BOLYAI SOCIETY
MATHEMATICAL STUDIES 14

BOLYAI SOCIETY MATHEMATICAL STUDIES

1. **Combinatorics, Paul Erdős is Eighty, Vol. 1**
D. Miklós, V.T. Sós, T. Szőnyi (Eds.)
2. **Combinatorics, Paul Erdős is Eighty, Vol. 2**
D. Miklós, V.T. Sós, T. Szőnyi (Eds.)
3. **Extremal Problems for Finite Sets**
P. Frankl, Z. Füredi, G. Katona, D. Miklós (Eds.)
4. **Topology with Applications**
A. Császár (Ed.)
5. **Approximation Theory and Function Series**
P. Vértesi, L. Leindler, Sz. Révész, J. Szabados, V. Totik (Eds.)
6. **Intuitive Geometry**
I. Bárány, K. Böröczky (Eds.)
7. **Graph Theory and Combinatorial Biology**
L. Lovász, A. Gyárfás, G. Katona, A. Recski (Eds.)
8. **Low Dimensional Topology**
K. Böröczky, Jr., W. Neumann, A. Stipsicz (Eds.)
9. **Random Walks**
P. Révész, B. Tóth (Eds.)
10. **Contemporary Combinatorics**
B. Bollobás (Ed.)
11. **Paul Erdős and His Mathematics I+II**
G. Halász, L. Lovász, M. Simonovits, V. T. Sós (Eds.)
12. **Higher Dimensional Varieties and Rational Points**
K. Böröczky, Jr., J. Kollár, T. Szamuely (Eds.)
13. **Surgery on Contact 3-Manifolds and Stein Surfaces**
B. Ozbagci, A. I. Stipsicz
14. **A Panorama of Hungarian Mathematics in the Twentieth Century, I**
J. Horváth (Ed.)

János Horváth (Ed.)

A Panorama of Hungarian Mathematics in the Twentieth Century I

Springer

JÁNOS BOLYAI MATHEMATICAL SOCIETY

János Horváth
Department of Mathematics
University of Maryland
College Park, MD 20742-4015
United States of America
e-mail: jhorvath@wam.umd.edu

Associate Editor: László Márki
Editorial Assistant: Gábor Sági

Mathematics Subject Classifictaion (2000):
01Axx, 26xx, 28xx, 30xx, 34xx, 35xx, 40xx, 41xx, 42xx, 43xx, 46xx, 47xx, 51xx, 52xx, 53xx, 54xx, 60xx, 62xx, 65xx, 91xx

Library of Congress Control Number: 2005931967

ISSN 1217-4696
ISBN 3-540-28945-3 Springer Berlin Heidelberg New York
ISBN 963 9453 04 8 János Bolyai Mathematical Society, Budapest

Springer is a part of Springer Science+Business Media
springeronline.com

Printed in Hungary

Cover design: Erich Kirchner, Heidelberg

Printed on acid-free paper 44/3142/db – 5 4 3 2 1 0

L Analyse Mathématique

M. Hadamard

Sur les fonctions bornées et intégrables

Note de M. Léopold Fejér, présentée par M. Picard

On sait que si la série $\sum_{n=0}^{\infty} a_n$ est convergente, la limite

$$\lim_{n=\infty} \frac{\sigma_0 + \sigma_1 + \cdots + \sigma_{n-1}}{n} \qquad (1)$$

existe et est égale à $\sum_{0}^{\infty} a_n$, où σ_{n-1} désigne la somme $\sum_{n=0}^{n-1} a_n$. Nous devons ranger, parmi les séries divergentes les plus simples, des séries, pour lesquelles existe au moins la limite (1). Une série divergente de telle sorte est la suivante:

$$\frac{1}{2} + \cos\vartheta + \cos 2\vartheta + \cdots + \cos(n-1)\vartheta + \cdots \qquad (2)$$

Ici on a

$$\frac{\sigma_0 + \sigma_1 + \cdots + \sigma_{n-1}}{n} = \frac{1}{2n} \frac{1 - \cos n\vartheta}{1 - \cos\vartheta} \qquad (3)$$

et donc (pour $\vartheta \gtrless 2k\pi$)

$$\lim_{n=\infty} \frac{\sigma_0 + \sigma_1 + \cdots + \sigma_{n-1}}{n} = 0$$

Nous voulons d'abord démontrer que la série de Fourier d'une fonction bornée et intégrable appartient à la classe des séries, pour lesquelles la limite (1) existe. Avec plus de précision, on a le théorème:

I. Soit $f(x)$ une fonction bornée et intégrable dans l'intervalle $(0, 2\pi)$ (c'est-à-dire de 0 à 2π en y comprenant les extrémités); alors la série de Fourier correspondante à $f(x)$

$$\frac{1}{2\pi}\int_0^{2\pi} f(\alpha)\,d\alpha + \sum_{n=1}^{\infty}\left\{\frac{1}{\pi}\int_0^{2\pi} f(\alpha)\cos n(\alpha - x)\,d\alpha\right\} \qquad (4)$$

peut être divergente, mais en tous les points x, pour les-

The manuscript of Lipót Fejér's Comptes Rendus note of 1900
(with the kind permission of the Académie des Sciences, Paris)

Contents

GEOMETRY

STOCHASTICS

Preface

I am often asked about the reason why mathematical research exploded in Hungary at the beginning of the twentieth century. My usual answer is, only half in jest: the two reasons are the personality of Lipót (Leopold) Fejér and the High School Mathematics Journal (Középiskolai Matematikai Lapok, abbreviated KöMaL).

This book will not answer the question because it would take a team of historians and sociologists to establish the causes of the scientific revolution that took place in Hungary during the first half of the twentieth century. Neither is this book a history of Hungarian mathematics in the twentieth century; that would be a work of several volumes written by professional historians of mathematics. Our goal is to present to the readers a sampling of beautiful mathematics created by Hungarians.

Hungary is a small country which became even smaller after the first World War. It could not support all the mathematical talent which suddenly appeared, which therefore had to find employment in other countries. The question therefore arises: whom do we consider to be a Hungarian mathematician? In this book the term "Hungarian mathematician" is applied to those who received their first impulse and started their research career in Hungary. This means most often that they obtained their Ph.D. degree from a Hungarian university, but this condition is not necessary: Alfréd Haar, who beyond the slightest doubt was a Hungarian mathematician, was a doctoral student of David Hilbert in Göttingen. In this book we use the Hungarian form of the first names of Hungarian mathematicians. We do this not only to emphasize their Hungarianness but also because most of them translated their names when publishing in a foreign language. For instance Frigyes Riesz published under the names Friedrich, Frédéric, Frederic, Federigo.

To keep the size of the work within reasonable bounds it was decided to deal with those researchers "whose activity can be considered essentially closed". This principle is not followed very rigorously: several older mathematicians had much younger collaborators (think Pál Erdős!) who can not go unmentioned. Also results, found by someone who satisfies the criterion, have often been sharpened and generalized by members of the younger generation; it would give a distorted picture if they were not mentioned. However, in the Biographies at the end of the volume we adhered to our principle – with a grain of salt.

From the Biographies the reader will detect a tragic side of twentieth century Hungarian mathematical life. Not only did two World Wars and the events related to them claim many victims, but several of our most outstanding mathematicians died from "natural causes" before their fiftieth birthday in possession of their full creative power.

The Bibliography lists only books. They are collected works of Hungarian mathematicians (and of some often-quoted non-Hungarians), a sampling of monographs authored or co-authored by Hungarians, and books referred to in the various chapters. References to the Bibliography are by numbers in square brackets: [ ]. Most authors of the chapters refer to journal articles within their text. However, some have their own reference list, to which they refer by some other means than numbers between square brackets (e.g., by numbers between { }, or letters, etc.).

I know from personal experience how risky it is to announce that a second volume will follow. Nevertheless, let me say that the János Bolyai Mathematical Society plans to publish another book in which the fields not covered in the present one but in which Hungarian mathematician had a very active role: logic, set theory, combinatorics, graph theory, number theory, algebra, will be discussed.

We have tried to use a uniform terminology in the various articles. Thus a map $f : A \to B$ such that $f(x) = f(y)$ implies $x = y$ is said to be *injective.* If for all $y \in B$ there is an $x \in A$ such that $f(x) = y$, then f is *surjective.* A real number α such that $\alpha \geq 0$ is said to be *positive*; if $\alpha > 0$, then it is *strictly positive.* If for $f : \mathbb{R} \to \mathbb{R}$ the relation $x < y$ implies $f(x) \leq f(y)$, then f is *increasing;* if $x < y$ implies $f(x) < f(y)$, then f is *strictly increasing.* Similarly for decreasing and strictly decreasing.

Finally, let me say that it gave me great pleasure to be connected during four years to the mathematics described in this volume, and to collaborate with the authors of the chapters. I want to express my sincere thanks to all authors, in particular to those who are not even Hungarian but are friends of Hungary and of Hungarian mathematics, and to all those who helped me during the years in my task as an editor. Last, but not least, I want to thank László Márki for his help in editing this volume and Gábor Ségi for technical assistance.

College Park, Maryland — János Horváth

July, 2005

BOLYAI SOCIETY
MATHEMATICAL STUDIES, 14

A Panorama of Hungarian Mathematics in the Twentieth Century, pp. 9–25.

Topology

MÁTYÁS BOGNÁR and ÁKOS CSÁSZÁR

Topology emerged as a separate branch of mathematics at the end of the nineteenth and the beginning of the twentieth century as a result of efforts to deal with convergence problems, to lay the foundations of real analysis and functional analysis, and to understand the geometric aspects of complex analysis. The main contributions were made in Western Europe, especially by Bernhard Riemann, Georg Cantor, Jules Henri Poincaré and Felix Hausdorff.

Hungarian research in topology started with nine articles published by Frigyes Riesz between 1904 and 1908 (*Qeuvres Complètes,* pp. 27–161). The first of these papers considers the Schoenflies theorem (the converse of the Jordan theorem on simple closed curves). In 1905 he published a paper on Borel's covering theorem for line segments in which he showed that Borel's assumption of countability of the covering can be omitted. Also in 1905, he published a paper on a theorem of L. Zoretti and a paper in which he attempted to give a description of R^n using a collection of n relations of order.

F. Riesz published two papers in Hungarian in 1906 and 1907, then their translation in German (Die Genesis des Raumbegriffes) also in 1907, in which he made an attempt to define something similar to the later concept (due to Hausdorff) of a topological space. Riesz's concept is based on the notion of an accumulation point. It appeared approximately at the same time as the investigations of Maurice Fréchet. A short description of the ideas of Riesz was contained in his talk at the International Congress of Mathematicians, held in Rome in 1908 (Stetigkeitsbegriff und abstrakte Mengenlehre).

Also in this talk he introduced a new structure, the so called "Verkettungstypus" (chaining type). The starting point is the observation that not every property of continuity can be described with the help of accumulation points. Such properties are for example the Cantor connectedness or the property of a planar domain to be bounded by a simple closed curve.

For that reason Riesz introduced this new structure based on the notion that any two subsets of the underlying set of the structure are either near to or far from each other (Are they "verkettet" or not?). Also he began to develop a theory of these structures.

The ideas of Riesz with respect to the "Verkettungstypus" were almost totally forgotten. As far as we are aware, before 1950 the only reference to this structure can be found in a paper of Tibor Radó and Paul Reichelderfer {24} where they write:

"The earliest suggestion of this kind of axiomatic treatment seems to be due to F. Riesz, who proposed to use (essentially) the concept of a pair of not mutually separated sets as a primitive concept (Stetigkeitsbegriff und abstrakte Mengenlehre)" (*Oeuvres Complètes I,* pp. 155–161).

The whole paper on this talk had little impact on the further development of topology before 1950.

It should be mentioned that a few decades later in 1951, independently of Riesz, V. A. Efremovich introduced the so called proximity spaces, which are in some respect similar to the "Verkettungstypus" of Riesz.

After these early papers Riesz abandoned publishing papers on topology except for a proof of the Jordan curve theorem (1938 in Hungarian and 1939 in French). He wrote exclusively on analysis. But his inspirations, letters and also some oral communications on topology were presented in the works of some other mathematicians.

First of all we should mention the only paper of Károly Kaluzsay which was published in Hungarian in 1915. Kaluzsay proved a three dimensional analogue of Schoenflies's theorem. This paper was quoted by Raymond Louis Wilder {28}. Here Wilder writes:

"I have recently come across an early attempt at the converse for the case of an H_2 in E_3 in a paper by K. Kaluzsay, A felületre vonatkozó Jordan tétel megfordítása, *Matematikai és Physikai Lapok,* vol. 24 (1915), pp. 101–141. Upon obtaining a translation of Kaluzsay's results, I have found the interesting fact, that these conditions (No. 5 apparently gives the uniform connectedness *im kleinen* of the complementary domains) closely approximate those which I gave in the Transaction paper just cited, except that instead of the condition on the Betti number, which I used, he assumed that (condition 3) any closed polygon in a complementary domain can be continuously deformed into a point, thus yielding only a special case."

According to Kaluzsay the inspiration of the problem treated in this paper came from F. Riesz.

Also to be mentioned is Dénes Kőnig's proof of the Helly–Radon theorem. D. Kőnig considers first a finite number of bounded closed convex sets and for passing from the finite case to the infinite one, he refers to an oral communication of a proof by F. Riesz. Thus in fact Riesz indirectly published a topological proof in 1921.

On the other hand, Béla Kerékjártó refers in his book [87] to a letter of Riesz's which contains a simple proof of Tietze's extension theorem. This proof of Riesz's figures in the book of Kerékjártó.

Beside Riesz there were a few other Hungarian mathematicians, widely known researchers on other chapters of mathematics than topology, also publishing in topology in this period.

György Pólya was a well-known analyst. However, in 1913 he published a construction of a Peano curve such that each point belongs at most to three values of the parameter (the existence of such a curve was claimed earlier).

Dénes Kőnig is known for his work in graph theory. But in 1918 he published a small book (in Hungarian) on the elements of analysis situs. The book contains a proof of the classification of closed and of bounded surfaces. For its clarity and suggestivity this book was very popular in Hungary for several years.

Kőnig published between 1911 and 1924 five papers which deal with the genus of systems of lines, with the combinatorial properties of surfaces and with one- and two-sidedness of manifolds of dimension higher than 2. Also he published a paper in 1922 in Acta Szeged, where he gave a generalization of a theorem of Borel.

It should be mentioned that Kőnig's investigations in graph theory are closely related to combinatorial topology as is referred in the introduction of his book [94].

Tibor Radó mainly worked in analysis, but he wrote an early paper in which he proved the triangulability of two-dimensional manifolds with a countable base in {22}:

"LEMMA 2. *If R is a two-dimensional manifold for which the countability axiom holds, then it can be triangulated.*" (Translation from German, p. 111).

Radó's proof is the first of this fundamental theorem of topology.

Later in 1936 he wrote two papers in connection with the topological index of a point under a continuous complex-valued function with respect to a directed continuous closed curve. In the second paper {23} he proves a lemma about the topological index with respect to some sequences of closed continuous curves and shows some corollaries of this lemma.

In a paper written together with J. W. T. Youngs and published in Acta Szeged in 1940, referring to the works of Moore, Alexandroff and Kerékjártó, the authors consider upper semicontinuous collections. The upper semicontinuity is the subject also of a paper co-authored with E. J. Mickle and published in the Proceedings of the *Amer. Math. Soc.* in 1950. The Borel transformations, i.e. the mappings, where the preimages of Borel sets are Borel sets, play an important role in this paper.

Tibor Radó also published a paper on semicontinuity of functions and functionals in the *Amer. Math. Monthly* in 1942.

Chiefly interesting is the paper of T. Radó and P. Reichelderfer "On cyclic transitivity" mentioned above. The authors suggest the desirability of a full axiomatic treatment of the theory of the structure of a general space using the notion of a "connected" set as an undefined concept. They refer also to section 2.1 of their paper, however this part of the paper was burned by the Germans. The editors of the *Fundamenta Mathematicae* announced on page 14:

"The manuscript of the work of Messrs. T. Radó and P. Reichelderfer was burned by the Germans and a large part of the already typeset pages destroyed. We were left by chance with these few sheets which had printed. ..." (Translation from French.)

We should mention that a condensed version of this paper was published in Duke Math. Journal but this version ends with section 1.23 and thus there is no reference to 2.1.

Between 1943 and 1949 T. Radó published three papers on Peano spaces and after 1950 five papers on algebraic topology: on singular homology, on chain homotopy and on general cohomology theory. It should be mentioned that in the monograph of T. Radó and P. Reichelderfeld [145], published in 1955, the subject of almost half of the book is topology, mainly cohomology theory.

The first paper of T. Radó, where the concept of the Čech cohomology group with integral coefficients occurs, is a paper co-authored with P. Reichelderfer written in 1949 (see {25}). The authors use this concept for the definition of the index in Euclidean n-space.

Two lemmas on metric spaces occur in a paper published in 1958 (E. J. Mickle and T. Radó {17}). One of them says that if A is a subset of a separable metric space M and $\emptyset \neq X$ is covering of A by proper closed spheres such that the diameters of these spheres have a finite upper bound, then there is a sequence $C_n = \gamma(a_n, r_n)$ of pairwise disjoint members of X such that $A \subset \bigcup_n \gamma(a_n, 5r_n)$, where $\gamma(a_n, s)$ is the closed sphere in M with centre a_n and radius s.

Gyula Pál mainly worked in analysis. However he also published some papers on topology. Three papers concern the construction of Jordan curves with given projections. In the third one {19} Gy. Pál proves that each locally connected planar continuum with at least two points is the orthogonal projection of a closed Jordan curve of the Euclidean 3-space.

Two papers are devoted to plane topology. In the second one {20} he describes the topic of a projected volume on topology.

Gyula Szőkefalvi Nagy mainly worked in geometry. However among his 150 papers there are also some papers belonging to topology. He published a paper in Hungarian and then in German {26} proving a theorem of Gauss on the double points of closed planar curves. The theorem says that under a successive numeration of the points of self intersection of a closed planar curve without singularities the serial number of each such point is once an even and once an odd number. The proof is based on the fact that two closed planar curves in relative general position have an even number of common points. Gy. Sz. Nagy's proof is also quoted in the book of Hans Rademacher and Otto Toeplitz {21}.

Gy. Sz. Nagy introduced in the paper mentioned above a new topological symbol connected with the cycles of the curves in question. He showed that planar curves with at most four double points can be fully characterized with these symbols.

Three papers of Gy. Sz. Nagy deal with closed curves on the sphere and on surfaces. In the first one {27} he considered cuts of self intersections of oriented closed curves on surfaces and proves the equality $h + k = n + 2 - 2r$ where n is the number of self intersection points of the curve C in question, r is a nonnegative integer, h is the number of components of the figure obtained from C after cutting it at each point of self intersection, and k is the number of components of the figure obtained from C making the complementary cut at each point of self intersection.

Pál Erdős was one of the most prolific and most travelled mathematicians ever. He worked in number theory and combinatorial analysis, and is considered the founder of discrete mathematics. Among his more than 1500 papers, there are also papers belonging to topology.

In his paper written in 1940 he states and proves the astonishing fact that the Menger–Urysohn dimension of the rational points of Hilbert space is 1 (*Ann. of Math.*, **41** (1940), 734–736).

One of the papers of Erdős published in 1944 concerns connected and biconnected sets. Some problems and questions related to these concepts are included there.

Also in 1945 and 1946 he published two papers on the Hausdorff dimension of some sets in Euclidean spaces.

In a paper co-authored with George Piranian (*Michigan Math. J.*, **5** (1958), 139–148) the authors have shown how the convergence field of certain regular Toeplitz matrices can serve as neighborhoods in the topologization of a quotient space of bounded sequences.

Erdős investigated Borel sets together with Arthur Stone (Proceedings of the *Amer. Mat. Soc.*, **25** (1970), 304–306). Erdős and Stone proved that the linear sum of two Borel subsets of the real line need not be Borel, even if one of them is compact and the other is a G_δ. However, after publication the authors have been informed of the earlier work of B. S. Sodnomov who proved in 1951 a theorem equivalent to the first theorem of Erdős and Stone.

Some papers of Erdős on topology are closely related to set theory.

In a paper co-authored with A. Tarski and published in 1943 the authors give a set theoretical equivalent of the statement that in every topological space of power $\leq 2^{\aleph_0}$ there exists a family of mutually disjoint open sets with a maximal power.

Two papers co-authored with András Hajnal and published in 1961 consider propositions of the form: Any product of $\aleph_\mu$ discrete λ-compact spaces is κ-compact, where μ, λ, κ are ordinals and for $\mu > 0$ $\aleph_\mu$ is the least cardinal number larger than all cardinals $\aleph_\delta$ with $\delta < \mu$, moreover a topological space is said to be μ-compact if each open cover has a subcover with cardinality less than $\aleph_\mu$. Using the generalized continuum hypothesis the authors show that the topological product of $\aleph_\kappa$ 1-compact (i.e. Lindelöf) discrete spaces is not necessarily κ-compact for any finite κ.

The subject of a paper of Erdős and Mary Ellen Rudin is the box product (1973). Erdős proves there that it is consistent with the usual axioms of set theory that the box product $\omega_k \times (\omega_0 + 1) \times (\omega_0 + 1) \times \ldots$ is ether normal or not normal for all integers $k > 1$, where ω_m is the initial ordinal number with the cardinality $\aleph_m$ for all nonnegative integers m.

Finally, one of the main results of an abstract and of a paper co-authorized with F. S. Cater and Fred Galvin (1976 and 1978) is that if κ and λ are infinite cardinals with $\lambda \leq \kappa^+$, where κ^+ is the least cardinal larger than κ, and X is a topological space with density $d(X) \geq 2$, where the density $d(Y)$ of a topological space Y is the minimum cardinality of a dense subset of Y, then the cofinality cfd $(X^\kappa)_{(\lambda)} \geq \mathrm{cf}\,\lambda$, where $(X^\kappa)_{(\lambda)}$ is the product of κ copies of X with λ-box topology, i.e. the basic open sets of this product are of the form $\prod_{\xi\in\kappa} U_\xi$, where U_ξ is open in X and the cardinality of the set $\{\xi;\, U_\xi \neq X\}$ is less than λ, moreover for any infinite cardinal number μ the cofinality cf μ of μ is the least cardinal α such that μ is the sum of α cardinals less then μ. In a certain sense this result generalizes the theorem of Gyula Kőnig that cf $2^\kappa > \kappa$.

György Alexits worked mainly in the theory of orthogonal series. However between 1932 and 1942 he was primarily interested in topology and wrote around 15 papers related to it. A part of them is devoted to Menger's theory of curves, another to locally connected continua.

One of his results is that each homogeneous rational curve is homeomorphic to the circle. This result was published only in Hungarian in {1}. It is closely related to a theorem of Mazurkiewicz which says that each locally connected homogeneous plane curve is homeomorphic to the circle.

His last paper on topology concerns spaces which are the union of a countable set of hereditarily locally connected continua. Among others, generalizing certain theorems of Gordon Thomas Whyburn, Alexits shows that a subset of such a space is totally disconnected if and only if it is zero dimensional, or if and only if each quasicomponent of the subset is a singleton, see {2}.

Alexits was also interested in defining some concepts of differential geometry (curvature, torsion) in metric and semimetric spaces using topological methods. One of the papers in this direction was co-authored with Jenő Egerváry.

*

The only Hungarian researcher of the first half of the century who devoted his activity mainly to topology was Béla Kerékjártó who wrote between 1919 and 1944 more than 70 scientific papers. Furthermore he is the author of the monograph [87]. He also planned to write the second volume of this books, but it has never been done.

The book itself is divided into two parts. The first deals with the topology of the plane, while the subject of the second part is the topology of surfaces. In the introductory chapter, which gives a view of the development of topology, also the notion of an n-manifold is formulated as follows:

An n-manifold is a connected Hausdorff space having a countable basis such that each member of this basis is homeomorphic to the interior of an n-ball.

Nowadays the existence of a countable basis is usually not assumed, but otherwise in general the formulation of Kerékjártó or some formulation equivalent to it is used for the notion of an n-manifold.

The main subject of the first part of the monograph are Jordan's theorem and numerous assertions closely related to this theorem. Here one can find the so called Kerékjártó theorem, which says that if $J_1, \dots, J_k$ are simple closed curves in the plane such that each pair of them has at least two common points, then the boundary of each complementary domain of $J_1 \cup \dots \cup J_k$ is a simple close curve (p. 87). This theorem is quoted in the book of M. H. A. Newman {18}.

The classification theorem of open surfaces, i.e. of the non-compact 2-manifolds, can be found in the second part. Kerékjártó's result in this relation was published earlier in the Jahresbericht der Deutschen Matematischen Vereinigung in 1922 (Bd. 31) with the heading: Haupsatz der Flächentopologie bei unendlich hohen Zusammenhang.

There is a fundamental difference between the open and the closed surfaces. In fact there is only a countable set of pairwise non-homeomorphic closed (i.e. compact) surfaces and they can be characterized in both the orientable and non-orientable cases with only one integer, e.g. with their genus. On the other hand there is a set of pairwise non-homeomorphic open surfaces of the cardinality of the continuum. This is true also in the case where we consider only simple surfaces, i.e. surfaces which are divided into two parts by each of their simple closed curves. One of the theorems of Kerékjártó says that each simple open surface is homomorphic to a domain of the plane, where the complementary set of the domain is totally disconnected. Moreover two open surfaces are homeomorphic if and only if these complementary sets are homeomorphic. But in the plane there can be found pairwise non-homeomorphic totally disconnected closed subsets of the cardinality of the continuum and the complement of each such subset is a domain in the plane, i.e. a simple open surface.

Kerékjártó formulated his scientific programme in his Privatdocent lecture held in Szeged on 15. December 1921 with the title "Az analízis és a geometria topológiai alapjairól" (On the topological fundaments of analysis and geometry). Here the intention formulated by Brouwer in his Privatdocent lecture in 1910 is emphasized, namely that complex analysis should be built with instruments of topology without metric elements such as length and area.

The principal direction of Kerékjártó's investigation is attached to this intention. He mainly investigates the properties of the topological transformations and groups of transformations of surfaces with special emphasis on continuous transformation groups.

One of the important periods of his investigations is the first half of the thirties. At that time he introduced the notion of regular map. He calls a bijective map τ of the plane onto itself regular at a point P if to each positive ε there is a positive δ such that if the distance of P and Q is less than δ then for each integer n the distance of $\tau^n(P)$ and $\tau^n(Q)$ is less than ε, where the distance is obtained from the spherical distance by a stereographic projection. The map τ itself is said to be regular if it is regular at each point P of the plane.

The notion of regularity can be defined in the same way in arbitrary metric spaces, and it can be transported to topological spaces as it is shown in a paper of Kerékjártó.

Now a fundamental statement of Kerékjártó says that the orientation preserving regular transformations without fixed points of the plane are equivalent to translations.

Several papers of Kerékjártó deal with regular topological transformations of various surfaces. One of them states that if $p > 1$ then each regular map of an orientable closed surface with genus p onto itself is periodic.

It should be mentioned that a somewhat strengthened version of the denial of Kerékjártó's regularity, the so called expensiveness, had a fundamental role in several branches of dynamical systems.

In the early forties Kerékjártó also investigates compact topological transformation groups. For example he examines the question of which closed surfaces with or without boundary have infinite compact transformation groups. He shows that there are only seven surfaces of this kind. These are: the sphere, the closed disc, the annulus, the torus, the projective plane, the Moebius strip and the Klein bottle.

Kerékjártó was the author also of the monographs "A geometria alapjairól" (Foundation of geometry) I. and II. (see [88]). The first volume was printed in 1934 and the second in 1944. The topological view appears in both volumes. For example in the introductory chapter of the second volume Kerékjártó mentions that each 2-times transitive continuous group of the topological transformations of the plane is equivalent to the group of similarity transformations of the Euclidean plane, where a group of topological transformations of a topological space is said to be n-times transitive if for any two n-tuples $(A_1, \ldots, A_n)$ and $(A'_1, \ldots, A'_n)$ of pairwise distinct points of the space there is precisely one element of the group which takes for $i = 1, \ldots, n$ the point A_i into A'_i. Also the topological concepts play an important role at the end of the second volume at the common characterization of the real and complex projective geometry.

We have to mention the simple proofs of Kerékjártó to some well known topological theorems of fundamental importance, e.g. the proof of the theorem of Schoenflies for the invariance of domains in the plane. The theorem says that if M and N are homeomorphic subsets of the plane then either both of them are open, or neither of them is open. Kerékjártó published his proof first in Hungarian, but the same proof appears in his monograph "Vorlesungen über Topologie". The proof is based on the fact that if Q is a square, and we consider a topological map of Q into the plane, then the image of the midsegments of Q cannot lie in the external residual domain of the image of the boundary of Q.

The proof of Kerékjártó, as well as the original proof of Schoenflies, requires an extension of the Jordan curve theorem, formulated by Schoenflies, however Kerékjártó's proof is quite different from that of Schoenflies.

To the Jordan curve theorem itself Kerékjártó gave two totally different proofs. One of them appeared also in his book "Vorlesungen über Topologie". More interesting is the second one which was published in 1930 in the Acta Szeged. The fundamental idea of this second proof is one of the simplest among the proofs of the theorem.

In the introductory chapter of the "Vorlesungen" Kerékjártó describes also the content of the second volume planned. However his early death prevented the implementing of his plans.

*

At present Hungarian research in topology is mainly inspired by the concept of syntopogeneous spaces due to Ákos Császár. They are a common generalization of topological spaces, uniform spaces introduced by André

Weil, and proximity spaces introduced by V. A. Efremovich. The first monograph about syntopogeneous spaces appeared in 1960 (Ákos Császár, [23]).

The productivity of two members of the present Hungarian school of topology is already closed by their early deaths.

János Czipszer as a publisher's reader of the first monograph of Ákos Császár gave the author valuable aid in preparing this edition. Also a basic theorem of this volume is due to J. Czipszer (Theorem 12.35, 12.37 in the second English edition). The second edition of the monograph (Ákos Császár, [24]) contains several results of Czipszer. The author writes in the preface of this edition:

"I have to express my warmest thanks to my collaborator J. Czipszer who, after having lent me valuable aid in the preparing the first edition, kindly communicated a series of his unpublished results which constitute an essential supplement to the theory of syntopogeneous structures. The material of Chapters 17, 18 and 19 is almost entirely due to him, and his ideas greatly contributed also to the subjects dealt with in the other chapters."

Among the 12 published papers of János Czipszer there are three dealing with topological questions. Two of them are co-authored with Ákos Császár. In the first paper {4} the authors consider curves without ramification points. Among others, they generalize one of Menger's theorem as follows: If X is a nondegenerate, connected, compact Hausdorff space which contains no point of infinite order and at most a finite number of points of order greater than 2, then X is the union of a finite collection of generalized arcs no one of which contains a non-end point of another.

In the second paper {5} the authors consider generalizations of the Stone–Weierstrass theorem. One of the main results says that to a family of bounded real-valued functions γ defined on a set E such that γ coincides with the set of all uniformly continuous (with respect to U) real-valued bounded functions if and only if γ is nonempty, it is closed with respect to uniform convergence and it satisfies the following conditions:

(a) $f \in \gamma \Longrightarrow f + \alpha \in \gamma$ for each $\alpha \in \mathbb{R}$,

(b) $f \in \gamma \Longrightarrow \alpha f \in \gamma$ for each $\alpha \geq 0$,

(c) $f, g \in \gamma \Longrightarrow \min(f, g) \in \gamma$ and $\max(f, g) \in \gamma$.

The third paper of J. Czipszer {3} is connected with the following problem proposed by G. Alexits:

Given a convergent sequence of real-valued functions defined over a complete separable metric space R, does there exist a dense subset of R on which the sequence converges locally uniformly?

A demonstration is given and produces an affirmative answer to the above question. Moreover it is shown that if the continuum hypothesis is assumed, then there exists a convergent sequence of real-valued functions defined over the real line $\mathbb{R}$ which does not converge locally uniformly on any subset of $\mathbb{R}$ having the power of the continuum.

Jenő Deák first published a few papers on analysis, then during the period from 1975 to his early death more than fifty papers on general topology, mostly devoted to the study of various topological structures.

His first papers on topology address the theory of directional structures, initiated in 1964 by Ervin Deák; the purpose of this theory was a topological characterization of the topological spaces homeomorphic to a subspace of the Euclidean space $\mathbb{R}^n$ $(n \in \mathbb{N})$. In order to reach this aim, let us say that a pseudo-direction $\mathcal{R}$ in a topological space X is a set of ordered pairs (G, F), where G is open and F is closed in X, and some natural conditions are satisfied. A pseudo-directional structure in X is a collection $\mathfrak{R}$ of pseudo-directions; $\mathfrak{R}$ is compatible with X if the sets G and $X \setminus F$ such that $(G, F) \in \mathcal{R} \in \mathfrak{R}$ constitute a subbase for the topology of X.

Now the main result of J. Deák in this theory is a stronger version of a theorem of E. Deák and says that a separable metrizable space X can be topologically embedded into $\mathbb{R}^n$ if and only if X admits a compatible pseudo-directional structure $\mathfrak{R}$ satisfying $|\mathfrak{R}| \leq n$, see {8}.

E. Deák has introduced a concept of dimension equal to the minimal cardinality of a compatible directional structure. J. Deák has presented, in a series of papers published between 1976 and 1980, a thorough discussion of this kind of dimension and of other similar concepts, partly introduced by him, gave a series of counter-examples and, in particular, he proved in {7} a generalization of a theorem of J. de Groot in dimension theory.

In {9} J. Deák develops further the investigations in which completely regular spaces are characterized by the existence of subbases possessing some special property. Continuing the studies of Péter Hamburger in this subject, he has presented several increasingly stronger results.

In {10} J. Deák gives an analysis of the ideas of Riesz and, in particular, he points out that the concept of a proximity, introduced in 1951 by V. A. Efremovich, can be brought in fact into a near connection with the chainings

(Verkettung) of Riesz, but the relation of the two concepts needs a careful discussion.

Extraordinarily intensive was the research work of J. Deák in the field of the theory of quasi-uniform spaces and bitopological spaces. As it is well-known, quasi-uniformities constitute a non-symmetric version of uniformities; i.e. a quasi-uniformity on a set X is a filter $\mathcal{U}$ on the product set $X \times X$ with the property that $(x, x) \in U$ for $x \in X$, $U \in \mathcal{U}$ and, for a given $U \in \mathcal{U}$, there is a $V \in \mathcal{U}$ such that $V^2 = V \circ V \subset U$. A bitopology on a set X is simply an ordered pair $(\mathcal{T}_1, \mathcal{T}_2)$ of topologies on X. The two kinds of structures are closely related because each quasi-uniformity $\mathcal{U}$ on X induces a topology $\mathcal{U}^{tp}$ for which the neighbourhoods of $x \in X$ are the sets $U(x)$ with $U \in \mathcal{U}$ so that the pair $(\mathcal{U}^{-tp}, \mathcal{U}^{tp})$ is a bitopology on X (of course, $\mathcal{U}^{-tp}$ denotes the topology induced by the quasi-uniformity $\mathcal{U}^{-1} = \{U^{-1} : U \in \mathcal{U}\}$).

As to bitopological spaces, a series of three papers (see {12}) presents (as a kind of a little monograph) a thorough discussion of separation axioms for bitopological spaces, then develops further results of R. E. Smithson on the relation of completely regular bitopological spaces and multifunctions, finally applies (pseudo-)directional structures to the characterization of completely regular bitopological spaces and their compactifications.

The problems belonging to the theory of extensions stand in the centre of the investigations of J. Deák on quasi-uniformities and bitopologies. In 1990 and 1991, he published several papers on such subjects; the most important is perhaps {11}. In these two papers the fundamental question is the following: we are given a quasi-uniform space $(X, \mathcal{U})$ inducing a bitopology $(\mathcal{U}^{-tp}, \mathcal{U}^{tp})$, further a set $Y \supset X$ and, on Y, a bitopology $(\mathcal{T}_{-1}, \mathcal{T}_1)$ such that $\mathcal{T}_{-1}|X = \mathcal{U}^{-tp}$ and $\mathcal{T}_1|X = \mathcal{U}^{tp}$; is there a quasi-uniformity $\mathcal{V}$ on Y such that $\mathcal{V}|(X \times X) = \mathcal{U}$ (i.e. $\mathcal{V}$ is an extension of $\mathcal{U}$) and $\mathcal{T}_{-1} = \mathcal{V}^{-tp}$, $\mathcal{T}_1 = \mathcal{V}^{tp}$ (i.e. $\mathcal{V}$ is compatible with the given bitopology)?

J. Deák presents a careful discussion of possible questions in connection with this problem: necessary or sufficient conditions for the existence of a compatible extension in general or satisfying further (e.g. cardinality) conditions, constructions for obtaining such extensions, etc. He also discusses a quite similar problem where, instead of quasi-uniformities, we consider syntopogenous structures in the sense of Császár.

A delicate question belonging to the same group of problems is the search for complete extensions, because the concept of completeness of a quasi-uniform space may be defined, and was in fact defined in the literature, in several more or less natural ways, and it is difficult to choose which is the most natural of them; probably the answer depends on the point of view the author wishes to emphasize. In any case, J. Deák's methods are useful for simplifying the situation in this rather complicated field of research (see {13}).

J. Deák published in 1993 a grandiose survey paper {14}. The paper gives a collection of results on extensions both of uniformities and of quasi-uniformities, in the latter case involving topologies or bitopologies, and contains not only a survey of the results in the literature but also a long list of new results and of open problems.

Still the questions related to extension problems are the subject of J. Deák's paper {15}; a quasi-metric on X is a function $d : X \times X \to [0, +\infty)$ satisfying $d(x, y) = 0 \Leftrightarrow x = y$ and $d(x, z) \leq d(x, y) + d(y, z)$. A quasi-metric induces a quasi-uniformity $\mathcal{U}$ if we set $U_{d,\varepsilon} = \{(x, y) : d(x, y) < \varepsilon\}$ for $\varepsilon > 0$ and then the topology $\mathcal{U}^{tp}$. The paper investigates the problem of the existence of a quasi-metric inducing a given topology on X and coinciding with a given quasi-metric on a subset of X.

A long series of further papers concerns questions on special properties of quasi-uniformities: weak symmetry properties, uniform local symmetry, doubly uniformly strict extensions, co-regularity, Cauchy-type properties, properties preserved by extensions, co-stability (the latter with co-author S. Romaguera). Through all this research work, J. Deák has deserved the rank of a leading researcher in the theory of quasi-uniform spaces.

While quasi-uniformities and topologies are particular cases of the general concept of a syntopogenous structure in the sense of Császár, other types of structures were considered in a further series of papers; these are particular cases of the general concept of a merotopy in the sense of Miroslav Katětov introduced in 1965. In an equivalent form due to Horst Herrlich, a merotopy on a set X is a set $\mathfrak{M}$ of covers of X with the property that a cover $\mathfrak{c}$ belongs to $\mathfrak{M}$ as soon as one of its refinements belongs to $\mathfrak{M}$ and, if $\mathfrak{c}_1$ and $\mathfrak{c}_2$ belong to $\mathfrak{M}$ then some common refinement of them also belongs to $\mathfrak{M}$; $\mathfrak{c}'$ is a refinement of $\mathfrak{c}$ if $C' \in \mathfrak{c}'$ implies the existence of $C \in \mathfrak{c}$ satisfying $C' \subset C$.

Special cases of merotopies are contiguities introduced by V. M. Ivanova and A. A. Ivanov in 1959 (sets of finite covers), Čech proximities introduced in the monograph of Eduard Čech in 1966 (two-element covers); less simple to explain but further special cases are filter merotopies (M. Katětov 1965) and in particular Cauchy structures (H. Keller 1968), Čech closures (E. Čech

1966), limitations (H. J. Kowalsky 1954). All these kinds of structures are connected by the fact that in the series

$$\text{merotopy} \to \text{filter merotopy} \to \text{contiguity} \to \text{Čech proximity}$$
$$\to \text{Čech closure} \to \text{limitation}$$

each kind of structure induces in a natural way a structure belonging to one of the later types in the series; we say that a richer structure induces another one. Also if $\mathcal{D}$ is a structure belonging to one type of the above list, given on a set X and $A \subset X$, there is a natural restriction $\mathcal{D}|A$ of $\mathcal{D}$ to A.

Now a very general problem that first appeared in a series of papers due to Császár and J. Deák is the following: we are given a structure $\mathcal{D}$ on a set X and subsets X_i of X ($i \in I$, where I is arbitrary, in general infinite, but possibly $I = \emptyset$) together with richer structures $\mathcal{C}_i$ given on each X_i. We look for a structure $\mathcal{C}$ of the same type inducing $\mathcal{D}$ and such that $\mathcal{C}|X_i = \mathcal{C}_i$; this is the problem of simultaneous extension of the type $(\mathcal{C}, \mathcal{D})$.

The series of papers co-authored with Császár (see {6}) contains investigations in this direction. In each case, necessary and/or sufficient conditions are given for the existence of a simultaneous extension and it is also discussed whether there is a coarsest or finest simultaneous extension. In later papers, J. Deák investigates other cases in the same spirit.

J. Deák also wrote a brilliant survey paper on the problem of simultaneous extensions (see {16}); his aim was to present the problem for mathematicians in general and not only for specialists in general topology.

References

[23] Császár, Ákos, *Fondements de la topologie générale,* Akadémiai Kiadó (Budapest–Gauthiers-Villars (Paris, 1960).

[24] Császár, Ákos, *Foundations of General Topology,* Pergamon Press (Oxford–London–New York–Paris, 1963).

[87] Kerékjártó, Béla, *Vorlesungen über Topologie. I. Flächentopologie,* Grundlehren der Mathematischen Wissenschaften in Einzeldarstellungen, Band VIII, Springer-Verlag (Berlin, 1923).

[88] Kerékjártó, Béla, *A geometria alapjairól. I. Az euklidesi geometria elemi felépítése,* A szerző kiadása a Magyar Tudományos Akadémia támogatásával, Szeged, 1937; II. Projektív geometria, Magyar Tudományos Akadémia (Budapest, 1944).

[94] Kőnig, Dénes, *Theorie der endlichen und unendlichen Graphen,* Kombinatorische Topologie der Streckenkomplexe, Akademische Verlagsgesellschaft M.B.H. (Leipzig, 1936)/ Chelsea Publishing Company (American Mathematical Society) (New York, 1950).

[145] Radó, Tibor – Reichelderfer, P. V., *Continuous Transformations in Analysis with an introduction to algebraic topology,* Die Grundlehren der Mathematischen Wissenschaften in Einzedarstellungen, Band LXXV, Springer-Verlag (Berlin–Göttingen–Heidelberg, 1955).

{1} Gy. Alexits, Homogén racionális görbékről, *Matematikai és Fizikai Lapok,* **44** (1937), 38–39.

{2} Gy. Alexits, Über verstreute Mengen, *Math. Annalen,* **118** (1942), 379–384.

{3} J. Czipszer, Sur une propriété générale des suites convergentes de fonctions arbitraires, *Acta Math. Acad. Sci. Hung.,* **10** (1959), 305–315.

{4} J. Czipszer and Á. Császár, Sur les courbes irramifiées, *Acta Mat. Acad. Sci. Hung.,* **9** (1958), 315–328.

{5} J. Czipszer and Á. Császár, Sur des critères généraux d'approximation uniforme, *Ann. Univ. Sci. Budapest Eötvös Sect. Math.,* **6** (1963), 17–26.

{6} Á. Császár and J. Deák, Simultaneous extensions of proximities, semi-uniformities, contiguities and merotopies I.–IV., *Math. Pannon.,* **1/2** (1990), 67–90, **2/1** (1991), 19–35, **2/2** (1991), 3–23, **3/1** (1992), 57–76.

{7} E. Deák, Applications of the Theory of Directional Structures, I, *Studia Sci. Math. Hungar.,* **15** (1980), 45–61.

{8} E. Deák, An embedding theorem for separable metrizable spaces, *Studia Sci. Math. Hungar.,* **18** (1983), 413–426.

{9} J. Deák, Preproximities and internal characterizations of complete regularity, *Studia Sci. Math. Hungar.,* **24** (1989), 147–177.

{10} J. Deák, On proximity-like relations introduced by F. Riesz in 1908, *Studia Sci. Math. Hungar.,* **25** (1990), 387–400.

{11} J. Deák, Extensions of quasi-uniformities for prescribed bitopologies I.–II., *Studia Sci. Math. Hungar.,* **25** (1990), 45–67, 69–91.

{12} J. Deák, On bitopological spaces I.–III., *Studia Sci. Math. Hungar.,* **25** (1990), 457–481, **26** (1991), 1–17, 19–33.

{13} J. Deák, On the coincidence of quasi-uniform completeness defined by filter pairs, *Studia Sci. Math. Hungar.,* **26** (1991), 411–413.

{14} J. Deák, A survey on compatible extensions, *Colloq. Math. Soc. János Bolyai,* **55** (1993), 127–175.

{15} J. Deák, Extending a quasi-metric, *Studia Sci. Math. Hungar.,* **28** (1993), 105–113.

{16} J. Deák, Simultaneous extensions, *Studia Sci. Math. Hungar.,* **32** (1996), 323–347.

{17} E. J. Mickle and T. Radó, On covering theorems, *Fund. Math.,* **45** (1958), 325–331.

{18} M. H. A. Newman, *Elements of the topology of plane sets of points,* Cambridge (1939).

{19} Gy. Pál, Térbeli Jordan Görbékről, *Math. és Phys. Lapok,* **24** (1915), 236–242.

{20} Gy. Pál, Zur Topologie der Ebene, *Acta Litt. ac Scient. Univ. Hung.,* **1** (1922–1923), 226–239.

{21} H. Rademacher and O. Toeplitz, *Von Zahlen und Figuren,* Berlin (1930).

{22} T. Radó, Über den Begriff der Riemannschen Fläche, *Acta Szeged,* **2** (1925), 101–121.

{23} T. Radó, Topological index, *Fund. Math.,* **27** (1936), 212–225.

{24} T. Radó and P. Reichelderfer, On cyclic transitivity, *Fundamenta Mathematicae,* **34** (1947), 14–29.

{25} T. Radó and P. Reichelderfer, On n-dimensional concept of bounded variation, absolute continuity and generalized Jacobian, *Proc. Nat. Acad. USA,* **35** (1949), 678–681.

{26} Gy. Szőkefalvi Nagy, Über ein topologisches Problem von Gauss, *Math. Zeitschr.,* **26** (1927), 579–592.

{27} Gy. Szőkefalvi Nagy, Über einen topologischen Satz, *Acta Szeged,* **4** (1929), 246–247.

{28} R. L. Wilder, Point sets in three and higher dimensions and their investigation by means of a unified Analysis Situs, *Bull. Amer. Math. Soc.,* **38** (1932), 649–692.

Mátyás Bognár
Ákos Császár

Eötvös Loránd University
Faculty os Natural Sciences
Department of Analysis
Pázmány P. sétány 1/C.
1117 Budapest
Hungary

csaszar@renyi.hu

Constructive Function Theory

BOLYAI SOCIETY
MATHEMATICAL STUDIES, 14

A Panorama of Hungarian Mathematics in the Twentieth Century, pp. 29–54.

Constructive Function Theory: I. Orthogonal Series

FERENC MÓRICZ*

1. The Riesz–Fischer Theorem

Let us consider the family of measurable functions defined on a Lebesgue measurable subset E of finite or infinite measure of the real line $\mathbb{R} := (-\infty, \infty)$. The functions may take real or complex values. The function space $L^2(E)$ consists of all measurable functions f whose squares $|f|^2$ are integrable in the Lebesgue sense. By the Schwarz inequality, f will then be integrable on the subsets of finite measure. Let us endow $L^2(E)$ with the inner product and norm

$$(f \mid g) := \int_E f(x)\overline{g(x)}\,dx \quad \text{and} \quad \|f\| := \sqrt{(f \mid f)},$$

respectively. Then $L^2(E)$ becomes a normed linear space whose norm is derived from the inner product. We say that a sequence $(f_n : n = 1, 2, \ldots)$ of functions in $L^2(E)$ converges in the mean to a function f in $L^2(E)$ if

$$\lim_{n\to\infty} \|f_n - f\| = 0.$$

Observe that the limit function f is uniquely determined, up to a set of measure zero. In fact, if

$$\lim_{n\to\infty} \|f_n - f\| = 0 \quad \text{and} \quad \lim_{n\to\infty} \|f_n - g\| = 0,$$

*This survey paper was partially supported by the Hungarian National Foundation for Scientific Research under Grants TS 044 782 and T 046 192.

then by the triangle inequality we have

$$\|f - g\| \le \|f_n - f\| + \|f_n - g\| \to 0 \quad \text{as} \quad n \to \infty.$$

Consequently, $\|f - g\| = 0$, which means that $f(x) - g(x) = 0$ almost everywhere (in abbreviation: a.e.).

The Riesz–Fischer theorem states that the classical Cauchy convergence criterion is valid in the case of this mean convergence notion. As a consequence, the space $L^2(E)$ is complete, and this fact turned out to be of basic importance in the theory of Hilbert spaces.

Riesz–Fischer theorem. *Given a sequence (f_n) of functions in $L^2(E)$, then in order that there exist a function in $L^2(E)$ to which it converges in the mean it is necessary and sufficient that*

$$\|f_m - f_n\| \to 0 \quad \text{as} \quad m, n \to \infty.$$

Frigyes Riesz and Ernst Fischer proved this theorem in 1907, independently of one another. Both of them published it in the Comptes Rendus Acad. Sci. Paris. However, Riesz submitted his paper two months earlier than Fischer. See [156, Vol. 1, C2, C5, pp. 378–381, 389–395], and also [203, Vol. 1, pp. 127–128, 377].

A system $(\phi_n : n = 1, 2, \ldots)$ of functions in $L^2(E)$ is called orthogonal if none of the ϕ_n vanishes a.e. and

$$(\phi_k \mid \phi_\ell) = 0 \quad \text{whenever} \quad k \neq \ell.$$

The system (ϕ_n) is called orthonormal (in abbreviation: ONS) if, besides the above condition of orthogonality, we also have

$$\|\phi_n\| = 1, \quad n = 1, 2, \ldots .$$

Clearly, each orthogonal system (ϕ_n) can be made normal by substituting $(\phi_n / \|\phi_n\|)$ for (ϕ_n).

Given a function f in $L^2(E)$, we form its expansion (one may say: generalized Fourier) coefficients with respect to the ONS (ϕ_n) as follows:

$$c_n := (f \mid \phi_n), \quad n = 1, 2, \ldots .$$

The series

$$\sum_{n=1}^{\infty} c_n \phi_n(x)$$

is called the orthogonal expansion (or we may say: generalized Fourier series) of f with respect to (ϕ_n). By the Bessel inequality, we have

$$\sum_{n=1}^{\infty} |c_n|^2 \leq \|f\|^2.$$

Since

$$\left\| \sum_{k=m}^{n} c_k \phi_k \right\|^2 = \sum_{k=m}^{n} |c_k|^2 \to 0 \quad \text{as} \quad m, n \to \infty,$$

by the Riesz–Fischer theorem, the series $\sum c_k \phi_k$ converges in the mean to some function g in $L^2(E)$. Since mean convergence implies weak convergence, we have

$$(g \mid \phi_\ell) = \lim_{n \to \infty} \sum_{k=1}^{n} c_k (\phi_k, \phi_\ell) = c_\ell := (f \mid \phi_\ell),$$

whence

$$(f - g \mid \phi_\ell) = 0, \quad \ell = 1, 2, \ldots;$$

that is, $f - g$ is orthogonal to every ϕ_ℓ of the given ONS.

The ONS $(\phi_n,\ n = 1, 2, \ldots)$ is called maximal (sometimes called complete) if it cannot be enlarged in the sense that whenever a new function not vanishing a.e., say ψ, is added to it, then the new system $(\psi, \phi_n :\ n = 1, 2, \ldots)$ is no longer orthogonal. To sum up the above reasoning, we obtain the following theorem: Given a complete ONS in $L^2(E)$, every function f in $L^2(E)$ admits an expansion convergent in the mean:

$$f = \sum_{n=1}^{\infty} (f \mid \phi_n) \phi_n.$$

We say that a set of functions in $L^2(E)$ is total (sometimes called closed) if it determines the entire space in the sense that every function f in $L^2(E)$ can be approximated arbitrarily closely, in the mean, by the linear combinations of the elements of this set. We may say briefly that a total set of functions in $L^2(E)$ spans the whole space $L^2(E)$. By the Bessel identity

$$\left\| f - \sum_{k=1}^{n} (f \mid \phi_k) \phi_k \right\|^2 = \|f\|^2 - \sum_{k=1}^{n} |(f \mid \phi_k)|^2,$$

it is easy to see that the notions of "maximal" and "total" systems (ϕ_n) are equivalent; furthermore, these are equivalent to the validity of the Parseval formula

$$\|f\|^2 = \sum_{n=1}^{\infty} \big|(f \mid \phi_n)\big|^2, \quad f \in L^2(E).$$

Combining these reasonings gives the following.

Riesz–Fischer theorem for orthogonal systems. *Given an orthonormal system $(\phi_n : n = 1, 2, \dots)$ in $L^2(E)$ (complete or not) and an arbitrary sequence $(a_n : n = 1, 2, \dots)$ of complex numbers such that*

$$\sum_{n=1}^{\infty} |a_n|^2 < \infty,$$

the orthogonal series $\sum a_n \phi_n$ converges in the mean to some function g in $L^2(E)$ such that the Fourier coefficients of g with respect to ϕ_n are these a_n:

$$(g \mid \phi_n) = a_n, \quad n = 1, 2, \dots .$$

It is this form as the Riesz–Fischer theorem was originally stated and proved by Frigyes Riesz. This theorem was one of those impressive theorems which first demonstrated the effectiveness of the integral introduced by Lebesgue in 1902.

We note that if the system (ϕ_n) is maximal, then the function g in the above theorem is uniquely determined. If the system (ϕ_n) is not maximal, then the expansion $\sum (g \mid \phi_n)\phi_n$ is the orthogonal projection of the function g onto the closed subspace spanned by the functions of the system (ϕ_n).

The Riesz–Fischer theorem explains why we usually start with a sequence (a_n) of numbers in the space ℓ^2, instead of a function f in the space $L^2(E)$ in Sections 5–7 below.

Frigyes Riesz played an outstanding role in the development of the powerful new theory of functional analysis, which emerged from the theory of real and complex functions, linear algebra and topology, etc. As a matter of fact, the emerging functional analysis preceded the present form of the modern linear algebra. For example, the notion of duality first became clear for Banach spaces. Up to the 1940's, the finite dimensional vector spaces were identified with their duals.

2. Riesz Typical Means

Almost everywhere convergence of an orthogonal series is a much more complicated question than its convergence in the mean. We shall discuss the problem of the pointwise convergence of the partial sums in Section 5 in connection with the work by Károly Tandori.

Instead of convergence of the partial sums of an orthogonal series, one may consider various summability methods to assign a reasonable sum (if any) to the given series. The so-called Riesz typical means were introduced by Marcel Riesz [158, Parts 5, 6, pp. 55–58, 59–61] in 1909 as follows.

Let $\lambda = (\lambda_n : n = 0, 1, \ldots)$ be a sequence of real numbers such that

$$0 \le \lambda_0 < \lambda_1 < \lambda_2 < \ldots \quad \text{and} \quad \lim_{n\to\infty} \lambda_n = \infty, \tag{2.1}$$

and let α be a positive number. Given a series $\sum_{n=0}^{\infty} u_n$ of real or complex numbers, the so-called Riesz typical means of type λ and of order α are defined by

$$R(\omega; \lambda, \alpha) := \sum_{\lambda_k \le \omega} \left(1 - \frac{\lambda_k}{\omega}\right)^{\alpha} u_k, \quad \omega > 0.$$

The series $\sum u_n$ is said to be summable to some L by the Riesz method of type λ and of order α if the finite limit

$$\lim_{\omega\to\infty} R(\omega; \lambda, \alpha) = L$$

exists. M. Riesz introduced these means for the study of the behaviour of Dirichlet series $\sum a_n e^{-\lambda_n z}$, where (a_n) is a given sequence of complex numbers and z is a complex variable.

In the particular case where $\omega := n + 1$ and $\lambda_n := n$, we get

$$R_n(\alpha) := R(n+1; (n), \alpha) = \sum_{k=0}^{n} \left(1 - \frac{k}{n+1}\right)^{\alpha} u_k,$$

which is fairly similar to the expression of the nth Cesàro mean of order α of the series $\sum u_n$ provided α is a positive integer:

$$C_n(\alpha) = \sum_{k=0}^{n} \left(1 - \frac{k}{n+1}\right)\left(1 - \frac{k}{n+2}\right) \ldots \left(1 - \frac{k}{n+\alpha}\right) u_k, \quad n = 0, 1, \ldots .$$

However, the parameter ω of the Riesz method approaches ∞ in a continuous way.

Let us consider another special case where $\omega := \lambda_n$ and $\alpha := 1$. Then

$$R(\lambda_n; \lambda, 1) := \frac{1}{\lambda_n} \sum_{k=0}^{n} (\lambda_n - \lambda_k) u_k = \frac{1}{\lambda_n} \sum_{k=0}^{n-1} (\lambda_{k+1} - \lambda_k) s_k,$$

where $s_k := \sum_{j=0}^{k} u_j$ is the kth partial sum. If, in addition, $\lambda_n := n$, then $R_n(1)$ is the first arithmetic mean of the partial sums.

Now, we return to the general case and introduce the notation

$$T(\omega; \lambda, \alpha) := \sum_{\lambda_k \leq \omega} (\omega - \lambda_k)^\alpha u_k,$$

while in case $\alpha = 0$ set

$$T_\lambda(\omega) := T(\omega; \lambda, 0) = \sum_{\lambda_k \leq \omega} u_k,$$

the latter may be called the partial sum function. Clearly, we have

$$R(\omega; \lambda, \alpha) = \frac{T(\omega; \lambda, \alpha)}{\omega^\alpha}.$$

We can express $T(\omega; \lambda, \alpha)$ in terms of a Riemann–Stieltjes integral as follows:

$$T(\omega; \lambda, \alpha) = \int_{\lambda_0}^{\omega} (\omega - \tau)^\alpha dT_\lambda(\tau) \quad \text{if} \quad \lambda_0 > 0,$$

while in case $\lambda_0 = 0$ we have to add the term $\omega^\alpha u_0$ to the right-hand side. By integration by parts, in either case we obtain

$$T(\omega; \lambda, \alpha) = \alpha \int_0^\omega (\omega - \tau)^{\alpha - 1} T_\lambda(\tau)\, d\tau.$$

Let us observe that the integral on the right-hand side coincides, up to a constant, with the Riemann–Liouville integral

$$(I^\alpha f)(\omega) := \frac{1}{\Gamma(\alpha)} \int_0^\omega (\omega - \tau)^{\alpha - 1} f(\tau)\, d\tau,$$

which is used for defining the fractional integral of f of order α, when it is applied to $f = T_\lambda$. Here $\Gamma(\alpha)$ is the Euler gamma function. From the identity

$$\Gamma(\alpha+1) = \alpha\Gamma(\alpha)$$

it follows that

$$R(\omega;\lambda,\alpha) = \frac{\Gamma(\alpha+1)}{\omega^\alpha}(I^\alpha T_\lambda)(\omega).$$

On the other hand, for the binomial coefficient

$$A_n^{(\alpha)} := \binom{n+\alpha}{n},$$

it follows from the Stirling formula that

$$\lim_{n\to\infty} \frac{A_n^{(\alpha)}}{n^\alpha} = \frac{1}{\Gamma\alpha+1)}, \qquad \alpha > 0.$$

Hence, for the nth Cesàro mean $C_n(\alpha)$ we obtain the following asymptotic equality:

$$C_n(\alpha) := \frac{S_n^{(\alpha)}}{A_n^{(\alpha)}} \approx \frac{\Gamma(\alpha+1)}{n^\alpha} S_n^{(\alpha)},$$

where

$$S_n^{(\alpha)} := \sum_{k=0}^{n} A_{n-k}^{(\alpha)} u_k = \sum_{k=0}^{n} A_{n-k}^{(\alpha-1)} s_k.$$

In the special case where α runs over the positive integers, the following recursive definition is equivalent to the above one:

$$S_n^{(j)} := \sum_{k=0}^{n} S_k^{(j-1)} \quad \text{for} \quad j = 1, 2, \dots, \quad \text{and} \quad S_n^{(0)} := s_n.$$

To sum up, the "integral of the partial sum function T_λ of fractional order α" occurs in the case of the Riesz typical means, while the "α times iteration of the partial sum s_n" comes in for the Cesàro means. This makes plausible the conjecture that the Cesàro method of summability of order α is equivalent to the Riesz method of summability of type $\lambda = (\lambda_n := n)$ and of order α whenever ω runs over the positive real numbers. This conjecture turned out to be true [158, Part 9, pp. 72–75], but its proof is rather complicated. (See the detailed proof for integer α in [68, §5.16, p. 113].)

On the other hand, the equivalence of the Cesàro and Riesz methods of summability is no longer true if ω runs over only the positive integers.

Marcel Riesz discovered the above summability method, named after him, in 1909, but he but he gave a detailed treatment of it only in 1915 in hiscitep [66] with G. H. Hardy.

The Riesz method of summability became of great significance in the theory of multiple Fourier series, as Jean-Pierre Kahane tells us in the Section "Commutative Harmonic Analysis". Here we confine ourselves to mention another result in the theory of orthogonal series. Given any orthogonal series

$$\sum_{n=1}^{\infty} a_n \phi_n(x) \quad \text{with} \quad \sum |a_n|^2 < \infty,$$

there exists a sequence $\lambda = (\lambda_n)$ satisfying the conditions in (2.1) such that the orthogonal series in question is summable a.e. by the Riesz method $R_n(\lambda_n; \lambda, 1)$.

Marcel Riesz achieved several significant results and each of them opened the way to new branches of analysis. Among others, the famous Riesz–Thorin convexity theorem started the development of the interpolation theory of linear operators, indispensable in modern harmonic analysis. M. Riesz also revealed the behavior of the harmonic conjugate to functions in the Lebesgue spaces L^p, where $1 < p < \infty$. Last but not least, the celebrated Riesz brothers' theorem on the unit circle surprisingly connected the theory of Fourier series with measure theory, thereby initiated a large number of deep results in both classical and abstract harmonic analysis.

3. The Haar Orthogonal System

An interesting and very useful orthonormal system was constructed by Alfréd Haar [62, B1, pp. 45–87] in his doctoral dissertation in 1909. The functions are defined on the interval $[0,1]$ and are conveniently labelled by two indices:

$$\chi_0^{(0)}(x); \chi_0^{(1)}(x); \chi_1^{(1)}(x), \chi_1^{(2)}(x); \ldots; \chi_n^{(1)}(x), \ldots, \chi_n^{(2^n)}(x); \ldots .$$

To go into details, the first of them, $\chi_0^{(0)}(x)$ is identically equal to 1; while for $n \geq 0$ and $1 \leq k \leq 2^n$, we set

$$\chi_n^{(k)}(x) := \begin{cases} 2^{n/2} & \text{for } x \in \left(\frac{k-1}{2^n}, \frac{k-1/2}{2^n}\right), \\ -2^{n/2} & \text{for } x \in \left(\frac{k-1/2}{2^n}, \frac{k}{2^n}\right), \\ 0 & \text{for } x \in \left(\frac{\ell-1}{2^n}, \frac{\ell}{2^n}\right) \text{ with } \ell \neq k,\ 1 \leq \ell \leq 2^n; \end{cases}$$

furthermore, at the points of discontinuity let $\chi_n^{(k)}(x)$ be equal to the arithmetic mean of the values assumed on the two adjacent intervals; and at the points $x = 0$ and $x = 1$ let $\chi_n^{(k)}(x)$ take the same value as on the interval $(0, 2^{-n-1})$ and $(1 - 2^{-n-1}, 1)$, respectively.

It is easy to see that the Haar system is an ONS on the interval $[0, 1]$. Let $f(x)$ be a function integrable on $[0, 1]$ in the Lebesgue sense. Its expansion with respect to the Haar system

$$f(x) \sim a_0^{(0)} \chi_0^{(0)}(x) + \sum_{n=0}^{\infty} \sum_{k=1}^{2^n} a_n^{(k)} \chi_n^{(k)}(x), \tag{3.1}$$

where

$$a_0^{(0)} := \int_0^1 f(t)\, dt, \quad a_n^{(k)} := \int_0^1 f(t) \chi_n^{(k)}(t)\, dt,$$

has remarkable properties of representing the function $f(x)$. The following theorem was theorem was proved citep Haar [62, B1, B2, pp. 45–87, 88–103].

Convergence theorem of Haar. (i) *If $f(x)$ is integrable in the Lebesgue sense on the interval $(0, 1)$, then its Haar expansion (3.1) converges to $f(x)$* a.e.

(ii) *The Haar expansion (3.1) converges to $f(x)$ at each point of continuity of $f(x)$, and converges uniformly on each interval on which $f(x)$ is uniformly continuous.*

Statement (ii) expresses one of the most interesting features of the Haar system that every continuous function on the interval $[0, 1]$ is uniformly approximated by the partial sums of its expansion with respect to the discontinuous functions of the Haar system.

Denote by $s_n^{(k)}(x)$ the partial sum of the series in (3.1) ending with the term $a_n^{(k)}\chi_n^{(k)}(x)$. Then by introducing the kernel

$$K_n^{(k)}(x,t) := \chi_0^{(0)}(x)\chi_0^{(0)}(t) + \chi_0^{(1)}(x)\chi_0^{(1)}(t) + \ldots + \chi_n^{(k)}(x)\chi_n^{(k)}(t),$$

we get the following representation:

$$s_n^{(k)}(x) = \int_0^1 f(t)K_n^{(k)}(x,t)\,dt.$$

The convergence theorem of Haar follows in a natural way from the nice properties of the kernel $K_n^{(k)}(x,t)$. It turns out that if x is not a dyadic rational number in $(0,1)$, then $K_n^{(k)}(x,t)$ differs from zero only on an interval $I_n^{(k)}$ (containing x) of length $\left|I_n^{(k)}\right| = 2^{-n}$ or 2^{-n-1}, on which it takes the value 2^n or 2^{n+1}, respectively. Therefore, we have

$$s_n^{(k)}(x) = \frac{1}{\left|I_n^{(k)}\right|}\int_{I_n^{(k)}} f(t)\,dt. \tag{3.2}$$

If x is a dyadic rational number in $(0,1)$, then

$$K_n^{(k)}(x,t) = \begin{cases} 2^n & \text{on } I_n^{(k)} := (x-2^{-n-1},x), \\ 2^{n-1} & \text{on } J_n^{(k)} := (x,x+2^{-n}), \\ 0 & \text{otherwise.} \end{cases}$$

Consequently, in this case we have

$$s_n^{(k)}(x) = \frac{1}{2\left|I_n^{(k)}\right|}\int_{I_n^{(k)}} f(t)\,dt + \frac{1}{2\left|J_n^{(k)}\right|}\int_{J_n^{(k)}} f(t)\,dt. \tag{3.3}$$

If $x=0$ or $x=1$, then the first or the second term on the right-hand side in (3.3) should be deleted, respectively.

If we let n tend to infinity, the intervals $I_n^{(k)}$ and $J_n^{(k)}$ contract into the point x and

$$\lim_{n\to\infty} s_n^{(k)}(x) = F'(x),$$

provided that the derivative of the integral

$$F(u) := \int_0^u f(t)\,dt$$

exists at the point $u = x$. As it is well known, the derivative $F'(x)$ exists at almost every x. So, statement (i) of the convergence theorem of Haar follows at once. The proof of statement (ii) also follows almost immediately from (3.2) and (3.3).

As a by-product of the above reasoning, we can conclude that the Haar system is maximal. Indeed, if each expansion coefficient of a function $f(x)$ equals zero in (3.1), then each partial sum $s_n^{(k)}(x) \equiv 0$. Thus, by statement (i) above, we necessarily have $f(x) = 0$ a.e. even in the case when $f(x)$ is only integrable. This proves the maximality of the Haar system not only in $L^2(0,1)$, but also in the larger space $L(0,1)$.

The Haar wavelet basis evolved from the classical Haar system, was the first prototype of the wavelet theory, one of the major events in Harmonic Analysis started in the 1980's. Signal processing, image compression, and many other areas of applied mathematics have been revolutionized because of wavelet theory. We note that the Nobel Prize winning physicist Dénes Gábor essentially discovered the notion of wavelets and applied them successfully in his physical computations already in 1946, decades before the rigorous mathematical theory.

In 1933, Alfréd Haar also solved the problem of the existence of an invariant measure on topological groups, thus giving one of the most powerful tools to all subsequent investigations of such groups. In the Section on "Noncommutative Harmonic Analysis", Jonathan Rosenberg will tell the details of this story.

4. The Saturation Problem for the Fejér Means

Let f be a 2π-periodic (real or complex-valued) function, integrable in the Lebesgue sense on the torus $[-\pi, \pi)$, in symbol: $f \in L_{2\pi}$. The Fourier series of f is given by

$$f(x) \sim \frac{1}{2}a_0 + \sum_{k=1}^{\infty}(a_k \cos kx + b_k \sin kx), \tag{4.1}$$

where

$$a_k := \frac{1}{\pi}\int_{-\pi}^{\pi} f(t)\cos kt\,dt, \quad b_k := \frac{1}{\pi}\int_{-\pi}^{\pi} f(t)\sin kt\,dt$$

are the Fourier coefficients. Denote by

$$s_n(f,x) := \frac{1}{2}a_0 + \sum_{k=1}^{n}(a_k \cos kx + b_k \sin kx)$$

the nth partial sum, and by

$$\begin{aligned}\sigma_n(f,x) &:= \frac{1}{n+1}\sum_{k=0}^{n} s_k(f,x)\\ &= \frac{1}{2}a_0 + \sum_{k=1}^{n}\left(1-\frac{k}{n+1}\right)(a_k \cos kx + b_k \sin kx)\end{aligned}$$

the first arithmetic, so-called Fejér mean, of series (4.1).

In 1900 Lipót Fejér proved that if a 2π-periodic function is continuous, then the means $\sigma_n(f,x)$ converge uniformly to $f(x)$. In 1905, Lebesgue proved that if $f \in L_{2\pi}$, then the means $\sigma_n(f,x)$ converge to $f(x)$ a.e. For more details see the Section "Commutative Harmonic Analysis" written by Jean-Pierre Kahane.

We recall that the conjugate series to (3.1) is defined by

$$\sum_{k=1}^{\infty}(a_k \sin kx - b_k \cos kx). \tag{4.2}$$

In 1911, I. I. Privalov proved that the first arithmetic means of series (4.2) (the so-called conjugate Fejér means) also converge a.e. for each $f \in L_{2\pi}$. This limit is called the conjugate function to f and denoted by $\tilde{f}$. This $\tilde{f}$ is not necessarily integrable in the Lebesgue sense. However, if $\tilde{f} \in L_{2\pi}$, then (4.2) is the Fourier series of $\tilde{f}$.

The conjugate function $\tilde{f}$ can be represented in the form of a Cauchy principal value integral as follows:

$$\begin{aligned}\tilde{f}(x) &= (\text{P.V.})\frac{1}{\pi}\int_{-\pi}^{\pi}\frac{f(x-t)}{2\tan\frac{t}{2}}\,dt\\ &=: \lim_{\varepsilon\downarrow 0}\frac{1}{\pi}\int_{\varepsilon}^{\pi}\frac{f(x-t)-f(x+t)}{2\tan\frac{t}{2}}\,dt,\end{aligned}$$

where the limit exists a.e. whenever $f \in L_{2\pi}$. This explains why $\tilde{f}$ is sometimes called the periodic Hilbert transform of f.

A 2π-periodic function f is said to satisfy the uniform Lipschitz condition of order $\alpha > 0$, in symbol: $f \in \mathrm{Lip}_{2\pi}\,\alpha$, if

$$\omega(f,\delta) := \sup\left\{\left|f(x_1) - f(x_2)\right| \,:\, |x_1 - x_2| \le \delta\right\} = \mathcal{O}(\delta^\alpha), \quad 0 \le \delta < 2\pi.$$

Only the case $0 < \alpha \le 1$ is interesting: if $\alpha > 1$, then $\omega(f,\delta)/\delta$ tends to zero with δ. Consequently, in this case $f'(x)$ exists and is zero everywhere, and f is constant. The function $\omega(f,\delta)$, $0 \le \delta < 2\pi$, is called the modulus of continuity of f. It is clear that a function f is uniformly continuous if and only if $\omega(f,\delta)$ tends to zero with δ. On the other hand, f belongs to $\mathrm{Lip}_{2\pi}\,1$ if and only if f is the antiderivative of a bounded function.

It is easy to check that if $f \in \mathrm{Lip}_{2\pi}\,\alpha$, then

$$\begin{aligned}\left\|\sigma_n(f) - f\right\|_C &:= \max_x \left|\sigma_n(f,x) - f(x)\right| \\ &= \begin{cases}\mathcal{O}(n^{-\alpha}) & \text{if } 0 < \alpha < 1, \\ \mathcal{O}(n^{-1}\ln n) & \text{if } \alpha = 1.\end{cases}\end{aligned}$$

In 1941, György Alexits {1} proved the following remarkable characterization.

Theorem of Alexits. *A necessary and sufficient condition for the relation*

$$\left\|\sigma_n(f) - f\right\|_C = \mathcal{O}(n^{-1}) \tag{4.3}$$

is that the conjugate function $\tilde{f} \in \mathrm{Lip}_{2\pi}\,1$.

This theorem becomes even more significant in the light of the following simple observation: If

$$\left\|\sigma_n(f) - f\right\|_C = o(n^{-1}), \tag{4.4}$$

then we have necessarily $f \equiv$ constant. In fact, since

$$\frac{1}{\pi}\int_{-\pi}^{\pi} \left[f(x) - \sigma_n(f,x)\right] \cos kx = ka_k/(n+1), \quad 1 \le k \le n,$$

relation (4.4) implies that the left-hand side here is $o(n^{-1})$, which means that $a_k = 0$ for $k \ge 1$. Analogously, we have $b_k = 0$ for $k \ge 1$. Consequently, $f \equiv \frac{1}{2}a_0$ as we have claimed.

Using the terminology introduced by Jean Favard, we may say that the Fejér means are saturated with saturation order $\mathcal{O}(n^{-1})$. The collection of functions f for which relation (4.3) is satisfied, is called the saturation class for the Fejér means. Now, the above theorem of Alexits says that the saturation class for the Fejér means consists of those functions f for which $\tilde{f} \in \mathrm{Lip}_{2\pi} 1$.

At this point, we have to correct a historical misunderstanding in the literature. Unfortunately, it had been a widespread belief that G. Alexits proved only the sufficiency part in the above theorem; that is, if $\tilde{f} \in \mathrm{Lip}_{2\pi} 1$, then (4.3) is satisfied. For example, A. Zygmund in his monograph [203, Vol. 1, p. 377] writes the following: "The sufficiency part... was proved by Alexits, the necessity by Zamansky." However, reading the text of {1} makes it clear that G. Alexits did prove both the necessity and the sufficiency part even in the case of the more general $(C, \alpha \geq 1)$ Cesàro means. We emphasize that Alexits' paper appeared 8 years earlier than that of Marc Zamansky {21}. What could have been the reason for this misunderstanding? We guess that the answer lies in the fact that Alexits' paper appeared in a Hungarian periodical in 1941, the third year of World War II.

The first international recognition of Alexits' paper was given by R. A. De Vore {3, pp. 59–60}, who wrote the following: "In 1941, G. Alexits in a now classical theorem gave a characterization of the saturation class for Fejér operators (the 'o' saturation theorem for Fejér operators goes back to Zygmund). This paper of Alexits was the beginning of the study of saturation of convolution operators. In 1949, J. Favard gave a general formulation of the phenomenon of saturation for convolution operators. Later, M. Zamansky initiated a systematic study of saturation for convolution with trigonometric polynomials. In particular, Zamansky was able to recover the Alexits theorem."

Last but not least, Alexits was one of the creators of the Hungarian orthogonal school in the middle of the twentieth century, who envisaged several far-reaching problems in the theory of orthogonal series and approximation theory. He had numerous disciples in Hungary as well as all over the world. His monograph [8] on orthogonal series appeared in German in 1960, translated into English the following year, and into Russian in 1963. It has became the standard reference book for researchers in the theory of orthogonal series.

Motivated by the notion of strong $(C, 1)$-summability introduced by G. H. Hardy and J. E. Littlewood in 1913, Alexits {2} introduced the notion of strong approximation by Fourier series in 1963, and by raising several problems in subsequent papers he laid the foundation of the so-called strong approximation, a new branch of approximation theory. The strong approximation by Fourier series as well as by general orthogonal series have been studied, while considered not only Fejér but also Cesàro and other means. We refer the reader to the monograph of László Leindler [108].

5. Almost Everywhere Convergence of Orthogonal Series

One of the main problems in the theory of orthogonal series is to characterize the pointwise behavior of the orthogonal series

$$\sum_{k=1}^{\infty} a_k \phi_k(x) \tag{5.1}$$

in terms of its coefficients a_k, where (a_k) is a sequence of real numbers and $\big(\phi_k(x)\big)$ is an arbitrary ONS on a finite interval, say $(0, 1)$ for the sake of simplicity.

By the Schwarz inequality and the Beppo Levi theorem, the condition

$$\sum_{k=1}^{\infty} |a_k| < \infty \tag{5.2}$$

implies even the absolute convergence of series (5.1) a.e. But this requirement is too strong for a more delicate study of convergence problems. On the other hand, the requirement

$$\sum_{k=1}^{\infty} a_k^2 < \infty \tag{5.3}$$

is indispensable, because in the special case where $\big(\phi_k(x)\big)$ is the familiar Rademacher system, the orthogonal series (5.1) converges a.e. if and only if (5.3) is satisfied. Thus, the useful convergence tests lie between conditions (5.2) and (5.3).

(i) *Sufficient condition: Rademacher* (1922), *Menshov* (1923). The most important convergence test was discovered nearly simultaneously by Hans Rademacher {13} and D. E. Menshov {8}.

Rademacher–Menshov theorem. *If*

$$\sum_{k=1}^{\infty} a_k^2 (\log k)^2 < \infty, \tag{5.4}$$

then the orthogonal series (5.1) *converges a.e.*

Here and in the sequel the logarithm is to the base 2, but any base greater than 1 would be appropriate.

It was D. E. Menshov {8} who first observed that condition (5.4) is the best possible in general.

Theorem of Menshov. *If* $\big(w(k) : k = 1, 2, \ldots\big)$ *is an increasing sequence of positive numbers with*

$$w(k) = o(\log k) \quad \textit{as} \quad k \to \infty,$$

then there exist a sequence (a_k) *of numbers and an* ONS $\big(\phi_k(x)\big)$ *such that*

$$\sum_{k=1}^{\infty} a_k^2 w^2(k) < \infty \tag{5.5}$$

and the orthogonal series (5.1) *diverges everywhere.*

(ii) *Necessary condition: Tandori* (1957). The breakthrough in the convergence (or one may say, in the divergence) theory of orthogonal series was achieved by Károly Tandori {14}, who proved the following deep complement of the Rademacher–Menshov theorem.

Divergence theorem of Tandori. *If*

$$|a_1| \geq |a_2| \geq \ldots \tag{5.6}$$

and

$$\sum_{k=1}^{\infty} |a_k|^2 (\log k)^2 = \infty, \tag{5.7}$$

then there exists an ONS $\big(\phi_k(x)\big)$ *depending on* (a_k) *such that the orthogonal series* (5.1) *diverges everywhere.*

It is easy to see that the theorem of Menshov above is a consequence of the divergence theorem of Tandori.

Combining the Rademacher–Menshov theorem with the divergence theorem of Tandori gives that if (a_k) satisfies (5.6), then condition (5.4) is not only sufficient, but also necessary that the orthogonal series (5.1) converge a.e. for all ONS $\big(\phi_k(x)\big)$.

Starting with {14}, which contains the divergence theorem, K. Tandori wrote a series of ten papers entitled "Über die orthogonalen Funktionen I–X", which is the longest of its kind ever published in the "Acta Scientiarum Mathematicarum". By constructing delicate counterexamples, he gave a great impetus to the further development of the theory of orthogonal series. His contribution is commensurable with the one D. E. Menshov exerted 30 years earlier. The majority of these results are collected in the monograph by G. Alexits [8], which serves as a by G. Alexits [8], which serves as a reference book even nowadays.

The divergence theorem of Tandori has several consequences. We cite here only one of them. It follows from the Rademacher–Menshov theorem and the familiar Kronecker lemma that if $\big(\lambda(k)\ :\ k=1,2,\dots\big)$ is an increasing sequence of positive numbers such that

$$\sum_{k=1}^{\infty}\frac{(\log k)^2}{\lambda^2(k)}<\infty, \tag{5.8}$$

then for every ONS $\big(\phi_k(x)\big)$ we have

$$\lim_{n\to\infty}\frac{1}{\lambda(n)}\sum_{k=1}^{n}\phi_k(x)=0 \quad \text{a.e.} \tag{5.9}$$

Now, from his divergence theorem K. Tandori {14} deduced that condition (5.8) is the best possible for the validity of (5.9).

Corollary. *If* $\big(\lambda(k)\big)$ *is an increasing sequence of positive numbers such that*

$$\sum_{k=1}^{\infty}\frac{(\log k)^2}{\lambda^2(k)}=\infty,$$

then there exists an ONS $\big(\phi_k(x)\big)$ *such that*

$$\limsup_{n\to\infty}\frac{1}{\lambda(n)}\left|\sum_{k=1}^{n}\phi_k(x)\right|=\infty \quad \text{a.e.}$$

In particular, set

$$\lambda(n) := \left\{ n(\log n)^3 (\log\log n)^{1+\varepsilon} \right\}^{1/2}.$$

Then (5.9) holds true for $\varepsilon > 0$, but it is no longer true for $\varepsilon = 0$. In a certain sense, (5.9) can be related to the "strong law of large numbers" well known in probability.

(iii) *Synthesis: Tandori* (1964). Denote by $\mathfrak{M}$ the class of those sequences $\mathfrak{a} = (a_k)$ of real numbers for which the orthogonal series (5.1) converges a.e. for all ONS $(\phi_k(x))$, where the set of measure zero of divergence points may depend on the particular ONS $(\phi_k(x))$. For each $N = 1, 2, \ldots$ introduce the quantity

$$I(a_1, a_2, \ldots, a_N) := \sup \int_0^1 \left(\max_{1 \le n \le N} \left| \sum_{k=1}^{n} a_k \phi_k(x) \right| \right)^2 dx,$$

where the supremum is taken over all ONS $(\phi_k(x))$. The following characterization of $\mathfrak{M}$ is due to K. Tandori {15}.

Characterization theorem of Tandori. *A sequence $\mathfrak{a} = (a_k)$ of numbers belongs to $\mathfrak{M}$ if and only if*

$$\|\mathfrak{a}\| := \lim_{N \to \infty} I^{1/2}(a_1, a_2, \ldots, a_N) < \infty. \tag{5.10}$$

Furthermore, the set $\mathfrak{M}$ is a reflexive Banach space with respect to the usual vector operations and the norm defined in (5.10).

The significance of this theorem is that it reduces the problem of a.e. convergence of general orthogonal series to the problem of evaluating the norm defined in (5.10). The Rademacher–Menshov theorem provides an upper estimate, while the divergence theorem of Tandori provides a lower estimate for this norm, the latter being valid only in the monotonicity case (5.6).

Corollary. *There exist positive constants C_1 and C_2 such that for every sequence $\mathfrak{a} = (a_k)$ of numbers we have*

$$\|\mathfrak{a}\|^2 \le C_1 \sum_{k=1}^{\infty} a_k^2 (\log 2k)^2;$$

and for every sequence $\mathfrak{a} = (a_k)$ *of numbers satisfying condition* (5.6), *we have*

$$\|\mathfrak{a}\|^2 \geq C_2 \sum_{k=1}^{\infty} a_k^2 (\log 2k)^2.$$

The space $\mathfrak{M}$ enjoys the following remarkable property. Let $\mathfrak{a} = (a_k)$ and $\mathfrak{b} = (b_k)$ be sequences of numbers for which

$$|a_k| \leq |b_k|, \quad k = 1, 2, \ldots .$$

If $\mathfrak{b} \in \mathfrak{M}$, then $\mathfrak{a} \in \mathfrak{M}$. If $\mathfrak{a} \notin \mathfrak{M}$, then $\mathfrak{b} \notin \mathfrak{M}$.

(iv) *Sign type* ONS: *Kashin* (1976). In the examples of divergent orthogonal series given by D. E. Menshov {9} and K. Tandori {14}, the orthonormal functions ϕ_k can be chosen to be uniformly bounded:

$$(5.11) \qquad \big|\phi_k(x)\big| \leq C, \qquad 0 \leq x \leq 1; \quad k = 1, 2, \ldots,$$

with some constant C.

The most important case is when $C = 1$. In this case, we necessarily have

$$\big|\phi_k(x)\big| = 1, \qquad 0 \leq x \leq 1; \quad k = 1, 2, \ldots;$$

and the ONS $\big(\phi_k(x)\big)$ is said to be sign type. Now, B. S. Kashin {6} improved the divergence theorem of Tandori as follows.

Divergence theorem of Kashin. *If conditions* (5.6) *and* (5.7) *are satisfied, then there exists a sign type* ONS $\big(\phi_k(x)\big)$ *depending on* (a_k) *such that the orthogonal series* (5.1) *diverges everywhere.*

Later K. Tandori {16} gave an extremely short proof of this surprising result.

Let $1 \leq C \leq \infty$ and denote by $\mathfrak{M}(C)$ the class of those sequences $\mathfrak{a} = (a_k)$ of real numbers for which the orthogonal series (5.1) converges a.e. for all ONS $\big(\phi_k(x)\big)$ satisfying condition (5.11). It is clear that if $1 < C_1 < C_2 < \infty$, then

$$\mathfrak{M}(\infty) \subseteq \mathfrak{M}(C_2) \subseteq \mathfrak{M}(C_1) \subseteq \mathfrak{M}(1).$$

Now, K. Tandori {17} proved that

$$\mathfrak{M}(C) = \mathfrak{M}(1), \quad 1 \leq C < \infty.$$

In other words, there is no difference between the class of all sign type ONS and the class of all uniformly bounded ONS as to the convergence a.e. of the orthogonal series (5.1). But the problem of whether $\mathfrak{M}(\infty) = \mathfrak{M}(1)$ is still open.

(v) *Rademacher–Menshov inequality revisited: Móricz and Tandori* (1996). The following sharpened form has been proved by Ferenc Móricz and Károly Tandori {11}.

Convergence theorem of Móricz and Tandori. *If for some* $0 < \varepsilon \le 2$ *we have*

$$\sum_{m=0}^{\infty} \sum_{k \in I_m} a_k^2 (\log k)^{\varepsilon} \left(\log \frac{2A_m^2}{a_k^2} \right)^{2-\varepsilon} < \infty, \tag{5.12}$$

where

$$I_m := \{2^m + 1, 2^m + 2, \ldots, 2^{m+1}\} \tag{5.13}$$

and

$$A_m^2 := \sum_{k \in I_m} a_k^2, \quad m = 0, 1, \ldots,$$

then the orthogonal series (5.1) *converges a.e.*

Denote by $\sum_\varepsilon$ the sum of the series occurring in (5.12). It is not difficult to show that if $0 \le \delta < \varepsilon \le 2$ and $\sum_\varepsilon < \infty$, then $\sum_\delta < \infty$ as well. Thus, the smaller ε is, the weaker is condition (5.12). Furthermore, if the sequence (a_k) satisfies condition (5.6) and $\sum_\varepsilon < \infty$ for some $0 \le \varepsilon \le 2$, then $\sum_\varepsilon < \infty$ for all $0 \le \varepsilon \le 2$.

It is interesting to point out that condition (5.12) for $\varepsilon = 0$ no longer guarantees the a.e. convergence of the orthogonal series (5.1). However, condition (5.12) for $\varepsilon = 0$ is necessary in order that the orthogonal series (5.1) converge a.e. for all ONS $(\phi_k(x))$. Even somewhat more was shown by Károly Tandori {18}: If the orthogonal series $\sum a_k \phi_k(x)$ converges a.e. for all ONS $(\phi_k(x))$, then

$$\sum_{k=1}^{\infty} a_k^2 \left(\log_+ \frac{1}{a_k^2} \right)^2 < \infty, \tag{5.14}$$

where

$$\log_+ u := \begin{cases} \log u & \text{if } u \ge 2 \\ 1 & \text{if } u < 2. \end{cases}$$

Obviously, (5.14) implies (5.12) for $\varepsilon = 0$. To sum up, condition (5.12) for some $\varepsilon > 0$ is sufficient, while for $\varepsilon = 0$ is necessary in order that the orthogonal series (5.1) converge a.e. for all ONS $\big(\phi_k(x)\big)$. Furthermore, these conditions are equivalent in the special case when (a_k) satisfies condition (5.6).

6. Cesàro Summability of Orthogonal Series

Similarly to trigonometric Fourier series, one can expect better convergence behavior of orthogonal series if ordinary convergence is replaced by $(C,1)$-summability. In fact, this is the case, but the improvement is less than that in the case of trigonometric series.

We recall that the first arithmetic or $(C,1)$-mean of the orthogonal series (5.1) with partial sum

$$s_n(x) := \sum_{k=1}^{n} a_k\phi_k(x)$$

is defined by

$$\sigma_n(x) := \frac{1}{n}\sum_{k=1}^{n} s_k(x) = \sum_{k=1}^{n}\left(1-\frac{k-1}{n}\right) a_k\phi_k(x), \qquad n = 1,2,\dots .$$

The whole theory of $(C,1)$-summability of orthogonal series is based on the following observations of A. N. Kolmogorov {7} and Stefan Kaczmarz {4}, respectively.

Theorem of Kolmogorov and Kaczmarz. *If condition* (5.3) *is satisfied, then*

$$\lim_{m\to\infty}\big\{ s_{2^m}(x) - \sigma_{2^m}(x)\big\} = 0 \quad \text{a.e.}$$

and

$$\lim_{m\to\infty}\ \max_{2^m<k<2^{m+1}} \big|\sigma_k(x) - \sigma_{2^m}(x)\big| = 0 \quad \text{a.e.}$$

As a corollary we obtain that, under condition (5.3), the orthogonal series (5.1) is $(C,1)$-summable a.e. if and only if the subsequence $\big(s_{2^m}(x) : m = 0,1,\dots\big)$ of the partial sums converges a.e. Now, it is a simple consequence of the Rademacher–Menshov theorem that if

$$\sum_{m=1}^{\infty}\Big(\sum_{k\in I_m} a_k^2\Big)(\log m)^2 < \infty, \tag{6.1}$$

where I_m is defined in (5.13), then the partial sums $s_{2^m}(x)$ of the orthogonal series (5.1) converge a.e. It is clear that (6.1) is equivalent to the condition

$$\sum_{k=4}^{\infty} a_k^2 (\log\log k)^2 < \infty. \tag{6.2}$$

Combining this observation with the theorem of Kolmogorov and Kaczmarz yields the following.

Theorem of Menshov and Kaczmarz. *If condition* (6.2) *is satisfied, then the orthogonal series* (5.1) *is* $(C, 1)$*-summable* a.e.

This theorem was first proved by D. E. Menshov {10} and S. Kaczmarz {5}. D. E. Menshov {10} also proved that condition (6.2) is the best possible.

Theorem of Menshov. *If* $\big(w(k) : k = 1, 2, \ldots\big)$ *is an increasing sequence of positive numbers such that*

$$w(k) = o(\log\log k) \quad \text{as} \quad k \to \infty,$$

then there exists a sequence (a_k) *of numbers and an ONS* $\big(\phi_k(x)\big)$ *such that condition* (5.5) *is satisfied and the orthogonal series* (5.1) *is nowhere* $(C, 1)$*-summable.*

K. Tandori {19} also sharpened this result of Menshov into a necessary and sufficient condition for certain sequences (a_k) of numbers as follows.

Theorem of Tandori. *If* (a_k) *is a sequence of numbers for which*

$$\sqrt{k}|a_k| \geq \sqrt{k+1}|a_{k+1}|, \quad k = 1, 2, \ldots$$

and

$$\sum_{k=4}^{\infty} a_k^2 (\log\log k)^2 = \infty,$$

then there exists an ONS $\big(\phi_k(x)\big)$ *depending on* (a_k) *such that the orthogonal series* (5.1) *is nowhere* $(C, 1)$*-summable.*

7. Unconditional Convergence of Orthogonal Series

One of the deepest results of K. Tandori {20} completely solved the problem of unconditional convergence of general orthogonal series. Since a general ONS $\big(\phi_k(x)\big)$ (unlike the classical trigonometric system, for example) has no a priori given arrangement of its terms, it is quite natural to study the question of convergence of the orthogonal series (5.1) in every rearrangement of its terms; that is, the convergence of the rearranged series

$$\sum_{\ell=1}^{\infty} a_{k(\ell)}\phi_{k(\ell)}(x), \tag{7.1}$$

where $\{k : \ell \to k(\ell) : \ell = 1, 2, \dots\}$ is a bijective mapping of the set of the positive integers (so-called permutation). By a.e. convergence of series (7.1), we mean the term "almost everywhere" in the sense that the set of measure zero of divergence points may vary with every rearrangement. Otherwise, the a.e. convergence in every rearrangement would reduce to the a.e. absolute convergence. Now, the aforementioned theorem of Tandori can be formulated as follows. Let

$$\nu_p := 2^{2^p} \quad \text{and} \quad J_p := \{\nu_p + 1, \nu_p + 2, \dots, \nu_{p+1}\}, \quad p = 0, 1, \dots .$$

We agree to say that the orthogonal series (5.1) converges unconditionally a.e. if the series in (7.1) converges a.e. for each permutation $\{k : \ell \to k(\ell) : \ell = 1, 2, \dots\}$.

Unconditional convergence theorem of Tandori. *Let*

$$|a_1^*| \ge |a_2^*| \ge \dots \ge |a_k^*| \ge \dots$$

be a decreasing rearrangement of the sequence (a_k) *of numbers. Then the orthogonal series* (5.1) *converges unconditionally* a.e. *if and only if*

$$\sum_{p=0}^{\infty} \Big\{ \sum_{k\in J_p} |a_k^*|^2 (\log k)^2 \Big\}^{1/2} < \infty.$$

The idea of studying unconditional convergence is due to Wladislaw Orlicz {12}, who proved the first basic result in this direction. His result, which is actually a consequence of the above theorem of Tandori, is formulated in terms of the so-called Weyl multiplier.

Corollary. *Let* $(\lambda(k) : k = 1, 2, \dots)$ *be an increasing sequence of positive numbers.*

(i) (due to Orlicz). *If*

$$\sum_{p=0}^{\infty} \frac{1}{\lambda(\nu_p)} < \infty, \quad \textit{or equivalently} \quad \sum_{m=1}^{\infty} \frac{1}{m\lambda(2^m)} < \infty, \tag{7.2}$$

and

$$\sum_{k=1}^{\infty} a_k^2 (\log k)^2 \lambda(k) < \infty, \tag{7.3}$$

then the orthogonal series (5.1) *converges unconditionally a.e.*

(ii) (due to Tandori). *On the other hand, if*

$$\sum_{p=0}^{\infty} \frac{1}{\lambda(\nu_p)} = \infty, \quad \textit{or equivalently} \quad \sum_{m=1}^{\infty} \frac{1}{m\lambda(2^m)} = \infty,$$

then there exist a sequence (a_k) *of numbers and an ONS* $(\phi_k(x))$ *such that condition* (7.3) *is satisfied and the orthogonal series* (5.1) *diverges a.e. in some rearrangement of its terms.*

For instance, if

$$\lambda(k) := (\log \log k)^{1+\varepsilon}, \quad k = 4, 5, \dots,$$

then condition (7.2) holds whenever $\varepsilon > 0$, but fails to hold when $\varepsilon = 0$. The above corollary can be reformulated as follows: The sequence $((\log k)^2 \lambda(k))$ is a Weyl multiplier for the unconditional convergence of orthogonal series for $\varepsilon > 0$, while it is not for $\varepsilon = 0$.

References

[8] Alexits, György, *Convergence Problems of Orthogonal Series,* International Series of Monographs on Pure and Applied Mathematics, Vol. 20. Pergamon Press (New York–Oxford–Paris, 1961).

[62] Haar, Alfréd, *Összegyűjtött Munkái = Gesammelte Arbeiten,* ed. Béla Sz.-Nagy, Akadémiai Kiadó (Budapest, 1959).

[66] Hardy, G. H. – Riesz, Marcel, *The General Theory of Dirichlet Series,* Cambridge Tracts in Mathematics and Mathematical Physics, No. 18, Cambridge, University Press (London, 1915), reprint: 1952.

[68] Hardy, G. H, *Divergent Series,* Oxford University Press (London, 1949).

[108] Leindler, László, *Strong Approximation by Fourier Series,* Akadémiai Kiadó (Budapest, 1985).

[156] Riesz, Frigyes, *Összegyűjtött Munkái = Œuvres complètes = Gesammelte Arbeiten,* ed. Á. Császár, Akadémiai Kiadó (Budapest, 1960).

[158] Riesz, Marcel, *Collected Papers,* ed. L. Gårding and L. Hörmander, Springer-Verlag (Berlin–Heidelberg–New York, 1988).

[203] Zygmund, Antoni, *Trigonometric Series,* Cambridge University Press (London–New York, 1959).

{1} G. Alexits, A Fourier-sor Cesàro közepeivel való approximáció nagyságrendjéről (On the order of approximation by the Cesàro means of Fourier series), *Mat. Fiz. Lapok,* **48** (1941), 410–421 — Sur l'ordre de grandeur de l'approximation d'une function périodique par les sommes de Fejér, *Acta Math. Acad. Sci. Hungar.*, **3** (1952), 29–42.

{2} G. Alexits, Sur les bornes de la théorie de l'approximation des fonctions continues par polynômes, *Magyar Tud. Akad. Kutató Int. Közl.,* **8** (1963), 329–340.

{3} R. A. De Vore, The approximation of continuous functions by positive linear operators, in: *Lect. Notes in Math.,* **293**, Springer (Berlin–Heidelberg–New York, 1972).

{4} S. Kaczmarz, Über die Reihen von allgemeinen Orthogonalfunktionen, *Math. Annalen,* **96** (1925), 148–151.

{5} S. Kaczmarz, Über die Summierbarkeit der Orthogonalreihen, *Math. Z.,* **26** (1927), 99–105.

{6} B. S. Kašin, On Weyl's multipliers for almost everywhere convergence of orthogonal series, *Analysis Math.,* **2** (1976), 249–266.

{7} A.N. Kolmogoroff, Une contribution à l'étude de la convergence des séries de Fourier, *Fund. Math.,* **5** (1924), 96–97.

{8} D. Menchoff, Sur les séries de fonctions orthogonales. I, *Fund. Math.,* **4** (1923), 82–105.

{9} D. E. Menchoff, Sur les séries de fonctions orthogonales bornées dans leur ensemble, *Recueil Math. Moscou (Mat. Sb.),* **3** (1938), 103–118.

{10} D. E. Menchoff, Sur les séries de fonctions orthogonales. II, *Fund. Math.,* **8** (1926), 56–108.

{11} F. Móricz and K. Tandori, An improved Menshov–Rademacher theorem, *Proc. Amer. Math. Soc.,* **124** (1996), 877–885.

{12} W. Orlicz, Zur Theorie der Orthogonalreihen, *Bulletin Internat. Acad. Sci. Polonaise Cracovie,* (1927), 81–115.

{13} H. Rademacher, Einige Sätze über Reihen von allgemeinen Orthogonalfunktionen, *Math. Annalen,* **87** (1922), 112–138.

{14} K. Tandori, Über die orthogonalen Funktionen. I, *Acta Sci. Math. (Szeged),* **18** (1957), 57–130.

{15} K. Tandori, Über die Konvergenz der Orthogonalreihen. II, *Acta Sci. Math. (Szeged),* **25** (1964), 219–232.

{16} K. Tandori, Einfacher Beweis eines Satzes von B. S. Kašin, *Acta Sci. Math. (Szeged),* **39** (1977), 175–178.

{17} K. Tandori, Über beschränkte orthonormierte Systeme, *Acta Math. Acad. Sci. Hungar.,* **31** (1978), 279–285.

{18} K. Tandori, Über die Divergenz der Orthogonalreihen, *Publicationes Math. Debrecen,* **8** (1961), 291–307.

{19} K. Tandori, Über die orthogonalen Funktionen. II (Summation), *Acta Sci. Math. (Szeged),* **18** (1957), 149–168.

{20} K. Tandori, Über die orthogonalen Funktionen. X (Unbedingte Konvergenz), *Acta Sci. Math. (Szeged),* **23** (1962), 185–221.

{21} M. Zamansky, Classes de saturation de certains procédés d'approximation des séries de Fourier, *Annales de l'Ecole Normale Supérieure,* **66** (1949), 19–93.

Ferenc Móricz
University of Szeged
Bolyai Institute
Aradi Vértanúk tere 1
6720 Szeged
Hungary

moricz@math.u-szeged.hu

BOLYAI SOCIETY
MATHEMATICAL STUDIES, 14

A Panorama of Hungarian Mathematics in the Twentieth Century, pp. 55–70.

Orthogonal Polynomials

JÓZSEF SZABADOS

The theory of orthogonal polynomials plays an important role in many branches of mathematics, such as approximation theory (best approximation, interpolation, quadrature), special functions, continued fractions, differential and integral equations. The notion of orthogonality originated from the theory of continued fractions, but later became an independent (and possibly more important) discipline. Among the contributors to the theory of orthogonal polynomials, we can find such outstanding mathematicians as Abel, Chebyshev, Fourier, Hermite, Laguerre, Laplace, Legendre, Markov, and Stieltjes, just to name a few. Beginning with Gábor Szegő, Hungarian mathematicians like Pál Erdős, Pál Turán, Géza Freud, Ervin Feldheim and others have made essential contributions to the flourishing theory of orthogonal polynomials in the last century. At this point I would like to mention two names who have made considerable efforts to propagate the work of the above mentioned Hungarian mathematicians: Richard Askey and Doron Lubinsky.

A considerable part of the material to be presented below is based on the classical book of Szegő [174]. Since its first publication in 1939, this monograph has reached three more editions, and until now is the most comprehensive, most quoted source on the subject. Another source of information concerning more recent developments in the theory of orthogonal polynomials is the monograph [47] of Géza Freud and a survey paper by Paul Nevai (J. Approx. Theory 48 (1986), pp. 3–167).

Problems connected with interpolation and mechanical quadrature on the roots of orthogonal polynomials are considered in another chapter written by Péter Vértesi.

Let α be a real valued increasing function on $\mathbf{R}$. We call such an α a *distribution function* if it assumes infinitely many values and the improper

Stieltjes integrals

$$c_n = \int_{\mathbf{R}} x^n \, d\alpha(x), \qquad n = 0, 1, \ldots \tag{1}$$

(called *moments*) exist and are finite. If α is absolutely continuous we write $d\alpha(x) = w(x)\,dx$, and call w a *weight function*. Let $\mathcal{P}_n$ and $\mathcal{P}_n^c$, $n = 0, 1, \ldots$ be the set of real and complex algebraic polynomials of degree at most n, respectively. Associated with each distribution function α, there is a unique sequence of *orthonormal polynomials* $p_n \in \mathcal{P}_n$, $\deg p_n = n$, $n = 0, 1, \ldots$ with positive leading coefficients γ_n and having the property

$$\int_{\mathbf{R}} p_n(x) p_m(x) \, d\alpha(x) = \delta_{mn}, \qquad m, n = 0, 1, \ldots .$$

Of course, the region of the above integration can be restricted to the *support* of the distribution α which, in general, may be a finite interval, the half line, or $\mathbf{R}$.

We list some important properties of orthogonal polynomials.

1. p_n, $n = 0, 1, \ldots$ are linearly independent.

2. All roots of p_n are real, simple, and lie in the interior of the support of α (which is a finite or infinite interval $[a, b]$).

3. With the help of the moments (1), the representation

$$p_n(x) = (D_{n-1} D_n)^{-1/2} \begin{vmatrix} c_0 & c_1 & c_2 & \ldots & c_n \\ c_1 & c_2 & c_3 & \ldots & c_{n+1} \\ \vdots & \vdots & \vdots & \ddots & \vdots \\ c_{n-1} & c_n & c_{n+1} & \ldots & c_{2n-1} \\ 1 & x & x^2 & \ldots & x^n \end{vmatrix}$$

holds, where $D_n = \det \left[c_{i+j}\right]_{i,j=0,\ldots,n}$, $n = 1, 2, \ldots$.

4. There is a three term recurrence relation

$$x p_n(x) = a_{n+1} p_{n+1}(x) + b_n p_n(x) + a_n p_{n-1}(x), \qquad n = 2, 3, \ldots, \tag{2}$$

where

$$a_n = \frac{\gamma_{n-1}}{\gamma_n} \quad \text{and} \quad b_n = \int_{\mathbf{R}} t p_n(t)^2 \, d\alpha(t).$$

5. The Christoffel–Darboux formula

$$K_n(x,y) := \sum_{k=0}^{n-1} p_k(x)p_k(y) = \frac{\gamma_{n-1}}{\gamma_n} \frac{p_n(x)p_{n-1}(y) - p_{n-1}(x)p_n(y)}{x-y} \tag{3}$$

holds.

Note also the following important property of K_n:

$$\lambda_n(x) := \frac{1}{K_n(x,x)} = \inf_{p(x)=1,\ p\in\mathcal{P}_{n-1}} \int_a^b p^2(t)\,d\alpha(t). \tag{4}$$

The latter is called the nth *Christoffel function* associated with $d\alpha$. Perhaps this is the most important notion in connection with orthogonal polynomials, since results concerning this function have a deep effect on various fields of approximation theory, harmonic and numerical analysis.

6. The Gauss–Jacobi quadrature formula

$$\int_{\mathbf{R}} p\,d\alpha = \sum_{k=0}^{n} \lambda_k p(x_k)$$

valid for every polynomial $p \in \mathcal{P}_{2n-1}$, where $x_k = x_{kn}$ are the roots of the orthogonal polynomial p_n, and

$$\lambda_k = \int_a^b \frac{p_n(x)}{p_n'(x_k)(x-x_k)}\,dx = \frac{1}{K_n(x_k,x_k)} > 0, \qquad k = 0,1,\ldots,$$

are the Cotes numbers.

With each function f such that $\int_{\mathbf{R}} f^2\,d\alpha < \infty$, we can associate the nth partial sum

$$S_n(d\alpha, f, x) := \sum_{k=0}^{n} f_k p_k(x), \tag{5}$$

of the orthogonal expansion, where

$$f_k := \int_{\mathbf{R}} f p_k\,d\alpha, \qquad k = 0,1,\ldots.$$

Of particular interest are the classical orthogonal polynomials, when

$$w(x) = \begin{cases} (1-x)^\alpha(1+x)^\beta \ (\alpha,\beta > -1) & \text{(Jacobi polynomials on } [-1,1], \\ x^\alpha e^{-x} \quad (\alpha > -1) & \text{(Laguerre polynomials on } [0,\infty), \\ e^{-x^2} & \text{(Hermite polynomials on } \mathbf{R}). \end{cases}$$

What makes these cases interesting is the existence of second order linear differential equations

$$(1-x^2)p_n''(x)+\left[\beta-\alpha-(\alpha+\beta+2)x\right]p_n'(x)+n(n+\alpha+\beta+1)p_n(x)=0,$$

$$xp_n''(x)+(\alpha+1-x)p_n'(x)+np_n(x)=0,$$

$$p_n''(x)-2xp_n'(x)+2np_n(x)=0,$$

respectively. These differential equations (which do not exist in the general case) make all considerations much easier. Hence our knowledge concerning the classical orthogonal polynomials is much more exhaustive.

The special case $\alpha=\beta=0$ of the above mentioned Jacobi polynomials, i.e. the Legendre polynomials, were generalized by Leopold Fejér in the following way. Let $f(z)=\sum_{n=0}^{\infty}a_nz^n$ be a generator function analytic in a neighborhood of the origin with all a_n real. Consider

$$\left|f(re^{i\theta})\right|^2=\sum_{n=0}^{\infty}p_n(\cos\theta)r^n,$$

where

$$p_n(\cos\theta)=\sum_{k=0}^{n}a_ka_{n-k}\cos(n-2k)\theta$$

is a polynomial of degree n of the variable $\cos\theta$. These are generalizations of the Legendre polynomials which are obtained in the special case $f(z)=(1-z)^{-1/2}$. More generally, for $f(z)=(1-z)^{-\lambda}$, $\lambda>-1/2$, we obtain the ultraspherical Jacobi polynomials (with the above notation, for $\alpha=\beta=\lambda-1/2$). Imposing proper monotony and asymptotic conditions on the sequence $\{a_n\}$, one can obtain polynomials p_n with different interesting properties (see Ch. 6.5 of Szegő's monograph [174]). The most intriguing question in this context is whether the polynomials p_n are orthogonal with respect to some weight function. Ervin Feldheim (Izv. Akad. Nauk SSSR, Ser. Math. **5** (1941), 241–248) and independently I. L. Lanzewizky showed that in case of orthogonality, these polynomials p_n can be rescaled to C_n to satisfy a three term recurrence relation of the form (2):

$$2x(1-\beta q^n)C_n(x)=(1-q^{n+1})C_{n+1}(x)+(1-\beta^2q^{n-1})C_{n-1}(x)$$

for $n\geq 1$, where

$$C_0(x)=1,\quad C_1(x)=2x\frac{1-\beta}{1-q},\qquad |\beta|,|q|<1.$$

(When $\beta = q^\lambda$ we obtain the ultraspherical polynomials mentioned above.) Although Feldheim was unable to obtain explicit representation for these polynomials, as well as for the weight function, this was a crucial step in characterizing these polynomials. Later the polynomials were found explicitly by Richard Askey who did pioneering work in the theory of so-called q-polynomials (cf. e.g. R. Askey, M. Ismail, Studies in Pure Math., P. Erdős editor, Birkhäuser, Basel, 1983, pp. 55–78.).

The notion of orthogonal polynomials was extended to complex domains mainly by the pioneering work of Szegő (cf. [174], Chs. XI and XVI). The most interesting case is the unit circle $|z| = 1$. Let μ be an increasing 2π periodic function which takes infinitely many values in any interval of length 2π. Then there is a uniquely determined sequence of complex polynomials $\phi_n(z) = \kappa_n z^n + \dots$, $\phi_n \in \mathcal{P}_n$, $\kappa > 0$ real numbers, such that

$$\frac{1}{2\pi}\int_{-\pi}^{\pi} \phi_m(z)\overline{\phi_n(z)}\, d\mu(\theta) = \delta_{mn}, \qquad z = e^{i\theta}; \quad m, n = 0, 1, \dots .$$

Szegő has shown that properties of these polynomials are somewhat simpler than those of the real orthogonal polynomials, to which they are related (in case of a finite interval). Such properties enable us to derive statements for real orthogonal polynomials from those of complex ones.

All roots of the complex orthonormal polynomials lie in the open unit circle. The problem of distribution of roots in the unit circle is a delicate question. Turán (J. Approx. Theory **29** (1980), 23–85.) raised the question whether the accumulation points of the roots of the orthogonal polynomials can fill up the whole unit disk. I (Acta Math. Acad. Sci. Hungar. **33** (1979), 197–210.) gave a partial answer to this question by constructing a weight function for any given $\varepsilon > 0$ such that the two dimensional Lebesgue measure of the above mentioned accumulation points greater than $\pi - \varepsilon$. The complete positive answer to Turán's problem was given later by M. P. Alfaro and L. Vigil (J. Approx. Theory **53** (1988), 195–197.).

Besides the three term recurrence relation (2), complex orthogonal polynomials on the unit circle obey the simpler two term forward-backward recurrence relations

$$\kappa_n z\phi_n(z) = \kappa_{n+1}\phi_{n+1}(z) - \phi_{n+1}(0)\phi^\star_{n+1}(z)$$

and

$$\kappa_n\phi_{n+1}(z) = \kappa_{n+1}\phi_n(z) + \phi_{n+1}(0)\phi^\star_n(z),$$

where $\phi_n^\star(z) := z^n\overline{\phi}_n(z^{-1})$ are the so-called *reciprocal polynomials,* where the bar indicates that the coefficients of the corresponding polynomial are conjugated (cf. Szegő [174]). These formulas play an important role in constructing a simpler algorithm for the prediction of a stationary time series (cf. [173], pp. 43–44).

In order to discover further properties of orthogonal polynomials, we have to consider the analog

$$K_n(z,u) := \sum_{k=0}^{n-1} \overline{\phi_k(z)}\phi_k(u) = \frac{\overline{\phi_n^\star(z)}\phi_n^\star(u) - \overline{\phi_n(z)}\phi_n(u)}{1-\overline{z}u}$$

of the Christoffel–Darboux formula (3). The analogue of the Christoffel function (4) is

$$\omega_n(d\mu,z) := K_n(z,z)^{-1} = \left[\sum_{k=0}^{n-1} |\phi_k(z)|^2\right]^{-1} \tag{6}$$

$$= \inf_{\substack{p\in\mathcal{P}_{n-1}^c\\ p(z)=1}} \frac{1}{2\pi}\int_0^{2\pi} |p(u)|^2 d\mu(t), \qquad u=e^{it}.$$

Let $0 \le g \in L_1$ be a positive measurable function in $[0,2\pi]$. The *Szegő function* associated with g is defined as

$$D(g,z) = \exp\left\{\frac{1}{4\pi}\int_0^{2\pi} \frac{u+z}{u-z}\log g(t)\,dt\right\}, \qquad u=e^{it}, \quad |z|<1. \tag{7}$$

$D(g,0) \neq 0$ in $|z|<1$, $D(g)$ is square integrable in the unit disk, $D(g,0)>0$, the radial limit $D(g,e^{it})$ exists, and $\left|D(g,e^{it})\right|^2 = g(t)$ almost everywhere. Szegő [174] proved that if μ is differentiable and $\log\mu' \in L_1$, then the asymptotic relation

$$\lim_{n\to\infty} \phi_n^\star(z) = D(\mu',z^{-1})$$

holds uniformly in every compact subset of $|z|<1$. Moreover,

$$\lim_{n\to\infty} z^{-n}\phi_n(z) = \overline{D(\mu',z^{-1})}, \qquad |z|>1,$$

and

$$\lim_{n\to\infty} \kappa_n = D(\mu',0)^{-1},$$

or equivalently

$$\lim_{n\to\infty}\int_0^{2\pi}\left[\left|\phi_n(z)\right|\mu'(t)^{1/2}-1\right]^2 dt = 0, \qquad z = e^{it}.$$

The latter result was later extended by Attila Máté, Paul Nevai, and Vilmos Totik to the case when $\log\mu' \in L_1$ is replaced by the weaker condition that $\mu' > 0$ almost everywhere.

As for the behavior of the orthogonal polynomials on the unit circle, Szegő and U. Grenander [59] showed that

$$\lim_{n\to\infty}\left[\phi_n(e^{it}) - e^{int}\overline{D^{-1}(\mu', e^{it})}\right] = 0 \tag{8}$$

uniformly in t, provided $0 < m \le \mu'(t) \le M < \infty$ and the modulus of continuity $\omega(\mu', h)$ of μ' is of order $\big(\log(1/h)\big)^{-\beta}$, $\beta > 1$. The latter condition was later weakened to $\int_0^\pi \frac{\omega(\mu',h)}{h}\,dh < \infty$ by Freud, who also gave similar conditions for the pointwise convergence in (8) (cf. [47]). The idea of localization of the asymptotic behavior of orthogonal polynomials is due to Tamás Frey (Mat. Sbornik **49** (1959), 133–180.).

The Szegő function (7) plays role in solving *Szegő's generalized extremal problem:* find

$$\omega(d\mu, z) := \lim_{n\to\infty}\omega_n(d\mu, z), \qquad |z| < 1$$

(note that

$$\omega(d\mu, z) = \begin{cases} \dfrac{\mu(z)}{2\pi}, & \text{if } |z| = 1,\\ 0, & \text{if } |z| > 1, \end{cases}$$

where $\mu(z)$ is the $d\mu$-measure of the point z).

Szegő found this limit function when μ is absolutely continuous. The case when μ is not necessarily absolutely continuous was proved by A. N. Kolmogorov and generalized by M. G. Krein for $L^p_{d\mu}$ measures ($p > 0$) in (6) instead of $L^2_{d\mu}$-measure.

In the above quoted work of Szegő and Grenander it is proved that

$$\omega(d\mu, z) = \left(1 - |z|^2\right)\left|D(\mu', z)\right|^2, \qquad |z| < 1$$

for every measure $d\mu$ on the unit circle.

The considerations resulting in the asymptotic formula (8) in the case of the unit circle led Szegő to the investigation of the asymptotic behavior of the Christoffel function (see (4)) in case of the interval $[-1,1]$. He proved that if the support of $d\alpha$ is in $[-1,1]$ and the derivative of the function

$$\frac{1}{|\sin t|\alpha'(\cos t)} \ge 1$$

is in some Lipschitz class, then (cf. Freud [47], p. 269)

$$\lambda_n^{-1}(x) = n - \frac{1}{2} + \operatorname{Re}\left[e^{i\xi}\frac{D'(\mu', e^{i\xi})}{D(\mu', e^{i\xi})}\right]$$

$$+\frac{1}{2\sin\xi}\operatorname{Im}\left[e^{(2n-1)i\xi}\frac{D(\mu', e^{i\xi})}{\overline{D(\mu', e^{i\xi})}}\right] + o(1), \qquad x = \cos\xi,$$

where

$$\mu(t) := (\operatorname{sgn} t)\big[\alpha(1) - \alpha(\cos t)\big], \qquad |t| \le \pi.$$

Purely in terms of orthogonal polynomials on the interval $[-1,1]$, a complete and general asymptotic of the Christoffel function can be found in a paper by Paul Nevai {1}. He proved that if $\log\alpha'(\cos t) \in L_1$ then

$$\lim_{n\to\infty} n\lambda_n(x) = \pi\alpha'(x)\left(1-x^2\right)^{1/2} \tag{9}$$

for almost all x in an interval where $1/\alpha'(x) \in L_1$. In addition, if $x \in (-1,1)$, α is absolutely continuous in a neighborhood of x and α' is continuous at x then (9) holds. The condition on α later was relaxed to $\log\alpha' \in L_1$ by Máté, Nevai and Totik (Annals of Math. **134** (1991), 433–453.).

Convergence problems of orthogonal expansions can be simplified by considering only Fourier series. This idea of *equiconvergence theorems* was developed by Alfréd Haar (Math. Annalen **78** (1917), 121–136.), and later generalized by Szegő (Math. Zeitschrift **12** (1921), 61–94.) and Freud ([47], p. 260.). The latter proved the following: Let $f \in L^2$ with respect to an absolutely continuous distribution function α whose support is in $[-1,1]$ such that its derivative has a positive polynomial minorant and is in a Lipschitz class. Further let $f^\star$ be that function which coincides with f in a neighborhood of f and zero otherwise. With the notation (5) and $d\alpha_0(x) = \left(1-x^2\right)^{-1/2} dx$ we have

$$\lim_{n\to\infty}\left[S_n(d\alpha, f, x) - S_n(d\alpha_0, f^\star, x)\right] = 0.$$

Note that the convergence properties of $S_n(d\alpha, f)$ depend only values of f taken in an arbitrarily small interval around the fixed point x. Under some additional conditions the above results can be made uniform in x.

The theory of orthogonal polynomials on infinite intervals is significantly different from that on a finite interval. Until the 1960's not much was known about this topic, except possibly the Hermite, Laguerre and a few other polynomials. Amazingly, while Szegő has made pioneering work in the theory on finite intervals, he was not interested in carrying over his ideas to infinite intervals. It was Freud who founded the now flourishing theory of orthogonal polynomials on the real line, and the corresponding representative polynomials are named after him. In the rest of this chapter we will describe this theory, except problems connected with interpolation and quadrature which are considered in the above mentioned work of Péter Vértesi.

Freud's aim was to extend the theory of best approximation, Jackson–Bernstein type estimates to the real axis. The natural way to do this was to explore properties of orthogonal polynomials, since the expectation was that orthogonal expansions, Cesàro and de la Vallée-Poussin means, may serve as near-best approximation.

What Freud did was to start from Hermite polynomials (orthogonal with respect to the weight e^{-x^2}), and generalize this weight. He considered weights of the form

$$w(x) = e^{-Q(x)} \tag{10}$$

where the even, twice differentiable function $Q > 0$ behaves like a polynomial at infinity, and $xQ'(x)$ is increasing. More exactly, we assume that

$$0 < \alpha = \liminf_{x\to\infty} \frac{\big(xQ'(x)\big)'}{Q'(x)} \le \limsup_{x\to\infty} \frac{\big(xQ'(x)\big)'}{Q'(x)} < \infty. \tag{11}$$

An important role is played by the number q_n which is defined (by Freud) as the unique solution of the equation

$$q_n Q'(q_n) = n, \qquad n = 1, 2 \ldots .$$

We mention that a closely related number a_n (which is of the same order of magnitude as q_n) was later determined independently by Mhaskar, Rahmanov and Saff as the unique positive solution of the integral equation

$$\frac{1}{\pi}\int_{-a_n}^{a_n} \frac{tQ'(t)\,dt}{\sqrt{a_n^2 - t^2}} = n, \qquad n = 1, 2, \ldots .$$

The number a_n describes the behavior of polynomials at infinity, namely the so-called infinite-finite range relation (this terminology was coined by D. Lubinsky)

$$\max_{|x|\le a_n} w(x)\big|p_n(x)\big| = \max_{x\in\mathbf{R}} w(x)\big|p_n(x)\big|$$

holds for every polynomial $p_n \in \mathcal{P}_n$. While this result tells us that the behavior of a weighted polynomial can be learned by studying it on a finite interval (whose length is independent of the polynomial and depends only on the weight and the degree), the number a_n plays the same role in considering L_p norms: we have

$$\text{(12)}\qquad \left(\int_{\mathbf{R}} w(x)\big|p_n(x)\big|^p\,dx\right)^{1/p} \le c\left(\int_{-a_n}^{a_n} w(x)\big|p_n(x)\big|^p\,dx\right)^{1/p},$$

for $p_n \in \mathcal{P}_n$, $p \ge 1$ where $c > 0$ are constants depending only on the weight. This infinite-finite range inequality was discovered by Freud for $p = 2$ and Q =polynomial, and later generalized for weights (10) with the property (11) by D. Lubinsky and others. As a further generalization, I (Advanced Problems in Constructive Approximation, Birkhäuser 2002, pp. 223–236.) considered weights with infinitely many zeros on $\mathbf{R}$. Let

$$0 < t_1 < t_2 < \dots, \qquad \lim_{k\to\infty} t_k = \infty$$

and

$$m_k > 0, \quad k = 1, \dots, \qquad \liminf_{k\to\infty} m_k > 0$$

be two sequences of real numbers satisfying the condition

$$\sum_{k=1}^{\infty} \frac{m_k}{t_k^{\varrho+\varepsilon}} < \infty, \quad \text{and} \quad \sum_{k=1}^{\infty} \frac{m_k}{t_k^{\varrho-\varepsilon}} = \infty$$

for some $\varrho \ge 0$ and all $\varepsilon > 0$. Now consider the weight

$$w(x) = e^{-Q(x)} \prod_{k=1}^{\infty} \left|1 - \left(\frac{x}{t_k}\right)^{2q}\right|^{m_k}, \qquad q = [\varrho/2] + 1,$$

where Q again satisfies (11). Here the infinite product converges uniformly in each compact subset of $\mathbf{R}$. Then (12) still holds for $p \ge 1$ provided $0 \le \varrho < \alpha$.

(12) is the key tool for investigating practically all problems arising in connection with orthogonal polynomials on infinite intervals.

The most important step in developing the theory of orthogonal polynomials on infinite intervals, just like in case of finite intervals, is to determine the exact order of magnitude of the Christoffel functions (4). Freud did this in the 1960's for the Hermite weight e^{-x^2}, using ad hoc methods. What he found was an interesting phenomenon. Namely, while

$$\max_{|x|\le(1-\varepsilon)\sqrt{2n}} e^{x^2}\lambda_n(x) \sim n^{-1/2} \tag{13}$$

for any fixed $0<\varepsilon<1$, at the same time

$$\max_{x\in\mathbf{R}} e^{x^2}\lambda_n(x) \sim n^{-1/6}. \tag{14}$$

In other words, the behavior of the Christoffel function (and, in fact, that of the orthogonal polynomials) is different near the critical point $a_n=\sqrt{2n}$. This "irregularity" caused problems in finding the optimal value of the weighted Lebesgue constant of Lagrange interpolation (for details see the chapter written by Péter Vértesi).

The key issue to overcome the difficulties in generalizing the above result e.g. for weights (10) satisfying (11) is the infinite-finite range inequality (12). Freud's original idea to generalize (13)–(14) for weights (10) when Q is restricted to polynomials of the form x^{2m} ($m\ge 1$ is an integer), was to use one-sided approximation for these weights, where it was essential that Q be a polynomial. Later he was able to remove this restriction and generalize (13)–(14) to

$$\min_{x\in\mathbf{R}}\frac{\lambda_n(x)}{w(x)}\ge\frac{ca_n}{n}\quad\text{and}\quad\max_{|x|\le c_1a_n}\frac{\lambda_n(x)}{w(x)}\le\frac{c_2a_n}{n}$$

for weights (10) satisfying (11). Freud's idea was to approximate w by polynomials in the sense

$$p_n^2(x)\sim w,\qquad |x|\le c_1a_n,\quad p_n\in\mathcal{P}_n.$$

The next important task is to find a good approximation for functions with the property that $\lim_{|x|\to\infty}f(x)w(x)=0$. Freud proved the weighted L_p-boundedness of the $(C,1)$-means of Fourier series, i.e.

$$\left\{\int_{\mathbf{R}}\left|\frac{1}{n}\sum_{k=0}^{n}S_k(w,f,t)w(t)\right|^p dt\right\}^{1/p}\le K\left\{\int_{\mathbf{R}}|f(t)w(t)|^p\,dt\right\}^{1/p},$$

$1 \le p \le \infty$, for weights (10) under some conditions which basically ensure that Q behaves like a polynomial at infinity. This boundedness easily implies that the de la Vallée-Poussin means

$$\frac{1}{n}\sum_{k=n+1}^{2n} S_k(w,f,x)$$

converge (in the weighted norm) in the order of best weighted approximation.

In order to obtain so-called *converse* theorems, that is statements about structural properties of functions deduced from the order of best approximation, one needs Bernstein–Markov type inequalities. These inequalities restrict the order of magnitude of the derivative of polynomials in terms of the weighted norm and the degree of the polynomial. The classical Markov–Bernstein inequality on the finite interval $[-1,1]$ states that

$$\max_{|x|\le 1}\left|p_n'(x)\right| \le \frac{cn^2}{1+n\sqrt{1-x^2}}\max_{|x|\le 1}\left|p_n(x)\right|$$

for all $p_n \in \mathcal{P}_n$. Freud proved the L_p-version of this for Hermite weights:

$$\left(\int_{\mathbf{R}} e^{-x^2}\left|p_n'(x)\right|^p dx\right)^{1/p} \le c\sqrt{n}\left(\int_{\mathbf{R}} e^{-x^2}\left|p_n(x)\right|^p dx\right)^{1/p}, \quad 1\le p\le\infty.$$

Later this was generalized to

(15)

$$\frac{\left(\int_{\mathbf{R}} e^{-|x|^\alpha}\left|p_n'(x)\right|^p dx\right)^{1/p}}{\left(\int_{\mathbf{R}} e^{-x^2}\left|p_n(x)\right|^p dx\right)^{1/p}} \le \begin{cases} cn^{1-1/\alpha} & \text{if } \alpha>1,\\ c\log n & \text{if } \alpha=1,\\ c & \text{if } 0<\alpha<1,\end{cases} \qquad 1\le p\le\infty$$

(for $\alpha \ge 2$ see Freud (J. Approx. Theory **19** (1977), 22–37.), for $1<\alpha<2$ see Levin and Lubinsky (J. Approx. Theory **49** (1987), 149–169.), and for $0<\alpha\le 1$ see Nevai and Totik (Constr. Approx. **2** (1986), 113–127.)). All these inequalities are sharp in the order of magnitude.

A further generalization of (15) is given by Levin and Lubinsky (SIAM J. Math. Anal. **21** (1990), 1065–182.) for weights (10)–(11) and $p=\infty$: then the right hand side of (15) becomes

$$c\int_1^{Q^{[-1]}(n)} \frac{Q(t)}{t^2}\,dt.$$

The sharpness of this "Markov factor" was proved by Kroó and Szabados (J. Approx. Theory **83** (1995), 41–64.).

Another class of polynomial inequalities is the so-called Nikolski-type inequalities. These compare L_p and L_q norms of polynomials. Concerning Freud weights $e^{-|x|^\alpha}$ with $\alpha > 0$, Nevai and Totik proved the following:

$$\frac{\left(\int_{\mathbf{R}} e^{-|x|^\alpha} |p_n(x)|^p \, dx\right)^{1/p}}{\left(\int_{\mathbf{R}} e^{-|x|^\alpha} |p_n(x)|^q \, dx\right)^{1/q}} \leq \begin{cases} cn^{1/\alpha} & \text{if } p \leq q, \\ cn^{1-1/\alpha} & \text{if } p > q, \ \alpha > 1, \\ c \log n & \text{if } p > q, \ \alpha = 1, \\ c & \text{if } p > q, \ 0 < \alpha < 1, \end{cases}$$

where $c > 0$ depends only on p and q. The proof uses the infinite-finite range inequality and estimates for the Christoffel function.

In the 1970's Freud made two remarkable conjectures concerning the recursion coefficients (2) and the greatest zero x_{1n} of the orthogonal polynomials with respect to the weight $w(x) = e^{-|x|^\alpha}$, $\alpha > 1$. These are the following:

$$\lim_{n\to\infty} n^{-1/\alpha} a_n = \left[\frac{\Gamma(\alpha/2)\Gamma(\alpha/2+1)}{\Gamma(\alpha+1)}\right]^{1/\alpha} \tag{16}$$

and

$$\lim_{n\to\infty} n^{-1/\alpha} x_{1n} = 2\left[\frac{\Gamma(\alpha/2)\Gamma(\alpha/2+1)}{\Gamma(\alpha+1)}\right]^{1/\alpha}. \tag{17}$$

Besides their beauty, these limit relations have a practical use in the theory of polynomial approximation. Using ad hoc methods, Freud was able to prove (16) for $\alpha = 2$, 4 and 6. His method breaks down for $\alpha \geq 8$. Eventually, A. Máté, P. Nevai and T. Zaslavsky (Trans. Amer. Math. Soc. **287** (1985), 495–505.) proved for any even integer α the asymptotic expansion

$$n^{-1/\alpha} a_n = \sum_{j=0}^{\infty} c_j n^{-2j},$$

where c_0 is the right hand side of (16).

As for (17), it is easy to see that (16) implies it (but not conversely). (17) in full generality ($\alpha > 1$ is a real number) was proved by E. A. Rakhmanov (Math. USSR Sbornik **47** (1984), 155–193.). A similar asymptotic relation for the kth root of the orthogonal polynomials was proved by Máté, Nevai, and Totik when α is an even integer.

Saff and Totik [160] established a very general result concerning the limit distribution of zeros of orthogonal polynomials with respect to certain measures. Let α be a finite Borel measure with support consisting of infinitely many points in $[0,1]$, and let the zeros of the corresponding nth orthogonal polynomial $p_{n,\alpha}$ be $x_1(\alpha,n),\dots,x_n(\alpha,n)$. In general, the limit distribution of these roots does not exist, so one should analyze the weak* limit of the "zero measures"

$$\nu(p_{n,\alpha}) = \frac{1}{n}\sum_{i=1}^{n}\delta_{x_i(\alpha,n)} \tag{18}$$

in the space of all unit Borel measures supported on $[0,1]$, where δ_t is the unit mass at t. The *carrier* of α is any Borel set whose complement has zero α-measure, and the *minimal carrier capacity* of α is $c_\alpha = \inf\{\operatorname{cap}(C) \mid C \text{ is a carrier of } \alpha\}$, where "cap" means logarithmic capacity. Now if $c_\alpha > 0$ and C is a minimal carrier of α, then any weak* limit ν of the zero distributions (18) satisfies

$$C \text{“}\subseteq\text{”} \operatorname{MAX} U^\nu \quad \text{and} \quad \operatorname{supp}(\nu) \subseteq \bar{C} \tag{19}$$

where "$\subseteq$" means inclusion except for a set of zero capacity, and $\operatorname{MAX} U^\nu$ is the set of maximum points of the logarithmic potential function

$$U^\nu(z) = \int_0^1 \log\frac{1}{|z-t|}\,d\nu(t)$$

associated with the limit distribution ν. Conversely, if $C \subseteq [0,1]$ is of positive capacity and $\mathcal{M}_C$ is the set of probability measures ν satisfying (19), then there is a measure α such that C is a minimal carrier of α and $\mathcal{M}_C = \{\nu \mid \nu \text{ is a weak}^\star \text{ limit point of the measures } \nu(p_{n,\alpha})\}$.

Finally, we mention that recently the notion of orthogonality e.g. on the interval $[-1,1]$ was extended to *varying weights,* mostly due to the work of Vilmos Totik. Let u be a measurable function satisfying the so-called Szegő condition

$$\int_{-1}^{1}\frac{\log u(t)}{\sqrt{1-t^2}}\,dt > -\infty,$$

and let w_n be a sequence of weights (in the sense discussed above). Then the kth orthonormal polynomial $p_{n,k}$ with positive leading coefficient is defined by

$$\int_{-1}^{1} p_{n,k}(x)p_{n,m}(x)w_n^{2n}(x)u^2(x)\,dx = \delta_{k,m}, \qquad k,n,m = 0,1,\ldots .$$

Totik [182, Ch. 14] gave asymptotics for these polynomials, and discussed their fundamental properties.

In this survey on orthogonal polynomials we tried to recite the most outstanding results achieved by Hungarian mathematicians. There are many other contributors (among them, of course, foreigners who were inspired mostly by the work of Szegő and Freud), but perhaps the mentioned results convince the reader that Hungarians were always in the forefront of research in this significant area of mathematics.

References

[47] Freud, Géza, *Orthogonale Polynome,* Akadémiai Kiadó (Budapest), Birkhäuser (Basel, 1969). *Orthogonal Polynomials,* Pergamon Press (London–Toronto–New York, 1971).

[59] Grenander, Ulf – Szegő, Gábor, *Toeplitz Forms and their Applications,* University of California Press (Berkeley and Los Angeles, 1958)/Chelsea (New York, 1984).

[160] Saff, Edward B. – Totik, Vilmos, *Logarithmic Potentials with External Fields,* Grundlehren der Mathematischen Wissenschaften, Vol. 316, Springer-Verlag (Berlin–New York, 1997).

[173] Szegő, Gábor, *Collected Papers,* ed. R. Askey, Birkhäuser (Boston–Basel–Stuttgart, 1982).

[174] Szegő, Gábor, *Orthogonal Polynomials,* American Mathematical Society Colloquium Publications, Vol. XXIII., 1939, revised 1959, 3rd edition 1967, 4th edition 1975.

[182] Totik, Vilmos, *Weighted Approximation with Varying Weight,* Springer Lecture Notes in Mathematics, No. 1569, Springer-Verlag (Berlin–New York, 1994).

{1} P. Nevai, Orthogonal Polynomials, *Memoirs Amer. Math. Soc.,* **213** (1979), 1–185.

József Szabados
Alfréd Rényi Institute of Mathematics
Hungarian Academy of Sciences
P.O.B. 127
1364 Budapest
Hungary
szabados@renyi.hu

BOLYAI SOCIETY
MATHEMATICAL STUDIES, 14

A Panorama of Hungarian Mathematics in the Twentieth Century, pp. 71–117.

Classical (Unweighted) and Weighted Interpolation

PÉTER VÉRTESI

1. Introduction

What is interpolation? "Perhaps it would be interesting to dig to the roots of the theory and to indicate its historical origin. Newton, who wanted to draw conclusions from the observed location of comets at equidistant times as to their location at arbitrary times arrived at the problem of determining a 'geometric' curve passing through arbitrarily many given points. He solved this problem by the interpolation polynomial bearing his name "(*Pál Turán* {128, p. 23}.)

Interpolation theory has been one of the favorite subjects of the twentieth century's Hungarian approximators. The backbone (mainly of classical interpolation) is the theory developed by *Lipót Fejér*, *Ervin Feldheim*, *Géza Grünwald*, *Pál Erdős* and *Pál Turán*.

One can find hundreds of papers dealing with different interpolatory processes (Lagrange-, Birkhoff (lacunary)-, Hermite–Fejér interpolation, etc.).

In the last 40 years or so there has developed a new branch in approximation theory: the so called *weighted approximation.*

During those years, even in this relatively new area, many interpolatory results were proved by the Hungarian school.

In Part A we quote the *classical* results while Part B considers the *weighted* ones. Since weighted approximation is relatively new, Part B is much shorter.

The interested readers may find many other details and results in the booklet of *Ervin Feldheim* {49} and in the book of *József Szabados* and *Péter Vértesi*, Interpolation of functions {111}. Generally, we concentrate on the Lagrange interpolation. Analogous results may be proved for the trigonometric and the complex case (see {111} again).

A
CLASSICAL CASE

2. Lagrange Interpolation. Lebesgue Function. Lebesgue Constant. Optimal Lebesgue Constant. Divergence of Interpolation

2.1. Let us begin with some definitions and notation. Let $C = C(I)$ denote the space of continuous functions on the interval $I := [-1,1]$, and let $\mathcal{P}_n$ denote the set of algebraic polynomials of degree at most n. $\|\cdot\|$ stands for the usual maximum norm on C. Let X be an *interpolatory matrix (array)*, i.e.,

$$X = \left\{ x_{kn} = \cos\vartheta_{kn}; \qquad k = 1, \dots, n; \qquad n = 0, 1, 2, \dots \right\},$$

with

$$-1 \le x_{nn} < x_{n-1,n} < \cdots < x_{2n} < x_{1n} \le 1 \tag{2.1}$$

and $0 \le \vartheta_{kn} \le \pi$, and consider the corresponding *Lagrange interpolation polynomial*

$$L_n(f, X, x) := \sum_{k=1}^{n} f(x_{kn})\ell_{kn}(X, x), \qquad n \in \mathbb{N}. \tag{2.2}$$

Here, for $n \in \mathbb{N}$,

$$\ell_{kn}(X, x) := \frac{\omega_n(X, x)}{\omega_n'(X, x_{kn})(x - x_{kn})}, \qquad 1 \le k \le n,$$

with

$$\omega_n(X,x) := \prod_{k=1}^{n}(x - x_{kn}),$$

are polynomials of exact degree $n-1$. They are called the *fundamental polynomials* associated with the *nodes* $\{x_{kn} : k = 1,\ldots,n\}$ obeying the relations $\ell_{kn}(X, x_{jn}) = \delta_{kj}$, $1 \le k,j \le n$.

The main question is: For what choices of the interpolation array X we can expect that (uniformly, pointwise, etc.) $L_n(f,X) \to f \ (n \to \infty)$?

Since, by the Čebishov alternation theorem ({95} Chap. 2, Theorem 9), the best uniform approximation, $P_{n-1}(f)$, to $f \in C$ from $\mathcal{P}_{n-1}$ interpolates f in at least n points, there exists, for each $f \in C$, an interpolation matrix, Y, for which

$$\big\| L_n(f,Y) - f \big\| = E_{n-1}(f) := \min_{P \in \mathcal{P}_{n-1}} \|f - P\|$$

goes to 0 as $n \to \infty$. However, for the *whole class* C, the situation is different.

To formulate the corresponding negative result, we quote some estimates and introduce further definitions.

By the classical Lebesgue estimate,

(2.3)

$$\begin{aligned}\big|L_n(f,X,x) - f(x)\big| &\le \big|L_n(f,X,x) - P_{n-1}(f,x)\big| + \big|P_{n-1}(f,x) - f(x)\big| \\ &\le \big|L_n(f - P_{n-1},X,x)\big| + E_{n-1}(f) \\ &\le \Big(\sum_{k=1}^{n}\big|\ell_{k,n}(X,x)\big| + 1\Big)E_{n-1}(f),\end{aligned}$$

therefore, with the notations

$$\lambda_n(X,x) := \sum_{k=1}^{n}\big|\ell_{kn}(X,x)\big|, \qquad n \in \mathbb{N}, \tag{2.4}$$

$$\Lambda_n(X) := \big\|\lambda_n(X,x)\big\|, \qquad n \in \mathbb{N}, \tag{2.5}$$

(*Lebesgue function* and *Lebesgue constant* (of Lagrange interpolation), respectively,) we have for $n \in \mathbb{N}$

$$\big|L_n(f,X,x) - f(x)\big| \le \big\{\lambda_n(X,x) + 1\big\}E_{n-1}(f) \tag{2.6}$$

and

$$\| L_n(f,X) - f \| \le \{ \Lambda_n(X) + 1 \} E_{n-1}(f). \tag{2.7}$$

"After ... the approximation theorem of *Karl Weierstrass*, it was hoped that there exists a (non-equidistant) system of nodes for which the Lagrange interpolation polynomials converge uniformly for every function continuous in $[-1,1]$. The mathematical world was awakened from this dream in 1914 by *Georg Faber* who showed that there is *no* such system." (*Turán* {128, p. 25})

Namely, he proved the then rather surprising lower bound

$$\Lambda_n(X) \ge \frac{1}{12} \log n, \qquad n \ge 1, \tag{2.8}$$

for any interpolation array X. Based on this result he obtained

Theorem 2.1 (Faber {40}). *For any fixed interpolation array X there exists a function $f \in C$ for which*

$$\varlimsup_{n\to\infty} \| L_n(f,X) \| = \infty. \tag{2.9}$$

2.2. The previous estimates show clearly the importance of the Lebesgue function, $\lambda_n(X,x)$, and the Lebesgue constant, $\Lambda_n(X)$. During the last 90 years, very general relations concerning their behaviour were proved and applied to obtain divergence theorems for $L_n(f,X)$.

First, we state the counterpart of (2.8). Namely, using an estimate of *L. Fejér* {45} (cf. {127, Section 4.12.6})

$$\Lambda_n(T) = \frac{2}{\pi} \log n + O(1),$$

one can see that the order $\log n$ in (2.8) is best possible (here T is the Čebishov matrix, i.e. $x_{kn} = \cos \frac{2k-1}{2n}\pi$).

A very natural problem, raised and answered in 1958 by *Erdős*, says that $\lambda_n(X,x)$ is "big" on a "large" set.

Theorem 2.2 (Erdős {36}). *For any fixed interpolation matrix $X \subset [-1,1]$, real $\varepsilon > 0$, and $A > 0$, there exists $n_0 = n_0(A,\varepsilon)$ so that the set*

$$\{ x \in \mathbb{R} : \lambda_n(X,x) \le A \text{ for all } n \ge n_0(A,\varepsilon) \}$$

has measure less than ε.

In 1978, *P. Erdős* and *J. Szabados* obtained a "best possible result in order" for $\lambda_n(X,x)$. Namely one has

Theorem 2.3 ({23}). *For any interpolatory matrix X and subinterval $[a,b]\subset[-1,1]$ there exists $c>0$ such that*

$$\int_a^b \lambda_n(X,x)\,dx \geq c(b-a)\log n, \qquad n\geq n_0(a,b).$$

The next statement, the more or less complete pointwise estimation is due to *P. Erdős* and *P. Vértesi* {31} from 1981.

Theorem 2.4. *Let $\varepsilon>0$ be given. Then, for any fixed interpolation matrix $X\subset[-1,1]$ there exist sets $H_n=H_n(\varepsilon,X)$ of measure $\leq\varepsilon$ and a number $\eta=\eta(\varepsilon)>0$ such that*

$$\lambda_n(X,x)>\eta\log n \tag{2.10}$$

if $x\in[-1,1]\setminus H_n$ and $n\geq 1$.

Closer investigation shows that (instead of the original $\eta=c\varepsilon^3$) $\eta=c\varepsilon$ can be attained ({134}). The behaviour of the Čebishov matrix, T, shows that (2.10) is the best possible in order.

2.3. Let us say some words about the *optimal Lebesgue constant.* In 1961, *P. Erdős*, improving a previous result of *P. Turán* and himself (see {24} and {38}), proved that

$$\left|\Lambda_n^* - \frac{2}{\pi}\log n\right| \leq c,$$

where

$$\Lambda_n^* := \min_{X\subset I}\Lambda_n(X), \qquad n\geq 1,$$

is the *optimal Lebesgue constant.* As a consequence of this result, the closer investigation of Λ_n^* attracted the attention of many mathematicians.

In 1978, *Ted Kilgore, Carl de Boor* and *Alan Pinkus* proved the so-called Bernstein-Erdős conjectures concerning the optimal interpolation array X (cf. {8}, {9}, {14} and {62}).

Using this result *P. Vértesi* {139} obtained the value of Λ_n^* *within the error* $o(1)$. Namely,

$$\Lambda_n^* = \frac{2}{\pi}\log n + \chi + O\left(\left(\frac{\log\log n}{\log n}\right)^2\right) \tag{2.11}$$

where $\chi = \frac{2}{\pi}(\gamma + \log\frac{4}{\pi}) = 0.521251\ldots$ and $\gamma = 0.577215\ldots$ is the Euler–Mascheroni constant (cf. 2.8.3).

2.4. One of the most talented approximators, *Géza Grünwald*, was a holocaust victim; he was killed in 1942 at the age 32. He was about 25 when, in two fundamental papers, he proved that the Lagrange interpolation can be very bad even for the good matrix $T = \left\{\cos\frac{2k-1}{2n}\pi\right\}$ (see {53}, {54}, {81}).

Theorem 2.5 (*Grünwald–Marcinkiewicz*)*. *There exists a function $f \in C$ for which*

$$\overline{\lim_{n\to\infty}}\left|L_n(f,T,x)\right| = \infty$$

for every $x \in [-1,1]$.

In their second joint paper, {21} *Erdős* and *Grünwald* sharpen this result. They construct a function $f \in C$ satisfying Theorem 2.5, where at the same time, the even function $f(\cos\vartheta)$ has a uniformly convergent Fourier series on $[0,\pi]$.

Marcinkiewicz {81} showed that for every x_0 there exists a continuous f for which

$$\overline{\lim_{n\to\infty}}\frac{1}{n}\sum_{k=1}^{n} L_k(f,T,x_0) = \infty. \tag{2.12}$$

In other words, the arithmetic means of Lagrange interpolating polynomials of a continuous function can diverge at a given point. This is in marked contrast to the celebrated theorem of Fejér {47} for Fourier series.

In their third joint paper, {20} *Erdős* and *Grünwald* claimed to prove a far-reaching generalization of (2.12), namely the existence of an $f \in C$ for which

$$\overline{\lim_{n\to\infty}}\frac{1}{n}\left|\sum_{k=1}^{n} L_k(f,T,x)\right| = \infty, \tag{2.13}$$

*At the same time the same statement was proved by the Polish mathematician, *Józef Marcinkiewicz.* We must note some other similarities between them. Both were born in 1910; both included the above theorem into their PhD dissertations; they were submitted in 1935; moreover *Marcinkiewicz* was also a victim of the war: as his teacher *Antoni Zygmund* writes: "On September 2 [1939], the second day of the war I came across him accidentally in the street in Wilno [Vilnius], already in military uniform... A few months later came the news that he was a prisoner of war and was asking for mathematical books. It seems that this was the last news about *Marcinkiewicz*" ({144, p. 4}).

for all $x \in [-1,1]$. However, as it was discovered later by *Erdős* himself, there is an oversight in the proof and the method only gives the result with the modulus sign *inside* the summation.

Only in {22}, where *Erdős* and *Gábor Halász* (who was born four years after the Erdős–Grünwald paper) were able to complete the proof and obtained the following.

Theorem 2.6. *Given a positive sequence* $\{\varepsilon_n\}$ *converging to zero however slowly, one can construct a function* $f \in C$ *such that for almost all* $x \in [-1,1]$

$$\frac{1}{n}\left|\sum_{k=1}^{n} L_k(f,T,x)\right| \geq \varepsilon_n \log\log n \tag{2.14}$$

for infinitely many n.

The right-hand side is optimal, for in the paper {39} *Erdős* has proved

Theorem 2.7.

$$\frac{1}{n}\left|\sum_{k=1}^{n} L_k(f,T,x)\right| = o(\log\log n)$$

for almost all x, *whenever* $f \in C$.

The proof was an ingenious combination of ideas from number theory, probability and interpolation; it is not by chance that the authors are *Erdős* and *Halász*!

2.5. After the result of *Grünwald* and *Marcinkiewicz* a natural problem was to obtain an analogous result for an *arbitrary* array X. In {37, p. 384}, *Erdős* wrote: "In a subsequent paper I hope to prove the following result:

Let $X \subset [-1,1]$ *be any point group [interpolatory array]. Then there exists a continuous function* $f(x)$ *so that for almost all* x

$$\overline{\lim_{n\to\infty}}\left|L_n(f,X,x)\right| = \infty."$$

After 4 years of work, *Erdős* and *Vértesi* proved the above result ({28}–{30}). *Erdős* writes in {29}: "[Here we prove the above] statement in full detail. The detailed proof turns out to be quite complicated and several unexpected difficulties had to be overcome."*

2.6. Another significant contribution of the Hungarian approximators to interpolation is the so called "fine and rough theory" (a name coined by *Erdős* and *Turán* in their basic joint paper {27} dedicated to *L. Fejér* on his 75th birthday in 1955).

In the class $\mathrm{Lip}\,\alpha$ $(0 < \alpha < 1)$ (we give the exact definitions a little later), a natural error estimate for Lagrange interpolation is

$$\|L_n(f,X) - f\| \le cn^{-\alpha}\Lambda_n(X)$$

(cf. (2.7)). *Erdős* and *Turán* raised the obvious question: *How sharp is this estimate in terms of the order of the Lebesgue constant as* $n \to \infty$? They themselves considered interpolatory arrays X where

$$\Lambda_n(X) \sim n^\beta \qquad (\beta > 0).$$

(In the class $\mathrm{Lip}\,\alpha$ this is the natural setting.) In the above paper {27} they prove essentially

Theorem 2.8. *Let X be as above. If $\alpha > \beta$, then we have uniform convergence in* $\mathrm{Lip}\,\alpha$. *If $\alpha \le \beta/(\beta+2)$, then for some $f \in \mathrm{Lip}\,\alpha$, Lagrange interpolation is divergent.*

These two cases comprise what is called the "rough theory", since *solely on the basis of the order* of $\Lambda_n(X)$ one can decide the convergence-divergence behavior. However,

Theorem 2.9. *If $\beta/(\beta+2) < \alpha \le \beta$ then anything can happen. That is, there is an interpolatory array Y_1 with $\Lambda_n(Y_1) \sim n^\beta$ and a function $f_1 \in \mathrm{Lip}\,\alpha$ such that $\overline{\lim}_{n\to\infty} \|L_n(f_1,Y_1)\| = \infty$, and another interpolation array Y_2 with $\Lambda_n(Y_2) \sim n^\beta$, such that $\lim_{n\to\infty} \|L_n(f,Y_2) - f\| = 0$ for every $f \in \mathrm{Lip}\,\alpha$.*

*In a personal letter *Erdős* wrote about the *main idea of the proof*: [First] "we should prove that for every fixed A and $\eta > 0$ there exists an M $(M = M(A,\eta))$ such that if we divide the interval $[-1,1]$ into M equal parts $I_1,\dots,I_M$ then

$$\sum\nolimits_k' |\ell_{k,n}(X,x)| > A, \qquad x \in I_r,$$

apart from a set of measure $\le \eta$. Here $\sum'$ means that k takes those values for which $x \notin I_r$".

That is, to decide the convergence-divergence behavior *we need more information than just the order of the Lebesgue constant.* The corresponding situation is called "fine theory".

This paper of *Erdős* and *Turán* has been very influential. It left open a number of problems and attracted the attention not only of the Hungarian school of interpolation (*Géza Freud, Ottó Kis, Melania Sallay, József Szabados, Péter Vértesi*), but also of others (including *R. J. Nessel, W. Dickmeis, E. van Wickeren*).

We mention three generalizations. Let $\omega_m(t)$ be an increasing continuous function for $t \geq 0$ with $\omega_m(0) = 0$, $\omega_m(t) > 0$ $(t > 0)$, $t^m/\omega_m(t) \leq T^m/\omega_m(T)$ $(t \leq T)$; $m \geq 1$ is a fixed integer. The function ω_m is *an* m-th *modulus of smoothness.* If $m = 1$, we write $\omega(t)$ (modulus of continuity).

With $\omega_m(t)$ we define the function-class $C(\omega_m)$ as

$$C(\omega_m) = \left\{ f \in C \text{ and } \omega_m(f,t) \leq c_m(f)\omega_m(t) \right\},$$

where, with $\Delta_h^m f(x) := \sum_{k=0}^m (-1)^{m-k} \binom{m}{k} f(x+kh)$,

$$\omega_m(f,t) = \sup_{\substack{x, x+mh \in [-1,1] \\ |h| \leq t}} \left| \Delta_h^m f(x) \right|,$$

is the m-th *modulus of smoothness* of f; if $m = 1$, $\omega(f,t)$ is the *modulus of continuity* of f. If $\omega(t) = t^\alpha$, $0 < \alpha \leq 1$, then by definition $C(\omega) \equiv \operatorname{Lip} \alpha$.

In his paper {69} *O. Kis* proved the following

Theorem 2.10. *For an arbitrary fixed interpolatory matrix X one can find an $f \in C(\omega_m)$ with*

$$\varlimsup_{n \to \infty} \frac{\left\| L_n(f,X,x) - f(x) \right\|}{\Lambda_n(X)\omega_m\big(d_n(X)\big)} > 1,$$

provided that

$$\lim_{t \to 0} \omega_m(t) t^{-m} = \infty \tag{2.15}$$

Here

$$d_n(X) = \min_{1 \leq k \leq n} (x_{kn} - x_{k+1,n}), \qquad n \geq 2.$$

Here is another generalization, a strong *pointwise-type* divergence result of *P. Vértesi* {111, Theorem 4.20}

Theorem 2.11. *Let X and $\omega(t)$ be given. If*

$$\underline{\lim}_{t\to 0}\,\omega(t)\big|\log t\big| > 0$$

then with an appropriate $f \in C(\omega)$

$$\overline{\lim}_{n\to\infty}\big|L_n(f,X,x) - f(x)\big| > 1$$

on a dense set of second category in $[-1,1]$.

To state a new and rather deep theorem of *G. Halász* {57}, we define, deviating somewhat from its previous meaning, the Lipschitz class Lip A of exponent $A = r + \alpha$ $(0 \le \alpha < 1)$, $r = [A]$, as the space of functions f for which $f^{(r)}$ exists everywhere. Moreover

$$\sup_{t_1\neq t_2}\big|f^{(r)}(t_1) - f^{(r)}(t_2)\big|/|t_1 - t_2|^{\alpha} < \infty,$$

in particular, Lip 0 consists of the bounded functions. We define the characteristic $D(A) = D(A,X)$ by

$$D(A,X) = \sup_{f\in \mathrm{Lip}\,A}\left\{a;\big|L_n(f,X,x) - f(x)\big| \le c\left(\frac{\sqrt{1-x^2}}{n} + \frac{1}{n^2}\right)^a\right\}$$

where $c > 0$ may depend on f and α but not on x and n.

It is clear that $-\infty \le D(A) \le A$. Moreover (cf. {57, Part 2}) we have

Theorem 2.12. *Let X be given. Then*

(i) *$D(A)$ is concave from below.*

(ii) *$D'(A) \ge \frac{1}{2}$ whenever $D(A)$ is finite.*

(iii) *$2D(A_1) + D(A_2) \le A_1 + 2A_2 + 2$ for any $A_1, A_2 \ge 0$.*

The trigonometric version of the above considerations is {57, Part 1.}. However, as it was proved by *J. Szabados* {109, Theorem}, the corresponding *trigonometric characteristic of D is fully described by the corresponding properties.*

2.7. The Faber-theorem (2.9) is a special case of a general statement proved by *S. M. Losinskii* and *F. I. Harsiladze* on *(linear) projection operators* (p.o.). (That means $\mathcal{L}_n$ is a linear bounded operator with $\mathcal{L}_n : C \to \mathcal{P}_{n-1}$ and $\mathcal{L}_n(f) \equiv f$ iff $f \in \mathcal{P}_{n-1}$). Namely, they proved that if

$$|||\mathcal{L}_n||| := \sup_{\|f\|\le 1} \big\| \mathcal{L}_n(f,x)\big\|, \qquad f \in C,$$

then

$$|||\mathcal{L}_n||| \ge \frac{\log n}{8\sqrt{\pi}}$$

($\mathcal{L}_n$ is a p.o.). If $\mathcal{L}_n = L_n(X)$ (Lagrange interpolation), then, obviously $\Lambda_n(X) = |||\mathcal{L}_n|||$.

In his paper {56}, *G. Halász* formulated some results on

$$\mathcal{L}_n(x) := \sup_{\|f\|\le 1} \big| \mathcal{L}_n(f,x)\big|, \qquad f \in C$$

(it generalizes the Lebesgue function $\lambda_n(X,x)$). Among others he states

Theorem 2.13. *For any sequence of projections* $\mathcal{L}_n$

(i) $\overline{\lim\limits_{n\to\infty}}\, \mathcal{L}_n(x) = \infty$ *on a set of positive measure in* $[-1,1]$;

(ii) $\lim\limits_{n\to\infty} \displaystyle\int_{-1}^{1} h\big(\log \mathcal{L}_n(x)\big) \log \mathcal{L}_n(x)\, dx = \infty$ *whenever*

$$I := \int_2^\infty \frac{h(x)}{x \log x}\, dx = \infty.$$

(iii) *If* $I < \infty$ *then there exists a sequence* $\mathcal{L}_n$ *such that*

$$\sup_n \int_{-1}^{1} h\big(\log \mathcal{L}_n(x)\big) \log \mathcal{L}_n(x)\, dx < \infty.$$

Let r be an integer, $r \ge 0$. We will be concerned with the investigation of the norm of the derivative $\mathcal{L}_n^{(r)}$ of the p.o. $\mathcal{L}_n$, i.e. $\mathcal{L}_n^{(r)}(f,x) = \frac{d^r}{dx^r}\mathcal{L}_n(f,x)$. If

$$\Big|\Big|\Big|\mathcal{L}_n^{[r]}\Big|\Big|\Big| := \sup_{\|f\|\le 1} \big\|\mathcal{L}_n^{(r)}(f,x)\big\|, \qquad f \in C$$

then, according *D. L. Berman* {7}, $\Big|\Big|\Big|\mathcal{L}_n^{[r]}\Big|\Big|\Big| \ge c_r n^{2r}$ $(r \ge 1)$. However, we can do better.

Motivated by the Nikolskii–Timan–Gopengauz phenomenon in polynomial approximation ({88}, say), let

$$\mathcal{L}_{n\mu}^{[r]}(x) := \sup_{|f(x)|\le(1-x^2)^\mu} \left|\mathcal{L}_n^{(r)}(f,x)\right|, \qquad f\in C, \qquad \mu\ge 0;$$

(see *N. S. Baiguzov* {2}, *L. Neckermann, P. O. Runck* {88} and for arbitrary $r\ge 3$ *J. Szabados* {115}). We can prove (see {115}, {130}):

Theorem 2.14. *For an arbitrary projection operator $\mathcal{L}_n$ and fixed $\mu\ge 0$, $r=0,1,2,\dots$,*

$$\int_{-1}^{1} \mathcal{L}_{n\mu}^{[r]}(x)\,dx \ge c(r,\mu)n^r\log n, \qquad n\ge 1.$$

By Theorem 2.13

$$\left(|||\mathcal{L}_n^{[r]}|||\ge\right)|||\mathcal{L}_{n\mu}^{[r]}||| := |||\mathcal{L}_{n\mu}^{[r]}(x)||| \ge c_1(r,\mu)n^r\log n. \tag{2.16}$$

Moreover as a nice application of the "additional points method", one can prove that the estimation (2.16) is the best possible in order (see {115}). Actually, the so called "additional point method" has a long history. Perhaps *Fejér* was the first who noticed that restricting the interpolation at the endpoints may improve its convergence behaviour ({44}, {43}). After some other initial results due to *E. Egerváry, P. Turán, P. Szász, G. Freud, N. S. Baiguzov* and others, *J. Szabados* was the first who *systematically applied the method of adding some new points to the original interpolatory matrix X to improve the behaviour of the interpolation.*

2.8. Remarks. 1. Let us mention two basic relations concerning the estimation of the Lebesgue function (see *P. Erdős, P. Turán* {26, p. 529} and *P. Erdős*, {36, p. 387}).

(a) *For an arbitrary interpolatory matrix $X\subset[-1,1]$*

$$\ell_{kn}(X,x)+\ell_{k+1,n}(X,x)\ge 1 \quad \text{if} \quad x\in[x_{k+1,n},x_{kn}], \qquad 1\le k\le n-1.$$

(b) *Let $y_1,y_2,\dots,y_t$ be any t ($t>t_0$) distinct numbers in $[-1,1]$ not necessarily in increasing order. Then for at least one u ($1\le u\le t$)*

$$\sum_{i=1}^{u-1}\frac{1}{|y_i-y_u|} > \frac{t\log t}{8}.$$

(The half-page proof is based on the inequality between the arithmetic and harmonic means.)

2. An improvement of Theorem 2.3 that settles "small" intervals whose lengths may depend on n is in {33} (see also {34}).

3. It may be instructive to compare some values of Λ_n^* with the Lebesgue constants $\Lambda_n(S)$ and $\Lambda_n(\widehat{T})$, respectively (see *Lev Brutman* {10, p. 122}, the values are of 7-digit precision) Here S is the matrix by which the estimations (2.11) were obtained; $\hat{T} = \left\{\cos \frac{2k-1}{2n}\pi / \cos \frac{\pi}{2n}\right\}$ is an extended Čebishov matrix.

n	Λ_n^*	$\Lambda_n(S)$	$\Lambda_n(\widehat{T})$
3	1.422 920	1.448 083	1.429 873
4	1.559 490	1.575 680	1.570 167
5	1.672 210	1.683 646	1.685 140
6	1.768 135	1.776 834	1.782 530
7	1.851 599	1.858 521	1.866 999
8	1.925 458	1.931 112	1.941 573
9	1.991 685	1.996 560	2.008 327
10	2.051 706	2.056 087	2.068 744
20	2.460 788	2.463 129	2.479 193
30	2.708 082	2.709 645	2.726 693
40	2.885 809	2.887 067	2.904 441
50	3.024 619	3.025 651	3.043 229
60	3.138 527	3.139 389	3.157 102
70	3.235 120	3.235 887	3.253 659
80	3.318 973	3.319 660	3.337 477
90	3.393 058	3.393 677	3.411 530
100	3.459 415	3.459 973	3.477 858
150	3.715 393	3.715 787	3.733 720
200	3.897 466	3.897 772	3.915 713

The above table shows that even for relatively small values of n, $\Lambda_n(S)$ is quite close to Λ_n^*; much closer than the corresponding values of $\Lambda_n(\widehat{T})$.

4. The optimal matrix and the corresponding Lebesgue constants are well-known in the trigonometric and complex cases (see {14}, {9}). For other generalizations see {111}, Chapters III and IV.

5. As *O. Kis* remarked in his paper {64}, there is a predecessor of the fundamental work {27}. Namely *S. M. Losinskii* in 1948 stated some analogous results in his short *Dokladi* paper {79}, but he never published the proofs. On the other hand, their verifications are in the exhausting paper {63}; for other developments see {111, Chapter I}.

6. The characterization of the "trigonometric $D(A)$" can be found in the papers {57} and *J. Szabados* {109}.

3. On the Convergence of the Interpolatory Processes

3.1. There are at least 4 simple possibilities to ensure convergence:

(a) raising the degree of the interpolatory polynomials (see Sections 3 and 4);
(b) using mean convergence (instead of the uniform one (see Section 5);
(c) restricting ourselves to a part (subclass) of C (see Sections 3.2–3.3);
(d) applying a combination of the fundamental functions $\ell_{kn}(X,x)$ (see Section 3.4).

3.2. Our first statement on "good" functions goes back to *L. Fejér* {42} and *László Kalmár* {61}.

Theorem 3.1. *Let f be analytic on $[-1,1]$ ($f \in A$, shortly). Then*

$$\lim_{n\to\infty} \| L_n(f,X,x) - f(x) \| = 0 \qquad \forall f \in A$$

iff the nodes $\{x_{kn}\}$ are uniformly distributed on $[-1,1]$

We say that the nodes $x_{kn} = \cos\vartheta_{kn}$, $1 \le k \le n$, $n \in \mathbb{N}$, are *uniformly distributed* on $[-1,1]$ if for every subinterval $I \subset [0,\pi]$,

$$\lim_{n\to\infty} \frac{N_n(I)}{n} = \frac{|I|}{\pi},$$

where $N_n(I)$ is the number of ϑ_{kn} in I (cf. (4.4)).

Exactly 30 years after Kalmár in his Ph.D. dissertation *O. Kis* {66, Theorem 5} proved the *trigonometric* version. Here is another statement on the convergence of the trigonometric interpolatory polynomials $T_n(g,T,t)$

belonging to the set T_n of trigonometric polynomials of order n, based on the interpolatory matrix $T = \{t_{kn},\ 0 \le k \le 2n,\ n \in \mathbb{N}\} \subset [0, 2\pi]$ for $g \in \widetilde{C}$ (g is 2π-periodic and continuous) (see {66, Theorem 6}). Let $B = \{g : g$ is 2π-periodic and analytic if $|\operatorname{Im} z| \le 2\log\left(1+\sqrt{2}\right)\}$.

Theorem 3.2. *For an arbitrary interpolatory* $T \subset [0, 2\pi)$

$$\lim_{n\to\infty} \left\| T_n(g, T, t) - g(t) \right\| = 0 \qquad g(t) \in \widetilde{C}$$

iff $g \in B$, *where* $\|\cdot\| = \max_{t\in\mathbb{R}} |\cdot|$.

3.3. Using the Lebesgue estimation (2.7), we obtained a convergence result if $f \in \operatorname{Lip}\alpha$ and $E_n(f) = o(n^{-\alpha})$ (see Section 2.6, too). Another, in a sense analogous statement (see the proof), is as follows. Let $f \in CBV$ (f is continuous and of bounded variation). Then (see {138}):

Theorem 3.3. *Let* $-1 < \gamma = \max(\alpha, \beta) < 1/2$ *be fixed. Then*

$$\lim_{n\to\infty} \left\| L_n^{(\alpha,\beta)}(f) - f \right\| = 0 \quad \text{if} \quad f \in CBV.$$

The result, in a sense, is the best possible.

Above, $L_n^{(\alpha,\beta)}$ is the Lagrange interpolation (2.2) based on the n roots of the Jacobi polynomials $P_n^{(\alpha,\beta)}(x)$, $\alpha, \beta > -1$ (see [174, Chapter 4]).

3.4. In a series of paper *O. Kis* generalized some convergent processes of *G. Grünwald, S. N. Bernstein* and others. Using fairly delicate considerations, he obtained some "best possible" estimates (see {65}).

Let $g \in \widetilde{C}$. Then for a fixed integer $k \ge 0$ the trigonometric polynomials

$$S_{nk}(g, x) := a_0 + \sum_{j=1}^{n-1} (a_j \cos jx + b_j \sin jx) + b_n \sin nx, \qquad n \ge 1$$

are *uniquely determined* by

$$S_{nk}\left(g, \frac{2i-1}{2n}\right) = \frac{1}{2^k} \sum_{j=0}^{k} \binom{k}{j} g\left(\frac{2i-2j+k-1}{2n}\pi\right), \qquad 1 \le i \le 2n.$$

Let us see some examples.

$$S_{n0}\left(g,\frac{2i-1}{2n}\pi\right)=g\left(\frac{2i-1}{2n}\pi\right),$$

$$S_{n1}\left(g,\frac{2i-1}{2n}\pi\right)=\frac{1}{2}\left\{g\left(\frac{i}{n}\pi\right)+g\left(\frac{i-1}{n}\pi\right)\right\},$$

$$S_{n2}\left(g,\frac{2i-1}{2n}\pi\right)=\frac{1}{4}\left\{g\left(\frac{2i+1}{2n}\pi\right)+2g\left(\frac{2i-1}{2n}\pi\right)+g\left(\frac{2i-3}{2n}\pi\right)\right\},$$

(usual trigonometric interpolation, Grünwald-type process and Bernstein process, respectively).

A natural setting is the investigation of

$$\lambda_{kn}(x):=\sup_{\substack{g\in\widetilde{C}\\ g\neq\text{const}}}\frac{\left|S_{kn}(g,x)-g(x)\right|}{\omega\left(g,\frac{\pi}{2n}\right)},\qquad \Lambda_{kn}:=\left\|\lambda_{kn}(x)\right\|.$$

The results are as follows:

Theorem 3.4. *We have*

$$\Lambda_{0n}=\frac{1}{2}+\frac{1}{2n}\sum_{i=1}^{n}\operatorname{ctg}\frac{2i-1}{4n}\qquad (\textit{H. Ehlich, K. Zeller});$$

$$\Lambda_{1n}=\begin{cases}1+\dfrac{1}{2n}\dfrac{1}{\sin\dfrac{\pi}{2n}}, & n=2,4,6\ldots;\\[2ex] 1+\dfrac{1}{2n}\operatorname{ctg}\dfrac{\pi}{2n}, & n=1,3,5\ldots;\end{cases}$$

$$\Lambda_{2n}=\frac{5}{4},\qquad n\geq 2;$$

$$\Lambda_{3n}\leq\frac{5}{4}+\frac{2}{3\pi},\qquad n=2,4,6\ldots;$$

$$\Lambda_{4n}=\frac{23}{16},\qquad n\geq 3.$$

3.5. Remarks. 1. The results in {42}, {61}, {66} say much more than the quoted theorems. The interested reader may consult the original paper or the book of *Dieter Gaier* {51}.

2. The statement of Theorem 3.3 is valid for functions satisfying the so-called one-sided Lip δ conditions. A new theorem deals with nodes corresponding to generalized Jacobi weights (see $\{16\}$).

3. There are many applications and generalizations of the idea in $\{65\}$ including algebraic, de la Vallée Poussin type procedures and saturation problems.

4. This part is restricted to the investigation of *uniform convergence*. There are, of course, many papers dealing with *pointwise* convergence. Most of them use tools closely connected to the theory of the orthogonal polynomials. The interested reader may consult the book of *G. Freud* [47] and the monograph of *Paul Nevai* (*Pál Névai*) $\{95\}$.

4. Hermite–Fejér Type and Other Convergent Interpolatory Processes

4.1. "After the discovery of *Faber* [cf. Theorem 2.1] the following question naturally arose. Does there exist a procedure different from Lagrange's interpolation which is efficient for the class C? Immediately after Faber's proof of his theorem *Fejér* discovered that the situation changes if we consider the Hermite interpolation that is the polynomial $\mathcal{H}_n(f, X, x)$ of degree at most $2n-1$ characterized by the properties

$$\left.\begin{aligned} &\mathcal{H}_n(f, X, x_{kn}) = f(x_{kn}), \qquad && 1 \le k \le n, \\ &\mathcal{H}_n'(f, X, x_{kn}) = y_{kn}, \qquad && 1 \le k \le n. \end{aligned}\right\} \tag{4.1}$$

These polynomials can be written as

$$\mathcal{H}_n(f, X, x) = \sum_{k=1}^{n} f(x_{kn}) h_{kn}(X, x) + \sum_{k=1}^{n} y_{kn} \mathfrak{h}_{kn}(X, x). \tag{4.2}$$

Here the fundamental functions of the first and second type satisfy the conditions

$$h_{kn}^{(i)}(x_{jn}) = \delta_{kj}\delta_{0i}, \ \mathfrak{h}_{kn}^{(i)}(x_{jn}) = \delta_{kj}\delta_{1i} \ (1 \le k, j \le n), \quad 0 \le i \le 1.$$

Fejér found the relations

$$h_{kn}(X,x) := v_{kn}(X,x)l_{kn}^2(X,x) \equiv$$

$$\equiv \left(1 - \frac{\omega_n''(X,x_{kn})}{\omega_n'(X,x_{kn})}(x - x_{kn})\right) l_{kn}^2(X,x), \qquad 1 \le k \le n$$

$$\mathfrak{h}_{kn}(X,x) = (x - x_{kn})l_{kn}^2(X,x), \qquad 1 \le k \le n."$$

(*P. Turán*, {128, p. 39}.)

In his fundamental paper {41, Theorem XI} *L. Fejér* proved for the matrix $T = \left\{\cos \frac{2k-1}{2n}\pi\right\}$:

Theorem 4.1. *Let $f \in C$. Then*

$$\lim_{n\to\infty} \left\| H_n(f,T,x) - f(x)\right\| = 0 \tag{4.3}$$

Here and later $H_n(f,X,x) = \sum_{k=1}^n f(x_{kn})h_{kn}(X,x)$ (i.e. $y_{kn} = 0$); this is the *classical Hermite–Fejér (HF) step-parabola* reminding us that the tangent lines to H_n at x_{kn} are parallel to the x-axis; if $y_{kn} = f'(x_{kn})$, then we write $\mathcal{H}_n(f,f',X,x)$ (above $1 \le k \le n$).

In 1932 *Gábor Szegő* [174, Theorem 14.6] generalized the previous result:

Theorem 4.2. *Supposing that $-1 < \alpha, \beta < 0$,*

$$\lim_{n\to\infty} \left\| H_n^{(\alpha,\beta)}(f,x) - f(x)\right\| = 0 \qquad \textit{whenever} \qquad f \in C.$$

Moreover, if $\gamma := \max(\alpha,\beta) \ge 0$, the result does not hold.

(Above, $H_n^{(\alpha,\beta)}$ stands for the H_n process based on the roots $x_{kn}^{(\alpha,\beta)}$ $(1 \le k \le n)$ of $P_n^{(\alpha,\beta)}(x)$.)

4.2. Let $\rho \ge 0$. If (the linear function) $v_{kn}(X,x) \ge \rho$ $(1 \le k \le n,\ n \in \mathbb{N},\ x \in [-1,1])$, then X is said to be *ρ-normal* if $\rho > 0$; when $\rho = 0$, X is *normal*. An easy calculation shows that $X^{(\alpha,\beta)} = \left\{x_{kn}^{(\alpha,\beta)}\right\}$ is $\rho = \min(-\alpha,-\beta)$ *normal* if $\alpha,\beta < 0$; the $X^{(0,0)}$ matrix (roots of the Legendre polynomials) forms a *normal* point-system (see [174, §14.5]).

The name ρ-normal (or normal) point-system was coined by *L. Fejér* {46, Part 5}. However, its real significance was revealed by *G. Grünwald* {55} in 1942:

Theorem 4.3. *Let X be ρ-normal. Then*

$$\lim_{n\to\infty} \|H_n(f,X,x)-f(x)\| = 0 \qquad \text{if} \qquad f\in C.$$

But, even in mathematics, there is *no* "free lunch": The price of the good convergence behaviour is *the saturation* of the process H_n. In {113} *J. Szabados* proved that $\|f - H_n(f,T)\| = o(n^{-1})$ iff f = const (at the same time, he gives the saturation class, too); a more general result of *Y. G. Shi* {104} says the following:

Let $f_k(x) = x^k$. Then we have

Theorem 4.4. *For an arbitrary interpolatory X*

$$\max_{k=1,2}\Big\{\|H_n(f_k,X,x)-f_k(x)\|\Big\} \neq o(n^{-1}).$$

4.3. The next natural problem was raised in {128, Problems XIX, XX}. Do the Hermite–Fejér step-parabolas have a rough convergence theory? (cf. Part 2.6). Now, if we use $\Lambda_{n2}(X) := \Big\|\sum_{k=1}^n |h_{kn}(X,x)|\Big\|$, the next surprising result can be proved (see *P. Vértesi* {111, Corollary 6.18}).

Theorem 4.5. *Using $\Lambda_{n2}(X)$ and $\mathrm{Lip}\,\alpha$, there is no rough convergence theory for the Hermite–Fejér step-parabolas either on the whole interval $[-1,1]$ or on a closed subinterval $[a,b]$.*

It is worthwhile to compare this result with Theorems 2.8 and 2.9.

4.4. *Fejér*'s result (4.3) shows that if the degree of the interpolation polynomial is about two times bigger than the number of interpolation points, then we can get convergence. *Erdős* raised the following question. Given $\varepsilon > 0$, suppose we interpolate at n nodes, but allow polynomials of degree at most $n(1+\varepsilon)$. Under what conditions will they converge for all continuous function?

The first answer was given by himself in {35}. Namely, he proved:

Theorem 4.6. *If the absolute values of the fundamental polynomials $\ell_{kn}(X,x)$ are uniformly bounded in $x\in[-1,1]$, k $(1\le k\le n)$ and $n\in\mathbb{N}$, then for every $\varepsilon>0$ and $f\in C$ there exists a sequence of polynomials $\varphi_n=\varphi_n(x)=\varphi_n(f,\varepsilon,x)$ with*

(i) $\deg\varphi_n \le n(1+\varepsilon)$,

(ii) $\varphi_n(x_{kn}) = f(x_{kn})$, $1 \le k \le n$, $n \in \mathbb{N}$,

(iii) $\lim_{n\to\infty} \|\varphi_n - f\| = 0$.

The answer for a more general system was given in the same paper and {32}.

The story is typically Erdősian. In {35}, *Erdős* stated an answer to the above problem, but instead of proving it, he just gave an indication that "the proof is a simple modification of Theorem 3". After some 45 years, as a result of the joint effort of *Erdős*, *András Kroó* and *Szabados*, the original statement concerning the above problem was completed, even in a slightly stronger form. The result is the following {32}:

Theorem 4.7. *For every $f \in C$ and $\varepsilon > 0$, there exists a sequence of polynomials $p_n(f)$ of degree at most $n(1+\varepsilon)$ such that*

$$p_n(f, x_{k,n}) = f(x_{k,n}), \qquad 1 \le k \le n,$$

and that

$$\|f - p_n(f)\| \le cE_{[n(1+\varepsilon)]}(f)$$

holds for some $c > 0$, if and only if

$$\limsup_{n\to\infty} \frac{N_n(I_n)}{n|I_n|} \le \frac{1}{\pi} \tag{4.4}$$

whenever I_n is a sequence of subintervals of I such that $\lim\limits_{n\to\infty} n|I_n| = \infty$ and

$$\lim_{n\to\infty} \left\{ n \min_{1\le k\le n-1} (\vartheta_{k+1,n} - \vartheta_{n,k}) \right\} > 0. \tag{4.5}$$

Here $N_n(I_n)$ is the number of the $\vartheta_{k,n}$ in $I_n \subset I$. Condition (4.4) ensures that the nodes are not too dense, and condition (4.5) says that adjacent nodes should not be too close.

4.5. Remarks. 1. First we call the reader's attention to the comprehensive bibliography on HF interpolation compiled by *H. H. Gonska* and *H-B. Knoop* {52} containing about 400 entries from the period 1914–1987. (Of course, dozens of new papers were (and will be) written after 1987.)

2. It is appropriate to make some historical remarks. In his paper {58} *Dunham Jackson* considered the discrete analogy of the famous Fejér means and proved that $J_n(g, t_{kn}) = g(t_{kn})$, $t_{kn} = 2k\pi/n$, $1 \le k \le n$, moreover

$$\lim_{n\to\infty} \|J_n(g) - g\| = 0 \qquad \text{for every} \qquad g \in \widetilde{C},$$

where

$$J_n(g,t) = \sum_{k=1}^{n} g(t_{kn}) \left(\frac{\sin n \dfrac{t-t_{kn}}{2}}{n \sin \dfrac{t-t_{kn}}{2}} \right)^2$$

(today they are called "Jackson polynomials").

As we know, *Bernstein* and *Fejér* {41, (85)} were the first to point out the property $J_n'(g, t_{kn}) = 0$, $1 \le k \le n$. For other details, see {52, p. 148} and {143, p. 21}.

3. The almost unbelievable popularity of the HF interpolation lies at least in 3 facts:

— simple form,

— easy to compute and (last but not least),

— it serves in many textbooks as a transparent proof of the Weierstrass approximation theorem.

4. The behaviour of $H_n^{(\alpha,\beta)}$ near at the endpoints ± 1 if $\max(\alpha,\beta) \ge 0$, was first investigated by *L. Fejér* {44}, {43}. Actually, in the previous, Hungarian version of {44}, he considered the Legendre roots only (i.e. when $\alpha = \beta = 0$), where the uniform convergence for the *whole interval* was ensured by the additional conditions $f(1) = f(-1) = \frac{1}{2}\int_{-1}^{1} f(x)\,dx$ (see {43}). A solution for arbitrary $\max(\alpha,\beta) \ge 0$ is given in *Szabados* {112}, {113}.

Here we quote a simple special case:

If $f' \in C$, *then* $\lim_{n\to\infty} \left\| H_n^{(1/2,1/2)}(f) - f \right\| = 0$ *whenever*

$$\int_{-1}^{1} \frac{x f(x)}{\sqrt{1-x^2}}\,dx = \int_{-1}^{1} \frac{(2x-1) f(x)}{\sqrt{1-x^2}}\,dx = 0.$$

Another approach is given by *Vértesi* {132} and *P. Nevai, P. Vértesi* {89}. Namely,

if $\alpha \ge 0$, $\beta > -1$, $f \in C$, *then*

$$\lim_{n\to\infty} \left\| H_n^{(\alpha,\beta)}(f) - f \right\|_{[-1+\varepsilon,1]} = 0 \quad (0 < \varepsilon < 2) \quad \textit{iff}$$

$$\begin{cases} \lim\limits_{n\to\infty} H_n^{(\alpha,\beta)}(f,1) = f(1) & \textit{and} \quad (\textit{if } \alpha \ge 1) \\ \left(H_n^{(\alpha,\beta)}(f,x) \right)^{(r)}_{x=1} = o(n^{2r}), & r = 1, 2, \ldots, [\alpha]. \end{cases}$$

For other details see {111, Ch. V/3}.

5. The previous theorem was proved using the idea of the so called quasi-Hermite–Fejér interpolation (qHFi): In their paper {17} *Jenő Egerváry* and *P. Turán* observed that if the HF step-parabolas are replaced by the polynomials of degree $2n+1$ (sic!) taking the values and zero derivatives at the Legendre nodes *and the values of the function* at ± 1, then the convergence of this so called qHF polynomials *becomes uniform in* $[-1,1]$ $(f \in C)$! I.e., adding two points with multiplicity one, we improve the convergence behaviour. This idea has many natural generalization (see the papers of *A. Schönhage*, *G. Freud* and others in {111, p. 199–200}.

6. Another generalization is the so called HF-type interpolation. Let $m \in \mathbb{N}$, $X \subset [-1,1]$ be given. If $f \in C$, then $I_{nm}(f, X, x) \in \mathcal{P}_{nm-1}$ is defined by

$$I_{nm}^{(t)}(f, X, x_{kn}) = f(x_{kn})\delta_{0t}, \qquad 1 \le k \le n,\ 0 \le t \le m-1.$$

(a) *If* m *is odd,* the processes will be denoted by L_{nm} (obviously $L_{n1} \equiv L_n$ (Lagrange interpolation)). As it turns out, they behave similarly to the Lagrange interpolation: One can prove a Faber type result if $n \to \infty$ (m is fixed); the corresponding Lebesgue constant, $\Lambda_{nm}(X) \ge c\log n$ for any X (see *Szabados* {116}; actually in this sophisticated paper the exact lower bound for other fundamental functions are given, too); the Lebesgue function $\lambda_{nm}(X,x) \ge c\log n$ on a "big" set (*Vértesi* {133}). Results on $I_{nm}(f, T, x)$ are in the papers *Terry M. Mills, P. Vértesi* {80} and *Simon J. Smith* {105}.

(b) *If* m *is even,* we use the notation H_{nm} (clearly, $H_{n2} = H_n$ (HF interpolation)) because the behaviour is similar to the HF process. Here is a convergence result: Using obvious notations, one can prove that *the following statements are equivalent.*

   (i) $\lim_{n\to\infty} \left\| H_{nm}^{(\alpha,\beta)}(f) - f \right\| = 0 \qquad \forall f \in C,$

   (ii) $-\frac{1}{2} - \frac{2}{m} \le \alpha, \beta < -\frac{1}{2} + \frac{1}{m}$ *and* $|\alpha - \beta| \le \frac{2}{m}$.

(see the survey paper *P. Vértesi* {135} and its references).

7. Let us say some words about *Grünwald*'s celebrated results (Theorem 4.3). First, today it may be proved using the result of *Pavel P. Korovkin* on positive linear operators (obviously, if X is ρ-normal, then $H_n(f, X)$ is a positive linear operator; cf. {70}). Secondly, as it turned out from the paper *János Balázs* {5}, the ρ-normality *is not necessary* to the good behaviour of $H_n(f, x)$ (cf. {44} and {131}).

8. There is a wide variety of the form of the error estimations. We refer to {111, Ch. V/2 Corr. 7.16}; another interesting question is the comparison of the process $L_n(f,x)$ and $H_n(f,X)$ (cf. {111, Ch. VI.}).

9. Around 1960, *Paul Butzer* raised the problem of proving the Jackson theorem by interpolatory processes. Many interesting papers were written by *G. Freud, O. Kis, M. Sallay* and others (see {111, Ch. II./ 5,6}).

10. After 1975, following a short paper of *J. Balázs*, the mathematical world rediscovered the simple but efficient rational interpolatory Shepard type operators. In the last 25 years dozens of papers were completed. Here we quote a nice result of the very talented young Hungarian mathematician *Gábor Somorjai**:

Let $f \in C[0,1]$, $\alpha > 2$ *real, and* let

$$S_n(f,x) := \frac{\sum_{k=0}^{n} f\left(\frac{k}{n}\right)\left|x - \frac{k}{n}\right|^{-\alpha}}{\sum_{k=0}^{n}\left|x - \frac{k}{n}\right|^{-\alpha}}.$$

We have

(i) $\left\|S_n(f) - f\right\|_{[0,1]} = o\left(\frac{1}{n}\right)$ *iff* $f = \text{const}$,

(ii) $\left\|S_n(f) - f\right\|_{[0,1]} = O\left(\frac{1}{n}\right)$ *iff* $f \in \text{Lip}\,1$.

(Cf. {106}; other relevant results are in the survey paper of *Bianca Della Vecchia* {15} and in {140}.)

5. Lacunary or Birkhoff Interpolation

5.1. In the classical Hermite interpolation we prescribe the consecutive derivatives of the interpolatory polynomials. Dropping this, we arrive at *lacunary* interpolation.

"While polynomials of the previous kind always exist, in Birkhoff's case, polynomials satisfying his conditions do not necessarily exist. Hence, we have the basic question:

*He died at the age 26, in 1978.

(a) existence,
(b) uniqueness,
(c) possibly, explicit representation,
(d) convergence,
(e) applications."

(*P. Turán* {128, p. 48}).

P. Turán and his collaborators, *János Surányi*, *J. Balázs* then *G. Freud* in a series of papers investigated the so-called $(0,2)$ interpolation on the roots of $P_n^{(-1,-1)}(x)$ (see {3}, {4}, {50} and {108}).

It turned out that for n =*even*, the $(0,2)$ interpolatory polynomial $R_n(f,x) = \sum_{k=1}^{n} f(x_{kn})r_{kn}(x) \in \mathcal{P}_{2n-1}$ satisfying

$$R_n(f,x_{kn}) = f(x_{kn}), \qquad R_n''(f,x_{kn}) = 0, \qquad 1 \le k \le n,$$

is uniquely defined ($f \in C$, the x_{kn} are the roots of $P_n^{(-1,-1)}$), moreover

Theorem 5.1. *If* $\omega_2(f,t) = o(t)$, *then*

$$\lim_{n\to\infty} \| R_n(f,x) - f(x) \| = 0, \qquad n = 2,4,6\ldots.$$

Furthermore, one can see that the condition $\omega_2(f,t) = o(t)$ is the best possible using that the corresponding $(0,2)$-Lebesgue constants satisfy $\| \sum_{k=1}^{n} |r_{kn}(x)| \| \ge cn$.

5.2. The theory of lacunary interpolation became very popular (again) not only in Hungary, but everywhere on the world:

This popularity resulted in the monograph of *G. G. Lorentz*, *Kurt Jetter* and *Sherman D. RiemenSchneider* {75}.

During the years many questions concerning the existence of the lacunary polynomials were answered. Among them we mention the relatively new paper of *A. A. Chak*, *A. Sharma* and *J. Szabados* {12}, solving the existence, uniqueness and representation on the roots of $P_n^{(\alpha,\beta)}(x)$, $\alpha,\beta \ge -1$.

5.3. The investigation of the $(0,m)$ *trigonometric* interpolation on equidistant nodes ($m \ge 2$) was initiated by *O. Kis* {67}. While *O. Kis* investigated the case $m = 2$, soon after, *A. Sharma* and *Arun K. Varma* settled the other values of m ({102}).

Their significant achievements were the simple explicit forms of the fundamental polynomials. Using these formulas and *J. Szabados* {111, Theorem 7.12}, we state the following:

Let $t_{kn} = \frac{2k\pi}{n}$, $k = 0, \pm1, \pm2, \ldots$. We are looking for a trigonometric polynomial $T_n(g,(0,m),t) \in \mathcal{T}_{n-1}$ ($g \in \widetilde{C}$, say) satisfying

$$T_n\big(g,(0,m),t_{kn}\big) = g(t_{kn}) \qquad T_n^{(m)}\big(g,(0,m),t_{kn}\big) = 0 \qquad k = 0, \pm1, \ldots$$

Theorem 5.2. *The above* $(0,m)$ *problem is uniquely solvable iff*

(i) m*=odd and* n *is arbitrary; or*

(ii) m *=even and* n *=odd.*

Moreover,

$$\big\|g(t) - T_n\big(g,(0,m),t\big)\big\| \le \frac{c}{n^m}\sum_{k=0}^{n}\frac{\widetilde{E}_k(g)}{(k+1)^{1-m}}$$

$$+\, c\big\{1+(-1)^m\big\}\, n\, \widetilde{E}_{[\frac{n}{4}]}(g).$$

In general the above conditions are also necessary (see {111, p. 252}*).*

Above, $\tilde{E}_n(g) := \min_{\tau\in\mathcal{T}_n} \|g - \tau\|$.

Let us note that when m is *odd* the second term does not appear and since the first term tends to 0 (as $n \to \infty$) the procedure is convergent for all $g \in \widetilde{C}$. When m is *even*, the condition, $\omega_2\big(g, \frac{1}{n}\big) = o\big(\frac{1}{n}\big)$ ensures uniform convergence.

5.4. As *Turán* suggested to him, *O. Kis* investigated the $(0,2)$ complex interpolation *at the roots of unity.* It turned out that existence and unicity always hold. Moreover, he proved the following (cf. {68}):

Theorem 5.3. *The corresponding Lebesgue constants (operator-norm) has the exact order* $\log n$.

The result is in striking contrast to Theorems 5.1 and 5.2, where the exact orders were n.*

The result was generalized by *A. S. Cavaretta, A. Sharma* and *R. S. Varga* {11} proving that in any case the $(0, m_1, m_2, \ldots, m_q)$ interpolation

*As *O. Kis* frequently told us, first *Turán* did not believe him: they sat down and *Turán* checked every small detail. And after some busy hours *Turán* was convinced...

on the roots of unity exists uniquely ($n \geq qm_q$) and the exact order of the corresponding operator norm is again $\log n$.

5.5. The next nice and surprising statement of *G. Somorjai* {107} shows that the above order $\log n$ is optimal.

Let $\Gamma = \{z : |z| = 1\}$ be the unit circle and let $C(T)$ be the Banach space of continuous functions on Γ endowed with the supremum norm $\|\cdot\|$. The closed subspace $AC \subset C(\Gamma)$ consists of the restriction to $C(\Gamma)$ of those functions which are analytic in $|z| < 1$ and continuous on $|z| \leq 1$. $B(C, A)$ will denote the space of bounded linear operators mapping $C(\Gamma)$ into AC endowed with the usual norm $\|\mathcal{L}\|_C = \sup_{f \in C(\Gamma), \|f\| \leq 1} \|\mathcal{L}(f, z)\|$ $(\mathcal{L} \in B(C, A))$. We shall say that *the operator $\mathcal{L} \in B(C, A)$ is determined on the set $H \subset \Gamma$* if $f \in C(\Gamma)$, $f|_H \equiv 0$ (i.e., the restriction of f to H is identically zero) implies $\mathcal{L}(f, z) \equiv 0$.

Theorem 5.4. *Let $H_n \subset \Gamma$ be closed sets of angular Lebesgue measure zero, and suppose that $\mathcal{L}_n \in B(C, A)$ are determined on H_n $(n = 1, 2, \dots)$. Then there is an $f \in AC$ for which*

$$\varlimsup_{n \to \infty} \|f(z) - \mathcal{L}_n(f, z)\| > 0.$$

In particular whatever is the set $\{z_{kn}\}_{k=1}^n \subset \Gamma$, *there do not exist discrete linear operators of the form*

$$\mathcal{L}_n(f, z) = \sum_{k=1}^{n} f(z_{kn}) a_{kn}(z), \qquad a_{kn}(z) \in AC, \qquad k = 1, \dots, n,$$

which would uniformly converge to every $f \in AC$.

5.6. In 1975 L. Pál {98} investigated a special Birkhoff interpolation which today called as *Pál-type interpolation.* His main idea was to prescribe the derivatives of the corresponding interpolation at points which are *different* from the nodes where the function values were given.

These simple idea was very fruitful in many cases. After getting the first convergence result in 1983 L. Szili {124}, in the last 20 years or so more than 50 papers have been written in this topic. A part of them consider regularity problem on quite general point-system,the other ones prove convergence theorems using special nodes-system.

While generally the "Lebesgue constants" tend to infinity,in their paper J. Szabados and A. K. Varma {118} obtained a process which converge for *arbitrary* continuous function.

Theorem 5.5. *For $n = 1, \ldots, n$ let x_{kn} $(k = 1, \ldots, n)$ resp. x^*_{jn} $(j = 1, \ldots, n-1)$ be the roots of $P_n^{(-1,-1)}$ resp. $P_n^{(1,1)}$. If $f \in C[-1,1]$ then there exists a unique polinomial $R_n(f, \cdot)$ of degree $\le 2n$ which satisfies*

$$R_n(f, x_{kn}) = f(x_{kn}) \quad (k = 1, \ldots, n),$$

$$R_n'(f, \pm) = R_n''(f, x^*_{jn}) = 0 \quad (j = 1, \ldots, n-1),$$

moreover we have

$$\lim_{n\to\infty} \big\| R_n(f, x) - f(x) \big\| = 0.$$

Other interesting results are in the short survey paper {119}.

5.7. Remarks. 1. If somebody uses a more systematic and detailed treatment of lacunary interpolation and tries to dig to the roots, the name of *György Pólya* must be mentioned (see the book {11}). Also, {111, Ch. VII} is a good source of some further results and proofs.

2. The $T_n\big((0,m)\big)$ process is *saturated with the order* n^{-m} (see {111, Theorem 7.14 and 7.15}).

3. One can consider $(0,2)$ interpolation on the infinite interval or weighted $(0,2)$ polynomials ({111, p. 234} and *J. Balázs* {11}).

6. On the Mean Convergence of Interpolation

6.1. The negative results in Part 2 (cf. Theorems 2.1, 2.5 and Part 2.5) motivate the fact that the attention turned to the *mean convergence* of interpolation. The first such result is due to *P. Erdős* and *P. Turán* {25} from 1937.

Theorem 6.1. *For an arbitrary weight w and $f \in C$,*

$$\lim_{n\to\infty} \int_{-1}^{1} \big\{ L_n(f, w, x) - f(x) \big\}^2 w(x)\, dx = 0. \tag{6.1}$$

Here and later w is a weight if $w \geq 0$ and $0 < \int_{-1}^{1} w < \infty$; $L_n(f, w)$ is the Lagrange interpolation with nodes at on the roots of the corresponding orthogonal polynomials (ONP) $p_n(w)$ (see [47] or [174]).

Using the Čebishov roots, *P. Erdős* and *Ervin Feldheim* proved much more {19}:

Theorem 6.2. *Let $f \in C$ and $p > 1$. Then*

$$\lim_{n = \to \infty} \int_{-1}^{1} \left| f(x) - L_n(f, T, x) \right|^p \frac{1}{\sqrt{1-x^2}} \, dx = 0.$$

6.2. However, as *E. Feldheim* {48} showed, one can have divergence type results in the L_p-metrics, too:

Namely, *for a suitable $f_1 \in C$,*

$$\overline{\lim_{n \to \infty}} \int_{-1}^{1} \left| L_n\big(f_1, X^{(1/2,1/2)}, x\big) - f_1(x) \right|^4 \sqrt{1-x^2} \, dx > 0. \tag{6.2}$$

The results (6.1) and (6.2), justify the problem of *Erdős*, *Turán* and *Freud* (see {25}, [47]): Investigate the expression

$$\left\| f - L_n(f, w) \right\|_{p,u} := \int_{-1}^{1} \left| f(x) - L_n(f, w, x) \right|^p u(x) dx.$$

Here $p > 0$, u and w are weights.

After the initial results of *Richard Askey* and *V. M. Badkov*, *P. Nevai* {94} proved the next fairly general

Theorem 6.3. *Assume that $w \in GJ$, $0 < p < \infty$, $u \geq 0$, $0 < \int_{-1}^{1} u(x) \log^+ u(x) dx < \infty$. Then*

$$\lim_{n \to \infty} \left\| \left| f - L_n(f, w) \right| u \right\|_p = 0 \qquad \forall f \in C$$

iff

$$\frac{u(x)}{\sqrt{w(x)\sqrt{1-x^2}}} \in L_p.$$

Here $w \in GJ$ if $w(x) = \prod_{k=0}^{v+1} |x - t_k|^{\Gamma_k}$, $\Gamma_k > -1$, $0 \leq k \leq v+1$ and $-1 \equiv t_{v+1} < t_v < \cdots < t_1 < t_0 \equiv 1$ are fixed.

6.3. Let us say some words about the proof. First we mention a result of *J. Marcinkiewicz* {82} on the trigonometric case. Namely, using the so called Marcinkiewicz–Zygmund inequalities

$$\text{(i)}\quad \left(\frac{2\pi}{2n+1}\sum_{k=0}^{2n}|T_n(t_{kn})|^p\right)^{1/p} \le c\|T_n\|_p, \quad 1\le p<\infty,$$

$$\text{(ii)}\quad \|T_n\|_p \le c_p\left(\frac{2\pi}{2n+1}\sum_{k=0}^{2n}|T_n(t_{kn})|^p\right)^{1/p}, \quad 1<p<\infty$$

$\left(t_{kn}=\frac{2k\pi}{2n+1},\, T_n\in\mathcal{T}_n\right)$, *one can prove that* for $p>1$ $\lim_{n\to\infty}\|I_n(g)-g\|_p=0$ *for every* $g\in C$.

Here I_n is the trigonometric interpolatory polynomial based on $\{t_{kn}\}$, $0\le k\le 2n$ (compare with Theorem 6.2).

So if we try to prove an "algebraic" mean convergence result, first, as did *Richard Askey*, we may try to find the analogue of (i) and (ii). The method works if both u and $w\in GJ$. On the other hand, for an *arbitrary weight* u, *Nevai*, using a different approach, considered Lagrange interpolation as a mapping from the space of bounded functions into the appropriate weighted L^p spaces (and *not* as a mapping from L^p into L^p). Using this philosophy and some delicate arguments, he proved the above result.

6.4. Theorem 6.1 is a reasonable motivation of the problem raised by *Turán* {128, Problem VIII}:

Does there exists a weight w *and* $f\in C$ *such that*

$$\overline{\lim_{n\to\infty}}\left\|f-L_n(f,w)\right\|_{p,w}=\infty$$

for every $p>2$?

The "yes" answer was conjectured by *R. Askey*, however the rather complicated proof solving a more general problem is due to *P. Nevai* from 1985 (see {1} and {97}).

Nevai's proof requires a lot of difficult and far-reaching statements on orthogonal polynomials. But as it turned out from a paper of *Y. G. Shi*, using a new approach, many considerations can be saved and at the same time more general results can be obtained. Among others he proves in {103, Corr. 14}

Theorem 6.4. *Let u and w be weights. If with a fixed $p_0 \geq 2$*

$$\left\| \frac{1}{\sqrt{w\sqrt{1-x^2}}} \right\|_{p,u} = \infty \qquad \textit{for every} \qquad p > p_0,$$

then there exists an $f \in C$ satisfying

$$\overline{\lim_{n\to\infty}} \left\| L_n(f,w) \right\|_{p,u} = \infty \qquad \textit{whenever} \qquad p > p_0.$$

6.5. Because of the good uniform convergence behaviour of the HFi, the investigation of its mean convergence is relatively new. In 1985, *Nevai* and *Vértesi* proved ({90})

Theorem 6.5. *Let $u, w \in J$. Then*

$$\lim_{n\to\infty} \left\| H_n(f,w) - f \right\|_{p,u} = 0 \qquad \textit{for all} \qquad f \in C \tag{6.3}$$

iff $w^{-p}u \in L_1$.

Let $u = w = 1$. By (6.3), $\int_{-1}^{1} \left| H_n^{(0,0)}(f) - f \right| dx \to 0$ for all $f \in C$; on the other hand if $f_1(x) = 1 - x$, then $\lim_{n\to\infty} \left\| f_1 - H_n^{(0,0)}(f_1) \right\| > 0$ ([174, (14.6.17)]), i.e. mean convergence *may improve the convergence behaviour of HFi, too.*

6.6. Remarks. 1. In paper {85} "very general" Jacobi weights have been considered and statements connected to Theorem 6.3 have been proved.

2. The main formal distinction between Theorem 6.4 and *Nevai*'s result in {97} is that *Nevai* must suppose $\log w(\cos t) \in L_1$; however *the ideas of the proofs are totally different:*

Shi uses some ideas of the *Erdős-Vértesi*'s paper, where $\lambda_n(X,x)$ was estimated ({31}, {134}). For other related results., see {84} and {141}.

3. Finally we mention two results: the first is due to *J. Prasad* and *A. K. Varma* {99}. *We have with $w(x) = 1/\sqrt{1-x^2}$*

$$\left\| H_n(f,w) - f \right\|_{p,w} \leq c\omega\left(f, \frac{1}{n}\right) \qquad \textit{if} \qquad f \in C;$$

i.e., the result is better than the well-known uniform estimation

$$\| H_n(f_1, w) - f_1 \| \le c \frac{\log n}{n}, \qquad f_1(x) = |x|.$$

The second one is due to *G. Mastroianni* and *P. Nevai* {83}, Theorem 3.2.

Let $u, w \in GJ$, $r \ge 0$, $0 < p < \infty$. *If* $X(w, r)$ *is an interpolatory matrix obtained by adding an appropriate number of points* to $X(w)$ *near* ± 1, *then under certain conditions*

$$\| f^{(l)} - L_n^{(l)}(f, X(w, r)) \|_{p,u} = O(n^{l-r}) \omega \left(f^{(r)}, \frac{1}{n} \right), \qquad 0 \le l \le r,$$

whenever $f^{(r)} \in C$.

I.e., mean convergence *eliminates the "*$\log n$*" factor in the above two theorems!*

The proof of the last statement heavily uses the additional points method.

4. For other related results you may consult {110}.

B
WEIGHTED CASE

7. Weighted Lagrange Interpolation, Weighted Lebesgue Function, Weighted Lebesgue Constant

7.1. Let f be a continuous function, say. If, instead of the interval $[-1,1]$, we try to approximate it on $\mathbb{R}=(-\infty,\infty)$, we have to deal with the obvious fact that polynomials (of degree ≥ 1) tend to infinity if $|x|\to\infty$. So to get a suitable approximation tool, we may try to *moderate their growth applying proper weights.*

If the weight $w(x)=e^{-Q(x)}$, $x\in\mathbb{R}$, satisfies

$$\lim_{|x|=\infty}\frac{Q(x)}{\log|x|}=\infty,$$

as well as some other mild restrictions and the Akhiezer–Babenko–Carleson–Dzrbasjan relation

$$\int_{-\infty}^{\infty}\frac{Q(x)}{1+x^2}\,dx=\infty,$$

then for $f\in C(w,\mathbb{R})$, where

$$C(w,\mathbb{R}):=\Big\{f;\ f \text{ is continuous on } \mathbb{R} \text{ and } \lim_{|x|\to\infty}f(x)w(x)=0\Big\},$$

we have, if $\|\cdot\|$ denotes now the supnorm on $\mathbb{R}$,

$$E_n(f,w):=\inf_{p\in\mathcal{P}_n}\big\|(f-p)w\big\|\equiv\inf_{p\in\mathcal{P}_n}\|fw-pw\|\to 0\qquad\text{as } n\to\infty.$$

So, instead of approximating $f\in C$ by $L_n(f,X)$ on $[-1,1]$, *we may estimate* $\{f(x)w(x)-L_n(f,w,X,x)\}$ *on the real line* $\mathbb{R}$ *for* $f\in C(w,\mathbb{R})$. Here $X\subset\mathbb{R}$,

$$t_k(x):=t_{kn}(w,X,x):=\frac{w(x)\omega_n(X,x)}{w(x_k)\omega_n'(X,x_k)(x-x_k)},\qquad 1\leq k\leq n,$$

and

$$L_n(f,w,X,x) := \sum_{k=1}^{n} \{ f(x_k)w(x_k) \} t_k(x), \qquad n \in \mathbb{N}.$$

The Lebesgue estimate now has the form

$$\text{(7.1)} \qquad \big| L_n(f,w,X,x) - f(x)w(x) \big| \le \{ \lambda_n(w,X,x) + 1 \} E_{n-1}(f,w)$$

where the *(weighted) Lebesgue function* is defined by

$$\text{(7.2)} \qquad \lambda_n(w,X,x) := \sum_{k=1}^{n} \big| t_k(w,X,x) \big|, \ x \in \mathbb{R}, \ n \in \mathbb{N}$$

(cf. Part 2.1); the existence of $r_{n-1}(f,w)$ for which $E_{n-1}(f,w) = \big\| (f - r_{n-1})w \big\|$ is well-known.

Formula (7.2) implies the natural definition of the *(weighted) Lebesgue constant*

$$\text{(7.3)} \qquad \Lambda_n(w,X) := \big\| \lambda_n(w,X,x) \big\|, \qquad n \in \mathbb{N}.$$

Estimation (7.1) and its immediate consequence

$$\big\| L_n(f,w,X) - fw \big\| \le \{ \Lambda_n(w,X) + 1 \} E_{n-1}(f,w), \qquad n \in \mathbb{N},$$

show that, analogous to the classical case, the investigation of $\lambda_n(w,X,x)$ and $\Lambda_n(w,X)$ is of fundamental importance to get convergence-divergence results for the weighted Lagrange interpolation.

To expect reasonable estimations, as it turns out, we need a considerable knowledge about the weight $w(x)$ and on the behaviour of the ONP $p_n(w^2,x)$ corresponding to the weight w^2.

7.2. As *Nevai* writes in his instructive monograph {92, Part 4.15}, about 40 years ago there was a great amount of information on orthogonal polynomials on infinite intervals, however as *G. Freud* realized in the sixties, there had been a complete lack of *systematic treatment* of the general theory; the results were of mostly ad hoc nature. And *G. Freud*, in the last 10 years of his life, laid down the basic tools of the systematic investigation.

During the years a great number from the approximators and/or orthogonalists joined *G. Freud* and his work, including many Hungarians. As a result, today our knowledge is more comprehensive and more solid than before. On the other hand, this branch is relatively young; a lot of new, exciting problems remain to be solved!

Let us return to *Freud*'s work. His first natural step was taking the Hermite polynomials as a prototype of ONP with weights whose support is noncompact. Later, he introduced $Q(x)$ (instead of the Hermitian $x^2/2$) and q_n. (It corresponds to the MRS number — see 7.3.)

Nowadays these "Hermite type" weights bear *Freud*'s name. To be more precise, here is a quite general definition of the so called *Freud-type weights.*

We say that $w(x) = e^{-Q(x)} \in \mathcal{F}$ if $Q : \mathbb{R} \to \mathbb{R}$ is even and differentiable in $\mathbb{R}$, Q'' is continuous in $(0, \infty)$, $Q' > 0$ in $(0, \infty)$ and for some $A, B > 1$

$$A \leq \frac{\left(xQ'(x)\right)'}{Q'(x)} \leq B, \qquad x \in (0, \infty).$$

Clearly, if $w \in \mathcal{F}$ then $w^2 \in \mathcal{F}$, too (see *A. L. Levin, Doron S. Lubinsky* {137}, say).

The simplest cases are the so-called *Freud weights* $w(x) = e^{-|x|^\alpha}$, $\alpha > 1$. Here $w \in \mathcal{F}$ with $A = B = \alpha$.

Let $w \in \mathcal{F}$ and denote by $\left\{ y_k = y_{kn}(w^2) \right\}_{k=1}^{n}$ the n different roots of the ONP $\left\{ p_n(w^2, x) \right\}_{n=0}^{\infty}$ (with respect to the weight $w^2 \in \mathcal{F}$). We index them in decreasing order as

$$(7.4) \qquad -\infty < y_{nn} < y_{n-1,n} < \ldots < y_{2n} < y_{1n} < \infty.$$

If $w \in \mathcal{F}$

$$(7.5) \qquad \Lambda_n\big(w, Y(w^2)\big) \sim n^{1/6}$$

where $Y(w^2) = \{y_{kn}(w^2);\ 1 \leq k \leq n,\ n \in \mathbb{N}\}$ (see *D. M. Matjila* and *J. Szabados* {86}, {117}).

However, one can do better (see {117, Theorem 1}).

Applying the "additional points method", *J. Szabados* {117, Theorem 1} improved (7.5) as follows.

Let $y_0 = y_{0n} > 0$ denote a point such that

$$\left| p_n(w^2, y_0) w(y_0) \right| = \left\| p_n(w^2) w \right\|.$$

Now if

$$(7.6) \qquad V(w^2) = \Big\{ \big\{ y_{kn}(w^2), 1 \leq k \leq n \big\} \cup \{y_{0n}, -y_{0n}\}, n \in \mathbb{N} \Big\},$$

one can prove the following:

Let $w \in \mathcal{F}$. Then

$$\Lambda_n\big(w, V(w^2)\big) \sim \log n. \tag{7.7}$$

7.3. To make the choice of $\pm y_0$ more clear, we introduce the so-called *Mhaskar-Rahmanov-Saff* (MRS) *number*. From now on let $I = (-1, 1)$ or $\mathbb{R}$ and $w = e^{-Q}$ where $Q : I \to \mathbb{R}$ is even and convex in I and has limit ∞ at the endpoints of I.

Let $u > 0$ be fixed; then the (unique) a satisfying

$$u = \frac{2}{\pi} \int_0^1 \frac{atQ'(at)}{\sqrt{1-t^2}}\, dt \tag{7.8}$$

is by definition the MRS *number.* It is denoted by $a_u(w)$. A very important and useful property of $a_n(w)$ is that

$$\begin{cases} \|r_n w\| = \max_{|x| \le a_n(w)} \big|r_n(x) w(x)\big|, \\ \|r_n w\| > \big|r_n(x) w(x)\big| \quad \text{for } |x| > a_n(w) \end{cases} \tag{7.9}$$

if $r_n \in \mathcal{P}_n$ $\quad(r_n \not\equiv 0$; $\|\cdot\|$ is the supnorm on $I)$ and that asymptotically (as $n \to \infty$) $a_n(w)$ is the smallest such number. Relation (7.9) may be formulated such that $r_n w$ "lives" on $[-a_n, a_n]$.

As *an example*, let $Q(x) = |x|^\alpha$. Then

$$a_n(w) = c(\alpha)\, n^{1/\alpha}, \qquad \alpha > 1,$$

and as one can see, y_{1n} and y_{0n} are "close" to a_n: namely, $|y_{0n} - y_{1n}| \le c\frac{a_n(w)}{n^{2/3}}$.

7.4. A very natural question is whether the order $\log n$ in (7.7) is optimal (see the Faber result (2.8)). *J. Szabados* {117, Theorem 2} verified this hint for the special Hermite weight $w(x) = e^{-x^2/2}$ (actually, he proved it also for other projection operators).

Generalizing the method and ideas of our common paper with *Erdős*, one can prove a statement on the weighted Lebesgue function $\lambda_n(w, X, x)$ (see *P. Vértesi* {136}).

Theorem 7.1. *Let $w \in \mathcal{F}$. If $\varepsilon > 0$ is an arbitrary fixed number, then for any interpolatory matrix $X \subset \mathbb{R}$ there exist sets $H_n = H_n(w, \varepsilon, X)$ with $|H_n| \le 2a_n(w)\varepsilon$ such that*

$$\lambda_n(w, X, x) \ge \frac{\varepsilon}{3840} \log n \tag{7.10}$$

if $x \in \big[-a_n(w), a_n(w)\big] \setminus H_n$, $n \ge n_1(\varepsilon)$.

This statement is a complete analogue of Theorem 2.4. Roughly speaking, it says that the weighted Lebesgue function is at least $c \log n$ on a "big part" of $[-a_n, a_n]$ for arbitrary fixed $X \subset (-\infty, \infty)$ and $w \in \mathcal{F}$.

7.5. The previous consideration can be developed for other weights. We mention some relations without going into the details.

We say that $w = e^{-Q}$ is an *Erdős weight* ($w \in \mathcal{E}$) if $Q(x) \approx \exp_k\big(|x|^\alpha\big)$, $\alpha > 1$ (see *A. L. Levin, D. S. Lubinsky* and *T. Z. Mthembu* {74} for the meaning of "$\approx$"). Using definitions and notations analogous the previous parts we have the following:

Let $w \in \mathcal{E}$. Then

$$a_n(w) = \{\log_k n\}^{1/\alpha}\big(1 + o(1)\big) \qquad \text{if} \qquad Q(x) = \exp_k |x|;$$

$$\begin{cases} \Lambda_n\big(w, Y(w^2)\big) \sim (nT_n)^{1/6}, & \text{where } \; T_n \nearrow \infty, \qquad T_n = o(n^2), \\ \Lambda_n\big(w, V(w^2)\big) \sim \log n \end{cases}$$

(see {74} and *Steve Damelin* {13}).

7.6. Analogous theorems can be proved if $w \in \mathcal{L}$ ($w(x) \approx x^\beta e^{-x^\alpha}$, $\beta > -1$, $\alpha > 1$, $x \in (0, \infty)$ (Laguerre type weights)); $w \in \mathcal{EXP}$ $\big(Q(x) \approx \exp_k\big((1-x^2)^{-\alpha}\big)$, $x \in (-1,1)\big)$; $w \in GSJ$ ($w(x) \approx \prod |x - t_k|$, where $T_n \sim n^2$, $x \in (-1,1)$) or $w \in M$ ($w = uv$ where $u \in \mathcal{F} \cup \mathcal{E} \cup \mathcal{L}$ and $v \in GSJ$). Further, we have {137}.

Theorem 7.2. *Let $w \in \mathcal{E} \bigcup \mathcal{L} \bigcup \mathcal{EXP} \bigcup GSJ \bigcup M$. Then the estimation analogous to (7.10) can be proved.*

7.7. Remarks. 1. First we formulate a useful relation analogous to 2.8.1

Let $(a, b) \subset \mathbb{R}$ *and* $w = e^{-Q} : (a, b) \to (0, \infty)$. *Assume that* Q' *exists and increasing in* (a, b). *Then for* $1 \le k \le n-1$,

$$t_{kn}(w, X, x) + t_{k+1,n}(w, X, x) \ge 1 \qquad \textit{if} \qquad x \in [x_{k+1,n}, x_{kn}]$$

for arbitrary interpolatory $X \subset (a, b)$ (see *D. S. Lubinsky* {130}).

2. There are a lot of problems to be solved: the weighted version of the *Grünwald–Marcinkiewicz, Erdős–Halász, Erdős–Vértesi, Erdős–Kroó–Szabados* results. Other ones can be formulated according to Part A.

8. Getting Convergence by Raising the Degree

8.1. Using Theorems 7.1 and 7.2, one can easily get Faber type results. To improve the behaviour of the weighted interpolation, one may try to apply the analogue of the HFi (cf. Part 4). However, as it turned out from the papers of *D. S. Lubinsky, P. Rabinowitz, J. Szabados* and others, the corresponding results are not quite satisfactory: one has to take the function class $C(w_2, \mathbb{R})$ to get convergence for $\|w_1(f - H_n(f, Y(w^2)))\|$ where $w_1(x) = o(w^2(x))$, $w^2(x) = o(w_2(x))$ $(|x| \to \infty,\ w \in \mathcal{F})$ instead of the "natural" $w_1 = w_2 = w^2$ (see {123} and its references).

8.2. Very recently *V. E. Sándor Szabó* {123} realized this "natural" settlement by taking a *Grünwald* type process. He proved

Theorem 8.1. *Let* $w \in \mathcal{F}$ *and* $f \in C(w^2, \mathbb{R})$. *Then with* $G_n(f, x) = \sum_{k=1}^{n} f(y_{kn}(w^2))\, \ell_{kn}^2(Y(w^2), x) \in \mathcal{P}_{2n-2}$

$$\lim_{n \to \infty} \|w^2(f - G(f))\| = 0.$$

Vértesi {129} refers to an *arbitrary matrix* X and proves the analogue of the Erdős theorem (see Theorem 4.6).

Theorem 8.2. *Let* $w \in \mathcal{F}$. *If* $|t_{kn}(w, X, x)| \le A$ *uniformly in* $x \in \mathbb{R}$, k *and* n, *then for every* $\varepsilon > 0$ *and to every* $f \in C(w^{1+\varepsilon}, \mathbb{R})$, *there exists a sequence of polynomials* $\varphi_\Delta(x) = \varphi_\Delta(f, \varepsilon, x) \in \mathcal{P}_\Delta$ *such that*

(i) $\Delta \le n(1 + \varepsilon + c\,\varepsilon n^{-2/3})$,

(ii) $\varphi_\Delta(x_{kn}) = f(x_{kn})$, $1 \le k \le n$, $n \in \mathbb{N}$,

(iii) $\|w^{1+\varepsilon}(f - \varphi_\Delta)\| \le c E_\Delta(f, w^{1+\varepsilon})$.

8.3. Remarks. 1. The main problem in proving Theorem 8.2 (which is a far-reaching generalization of Theorem 8.1) is that now $\| t_{kn}(w,X,x)\| \le A$ *does not* involve the nice property $\vartheta_{k-1,n} - \vartheta_{kn} \sim \frac{1}{n}$ (which holds true at the classical case whenever $\| \ell_{kn}(X,x)\| \le A$).

2. Statements analogous to Theorem 8.2 for other weights have been proved in {126}.

9. Mean Convergence

9.1. The results of this part originally were formulated for the classical $L_n(f,X,x) = \sum f(x_{kn})\ell_{kn}(X,x)$ Lagrange interpolatory case. However by

$$|f(x) - L_n(f,X,x)|^p\, w^p(x) = |f(x)w(x) - L_n(f,w,X,x)|^p$$

they can be transformed into the weighted case. We shall use both formulations. First a counterpart of Theorem 6.1 due to *Shohat* ({77, ref. {65}}):

Theorem 9.1. *Let $f^2 \in C(w^2,\mathbb{R})$. Then*

$$\lim_{n\to\infty} \| fw - L_n(f,w,Y(w^2))\|_{L_2(\mathbb{R})} = 0.$$

Here and later $Y(w^2)$ (or $V(w^2)$) is analogous to (7.4) (or (7.6)); $\|h\|_{L_p(\mathbb{R})} = \left(\int_{\mathbb{R}} |h|^p\right)^{1/p}$.

9.2. In many respects the Hermite weight on $\mathbb{R}$ corresponds to the Čebishov weight $1/\sqrt{1-x^2}$ on $(-1,1)$. In spite of this, a result similar to (6.2) does not hold. For simplicity, we quote a rather special case of a paper of *Lubinsky* and *Matjila*, from 1995 {77, ref. {63}}.

Theorem 9.2. *Let $w \in \mathcal{F}$, $1 < p \le 4$, $0 < \alpha \le 1$. Then*

$$\lim_{n\to\infty} \left\| \left(f - L_n(f,Y(w^2))\, w_{-\Delta}\right)\right\|_{L_p(\mathbb{R})} = 0$$

for every $f \in C(w_\alpha,\mathbb{R})$ iff $\Delta > -\alpha + 1/p$. Here $w_\delta(x) = w(x)\left(1+|x|\right)^\delta$.

9.3. The situation is more satisfactory if we use the matrix $V(w^2)$ instead of $Y(w^2)$. We quote two recent results of *D. S. Lubinsky* and *G. Mastroianni* {76}, {78}.

Theorem 9.3. *Let $w \in \mathcal{F}$ and $1 < p < \infty$. Then for every $f \in C(w, \mathbb{R})$,*

$$\lim_{n\to\infty} \left\| fw - L_n\big(f, w, V(w^2)\big) \right\|_{L_p(\mathbb{R})} = 0.$$

The analogous statement holds true if $w \in \mathcal{EXP}$.

The basic ideas are analogous to the ones in the previous theorem. However, the main emphasis was to get a general Marcinkiewicz–Zygmund inequality using the fairly sophisticated König method.

9.4. Finally here is a result of *Nevai* {91} analogous to Theorem 6.4.

Theorem 9.4. *Let $w(x) = \exp\big(-|x|^m\big)$, $m > 0$, even. Let u (≥ 0) and $\int_{\mathbb{R}} u < \infty$. If $0 < p < \infty$ and*

$$\int_{\mathbb{R}} \left[\big(w(t)\big)^{1/2}\big(1 + |t|\big)\right]^{-p} u(t)\, dt = \infty,$$

then there exists a function f supported on a finite interval such that

$$\overline{\lim_{n\to\infty}} \int_{\mathbb{R}} \big|L_n\big(f, X(w), t\big)\big|^p\, u(t)\, dt = \infty.$$

References

[47] Freud, Géza, *Orthogonale Polynome,* Akadémiai Kiadó (Budapest), Birkhäuser (Basel, 1969). *Orthogonal Polynomials,* Pergamon Press (London–Toronto–New York, 1971).

[174] Szegő, Gábor, *Orthogonal Polynomials,* American Mathematical Society Colloquium Publications, Vol. XXIII., 1939, revised 1959, 3rd edition 1967, 4th edition 1975.

{1} R. Askey, Mean convergence of orthogonal series and Lagrange interpolation, *Acta Math. Acad. Sci. Hungar.*, **23** (1972), 71–85.

{2} N. S. Baiguzov, Some estimates for the derivates of algebraic polynomials and an application to numerical differentiation, *Mat. Zametki,* **5** (1969), 183–194 (in Russian).

{3} J. Balázs and P. Turán, Notes on interpolation, II, *Acta Math. Acad. Sci. Hungar.*, **8** (1957), 201–215.

{4} J. Balázs and P. Turán, Notes on interpolation, III, *Acta Math. Acad. Sci. Hungar.*, **9** (1958), 195–214.

{5} J. Balázs, On the convergence of Hermite–Fejér interpolation process, *Acta Math. Acad. Sci. Hungar.*, **9** (1958), 259–267.

{6} J. Balázs, Weighted $(0, 2)$ interpolation on the ultrasherical nodes, *MTA III. Osztálya Közl.*, **11(3)** (1961), 305–338.

{7} D. L. Berman, On some linear operators, *Dokl. Akad. Nauk SSSR.*, **73** (1950), 249–252 (Russian).

{8} L. Brutman and A. Pinkus, On the Erdős conjecture concerning minimal norm interpolation on the unit circle, *SIAM J. Numer. Anal.*, **17** (1980), 373–375.

{9} L. Brutman and A. Pinkus, On the Erdős conjecture concerning minimal norm interpolation on the unit circle, *SIAM J. Numer. Anal.*, **17** (1980), 373–375.

{10} L. Brutman, Lebesgue functions for polynomial interpolation. A survey, *Ann. of Numer. Math.*, **4** (1997), 111–127.

{11} A. S. Cavaretta, Jr., A. Sharma and R. S. Varga, Hermite Birkhoff interpolation in the n-th roots of unity, *Trans. Amer. Math. Soc.*, **259** (1980), 621–628.

{12} A. M. Chak, A. Sharma and J. Szabados, On a problem of P. Turán, *Studia Sci. Math. Hungar.*, **15** (1980), 441–455.

{13} S. Damelin, The Lebesgue function and Lebesgue constant of Lagrange interpolation for Erdős weights, *J. Approx. Theory,* to appear.

{14} C. de Boor and A. Pinkus, Proof of the conjectures of Bernstein and Erdős concerning the optimal nodes for polynomial interpolation, *J. Approx. Theory,* **24** (1978), 289–303.

{15} B. della Vecchia, Direct and converse results by rational operators, *Constr. Approx.*, **12** (1996), 271–285.

{16} B. della Vecchia, G. Mastroianni and P. Vértesi, *One-sided convergence of Lagrange interpolation based on generalized Jacobi weights,* to appear.

{17} E. Egerváry and P. Turán, Notes on interpolation, V., *Acta Math. Acad. Sci. Hungar.*, **9** (1958), 259–267.

{18} S. A. N. Eneduanya, On the convergence of interpolation polynomials, *Anal. Math.*, **11** (1985), 13–22.

{19} P. Erdős and E. Feldheim, Sur le mode de convergence pour l'interpolation de Lagrange, *C. R. Acad. Sci. Paris,* **203** (1936), 913–915.

{20} P. Erdős and G. Grünwald, Über die aritmetischen Mittelwerte der Lagrangeschen Interpolationspolynome, *Studia Math.*, **7** (1938), 82–95.

{21} P. Erdős and G. Grünwald, Über einen Faber'schen Satz, *Ann. Math.*, **39** (1938), 257–261.

{22} P. Erdős and G. Halász, On the arithmetic means of Lagrange interpolation, in: *Approximation Theory* (J. Szabados, K. Tandori, eds.), Colloq. Math. Soc. J. Bolyai, **58** (1991), pp. 263–274.

{23} P. Erdős and J. Szabados, On the integral of the Lebesgue function of interpolation, *Acta Math. Acad. Sci. Hungar.*, **32** (1978), 191–195.

{24} P. Erdős and P. Turán, An extremal problem in the theory of interpolation, *Acta Math. Acad. Sci. Hungar.*, **12** (1961), 221–234.

{25} P. Erdős and P. Turán, On interpolation, I.Quadrature and mean convergence in the Lagrange interpolation, *Ann.of Math.*, **38** (1937), 142–155.

{26} P. Erdős and P. Turán, On interpolation, III., *Ann. of Math.*, **41** (1940), 510–553.

{27} P. Erdős and P. Turán, On the role of the Lebesgue function the theory of Lagrange interpolation, *Acta Math. Acad. Sci. Hungar.*, **6** (1955), 47–66.

{28} P. Erdős and P. Vértesi, Correction of some misprints in our paper: "On the almost everywhere divergence of Lagrange interpolation polynomials for arbitrary systems of nodes" [*Acta Math. Acad. Sci. Hungar.*, **36**, no. 1–2 (1980), 71–89], *Acta Math. Acad. Sci. Hungar.* **38** (1981), 263.

{29} P. Erdős and P. Vértesi, On the almost everywhere divergence of Lagrange interpolation of polynomials for arbitrary systems of nodes, *Acta Math. Acad. Sci. Hungar.*, **36** (1980), 71–89.

{30} P. Erdős and P. Vértesi, On the almost everywhere divergence of Lagrange interpolation, in: *Approximation and Function Spaces* (Gdańsk, 1979), North-Holland (Amsterdam–New York, 1981), pp. 270–278.

{31} P. Erdős and P. Vértesi, On the Lebesgue function of interpolation, in: *Functional Analysis and Approximation* (P. L. Butzer, B. Sz.-Nagy, E. Görlich, eds.), ISNM, **60**, Birkhäuser (1981), pp. 299–309.

{32} P. Erdős, A. Kroó and J. Szabados, On convergent interpolatory polynomials, *J. Approx. Theory,* **58** (1989), 232–241.

{33} P. Erdős, J. Szabados and P. Vértesi, On the integral of Lebesgue function of interpolation. II, *Acta Math. Hungar.*, **68** (1995), 1–6.

{34} P. Erdős, J. Szabados, A. K. Varma and P. Vértesi, On an interpolation theoretical extremal problem, *Studia Sci. Math. Hungar.*, **29** (1994), 55–60.

{35} P. Erdős, On some convergence properties in the interpolation polynomials, *Ann.of Math.,* **44** (1943), 330–337.

{36} P. Erdős, Problems and results on the theory of interpolation, I, *Acta Math. Acad. Sci. Hungar.*, **9** (1958), 381–388.

{37} P. Erdős, Problems and results on the theory of interpolation, I, *Acta Math. Acad. Sci. Hungar.*, **9** (1958), 381–388.

{38} P. Erdős, Problems and results on the theory of interpolation, II, *Acta Math. Acad. Sci. Hungar.*, **12** (1961), 235–244.

{39} P. Erdős, Some theorems and remarks on interpolation, *Acta Sci. Math. (Szeged),* **12** (1950), 11–17.

{40} G. Faber, Über die interpolatorische Darstellung steiger Funktionen, *Jahresber. der Deutschen Math. Verein.*, **23** (1914), 190–210.

{41} L. Fejér, Über Interpolation, *Göttinger Nachrichten* (1916), 66–91.

{42} L. Fejér, Interpolation und konforme Abbildung, *Gött. Nachr.* (1918), 319–331.

{43} L. Fejér, Interpolation-ról (első közlemény), *Mat. és Term. Értesítő,* **34** (1916), 209–229.

{44} L. Fejér, Lagrangesche interpolation und die zugehörigen konjugierten Punkte, *Math. Ann.*, **106** (1932), 1–55.

{45} L. Fejér, Lebeguesche Konstanten und divergente Fourierreihen, *J. Reine Angew, Math.*, **138** (1910), 22–53.

{46} L. Fejér, On the characterization of some remarkable systems point of interpolation by means of conjugate points, *American Math. Monthly,* **41** (1934), 1–14.

{47} L. Fejér, Sur les fonctions bornées et integrables, *C. R. Acad. Sci. Paris,* **131** (1900), 984–987.

{48} E. Feldheim, Sur le mode de convergence dans l'interpolation de Lagrange, *Dokl. Nauk. USSR,* **14** (1937), 327–331.

{49} E. Feldheim, Théorie de la convergence des procédés d'interpolation et de quadrature mécanique, *Mémorial des Sciences Mathématiques,* **95** (1939), 1–9, Paris, Gauthier–Villars.

{50} G. Freud, Bemarkung über die Konvergenz eines interpolations verfahren von P. Turán, *Acta Math. Acad. Sci. Hungar.*, **9** (1958), 337–341.

{51} D. Gaier, *Lectures on Complex Approximation,* Birkhäuser (1987), pp. 196+IX.

{52} H. H. Gonska and H.-B. Knoop, On Hermite–Fejér Interpolation; A Biography (1914–1987), *Studia Sci. Math. Hungar.*, **25** (1990), 147–198.

{53} G. Grünwald, Über Divergenzerscheinungen der Lagrangeschen Interpolationspolynome stetiger Funktionen, *Ann. Math.*, **37** (1936), 908–918.

{54} G. Grünwald, Über Divergenzerscheinungen der Lagrangeschen Interpolationspolynome, *Acta Sci. Math. (Szeged),* **7** (1935), 207–221.

{55} G. Grünwald, On the theory of interpolation, *Acta Math.*, **75** (1942), 219–245.

{56} G. Halász, On projections into the space of trigonometric polynomials, *Acta Sci. Math. (Szeged),* **57** (1993), 353–366.

{57} G. Halász, The "coarse and fine theory of interpolation" of Erdős and Turán in a boarder view, *Constr. Approx.*, **8** (1992), 169–185.

{58} D. Jackson, A formula of trigonometric interpolation, *Rendiconti der circolo matematico di Palermo,* **37** (1914), 371–375.

{59} I. Joó, On Pál interpolation, *Ann. Univ. Sci. Budapest. Sect. Math.*, **37** (1994), 247–262.

{60} I. Joó and V. E. S. Szabó, A generalization of Pál interpolation process, *Acta Sci. Math. (Szeged),* **60** (1995), 429–438.

{61} A. Kalmár, Az interpolációról, *Math. és Phys. Lapok,* **33** (1926), 120–149 (in Hungarian).

{62} T. A. Kilgore, A characterization of the Lagrange interpolating projection with minimal Tchebycheff norm, *J. Approx. Theory,* **24** (1978), 273–288.

{63} O. Kis and J. Szabados, On the convergence of Lagrange interpolation, *Acta Math. Acad. Sci. Hungar.*, **16** (1965), 389–430 (in Russian).

{64} O. Kis, Convergence of interpolation in some function spaces, *MTA Mat. Kut. Int. Közl.*, **7** (1962), 95–111 (Russian).

{65} O. Kis, On certain interpolatory processes, II, *Acta Math. Acad. Sci. Hungar.*, **26** (1973), 171–190 (in Russian).

{66} O. Kis, On the convergence of the trigonometrical and harmonical interpolation, *Acta Math. Acad. Sci. Hungar.*, **7** (1956), 173–200 (in Russian).

{67} O. Kis, On trigonometric $(0, 2)$ interpolation, *Acta Math. Acad. Sci. Hungar.*, **11** (1960), 255–276 (in Russian).

{68} O. Kis, Remarks on interpolation, *Acta Math. Acad. Sci. Hungar.*, **11** (1960), 49–64 (in Russian).

{69} O. Kis, Remarks on the error of interpolation, *Acta Math. Acad. Sci. Hungar.*, **20** (1969), 339–346 (in Russian).

{70} P. P. Korovkin, *Linear Operators and Approximation Theory,* Hindustan (Delhi, 1960).

{71} M. Lénárd, On $(0; 1)$ Pál-type interpolation with boundary conditions, *Publ. Math. Debrecen,* **55 (3-4)** (1999), 465–478.

{72} M. Lénárd, Birkhoff quadrature formulae based on the zeros of Jacobi polynomials, *Mathematical and Computer Modelling,* **38** (2003), 917–927.

{73} A. L. Levin and D. S. Lubinsky, Christoffel functions, orthogonal polynomials, and Nevai's conjecture for Freud weights, *Constr. Approx.*, **8** (1992), 463–535.

{74} A. L. Levin, D. S. Lubinsky and T. Z. Mthembu, Christoffel functions and orthogonal polynomials for Erdős weights on $(-\infty, \infty)$, *Rendiconti di Matematica di Roma,* **14** (1994), 199–289.

{75} G. G. Lorentz, K. Jetter and S. D. Riemenschneider, Birkhoff interpolation, in: *Encyclopedia of Mathematics and its Applications,* **19**, Addison-Wesley (London–Amsterdam–Sydney–Tokyo, 1983).

{76} D. S. Lubinsky and G. Mastroianni, *Mean convergence of extended Lagrange interpolation with Freud weights,* to appear.

{77} D. S. Lubinsky, An update on orthogonal polynomials and weighted approximation on the real line, *Acta Appl. Math.*, **33** (1993), 121–164.

{78} D. S. Lubinsky, On converse Marcinkiewicz–Zygmund inequalities in L^p $p > 1$ constr., *Approx.*, **15** (1999), 577–610.

{79} S. M. Lozinskii, The spaces $\tilde{C}_\omega^*$ and $\tilde{C}_\omega^*$ and the convergence of interpolation processes in them, *Dokl. Akad. Nauk SSSR (N.S.),* **59** (1948), 1389–1392 (in Russian).

{80} T. M. Mills and P. Vértesi, An extension of the Grünwald-Marcinkiewicz interpolation theorem, *Bull. Austral Math. Soc.*, **63** (201), 299–320.

{81} J. Marcinkiewicz, Sur la divergence des polynômes d'interpolation, *Acta Sci. Math. (Szeged),* **8** (1937), 131–135.

{82} J. Marczinkiewicz, On interpolation, I., *Studia Math.,* **6** (1936), 1–17 (in French).

{83} G. Mastroianni and P. Nevai, Mean convergence of derivatives of Lagrange interpolation, *J. Comput. Appl. Math.,* **34(3)** (1991), 385–396.

{84} G. Mastroianni and P. Vértesi, Mean convergence of Lagrange interpolation on arbitrary system of nodes, *Acta Sci. Math. (Szeged),* **57** (1993), 425–436.

{85} G. Mastroianni and P. Vértesi, Some applications of generalized Jacobi weights, *Acta Math. Acad. Sci. Hungar.*, **77** (1997), 323–357.

{86} D. M. Matjila, Bounds for the weighted Lebesgue function for Freud weights on a larger interval, *J. Comp. Appl. Math.,* **65** (1995), 293–298.

{87} T. M. Mills, Some techniques in approximation theory, *Math. Sci.,* **5** (1980), 105–120.

{88} L. Neckermann and P. O. Runck, Über Approximationseigenschaften differenzierter Lagrangescher Interpolationspolynome mit Jacobischen Abszissen, *Numer. Math.,* **12** (1968), 1959–1969.

{89} P. Nevai and P. Vértesi, Convergence of Hermite–Fejér interpolation at zeros of generalized polynomials, *Acta Sci. Math. Szeged,* **53** (1989), 77–98.

{90} P. Nevai and P. Vértesi, Mean convergence of Hermite–Fejér interpolation, *Math. Anal. Appl.,* **105** (1985), 26–58.

{91} P. Nevai, *Necessary conditions for weighted mean convergence of Lagrange interpolation associated with exponential weights,* in preparation.

{92} P. Nevai, Géza Freud, orthogonal polynomials and Christoffel functions: A case study, *J. Approx. Th.,* **48** (1986), 3–167.

{93} P. Nevai, Lagrange interpolation at zeros of orthogonal polynomials, in: *Approximation Theory II.,* Academic Press (1976), pp. 163–201.

{94} P. Nevai, Mean convergence of Lagrange interpolation. III, *Trans. Amer. Math. Soc.,* **282** (1984), 669–698., and *J. Approx. Theory,* **60** (1990), 360–363.

{95} P. Nevai, Orthogonal Polynomials, *Mem. Amer. Math. Soc.,* **213** (Amer. Mathematical Soc. Providence RI, 1979).

{96} P. Nevai, Remarks on interpolation, *Acta Math. Acad. Sci. Hungar.,* **25** (1974), 129–144 (in Russian).

{97} P. Nevai, Solution of Turán's problem on divergence of Lagrange interpolation in L^p with $p > 2$, *J. Approx Theory,* **43** (1985), 190–193.

{98} L. G. Pál, A new modification of the Hermite–Fejér interpolation, *Anal. Math.,* **1** (1975), 197–205.

{99} J. Prasad and A. K. Varma, *An analogue of a problem of P. Erdős and E. Feldheim on L_p convergence of interpolating process,* to appear.

{100} Z. F. Sebestyén, Pál-type interpolation on the roots of Hermite polynomials, *Pure Math. Appl.*, **9(3–4)** (1998), 429–439.

{101} Z. F. Sebestyén, Supplement to the Pál type $(0; 0, 1)$ lacunary interpolation *Anal. Math.*, **25(2)** (1999), 147–154.

{102} A. Sharma and A. K. Varma, Trigonometric interpolation, *Duke Math J.*, **32** (1965), 341–357.

{103} Y. G. Shi, Bounds and inequalities for general orthogonal polynomials on finite intervals, *J. Approx. Theory,* **73** (1993), 303–319.

{104} Y. G. Shi, On critical order of Hermite–Fejér type interpolation, in: *Progress in Approximation Theory,* Academic Press (1991), pp. 761–766.

{105} S. J. Smith, Generalized Hermite–Fejér interpolation polynomials, *Expo. Math.*, **18** (2000), 389–404.

{106} G. Somorjai, On a saturation problem, *Acta Math. Acad. Sci. Hungar.*, **32** (1978), 377–381.

{107} G. Somorjai, On discrete linear operators in the function space A, in: *Constructive Function Theory '77* (Sofia, 1980) pp. 489–496.

{108} J. Surányi and P. Turán, Notes on interpolation, I., *Acta Math. Acad. Sci. Hungar.*, **6** (1955), 67–80.

{109} J. Szabados, The exact error of trigonometric interpolation for differentiable functions, *Constr. Approx.*, **8** (1992), 203–210.

{110} J. Szabados and P. Vértesi, A survey on mean convergence of interpolatory process, *J. Comp. App. Math.*, **43** (1992), 3–18.

{111} J. Szabados and P. Vértesi, Interpolation of Functions, in: *World Scientific* (1990), pp. 1–305+I–XII.

{112} J. Szabados, On Hermite–Fejér interpolation for the Jacobi abscissas, *Acta Math. Acad. Sci. Hungar.*, **23** (1972), 449–464.

{113} J. Szabados, Optimal order of convergence of Hermite–Fejér interpolation for general system of nodes, *Acad. Sci. Math. (Szeged),* **57** (1993), 463–470.

{114} J. Szabados, On the convergence of quadrature procedures in certain classes of functions, *Acta Math. Acad. Sci. Hungar.*, **18** (1967), 97–111.

{115} J. Szabados, On the convergence of the derivatives of projection operators, *Analysis,* **7** (1987), 349–357.

{116} J. Szabados, On the order of magnitude oof fundamental polynomials of Hermite interpolation, *Acta Math. Acad. Sci. Hungar.*, **61** (1993), 357–368.

{117} J. Szabados, Weighted Lagrange and Hermite–Fejér interpolation on the real line, *J. of Inequal. and Appl.*, **1** (1997), 99–123.

{118} J. Szabados and A. K. Varma, On a convergent Pál-type $(0, 2)$ interpolation process, *Acta Math. Hungar.*, **66** (1995), 301–326.

{119} V. E. S. Szabó, On Pál-type interpolation processes, *Approximation and optimization,* Vol. II (Cluj-Napoca, 1996), 221–226, Transilvania, Cluj-Napoca, 1997.

{120} V. E. S. Szabó, A generalization of Pál interpolation processes. II, *Acta Math. Hungar.*, **74 (1-2)** (1997), 19–29.

{121} V. E. S. Szabó, A generalization of Pál interpolation processes. III. The Hermite case, *Acta Math. Hungar.*, **74 (3)** (1997), 191–201.

{122} V. E. S. Szabó, A generalization of Pál interpolation processes. IV. The Jacobi case, *Acta Math. Hungar.*, **74 (4)** (1997), 287–300.

{123} V. E. S. Szabó, *Weighted interpolation, I.*, to appear.

{124} L. Szili, An interpolation process on the roots of the integrated Legendre polynomials, *Anal. Math.*, **9** (1983), 235–245.

{125} L. Szili, A convergence theorem for the Pál method of interpolation on the roots of Hermite polynomials, *Anal. Math.*, **11** (1985), 75–84.

{126} L. Szili and P. Vértesi, An Erdős type convergence process in weighted interpolation, II, *Acta Math. Acad. Sci. Hungar.*, **98** (2003), 129–162.

{127} A. F. Timan, *Approximation Theory of Functions of a Real Variable* (Moscow, 1960) (in Russian).

{128} P. Turán, On some open problems of approximation theory, *J. Approx. Theory,* **29** (1980), 23–85.

{129} P. Vértesi, An Erdős type convergence process in weighted interpolation. I. (Freud type weights), *Acta Math. Hungar.*, **91** (2000), 195–215.

{130} P. Vértesi, Derivatives of projection operators, *Analysis,* **9** (1989), 145–156.

{131} P. Vértesi, Hermite–Fejér type interpolations, II, *Acta Math. Acad. Sci. Hungar.*, **33** (1979), 333-343.

{132} P. Vértesi, Hermite–Fejér type interpolations, IV., *Acta Math. Acad. Sci. Hungar.*, **39** (1982), 85–93.

{133} P. Vértesi, Lebesgue function type sums of Hermite interpolations, *Acta Math. Acad. Sci. Hungar.*, **64** (1994), 341–349.

{134} P. Vértesi, New estimation for the Lebesgue function of Lagrange interpolation, *Acta Math. Acad. Sci. Hungar.*, **40** (1982), 21–27.

{135} P. Vértesi, On ρ-normal pointsystems, in: *A survey Approximation Theory* (edited by M. W. Müller et. al.), Math. Res., vol. 86., Proc. IDoMAT 95, pp. 317–327.

{136} P. Vértesi, On the Lebesgue function of weighted Lagrange interpolation, I., *Constr. Approx.*, **15** (1999), 355–367.

{137} P. Vértesi, On the Lebesgue function of weighted Lagrange interpolation, II., *J. Austral Math. Soc. (Series A),* **65** (1998), 145–162.

{138} P. Vértesi, One sided convergence conditions for Lagrange interpolation, *Acta Sci. Math. (Szeged),* **45** (1983), 419–428.

{139} P. Vértesi, Optimal Lebesgue constant for Lagrange interpolation, *SIAM J. Numer. Anal.,* **27** (1990), 1322–1331.

{140} P. Vértesi, Saturation of the Shepard operators, *Acta Math. Acad. Sci. Hungar.*, **72** (1996), 307–317.

{141} P. Vértesi, Turán-type problems on mean convergence, I–II., *Acta Math. Acad. Sci. Hungar.*, **65** (1994), 115–139., and (1994), 237–242.

{142} A. Zygmund, *Trigonometric Series I,* Cambridge University Press (1959).

{143} A. Zygmund, *Trigonometric Series II,* Cambridge University Press (1959).

{144} A. Zygmund, Jozef Marcinkiewicz, in: *Collected works of Jozef Marcinkiewicz,* pp. 4.

Péter Vértesi
Alfréd Rényi Institute of Mathematics
Hungarian Academy of Sciences
P.O.B. 127
1364 Budapest
Hungary

veter@renyi.hu

BOLYAI SOCIETY
MATHEMATICAL STUDIES, 14

A Panorama of Hungarian Mathematics in the Twentieth Century, pp. 119–156.

Extremal Properties of Polynomials

TAMÁS ERDÉLYI

This article focuses on those problems about extremal properties of polynomials that were considered by the Hungarian mathematicians Lipót Fejér, Mihály Fekete, Marcel Riesz, Alfréd Rényi, György Pólya, Gábor Szegő, Pál Erdős, Pál Turán, Géza Freud, Gábor Somorjai, and their associates, who lived and died mostly in the twentieth century. It reflects my personal taste and is far from complete even within the subdomains we focus on most, namely inequalities for polynomials with constraints, Müntz polynomials, and the geometry of polynomials. There are separate chapters of this book devoted to orthogonal polynomials, interpolation, and function series, so here we touch these issues only marginally.

1. Markov- and Bernstein-Type Inequalities

Let $\|f\|_A$ denote the supremum norm of a function f on A. The Markov inequality asserts that

$$\|p'\|_{[-1,1]} \le n^2 \|p\|_{[-1,1]}$$

holds for every polynomial p of degree at most n with complex coefficients. The inequality

$$|p'(y)| \le \frac{n}{\sqrt{1-y^2}} \|p\|_{[-1,1]}$$

holds for every polynomial p of degree at most n with complex coefficients and for every $y \in (-1,1)$, and is known as Bernstein inequality. Various analogues of the above two inequalities are known in which the underlying intervals, the maximum norms, and the family of functions are replaced by more general sets, norms, and families of functions, respectively. These inequalities are called Markov- and Bernstein-type inequalities. If the norms are the same on both sides, the inequality is called Markov-type, otherwise it is called Bernstein-type (this distinction is not completely standard). Markov- and Bernstein-type inequalities are known on various regions of the complex plane and n-dimensional Euclidean space, for various norms such as weighted L_p norms, and for many classes of functions such as polynomials with various constraints and exponential sums of n terms, just to mention a couple. Markov- and Bernstein-type inequalities have their own intrinsic interest. In addition, they play a fundamental role in approximation theory.

The inequality

$$\left\|p^{(m)}\right\|_{[-1,1]} \leq T_n^{(m)}(1) \cdot \|p\|_{[-1,1]}$$

for every (algebraic) polynomial p of degree at most n with complex coefficients was first proved by V.A. Markov in 1892. Here, and in the sequel T_n denotes the Čebyshov polynomial of degree n defined by

$$T_n(x) := \frac{1}{2}\left(\left(x + \sqrt{x^2-1}\,\right)^n + \left(x - \sqrt{x^2-1}\,\right)^n\right)$$

(equivalently, $T_n(\cos\theta) := \cos(n\theta)$, $\theta \in \mathbb{R}$). V. A. Markov was the brother of the more famous A. A. Markov who proved the above inequality for $m = 1$ in 1889 by answering a question raised by the prominent Russian chemist, D. I. Mendeleev. Sergei N. Bernstein presented a shorter variational proof of V. A. Markov's inequality in 1938. The simplest known proof of Markov's inequality for higher derivatives is due to Duffin and Schaeffer {15}, who gave various extensions as well.

Let $\mathbb{T} := \mathbb{R} \pmod{2\pi}$. The inequality

$$\|t'\|_{\mathbb{T}} \leq n\|t\|_{\mathbb{T}}$$

for all (real or complex) trigonometric polynomials of order n is also called the Bernstein inequality. It was proved by Bernstein in 1912 with $2n$ in place of n. See also {41}. The sharp inequality appears first in a paper of Fekete in 1916 who attributes the proof to Fejér. Bernstein attributes the proof to Edmund Landau. A clever proof based on zero-counting may be found in many books dealing with approximation theory. In books Markov's inequality for the first derivative is then deduced as a combination of Bernstein's inequality and an inequality due to Issai Schur:

$$\|p\|_{[-1,1]} \leq (n+1) \max_{x\in[-1,1]} \left|p(x)\sqrt{1-x^2}\right|$$

for every polynomial p of degree at most n with real coefficients.

Bernstein used his inequality to prove inverse theorems of approximation. Bernstein's method is presented in the proof of the next theorem, which is one of the simplest cases. However, several other inverse theorems of approximation can be proved by straightforward modifications of the proof of this result. That is why Bernstein- and Markov-type inequalities play a significant role in approximation theory. Direct and inverse theorems of approximation and related matters may be found in many books on approximation theory, including [111], {14}, and {56}.

Let $\mathcal{T}_n$ be the collection of all trigonometric polynomials of order at most n with real coefficients. Let Lip_α, $\alpha \in (0,1]$, denote the family of all real-valued functions g defined on $\mathbb{T}$ satisfying

$$\sup\left\{\frac{|g(x)-g(y)|}{|x-y|^\alpha} : x \neq y \in \mathbb{T}\right\} < \infty.$$

For $f \in C(\mathbb{T})$, let

$$E_n(f) := \inf\left\{\|t-f\|_{\mathbb{T}} : t \in \mathcal{T}_n\right\}.$$

An example for a direct theorem of approximation is the following. *Suppose f is m times differentiable on $\mathbb{T}$ and $f^{(m)} \in \mathrm{Lip}_\alpha$ for some $\alpha \in (0,1]$. Then there is a constant C depending only on f so that*

$$E_n(f) \leq Cn^{-(m+\alpha)}, \qquad n = 1, 2, \ldots .$$

A proof may be found in [111], for example. The inverse theorem of the above result can be formulated as follows. *Suppose $m \geq 1$ is an integer, $\alpha \in (0,1)$, and $f \in C(\mathbb{T})$. Suppose there is a constant $C > 0$ depending only on f such that*

$$E_n(f) \leq Cn^{-(m+\alpha)}, \qquad n = 1, 2, \ldots .$$

Then f is m times continuously differentiable on $\mathbb{T}$ and $f^{(m)} \in \mathrm{Lip}_\alpha$.

We outline the proof of the above inverse theorem. We show only that f is m times continuously differentiable on $\mathbb{T}$. The rest can be proved similarly, but its proof requires more technical details. See, for example, George G. Lorentz's book [111]. For each $k \in \mathbb{N}$, let $Q_{2^k} \in \mathcal{T}_{2^k}$ be chosen so that

$$\|Q_{2^k} - f\|_{\mathbb{T}} \leq C\,2^{-k(m+\alpha)}.$$

Then

$$\|Q_{2^{k+1}} - Q_{2^k}\|_{\mathbb{T}} \le 2C\, 2^{-k(m+\alpha)}.$$

Now

$$f(\theta) = Q_{2^0}(\theta) + \sum_{k=1}^{\infty}(Q_{2^{k+1}} - Q_{2^k})(\theta), \qquad \theta \in \mathbb{T},$$

and by Bernstein's inequality

$$\begin{aligned}
&\left|Q_{2^0}^{(j)}(\theta)\right| + \sum_{k=0}^{\infty}\left|(Q_{2^{k+1}} - Q_{2^k})^{(j)}(\theta)\right| \\
&\le \|Q_1\|_{\mathbb{T}} + \sum_{k=0}^{\infty}\left(2^{k+1}\right)^j \left\| Q_{2^{k+1}} - Q_{2^k}\right\|_{\mathbb{T}} \\
&\le \|Q_1\|_{\mathbb{T}} + \sum_{k=0}^{\infty}\left(2^{k+1}\right)^j 2C\, 2^{-k(m+\alpha)} \\
&\le \|Q_1\|_{\mathbb{T}} + 2^{j+1}C\sum_{k=0}^{\infty}\left(2^{j-m-\alpha}\right)^k < \infty
\end{aligned}$$

for every $\theta \in \mathbb{T}$ and $j = 0, 1, \ldots, m$, since $\alpha > 0$. Now we can conclude that $f^{(j)}(\theta)$ exists and

$$f^{(j)}(\theta) = Q_1^{(j)}(\theta) + \sum_{k=0}^{\infty}(Q_{2^{k+1}} - Q_{2^k})^{(j)}(\theta)$$

for every $\theta \in \mathbb{T}$ and $j = 0, 1, \ldots, m$. The fact that $f^{(m)} \in C(\mathbb{T})$ can be seen by the Weierstrass M-test. This finishes the proof.

For Erdős, Markov- and Bernstein-type inequalities had their own intrinsic interest and he explored what happens when the polynomials are restricted in certain ways. It had been observed by Bernstein that Markov's inequality for monotone polynomials is not essentially better than for arbitrary polynomials. Bernstein proved that *if n is odd, then*

$$\sup_p \frac{\|p'\|_{[-1,1]}}{\|p\|_{[-1,1]}} = \left(\frac{n+1}{2}\right)^2,$$

where the supremum is taken over all polynomials p of degree at most n with real coefficients which are monotone on $[-1,1]$. This is surprising, since one would expect that if a polynomial is this far away from the "equioscillating" property of the Čebyshov polynomial T_n, then there should be a more significant improvement in the Markov inequality. In the short paper {20}, Erdős gave a class of restricted polynomials for which the Markov factor n^2 improves to cn. He proved that there is an absolute constant c such that

$$|p'(y)| \leq \min\left\{\frac{c\sqrt{n}}{(1-y^2)^2}, \frac{en}{2}\right\}\|p\|_{[-1,1]}, \qquad y \in (-1,1),$$

for every polynomial of degree at most n that has all its zeros in $\mathbb{R} \setminus (-1,1)$. This result motivated several people to study Markov- and Bernstein-type inequalities for polynomials with restricted zeros and under some other constraints. Generalizations of the above Markov- and Bernstein-type inequality of Erdős have been extended in many directions by many people including G. G. Lorentz, John T. Scheick, József Szabados, Arun Kumar Varma, Attila Máté, Quazi Ibadur Rahman, and Narendra K. Govil. Many of these results are contained in the following essentially sharp result, due to Peter Borwein and Tamás Erdélyi {7}: *there is an absolute constant c such that*

$$|p'(y)| \leq c\min\left\{\sqrt{\frac{n(k+1)}{1-y^2}}, n(k+1)\right\}\|p\|_{[-1,1]}, \qquad y \in (-1,1),$$

for every polynomial p of degree at most n with real coefficients that has at most k zeros in the open unit disk.

Let K_α be the open diamond of the complex plane with diagonals $[-1,1]$ and $[-ia,ia]$ such that the angle between $[ia,1]$ and $[1,-ia]$ is $\alpha\pi$. A challenging question of Erdős, that Gábor Halász {45} answered in 1996, is: how large can the quantity

$$\frac{\|p'\|_{[-1,1]}}{\|p\|_{[-1,1]}}$$

be, assuming that p is a polynomial of degree at most n which has no zeros in a diamond K_α, $\alpha \in [0,2)$? He proved that *if $\alpha \in [0,1)$ then there are constants $c_1 > 0$ and $c_2 > 0$ depending only on α such that*

$$c_1 n^{2-\alpha} \leq \sup_p \frac{|p'(1)|}{\|p\|_{[-1,1]}} \leq \sup_p \frac{\|p'\|_{[-1,1]}}{\|p\|_{[-1,1]}} \leq c_2 n^{2-\alpha},$$

where the supremum is taken for all polynomials p of degree at most n (with either real or complex coefficients) having no zeros in K_α. He also showed that there is an absolute constant $c_2 > 0$ such that

$$\|p'\|_{[-1,1]} \leq c_2 n \log n \|p\|_{[-1,1]}$$

holds for all polynomials p of degree at most n with complex coefficients having no zeros in K_1, while for every $\alpha \in (1,2)$ there is a constant c_2 depending on α such that

$$\|p'\|_{[-1,1]} \leq c_2 n \|p\|_{[-1,1]}$$

for all polynomials p of degree at most n with complex coefficients having no zeros in K_α.

Halász reduced the proof to the following result of Szegő {82}: *the inequality*

$$|p'(0)| \leq c_\alpha n^{2\alpha} \|p\|_{D_\alpha}$$

holds for every polynomial p of degree at most n with complex coefficients, where c_α is a constant depending only on α, and

$$D_\alpha := \{z \in \mathbb{C} : |z| \leq 1, \ |\arg(z)| \leq \pi(1-\alpha)\}, \qquad \alpha \in (0,1].$$

The identity

$$t'(\theta) = \sum_{\nu=1}^{2n} (-1)^{\nu+1} \lambda_\nu t(\theta + \theta_\nu)$$

with

$$\lambda_\nu := \frac{1}{n} \frac{1}{\left(2 \sin\left(\frac{1}{2}\theta_\nu\right)\right)^2}, \qquad \theta_\nu := \frac{2\nu - 1}{2n}\pi, \qquad \nu = 1, 2, \ldots, 2n,$$

for all trigonometric polynomials t of order at most n was established by M. Riesz {70} and it is called as the Riesz Interpolation Formula. Here, choosing $t(\theta) := \sin(n\theta)$ and the point $\theta = 0$, we obtain that $\sum_{\nu=1}^{2n} \lambda_\nu = n$. The above identity can be used to prove not only Bernstein's inequality, but an L_p version of it for all $p \geq 1$. Namely, combining the triangle inequality and Hölder's inequality in the Riesz Interpolation Formula, and then integrating both sides, we obtain

$$\int_0^{2\pi} |t'(\theta)|^p \, d\theta \leq n^p \int_0^{2\pi} |t(\theta)|^p \, d\theta$$

for all trigonometric polynomials t of order at most n. It is interesting to note that the sharp L_p version of Bernstein's inequality with Bernstein factor n for all $0 < p < 1$ was established only much later, in 1981, by Vitali V. Arestov {1}. It followed the paper {61} by Máté and Nevai, where the $11^{1/p}n$ Bernstein factor was proved. A short and elegant proof of Arestov's result due to Manfred Golitschek and G. G. Lorenz is presented in {14, pages 104–109}.

For *real* trigonometric polynomials t the inequality

$$t'(\theta)^2 + n^2 t(\theta)^2 \leq n^2 \|t\|_{\mathbb{R}}^2, \qquad \theta \in \mathbb{R},$$

holds and is known as the Bernstein–Szegő inequality. Various extensions and generalizations of this have been established throughout the century.

There is a Bernstein inequality on the unit circle ∂D of the complex plane. It states that

$$\|p'\|_{\partial D} \leq n \, \|p\|_{\partial D}$$

for all polynomial p of degree at most n with complex coefficients.

It was conjectured by Erdős and proved by Péter Lax {53} in 1944 that

$$\|p'\|_{\partial D} \leq \frac{n}{2} \, \|p\|_{\partial D}$$

for every polynomial p of degree at most n with complex coefficients having no zeros in D.

The question about the right Bernstein factor on the unit circle (between $n/2$ and n) is unsettled in the case when we know that there are k zeros inside the open unit disk and $n - k$ zeros are outside it.

A technical detail related to the proof of the Bernstein–Szegő inequality is known as the Riesz Lemma after Marcel Riesz, see {8}, for instance. It states that *if t is a real trigonometric polynomial of order n and for an $\alpha \in \mathbb{R}$ $t(\alpha) = \|t\|_{\mathbb{R}} = 1$, then*

$$t(\theta) \geq \cos\big(n(\theta - \alpha)\big), \qquad \theta \in \left[\alpha - \frac{\pi}{2n}, \alpha + \frac{\pi}{2n}\right].$$

In particular t does not vanish in $\left(\alpha - \frac{\pi}{2n}, \alpha + \frac{\pi}{2n}\right)$.

Géza Freud paid serious attention to Markov–Bernstein-type inequalities on the real line associated with $w_\alpha(x) := \exp\big(-|x|^\alpha\big)$, $\alpha > 0$ in L_p norm. After Freud the name *Freud weight* has become common to refer to the weights w_α and their generalizations. Freud handled the Hermite weight, case $\alpha = 2$, coupled with the assumption $1 \le p \le \infty$. However, it was his student, Paul Nevai, together with Eli Levin, Doron S. Lubinsky, and Vilmos Totik who put the right pieces together to obtain

$$\|Q'w_\alpha\|_p \le Cn^{1-1/\alpha}\|Qw_\alpha\|_p, \qquad \alpha > 1,$$

$$\|Q'w_\alpha\|_p \le C\log n\|Qw_\alpha\|_p, \qquad \alpha = 1,$$

and

$$\|Q'w_\alpha\|_p \le C\|Qw_\alpha\|_p, \qquad 0 < \alpha < 1,$$

for every polynomial Q of degree at most n with real coefficients and for every $0 < p \le \infty$, where $C = C(\alpha, p)$ is a constant depending only on α and p, and

$$\|f\|_p^p := \int_{\mathbb{R}} |f(t)|^p\,dt \qquad \text{and} \qquad \|f\|_\infty := \|f\|_{\mathbb{R}}.$$

In their proof the idea of an infinite-finite range inequality played a significant role. This also goes back to Freud who observed that

$$\big\|Q(x)\exp(-x^2)\big\|_{\mathbb{R}} \le Cn^{1/2}\big\|Q(x)\exp(-x^2)\big\|_{I_n}$$

for every polynomial Q of degree at most n with real coefficients, where

$$I_n := \big[-(3/2)n^{1/2}, (3/2)n^{1/2}\big],$$

and C is an absolute constant.

There is an elementary paper of Szegő {84} dealing with weighted Markov- and Bernstein-type inequalities on $[0,\infty)$ with respect to the Laguerre weight e^{-x} on $[0,\infty)$, which proves

$$\big\|p'(x)e^{-x}\big\|_{[0,\infty)} \le (8n+2)\big\|p(x)e^{-x}\big\|_{[0,\infty)}$$

for every polynomial p of degree at most n with real coefficients. A sharp L_2 version of the above inequality was the topic of Turán's paper {87}. Turán has also found the extremal polynomials in this L_2 case.

An interesting inequality of Turán {86}) states that

$$\frac{\|p'\|_{[-1,1]}}{\|p\|_{[-1,1]}} > \frac{1}{6}\sqrt{n}$$

for every polynomial p of degree n having all its zeros in $[-1,1]$. He also showed that

$$\|p'\|_{\overline{D}} \geq \frac{n}{2}\|p\|_{\overline{D}}$$

if p has each of its zeros in the closed unit disk $\overline{D}$ of the complex plane.

Turán posed some problems about bounding the derivative of a polynomial if its modulus is bounded by a certain curve. One of his problems asked for the right upper bound for $\|p'\|_{[-1,1]}$ for polynomials p of degree at most n with real coefficients satisfying $|p(x)| \leq (1-x^2)^{1/2}$ for every $x \in [-1,1]$. This problem has been solved by Q. I. Rahman {69} who established the inequality $\|p'\|_{[-1,1]} \leq 2(n-1)$ for all such polynomials.

2. Müntz Polynomials and Exponential Sums

James A. Clarkson and Erdős wrote a seminal paper on the density of Müntz polynomials. C. Herman Müntz's classical theorem characterizes sequences $\Lambda := (\lambda_j)_{j=0}^{\infty}$ with

$$(1) \qquad 0 = \lambda_0 < \lambda_1 < \lambda_2 < \cdots$$

for which the Müntz space $M(\Lambda) := \operatorname{span}\{x^{\lambda_0}, x^{\lambda_1}, \ldots\}$ is dense in $C[0,1]$. Here the span denotes the collection of all finite linear combinations of the functions $x^{\lambda_0}, x^{\lambda_1}, \ldots$ with real coefficients, and $C(A)$ is the space of all real-valued continuous functions on $A \subset [0,\infty)$ equipped with the uniform norm. If $A := [a,b]$ is a finite closed interval, then the notation $C[a,b] := C\big([a,b]\big)$ is used.

Müntz's Theorem. *Suppose* $\Lambda := (\lambda_j)_{i=0}^{\infty}$ *is a sequence satisfying* (1). *Then* $M(\Lambda)$ *is dense in* $C[0,1]$ *if and only if* $\sum_{j=1}^{\infty} 1/\lambda_j = \infty$.

The original Müntz Theorem proved by C. Müntz {63} in 1914, by Ottó Szász {76} in 1916, and anticipated by Bernstein was only for sequences of exponents tending to infinity. Szász proved more than Müntz. He did not assume (1). He assumed only that the numbers λ_k are arbitrary complex with $\lambda_0 = 1$, $\operatorname{Re}\lambda_k > 0$, $k = 1,2,\ldots$, and the exponents λ_k are distinct. He gave a necessary and sufficient conditions which characterize denseness in the case when $\operatorname{Re}\lambda_k \geq c > 0$, $k = 1,2,\ldots$.

The point 0 is special in the study of Müntz spaces. Even replacing $[0,1]$ by an interval $[a,b] \subset [0,\infty)$ in Müntz's Theorem is a non-trivial issue. Such an extension is, in large measure, due to James A. Clarkson and Erdős {13} and Laurent Schwartz [164]. In {13}, Clarkson and Erdős showed that Müntz's Theorem holds on any interval $[a,b]$ with $a \geq 0$. That is, *for any increasing positive sequence* $\Lambda := (\lambda_j)_{j=0}^{\infty}$ *and any* $0 < a < b$, $M(\Lambda)$ *is dense in* $C[a,b]$ *if and only if* $\sum_{j=1}^{\infty} 1/\lambda_j = \infty$. Moreover, they showed that under the assumptions $\sum_{j=1}^{\infty} 1/\lambda_j < \infty$ and

$$\inf\{\lambda_{j+1} - \lambda_j \,:\, j = 0,1,2,\ldots\} > 0$$

every function $f \in C[a,b]$ from the uniform closure of $M(\Lambda)$ on $[a,b]$ is of the form

$$f(x) = \sum_{j=0}^{\infty} a_j x^{\lambda_j}, \qquad x \in [a,b). \tag{2}$$

In particular, f can be extended analytically throughout the open disk centered at 0 with radius b, cut by $(-\infty,0]$.

Erdős considered this result his best contribution to complex analysis. Later, by different methods, L. Schwartz extended some of the Clarkson–Erdős results to the case when the exponents λ_i are arbitrary distinct positive real numbers. For example, in that case, under the assumption $\sum_{j=1}^{\infty} 1/\lambda_j < \infty$ every function $f \in C[a,b]$ from the uniform closure of $M(\Lambda)$ on $[a,b]$ can still be extended analytically throughout the region

$$\{z \in \mathbb{C} \setminus (-\infty,0] \,:\, |z| < b\},$$

although such an analytic extension does not necessarily have a representation given by (2). The Clarkson–Erdős results were further extended by P. Borwein and T. Erdélyi {10} from the interval $[0,1]$ to compact subsets of $[0,\infty)$ with positive Lebesgue measure. That is, *if* $\Lambda := (\lambda_j)_{j=0}^{\infty}$ *is an increasing sequence of positive real numbers with* $\lambda_0 = 0$ *and* $A \subset [0,\infty)$ *is a compact set with positive Lebesgue measure, then* $M(\Lambda)$ *is dense in* $C(A)$ *if and only if* $\sum_{j=1}^{\infty} 1/\lambda_j = \infty$. This result had been expected by Erdős and others for a long time.

Somorjai {74} and Joseph Bak and Donald J. Newman {2} proved that

$$R(\Lambda) := \{p/q \,:\, p,q \in M(\Lambda)\}$$

is always dense in $C[0,1]$. This surprising result says that while the set $M(\Lambda)$ of Müntz polynomials may be far from dense, the set $R(\Lambda)$ of Müntz rationals is always dense in $C[0,1]$ no matter what the underlying sequence Λ is. Newman was truly impressed by Somorjai's result. In the light of Somorjai's theorem, Newman, in 1978 [123, p. 50] raises "the very sane, if very prosaic question": are the functions

$$\prod_{j=1}^{k}\left(\sum_{i=0}^{n_j} a_{i,j}x^{i^2}\right), \qquad a_{i,j}\in\mathbb{R}, \quad n_j\in\mathbb{N},$$

dense in $C[0,1]$ for some fixed $k\geq 2$? In other words does the "extra multiplication" have the same power that the "extra division" has in the Bak–Newman–Somorjai result? Newman speculated that it did not. P. Borwein and T. Erdélyi proved this conjecture in {10} in a generalized form.

Müntz–Jackson type theorems via interpolation have been considered in {60} by László Márki, Somorjai, and Szabados.

The main results of {8, Section 4.2} and {16} are the following.

Full Müntz Theorem in $C[0,1]$. *Suppose $(\lambda_j)_{j=1}^{\infty}$ is a sequence of distinct real numbers greater than 0. Then* span $\{1, x^{\lambda_1}, x^{\lambda_2}, \ldots\}$ *is dense in $C[0,1]$ if and only if*

$$\sum_{j=1}^{\infty}\frac{\lambda_j}{\lambda_j^2+1}=\infty.$$

Moreover, if

$$\sum_{j=1}^{\infty}\frac{\lambda_j}{\lambda_j^2+1}<\infty,$$

then every function from the $C[0,1]$ closure of span $\{1, x^{\lambda_1}, x^{\lambda_2}, \ldots\}$ *is infinitely many times differentiable on $(0,1)$.*

Full Müntz Theorem in $L_p(A)$. *Let $A\subset[0,1]$ be a compact set with positive lower density at 0. Let $p\in(0,\infty)$. Suppose $(\lambda_j)_{j=1}^{\infty}$ is a sequence of distinct real numbers greater than $-(1/p)$. Then* span $\{x^{\lambda_1}, x^{\lambda_2}, \ldots\}$ *is dense in $L_p(A)$ if and only if*

$$\sum_{j=1}^{\infty}\frac{\lambda_j+(1/p)}{\left(\lambda_j+(1/p)\right)^2+1}=\infty.$$

Moreover, if

$$\sum_{j=1}^{\infty}\frac{\lambda_j+(1/p)}{\left(\lambda_j+(1/p)\right)^2+1}<\infty,$$

then every function from the $L_p(A)$ closure of $\text{span}\,\{x^{\lambda_1}, x^{\lambda_2}, \dots\}$ *can be represented as an analytic function on* $\{z \in \mathbb{C} \setminus (-\infty, 0] : |z| < r_A\}$ *restricted to* $A \cap (0, r_A)$, *where*

$$r_A := \sup \{y \in \mathbb{R} : m(A \cap [y, \infty)) > 0\}$$

($m(\cdot)$ denotes the one-dimensional Lebesgue measure).

These improve and extend earlier results of Müntz {63}, Szász {76}, and Clarkson and Erdős {13}. Related issues about the denseness of $\text{span}\,\{x^{\lambda_1}, x^{\lambda_2}, \dots\}$ are also considered.

Based on the work of Edmond Laguerre {52} and Pólya {66} the following is known.

Let $\mathcal{A}_1$ denote the class of entire functions f of the form

$$f(z) = Cz^m e^{-c^2z^2+az} \prod_{k=1}^{\infty} (1 + \alpha_k z) e^{-\alpha_k z}, \qquad z \in \mathbb{C},$$

where $C, c, a, \alpha_k \in \mathbb{R}$, m is a nonnegative integer, and $\sum_{k=1}^{\infty} \alpha_k^2 < \infty$. Then $\mathcal{A}_1$ is the collection of the analytic extensions of those functions defined on $\mathbb{R}$ which may be obtained as the uniform limit, on every compact subset of $\mathbb{R}$, of polynomials having only real zeros.

Let $\mathcal{A}_2$ denote the class of entire functions f of the form

$$f(z) = Cz^m e^{-az} \prod_{k=1}^{\infty} (1 - \alpha_k z), \qquad z \in \mathbb{C},$$

where $C \in \mathbb{R}$, $a > 0$, m is a nonnegative integer, $\alpha_k \geq 0$, and $\sum_{k=1}^{\infty} \alpha_k < \infty$. Then $\mathcal{A}_2$ is the collection of the analytic extensions of those functions defined on $\mathbb{R}$ which may be obtained as the uniform limit, on every compact subset of $\mathbb{R}$, of polynomials having only positive zeros.

The functions of the classes $\mathcal{A}_1$ and $\mathcal{A}_2$ are sometimes called the Pólya–Laguerre functions.

In {68} Pólya posed the question: for which sequences $0 < \beta_1 < \beta_2 < \cdots$ are the linear combinations of the functions

$$\cos(\beta_k t), \quad \sin(\beta_k t), \qquad k = 1, 2, \dots,$$

complete in $C[0, 2\pi]$? Pólya himself conjectured that

$$\limsup_{k \to \infty} \frac{\beta_k}{k} < 1$$

is sufficient. Szász {77} proved Pólya's conjecture.

The following pretty results of Fejér may be found in {8} (see also {40}): *Let*

$$p(z) := \sum_{k=0}^{n} a_k z^{\lambda_k}, \qquad a_k \in \mathbb{C}, \quad a_0 a_1 \neq 0.$$

Then p has at least one zero $z_0 \in \mathbb{C}$ so that

$$|z_0| \le \left(\frac{\lambda_2 \lambda_3 \cdots \lambda_n}{(\lambda_2 - \lambda_1)(\lambda_3 - \lambda_1) \cdots (\lambda_n - \lambda_1)} \right)^{1/\lambda_1} \left| \frac{a_0}{a_1} \right|^{1/\lambda_1}.$$

From the above result the following beautiful consequence follows easily: *Suppose*

$$f(z) = \sum_{k=0}^{\infty} a_k z^{\lambda_k}, \qquad a_k \in \mathbb{C}$$

is an entire function so that $\sum_{k=1}^{\infty} 1/\lambda_k < \infty$, that is, the entire function f satisfies the Fejér gap condition. Then there is a $z_0 \in \mathbb{C}$ so that $f(z_0) = 0$. Hence for every $a \in \mathbb{C}$ there is a z_0 such that $f(z_0) = a$, that is f has no Picard exceptional value.

Important results of Turán are based on the following observations: *Let*

$$g(\nu) := \sum_{j=1}^{n} b_j z_j^{\nu}, \qquad b_j, z_j \in \mathbb{C}.$$

Suppose

$$\min_{1 \le j \le n} |z_j| \ge 1, \qquad j = 1, 2, \ldots, n.$$

Then

$$\max_{\nu = m+1, \ldots, m+n} |g(\nu)| \ge \left(\frac{n}{2e(m+n)} \right)^n |b_1 + b_2 + \cdots + b_n|$$

for every positive integer m.

A consequence of the preceding is the famous Turán Lemma: *if*

$$f(t) := \sum_{j=1}^{n} b_j e^{\lambda_j t}, \qquad b_j, \lambda_j \in \mathbb{C}.$$

and

$$\min_{1 \le j \le n} \operatorname{Re}(\lambda_j) \ge 0,$$

then

$$|f(0)| \le \left(\frac{2e(a+d)}{d}\right)^n \|f\|_{[a,a+d]}$$

for every $a > 0$ and $d > 0$.

Another consequence of this is the fact that *if*

$$p(z) := \sum_{j=1}^n b_j z^{\lambda_j}, \qquad b_j \in \mathbb{C}, \quad \lambda_j \in \mathbb{R}, \quad z = e^{i\theta},$$

then

$$\max_{|z|=1} |p(z)| \le \left(\frac{4e\pi}{\delta}\right)^n \max_{\substack{|z|=1 \\ \alpha \le \arg(z) \le \alpha+\delta}} |p(z)|$$

for every $0 \le \alpha < \alpha + \delta \le 2\pi$.

Turán's inequalities above and their variants play a central role in the book of Turán {88}, where many applications are also presented. The main point in these inequalities is that the exponent on the right-hand side is only the number of terms n, and so it is independent of the numbers λ_j. An inequality of type

$$\max_{|z|=1} |p(z)| \le c(\delta)^{\lambda_n} \max_{\substack{|z|=1 \\ \alpha \le \arg(z) \le \alpha+\delta}} |p(z)|,$$

where $0 \le \lambda_1 < \lambda_2 < \cdots < \lambda_n$ are integers and $c(\delta)$ depends only on δ, could be obtained by a simple direct argument, but it is much less useful than Turán's inequality. Fedor Nazarov has a paper {64} devoted to Turán-type inequalities for exponential sums, and their applications to various uniqueness theorems in harmonic analysis of the uncertainty principle type. The author derives an estimate for the maximum modulus of an exponential sum

$$\sum_{k=1}^n c_k e^{\lambda_k t}, \qquad c_k, \lambda_k \in \mathbb{C},$$

on an interval $I \subset \mathbb{R}$ in terms of its maximum modulus on a measurable set $E \subset I$ of positive Lebesgue measure:

$$\sup_{t \in I} |p(t)| \le e^{\max|\operatorname{Re}\lambda_k| m(I)} \left(\frac{Am(I)}{m(E)}\right)^{n-1} \sup_{t \in E} |p(t)|.$$

In {9} a subtle Bernstein-type extremal problem related to Turán's result is solved by establishing the equality

$$\sup_{0 \neq f \in \widetilde{E}_{2n}} \frac{|f'(0)|}{\|f\|_{[-1,1]}} = 2n - 1,$$

where

$$\widetilde{E}_{2n} := \left\{ f \, : \, f(t) = a_0 + \sum_{j=1}^{n} \left(a_j e^{\lambda_j t} + b_j e^{-\lambda_j t} \right), \quad a_j, b_j, \lambda_j \in \mathbb{R} \right\}.$$

This settles a conjecture of G. G. Lorentz and others and it is surprising to be able to provide a sharp solution. It follows fairly simply from the above that

$$\frac{1}{e-1} \frac{n-1}{\min\{y-a, b-y\}} \leq \sup_{0 \neq f \in E_n} \frac{|f'(y)|}{\|f\|_{[a,b]}} \leq \frac{2n-1}{\min\{y-a, b-y\}}$$

for every $y \in (a,b)$, where

$$E_n := \left\{ f \, : \, f(t) = a_0 + \sum_{j=1}^{n} a_j e^{\lambda_j t}, \quad a_j, \lambda_j \in \mathbb{R} \right\}.$$

3. Geometric Properties of Polynomials

There are a number of contributions by Hungarian mathematicians exploring the relation between the coefficients of a polynomial and the number of its zeros in certain regions of the complex domain. I find the following result of Erdős and Turán {38} especially attractive. It states that *if* $p(z) = \sum_{j=0}^{n} a_j z^j$ *has* m *positive real zeros, then*

$$m^2 \leq 2n \log \left(\frac{|a_0| + |a_1| + \cdots + |a_n|}{\sqrt{|a_0 a_n|}} \right).$$

This result was originally due to I. Schur. Erdős and Turán rediscovered it with a short proof that we outline now:

Step 1. We utilize the following observation due to Szász, see page 173 of {8}. Let $\gamma, \lambda_0, \lambda_1, \dots, \lambda_n$ be distinct real numbers greater than $-1/2$. Then the $L_2[0,1]$ distance d_n from x^γ to span $\{x^{\lambda_0}, x^{\lambda_1}, \dots, x^{\lambda_n}\}$ is given by

$$d_n = \frac{1}{\sqrt{2\gamma+1}} \prod_{j=0}^{n} \left| \frac{\gamma - \lambda_j}{\gamma + \lambda_j + 1} \right| .$$

Step 2. Let

$$p(z) = a_n \prod_{k=1}^{n} (z - r_k e^{i\theta_k})$$

and

$$q(z) := \prod_{k=1}^{n} (z - e^{i\theta_k}).$$

Note that for $|z| = 1$,

$$\frac{|z - re^{i\theta}|^2}{r} \geq |z - e^{i\theta}|^2 .$$

Use this to deduce that

$$|q(z)|^2 \leq \frac{|p(z)|^2}{|a_0 a_n|} \leq \left(\frac{|a_0| + |a_1| + \cdots + |a_n|}{\sqrt{|a_0 a_n|}} \right)^2$$

whenever $|z| = 1$.

Step 3. Let ∂D be the unit circle of the complex plane. Since p has m positive real roots, q has m roots at 1. Use the change of variables $x := z + z^{-1}$ applied to $z^n q(z^{-1}) q(z)$ to show that

$$\begin{aligned}
\|q\|_{\partial D}^2 &\geq \min_{\{b_k\}} \left\| (z-1)^m (z^{n-m} + b_{n-m-1} z^{n-m-1} + \cdots + b_1 z + b_0) \right\|_{\partial D}^2 \\
&\geq \min_{\{c_k\}} \left\| x^m (x^{n-m} + c_{n-m-1} x^{n-m-1} + \cdots + c_1 x + c_0) \right\|_{[0,4]} \\
&= 4^n \min_{\{d_k\}} \left\| x^m (x^{n-m} + d_{n-m-1} x^{n-m-1} + \cdots + d_1 x + d_0) \right\|_{[0,1]} \\
&\geq \frac{4^n}{\sqrt{2n+1} \binom{2n}{n+m}},
\end{aligned}$$

where the last inequality follows from Step 1.

Step 4. Now one can easily show that

$$\log\left(\frac{4^n}{\sqrt{2n+1}\binom{2n}{n+m}}\right) \geq m^2/n$$

for $1 < m < n$, and the proof of Erdős and Turán is finished.

Another beautiful result of Erdős and Turán {38} states that if the zeros of $p(z) = \sum_{j=0}^n a_j z^j$ are denoted by

$$z_k = r_k e^{i\varphi_k}, \qquad k = 1, 2, \dots, n,$$

then for every $0 < \alpha < \beta \leq 2\pi$ we have

$$\left| \sum_{k \in I(\alpha,\beta)} 1 - \frac{\beta - \alpha}{2\pi} n \right| \leq 16\sqrt{n \log R},$$

where

$$R := \frac{|a_0| + |a_1| + \cdots + |a_n|}{\sqrt{|a_0 a_n|}}$$

and

$$I(\alpha, \beta) := \left\{ k \in \{1, 2, \dots, n\} \, : \, \varphi_k \in [\alpha, \beta] \right\}.$$

André Bloch and Pólya {68} proved that the average number of real zeros of a polynomial from

$$\mathcal{F}_n := \left\{ p \, : \, p(z) = \sum_{k=0}^n a_k z^k, \; a_k \in \{-1, 0, 1\} \right\}.$$

is at most $c\sqrt{n}$. They also proved that a polynomial from $\mathcal{F}_n$ cannot have more than

$$\frac{cn \log \log n}{\log n}$$

real zeros. This quite weak result appears to be the first on this subject. Schur {73} and by different methods Szegő {83} and Erdős and Turán {38} improve this to $c\sqrt{n \log n}$ (see also {8}). Their results are more general, but in this specialization not sharp. In {12} the right upper bound $c\sqrt{n}$ is found for the number of real zeros of polynomials from a much larger class, the class of all polynomials of the form

$$p(x) = \sum_{j=0}^n a_j x^j, \qquad |a_j| \leq 1, \quad |a_0| = |a_n| = 1, \quad a_j \in \mathbb{C}.$$

In fact our method is able to give $c\sqrt{n}$ as an upper bound for the number of zeros of a polynomial p of degree at most n with $|a_0| = 1$, $|a_j| \le 1$, inside any polygon with vertices on the unit circle (of course, c depends on the polygon). This is discussed in {11}.

Bloch and Pólya {4} also prove that there are polynomials $p \in \mathcal{F}_n$ with

$$\frac{cn^{1/4}}{\sqrt{\log n}}$$

distinct real zeros of odd multiplicity. (Schur {73} claims they do it for polynomials with coefficients only from $\{-1, 1\}$, but this appears to be incorrect.)

A surprising theorem of Szegő {80} states that *if*

$$f(x) := \sum_{k=0}^{n} a_k \binom{n}{k} x^k, \qquad a_n \neq 0,$$

$$g(x) := \sum_{k=0}^{n} b_k \binom{n}{k} x^k, \qquad b_n \neq 0,$$

and

$$h(x) := \sum_{k=0}^{n} a_k b_k \binom{n}{k} x^k,$$

f has all its zeros in a closed disk $\overline{D}$, and g has zeros $\beta_1, \dots, \beta_n$, then all the zeros of h are of the form $-\beta_j \gamma_j$ with $\gamma_j \in \overline{D}$. An interesting consequence of this is the fact that *if a polynomial p of degree n has all its zeros in $\overline{D}_1 := \left\{z \in \mathbb{C} : |z| \le 1\right\}$, then the polynomial q defined by $q(x) := \int_0^x p(t)\, dt$ has all its zeros in $\overline{D}_2 := \left\{z \in \mathbb{C} : |z| \le 2\right\}$.*

In {18}, Erdős proved that the arc length from 0 to 2π of a real trigonometric polynomial f of order at most n satisfying $\|f\|_{\mathbb{R}} \le 1$ is maximal for $\cos n\theta$. An interesting question he posed quite often is the following: *Let $0 < a < b < 2\pi$. Is it still true that the variation and arc-length of a real trigonometric polynomial with $\|f\|_{\mathbb{R}} \le 1$ in $[a, b]$ is maximal for $\cos(n\theta + \alpha)$ for a suitable α?* The following related conjecture of Erdős was open for quite a long time: *Is it true that the arc length from -1 to 1 of a real algebraic polynomial p of degree at most n with $\|p\|_{[-1,1]}$ is maximal for the*

Čebyshov polynomial T_n? This was proved independently by Gundorph K. Kristiansen {51} and by Borislav Bojanov {5}.

A well-known theorem of Piotr L. Čebyshov states that *if p is a real algebraic polynomial of degree at most n and $z_0 \in \mathbb{R}\setminus[-1,1]$, then $|p(z_0)| \le |T_n(z_0)| \cdot \|p\|_{[-1,1]}$, where T_n is the Čebyshov polynomial of degree n.* The standard proof of this is based on zero counting which can no longer be applied if z_0 is not real. By letting $z_0 \in \mathbb{C}$ tend to a point in $(-1,1)$, it is fairly obvious that this result cannot be extended to all $z_0 \in \mathbb{C}$. However, a surprising result of Erdős {23} shows that Čebyshov's inequality can be extended to all $z_0 \in \mathbb{C}$ outside the open unit disk.

Erdős and Turán were probably the first to discover the power and applicability of an almost forgotten result of Evgenii Ja. Remez. The so-called Remez inequality is not only attractive and interesting in its own right, but it also plays a fundamental role in proving various other things about polynomials. Let $\mathcal{P}_n$ be the set of all algebraic polynomials of degree at most n with real coefficients. For a fixed $s \in (0,2)$, let

$$\mathcal{P}_n(s) := \Big\{p \in \mathcal{P}_n \,:\, m\Big(\big\{x \in [-1,1] \,:\, |p(x)| \le 1\big\}\Big) \ge 2-s\Big\},$$

where $m(\cdot)$ denotes linear Lebesgue measure. The Remez inequality concerns the problem of bounding the uniform norm of a polynomial p of degree n on $[-1,1]$ given that its modulus is bounded by 1 on a subset of $[-1,1]$ of Lebesgue measure at least $2-s$. That is, *how large can $\|p\|_{[-1,1]}$ (the uniform norm of p on $[-1,1]$) be if $p \in \mathcal{P}_n(s)$?* The answer is given in terms of the Čebyshov polynomials T_n. We have

$$\|p\|_{[-1,1]} \le T_n\left(\frac{2-s}{2+s}\right)$$

for all $p \in \mathcal{P}_n(s)$, and the extremal polynomials for the above problem are the Čebyshov polynomials $\pm T_n\big(h(x)\big)$, where h is a linear function which maps $[-1,1-s]$ or $[-1+s,1]$ onto $[-1,1]$. See page 228 of {8}.

One of the applications of the Remez inequality by Erdős and Turán {37} deals with orthogonal polynomials. Let w be an integrable weight function on $[-1,1]$ that is strictly positive almost everywhere. Denote the sequence of the associated orthonormal polynomials by $(p_n)_{n=0}^{\infty}$. The leading coefficient of p_n is denoted by $\gamma_n > 0$. Then a theorem of Erdős and Turán {37} states that

$$\lim_{n\to\infty} \big[p_n(z)\big]^{1/n} = z + \sqrt{z^2-1}$$

holds uniformly on every closed subset of $\mathbb{C} \setminus [-1,1]$.

Erdős and Turán {35} established a number of results on the spacing of zeros of orthogonal polynomials. One of these is the following. *Let* w *be an integrable weight function on* $[-1,1]$ *with* $\int_{-1}^{1} \left(w(x)\right)^{-1} dx =: M < \infty$, *and let*

$$(1 >)\, x_{1,n} > x_{2,n} > \cdots > x_{n,n}\, (> -1)$$

be the zeros of the associated orthonormal polynomials p_n *in decreasing order. Let*

$$x_{\nu,n} = \cos\theta_{\nu,n}, \qquad 0 < \theta_{\nu,n} < \pi, \qquad \nu = 1, 2, \ldots, n.$$

Let $\theta_{0,n} := 0$ *and* $\theta_{n+1,n} := \pi$. *Then there is a constant* K *depending only on* M *such that*

$$\theta_{\nu+1,n} - \theta_{\nu,n} < \frac{K \log n}{n}, \qquad \nu = 0, 1, \ldots, n.$$

Pólya {67} proved that if

$$p(x) = \sum_{k=0}^{n} a_k x^k,$$

then

$$|a_n| \leq \frac{1}{2} \left(\frac{4}{m(E)}\right)^n \sup_{x \in E} |p(x)|$$

for every measurable set $E \subset \mathbb{R}$, $0 < m(E) < \infty$. Equality holds if and only if E is an interval $[a, a+\lambda]$ and $P(x) = AT_n\big(2(x-a)/\lambda - 1\big)$, where $a, A \in \mathbb{R}$ and $\lambda > 0$. Here, as before, T_n denotes the Čebyshov polynomial of degree n.

Let $H \subset \mathbb{C}$ be a compact set. Let

$$\mu_n = \inf\left(\max_{z \in H} |p(z)|\right)$$

where the infimum is taken for all monic polynomials p of degree n with complex coefficients. Let

$$\widetilde{\mu}_n = \inf\left(\max_{z \in H} |p(z)|\right),$$

where the infimum is taken for all monic polynomials p of degree n with complex coefficients and with all zeros in H. The numbers

$$\mu := \mu(H) = \lim_{n\to\infty} \mu_n^{1/n} \qquad \text{and} \qquad \widetilde{\mu} := \widetilde{\mu}(H) = \lim_{n\to\infty} \widetilde{\mu}_n^{1/n}$$

exist and are called the Čebyshov constant and modified Čebyshov constant, respectively.

Let

$$d(z_1, z_2, \ldots, z_n) = \prod_{1\le k<j\le n} |z_k - z_j|$$

and

$$d_n := \sup \big(d(z_1, z_2, \ldots, z_n) \big)^{\frac{2}{n(n-1)}},$$

where the supremum is taken for all $z_1, z_2, \ldots, z_n \in H$. The points $z_1, z_2, \ldots, z_n$ for which the above supremum is achieved are called nth *Fekete points.* Then the value

$$d(H) := \lim_{n\to\infty} d_n$$

exists and is called the transfinite diameter (Fekete constant) of H.

The logarithmic energy $I(\mu)$ of a $\mu \in \mathcal{M}(H)$ is defined as

$$I(\mu) := \int_H \int_H \log \frac{1}{|z-t|} \, d\mu(z) \, d\mu(t),$$

and the energy V of H by

$$V := \inf \{ I(\mu) \, : \, \mu \in \mathcal{M}(H) \},$$

where $\mathcal{M}(H)$ is the collection of all positive Borel measures with $\mu(H) = 1$ and with support in H. Then V turns out to be finite or $+\infty$ and in the finite case there is a unique measure $\mu = \mu_H \in \mathcal{M}(H)$ for which the infimum defining V is attained. This μ_H is called the equilibrium distribution or measure of a compact set H. The quantity $\operatorname{cap}(H) := e^{-V}$ is called the logarithmic capacity of H.

Fekete {42} and Szegő {81} proved that the the Čebyshov constants, the transfinite diameter and the logarithmic capacity of a compact set $H \subset \mathbb{C}$ are equal, that is,

$$\mu(H) = \widetilde{\mu}(H) = d(H) = \operatorname{cap}(H)$$

for every compact subset H of the complex plane.

Tamás Kővári has two papers, {48} (written jointly with Pommerenke) and {47}, on the distribution of Fekete points.

Approximation of functions $f \in C(H)$ by polynomials with integer coefficients on a compact set $H \subset \mathbb{C}$ has been considered by Fekete. Several papers appeared on the subject, the two most important of them are {42} and {44}. His typical results include that a function $f \in C[a,b]$, $b-a \geq 4$, is approximable from the collection of polynomials with integer coefficients if and only if f is a polynomial with integer coefficients. Also, a function $f \in C[0,1]$ is approximable from the collection of polynomials with integer coefficients if and only if $f(0)$ and $f(1)$ are both integers.

Another important result of Fekete {42} is an extension of a sresult of David Hilbert related to the so-called Integer Chebyshev Problem. An integer Chebyshev polynomial Q_n for a compact subset $E \subset \mathbb{C}$ is a polynomial Q_n of degree at most n with integer coefficients such that $\|Q_n\|_E = \inf_{P_n} \|P_n\|_E$, where the infimum is taken for all not identically zero polynomials P_n of degree at most n with integer coefficients (it is easy to see that at least one such Q_n exists). The integer Chebyshev constant for a compact subset $E \subset \mathbb{C}$ is defined by $t_{\mathbb{Z}}(E) := \lim_{n\to\infty} \|Q_n\|_E^{1/n}$ (it is a routine argument to show that the limit exists). In the above notation Fekete's result simply reads as

$$t_{\mathbb{Z}}(E) \leq \sqrt{\operatorname{cap}(E)},$$

and contains Hilbert's result as a special case when the set E is a closed interval of the real line. The Integer Chebyshev Problem has continued to attract a large number of well known mathematicians later in the century. Yet, there are many unanswered questions about it posed in papers appearing in general mathematics journals in the XXI-st century.

For a prime p, the polynomials

$$f_p(z) := \sum_{k=1}^{p-1} \left(\frac{k}{p}\right) z^k$$

are named after Fekete and have a variety of remarkable properties (the coefficients are Legendre symbols).

Erdős and Freud {30} worked together on orthogonal polynomials with regularly distributed zeros. Let α be a positive measure on $(-\infty,\infty)$ for which all the moments

$$\mu_m := \int_{-\infty}^{\infty} x^m \, d\alpha(x), \qquad m = 0, 1, \dots,$$

exist and are finite. Denote the sequence of the associated orthonormal polynomials by $(p_n)_{n=0}^{\infty}$. Let $x_{1,n} > x_{2,n} > \cdots > x_{n,n}$ be the zeros of of p_n in decreasing order. Let $N_n(\alpha, t)$ denote the number of positive integers k for which

$$x_{k,n} - x_{n,n} \geq t(x_{1,n} - x_{n,n}).$$

The distribution function β of the zeros is defined, when it exists, as

$$\beta(t) = \lim_{n\to\infty} n^{-1} N_n(\alpha, t), \qquad 0 \leq t \leq 1.$$

Let

$$\beta_0(t) = \frac{1}{2} - \frac{1}{\pi} \arcsin(2t - 1).$$

A positive measure α for which the array $x_{k,n}$ has the distribution function $\beta_0(t)$ is called an arc-sine measure. If $d\alpha(x) = w(x)\,dx$ is absolutely continuous and α is an arc-sine measure, then w is called an arc-sine weight. One of the theorems of Erdős and Freud {30} states that *the condition*

$$\limsup_{n\to\infty} (\gamma_{n-1})^{1/(n-1)} (x_{1,n} - x_{n,n}) \leq 4$$

implies that α is arc-sine and

$$\lim_{n\to\infty} (\gamma_{n-1})^{1/(n-1)} (x_{1,n} - x_{n,n}) = 4, \tag{3}$$

where, as before, γ_{n-1} is the leading coefficient of p_{n-1}.

They also show that the weights $w_\alpha(x) := \exp\big(-|x|^\alpha\big)$, $\alpha > 0$, are not arc-sine. It is further proved by a counter-example that even the stronger sufficient condition (3) in the above-quoted result is not necessary in general to characterize arc-sine measures. As the next result of their paper shows, the case is different if w has compact support. Namely they show that a *weight w, the support of which is contained in $[-1, 1]$, is arc-sine on $[-1, 1]$ if and only if*

$$\limsup_{n\to\infty} (\gamma_n)^{1/n} \leq 2.$$

A set $A \subset [-1, 1]$ is called a determining set if all weights w with support contained in $[-1, 1]$, the restricted support $\{x : w(x) > 0\}$ of which contain A, are arc-sine on $[-1, 1]$. A set $A \subset [-1, 1]$ is said to have minimal capacity c if for every $\varepsilon > 0$ there exists a $\delta(\varepsilon) > 0$ such that for every $B \subset [-1, 1]$ having Lebesgue measure less than $\delta(\varepsilon)$ we have $\operatorname{cap}(A \backslash B) > c - \varepsilon$. Another remarkable result of this paper by Erdős and Freud is that a *measurable set $A \subset [-1, 1]$ is a determining set if and only if it has minimal capacity $1/2$.*

Erdős' paper {32} with Fritz Herzog and George Piranian on the geometry of polynomials is seminal. In this paper, they proved a number of interesting results and raised many challenging questions. Although quite a few of these have been solved by Christian Pommerenke and others, many of them are still open. Erdős liked this paper very much. In his talks about polynomials, he often revisited these topics and mentioned the unsolved problems again and again. A taste of this paper is given by the following results and still unsolved problems from it. As before, associated with a monic polynomial

$$f(z) = \prod_{j=1}^{n} (z - z_j), \qquad z_j \in \mathbb{C}, \tag{4}$$

let

$$E = E(f) = E_n(f) := \{z \in \mathbb{C} : |f(z)| \le 1\}.$$

One of the results of Erdős, Herzog, and Piranian tells us that *the infimum of $m\big(E(f)\big)$ is 0, where the infimum is taken over all polynomials f of the form (4) with all their zeros in the closed unit disk (n varies and m denotes the two-dimensional Lebesgue measure).* Another result is the following. *Let F be a closed set of transfinite diameter less than 1. Then there exists a positive number $\rho(F)$ such that, for every polynomial of the form (4) whose zeros lie in F, the set $E(f)$ contains a disk of radius $\rho(F)$.* There are results on the number of components of E, the sum of the diameters of the components of E, some implications of the connectedness of E, some necessary assumptions that imply the convexity of E. An interesting conjecture of Erdős states that *the length of the boundary of $E_n(f)$ for a polynomial f of the form (4) is $2n + O(1)$.* This problem seems almost impossible to settle. The best result in this direction is $O(n)$ by P. Borwein {6} that improves an earlier upper bound $74n^2$ given by Pommerenke.

One of the papers where Erdős revisits this topic is {33} written jointly with Elisha Netanyahu. The result of this paper states that *if the zeros $z_j \in \mathbb{C}$ are in a bounded, closed, and connected set whose transfinite diameter is $1 - c$ $(0 < c < 1)$, then $E(f)$ contains a disk of positive radius ρ depending only on c.*

I have a postcard from Erdős asking for a proof of the fact that the diameter of $E(f)$ is always at least 2 for monic polynomials f (of the form (4)). "This ought to be trivial, but I do not have a proof" is commented by Erdős on the card. Based on a result of Pommerenke, a proof is presented in {8, p. 354}.

János Erőd {39} attributes the following interesting result to Erdős and Turán and presents its proof in his paper. *If*

$$(5) \qquad f(x) = \pm \prod_{j=1}^{n} (x - x_j), \qquad -1 \le x_1 \le x_2 \le \cdots \le x_n \le 1,$$

and f is convex between x_{k-1} and x_k for an index k, then

$$x_k - x_{k-1} \le \frac{16}{\sqrt{n}}.$$

It is not clear to me whether or not Erdős and Turán published this result. In any case, Erőd proves more, namely he shows that in the Erdős–Turán inequality the constant 16 can be replaced by

$$c_n = 2\sqrt{\frac{n}{2n-3}} \quad \text{if } n \text{ is even, and} \qquad c_n = \frac{2n}{n-1}\sqrt{\frac{n-2}{2n-3}} \quad \text{if } n \text{ is odd}$$

(so $\lim_{n\to\infty} c_n = \sqrt{2}$ in both cases).

An elementary paper of Erdős and Tibor Grünwald (Gallai) {31} deals with some geometric properties of polynomials with only real zeros. One of their results states that *if f is a polynomial of the form* (5), *then*

$$\int_{x_k}^{x_{k+1}} \left| f(x) \right| dx \le \frac{2}{3} (x_{k+1} - x_k) \max_{x \in [x_k, x_{k+1}]} \left| f(x) \right|.$$

Some extensions of the above are proved in {19}. In this paper Erdős raised a number of questions. For example, he conjectured that if t is a real trigonometric polynomial with only real zeros and with $\|t\|_{\mathbb{R}} \le 1$, then

$$\int_0^{2\pi} \left| t(\theta) \right| d\theta \le 4.$$

Concerning polynomials p of degree at most n with all their zeros in $(-1, 1)$ and with $\|p\|_{[-1,1]} = 1$, Erdős conjectured that *if $x_k < x_{k+1}$ are two consecutive zeros of p, then*

$$\int_{x_k}^{x_{k+1}} \left| p(x) \right| dx \le d_n (x_{k+1} - x_k),$$

where

$$d_n := \frac{1}{y_{k+1} - y_k} \int_{y_k}^{y_{k+1}} \left| T_n(y) \right| dy,$$

T_n is the usual Čebyshov polynomial of degree n, and $y_k < y_{k+1}$ are two consecutive zeros of T_n. (Note that d_n is independent of k and that $\lim_{n\to\infty} d_n = 2/\pi$.) These conjectures and more have all been proved in 1974, see Edward B. Saff and T. Sheil-Small {71} and Kristiansen {49}. {49} contains an error. This was corrected in {50}. See also {75}.

In {36} Erdős and Turán proved the following. Let $\left\{\zeta_\nu^{(n)}\right\}$ be a triangular sequence of numbers such that

$$1 \ge \zeta_1^{(n)} > \zeta_2^{(n)} > \cdots > \zeta_n^{(n)} \ge -1.$$

Let

$$\zeta_\nu^{(n)} = \cos\phi_\nu^{(n)}, \qquad 0 \le \phi_\nu^{(n)} \le \pi, \qquad \text{and} \qquad \omega_n(\zeta) = \prod_{\nu=1}^{n} \left(\zeta - \zeta_\nu^{(n)}\right).$$

For $(\alpha,\beta) \subset (0,\pi)$, let $N_n(\alpha,\beta)$ denote the number of $\phi_\nu^{(n)}$ in (α,β). Suppose $\left|\omega_n(\zeta)\right| < 2^{-n}A(n)$ on $(-1,1)$ for every n. Then for every subinterval (α,β) of $(0,\pi)$ one has

$$\left|N_n(\alpha,\beta) - \frac{\beta-\alpha}{\pi}n\right| < \frac{8}{\log 3}\left(n\log A(n)\right)^{1/2}.$$

Extending the results of his paper {36} with Turán, Erdős {21} proved that *if there are absolute constants* $c_1, c_2 > 0$ *and a function* f *such that*

$$\frac{c_2 f(n)}{2^n} \le \max_{\zeta_{\nu+1}^{(n)} \le \zeta \le \zeta_\nu^{(n)}} \left|\omega_n(\zeta)\right| \le \frac{c_1 f(n)}{2^n}, \qquad \nu = 0,1,\ldots,n,$$

then for $(\alpha,\beta) \subset (0,\pi)$,

$$N_n(\alpha,\beta) = \frac{\beta-\alpha}{\pi}n + O\big((\log n)\big(\log f(n)\big)\big).$$

This result has been extended by various people in many directions. See, for example, Totik {85}.

Erdős {28} gives an extension of some results of Bernstein and Antoni Zygmund. Bernstein had asked the question whether one can deduce boundedness of $\left|P_n(x)\right|$ on $[-1,1]$ for polynomials p_n of degree at most n if one knows that $\left|P_n(x)\right| \le 1$ for $m > (1+c)n$ values of x with some $c > 0$. His answer was affirmative. He showed that *if* $\left|P_n\big(x_i^{(m)}\big)\right| \le 1$ *for* *all zeros $x_i^{(m)}$ of the mth Čebyshov polynomial T_m with $m > (1+c)n$, then $|P_n(x)| \le A(c)$ for all $x \in [-1,1]$, with $A(c)$ depending only on c.* Zygmund had shown that the same conclusion is valid if T_m is replaced by the mth Legendre polynomial L_m. Erdős established a necessary and sufficient condition to characterize the system of nodes

$$-1 \le x_1^{(m)} < x_2^{(m)} < \cdots < x_n^{(m)} \le 1$$

for which

$$\left|P_n\left(x_i^{(m)}\right)\right| \le 1, \qquad i = 1, 2, \ldots, m, \qquad m > (1+c)n,$$

imply $|P_n(x)| \le A(c)$ for all polynomials P_n of degree at most n and for all $x \in [-1,1]$, with $A(c)$ depending only on c. His result contains both that of Bernstein and of Zygmund as special cases. Note that such an implication is impossible if $m \le n+1$, by a well-known result of Georg Faber.

Erdős wrote a paper {22} on the coefficients of the cyclotomic polynomials. The cyclotomic polynomial F_n is defined as the monic polynomial whose zeros are the primitive nth roots of unity. It is well known that

$$F_n(x) = \prod_{d|n} \left(x^{n/d} - 1\right)^{\mu(d)},$$

where μ is the Möbius function. For $n < 105$, all coefficients of F_n are ± 1 or 0. For $n = 105$, the coefficient 2 occurs for the first time. Denote by A_n the maximum over the absolute values of the coefficients of F_n. Schur proved that $\limsup A_n = \infty$. Emma Lehmer proved that $A_n > cn^{1/3}$ for infinitely many n. In his paper {22}, Erdős proved that *for every k, $A_n > n^k$ for infinitely many n.* This is implied by his even sharper theorem saying that

$$A_n > \exp\left[c\,(\log n)^{4/3}\right]$$

for $n = 2 \cdot 3 \cdot 5 \cdot \cdots \cdot p_k$ with k sufficiently large, where p_k denotes the kth prime number. Recent improvements and generalizations of this can be explored in {57}, {58}, and {59}.

Erdős {25} has a note on the number of terms in the square of a polynomial. Let

$$f_k(x) = a_0 + a_1 x^{n_1} + \cdots + a_{k-1} x^{n_{k-1}}, \qquad 0 \ne a_i \in \mathbb{R},$$

be a polynomial with k terms. Denote by $Q(f_k)$ the number of terms of f_k^2. Let $Q_k := \min Q(f_k)$, where the minimum is taken over all f_k of the above form. László Rédei posed the problem whether $Q_k < k$ is possible. Rényi, László Kalmár, and Rédei proved that, in fact, $\liminf_{k\to\infty} Q_k/k = 0$, and also that $Q(29) \le 28$. Rényi further proved that

$$\lim_{n\to\infty} \frac{1}{n} \sum_{k=1}^{n} \frac{Q_k}{k} = 0.$$

He also conjectured that $\lim_{k\to\infty} Q_k/k = 0$. In his short note {25}, Erdős proves this conjecture. In fact, he shows that *there are absolute constants* $c_1 > 0$ *and* $0 < c_2 < 1$ *such that* $Q_k < c_2 k^{1-c_1}$. Rényi conjectured that $\lim_{k\to\infty} Q_k = \infty$. He also asked whether or not Q_k remains the same if the coefficients are complex. These questions remained open at the time of the writing of this paper.

Erdős {27} proved a significant result related to his conjecture about polynomials with ± 1 coefficients. He showed that if

$$f_n(\theta) := \sum_{k=1}^{n} (a_k \cos k\theta + b_k \sin k\theta)$$

is a trigonometric polynomial with real coefficients,

$$\max_{1\le k\le n} \left\{ \max \left\{ |a_k|, |b_k| \right\} \right\} = 1 \qquad \textit{and} \qquad \sum_{k=1}^{n} (a_k^2 + b_k^2) = An,$$

then there exists a $c = c(A) > 0$ *depending only on* A *for which* $\lim_{A\to 0} c(A) = 0$ *and*

$$\max_{0\le\theta\le 2\pi} \left| f_n(\theta) \right| \ge \frac{1 + c(A)}{\sqrt{2}} \left(\sum_{k=1}^{n} (a_k^2 + b_k^2) \right)^{1/2}.$$

Closely related to this is a problem for which Erdős offered \$100 and which has become one of my favorite Erdős problems (see also Problem 22 in {26}): *Is there an absolute constant* $\varepsilon > 0$ *such that the maximum modulus on the unit circle of any polynomial* $p(x) = \sum_{j=0}^{n} a_j x^j$ *with each* $a_j \in \{-1, 1\}$ *is at least* $(1+\varepsilon)\sqrt{n}$? Erdős conjectured that there is such an $\varepsilon > 0$. Even the weaker version of the above, with $(1+\varepsilon)\sqrt{n}$ replaced by $\sqrt{n} + \varepsilon$ with an absolute constant $\varepsilon > 0$, looks really difficult. (The lower bound $\sqrt{n+1}$ is obvious by the Parseval formula.) These problems are unsettled to this date.

Let D be the open unit disk of the complex plane. Let ∂D be the unit circle of the complex plane. Let

$$\mathcal{K}_n := \left\{ p : p(z) = \sum_{k=0}^{n} a_k z^k,\ a_k \in \mathbb{C},\ |a_k| = 1 \right\}.$$

The class $\mathcal{K}_n$ is often called the collection of all (complex) unimodular polynomials of degree n. Given a sequence (ε_{n_k}) of strictly positive numbers tending to 0, we say that a sequence (P_{n_k}) of polynomials $P_{n_k} \in \mathcal{K}_{n_k}$ is (ε_{n_k})-ultraflat if

$$(1 - \varepsilon_{n_k})\sqrt{n_k + 1} \le \left| P_{n_k}(z) \right| \le (1 + \varepsilon_{n_k})\sqrt{n_k + 1}, \qquad z \in \partial D,$$

or equivalently

$$\max_{z \in \partial D} \left| \left| P_{n_k}(z) \right| - \sqrt{n_k + 1} \right| \le \varepsilon_{n_k} \sqrt{n_k + 1}.$$

The existence of an ultraflat sequence of unimodular polynomials (for some sequence (ε_{n_k}) of positive real numbers tending to 0) seemed very unlikely, in view of an extended version of the above mentioned Erdős conjecture to polynomials $P \in \mathcal{K}_n$ with $n \ge 1$.

Yet, refining a method of Thomas W. Körner, Jean-Pierre Kahane {46} proved that there exists a sequence (P_n) with $P_n \in \mathcal{K}_n$ which is (ε_n)-ultraflat, where

$$\varepsilon_n = O\left(n^{-1/17} \sqrt{\log n}\,\right).$$

Thus this extended version of the above mentioned Erdős conjecture was disproved for the classes $\mathcal{K}_n$.

The structure of ultraflat sequences of unimodular polynomials is beautiful. The following uniform distribution theorem for the angular speed, conjectured by Saffari {72}, is proved in {17}. *Suppose (P_n) is an ultraflat sequence of unimodular polynomials $P_n \in \mathcal{K}_n$. We write*

$$P_n(e^{it}) = R_n(t) e^{i\alpha_n(t)}, \qquad R_n(t) = \left| P_n(e^{it}) \right|, \qquad t \in \mathbb{R}.$$

It is a simple exercise to show that α_n can be chosen so that it is differentiable in t on $\mathbb{R}$. Then in the interval $[0, 2\pi]$, the distribution of the normalized angular speed $\alpha_n'(t)/n$ converges to the uniform distribution as $n \to \infty$. That is, we have

$$m\left(\left\{ t \in [0, 2\pi] : 0 \le \alpha_n'(t) \le nx \right\} \right) = 2\pi x + \gamma_n(x)$$

for every $x \in [0,1]$, *where* $\lim_{n\to\infty} \gamma_n(x) = 0$ *for every* $x \in [0,1]$. *As a consequence,* $\left|P_n'(e^{it})\right|/n^{3/2}$ *also converges to the uniform distribution as* $n \to \infty$. *That is, we have*

$$m\Big(\big\{t \in [0,2\pi] \,:\, 0 \le \left|P_n'(e^{it})\right| \le n^{3/2}x\big\}\Big) = 2\pi x + \gamma_n(x)$$

for every $x \in [0,1]$, *where* $\lim_{n\to\infty} \gamma_n(x) = 0$ *for every* $x \in [0,1]$ *(*$m(\cdot)$ *denotes the one-dimensional Lebesgue measure). In both statements the convergence of* $\gamma_n(x)$ *is uniform on* $[0,1]$.

For higher derivatives, the following result is proved in {17}. *Suppose* (P_n) *is an ultraflat sequence of unimodular polynomials* $P_n \in \mathcal{K}_n$. *Then*

$$\left(\frac{\left|P_n^{(m)}(e^{it})\right|}{n^{m+1/2}}\right)^{1/m}$$

converges to the uniform distribution as $n \to \infty$. *More precisely, we have*

$$m\Big(\big\{t \in [0,2\pi] \,:\, 0 \le \left|P_n^{(m)}(e^{it})\right| \le n^{m+1/2}x^m\big\}\Big) = 2\pi x + \gamma_{m,n}(x)$$

for every $x \in [0,1]$, *where* $\lim_{n\to\infty} \gamma_{m,n}(x) = 0$ *for every fixed* $m = 1,2,\dots$ *and* $x \in [0,1]$. *For every fixed* $m = 1,2,\dots$, *the convergence of* $\gamma_{n,m}(x)$ *is uniform on* $[0,1]$.

Several topics from Erdős's problem paper {29} have already been discussed before. Here is one more interesting group of problems. Let $(z_k)_{k=1}^\infty$ be a sequence of complex numbers of modulus 1. Let

$$A_n := \max_{|z|=1} \prod_{k=1}^n |z - z_k|.$$

What can one say about the growth of A_n? Erdős conjectured that $\limsup A_n = \infty$. In my copy of {29} that Erdős gave me, there are some handwritten notes (in Hungarian) saying the following. "Wagner proved that $\limsup A_n = \infty$. It is still open whether or not $A_n > n^c$ or $\sum_{k=1}^n A_k > n^{1+c}$ happens for infinitely many n (with an absolute constant $c > 0$). These are probably difficult to answer." See {89}.

Erdős was famous for anticipating the "right" results. "This is obviously true; only a proof is needed" he used to say quite often. Most of the times, his conjectures turned out to be true. Some of his conjectures failed for the more or less trivial reason that he was not always completely precise with the formulation of the problem. However, it happened only very rarely that he was essentially wrong with his conjectures. If someone proved something that was in contrast with Erdős' anticipation, he or she could really boast to have proved a really surprising result. Erdős was always honest with his conjectures. If he did not have a sense about which way to go, he formulated the problem "prove or disprove". Erdős turned even his "ill fated" conjectures into challenging open problems. The following quotation is a typical example of how Erdős treated the rare cases when a conjecture of his was disproved. It is from his problem paper {29} entitled "Extremal problems on polynomials". For this quotation, we need to recall the following notation. Associated with a monic polynomial $f(z) = \prod_{j=1}^{n}(z - z_j)$, where z_j are complex numbers, let $E_n(f) := \{z \in \mathbb{C} : |f(z)| \le 1\}$. In his problem paper Erdős writes (in terms of the notation employed here): "In [7] we made the ill fated conjecture that the number of components of $E_n(f)$ with diameter greater than $1 + c$ $(c > 0)$ is less than δ_c, δ_c bounded. Pommerenke [14] showed that nothing could be farther from the truth, in fact he showed that for every $\varepsilon > 0$ and $k \in \mathbb{N}$, there is an $E_n(f)$ which has more than k components of diameter greater than $4 - \varepsilon$. Our conjecture can probably be saved as follows: Denote by $\Phi_n(c)$ the largest number of components of diameter greater than $1 + c$ $(c > 0)$ which $E_n(f)$ can have. Surely, for every $c > 0$, $\Phi_n(c) = o(n)$, and hopefully $\Phi_n(c) = o(n^\varepsilon)$ for every $\varepsilon > 0$. I have no guess about a lower bound for $\Phi_n(c)$, also I am not sure whether the growth of $\Phi_n(c)$, $(1 < c < 4)$ depends on c very much."

If $p(x) = \sum_{j=0}^{n} a_j x^j$, then we introduce $l_1(p) := \sum_{j=0}^{n} |a_j|$. An interesting problem of Erdős and György Szekeres is to minimize $l_1(p)$ over all polynomials

$$p(x) = \prod_{k=1}^{N}(1 - x^{\alpha_k}),$$

where N is a fixed positive integer, while the positive integers $\alpha_1, \alpha_2, \ldots, \alpha_N$ vary. Let $E_N^* := \min l_1(p)$, where the minimum is taken for all p of the above form. It is conjectured by Erdős and Szekeres that $E_N^* \ge N^K$ for any fixed K and sufficiently large N. Erdős and Szekeres {34} proved a sub-exponential upper bound for E_N^*. The best known upper bound for E_N^* today is $\exp\left(O(\log n)^4\right)$ given by A. S. Belov and Sergei V. Konyagin {3}, that improves weaker bounds given earlier by Andrew Odlyzko and Mihail N. Kolountzakis.

For a fixed $z \in \mathbb{C}$ and a Borel measure μ on $[-\pi, \pi)$ the Christoffel function $\omega_n(\mu, z)$ is defined to be the minimum of

$$\frac{1}{2\pi} \int_{-\pi}^{\pi} \left|p(e^{i\theta})\right|^2 d\mu(\theta)$$

taken over all polynomials p of degree less than n with $p(z) = 1$. Szegő {78} studied $\omega_n(\mu, 0)$ for absolutely continuous measures μ in 1915. Later, in 1922, Szegő {79} showed that

$$\lim_{n\to\infty} n\omega_n(\mu, e^{it}) = \mu'(t), \qquad t \in (-\pi, \pi),$$

assuming that μ is absolutely continuous and $\mu' > 0$ is twice continuously differentiable. This result is very important for applications in orthogonal polynomials, probability theory, and statistics (linear prediction), and other areas. It gives a useful and numerically adaptable method of computing the weight function for orthogonal polynomials. In their paper {62}, Máté, Nevai, and Totik show that

$$\lim_{n\to\infty} n\omega_n(\mu, e^{it}) = \mu'(t)$$

holds almost everywhere on every interval $I \subset [-\pi, \pi)$ for which

$$\int_I \log \mu'(\theta)\, d\theta > -\infty.$$

In an earlier paper {61} Máté and Nevai showed that

$$\int_{-\pi}^{\pi} \log \mu'(\theta)\, d\theta > -\infty$$

implies

$$2e^{-1}\mu'(t) \le \liminf_{n\to\infty} n\omega_n(\mu, e^{it}) \le \limsup_{n\to\infty} n\omega_n(\mu, e^{it}) = \mu'(t)$$

for almost every real t.

References

[111] Lorentz, George G., *Approximation of Functions,* Holt, Rinehart and Winston (New York, 1966); 2nd edition, Chelsea (New York, 1986).

[123] Newman, D. H., *Approximation with Rational Functions.* CBMS Regional Conference Series in Mathematics, Vol. 41, American Mathematical Society (Providence, RI, 1979).

[164] Schwartz, Laurent, *Étude des Sommes d'Exponentielles,* Hermann (Paris, 1959).

[173] Szegő, Gábor, *Collected Papers,* ed. R. Askey, Birkhäuser (Boston–Basel–Stuttgart, 1982).

{1} V. V. Arestov, On integral inequalities for trigonometric polynomials and their derivatives, *Izv.,* **45**, 3–22.

{2} J. Bak and D. J. Newman, Rational combinations of x^{λ_k}, $\lambda_k \geq 0$ are always dense in $C[0,1]$, *J. Approx. Theory,* **23** (1978), 155–157.

{3} A. S. Belov and S. V. Konyagin, On estimates of the constant term of a nonnegative trigonometric polynomial with integral coefficients, *Mat. Zametki,* **59** (1996), 627–629.

{4} A. Bloch and G. Pólya, On the roots of certain algebraic equations, *Proc. London Math. Soc,* **33** (1932), 102–114; [Pólya Collected Papers] II, pp. 361–338.

{5} B. D. Bojanov, Proof of a conjecture of Erdős about the longest polynomial, *Proc. Amer. Math. Soc.,* **84** (1982), 99–103.

{6} P. Borwein, The arc length of the lemniscate ($|p(z)| = 1$), *Proc. Amer. Math. Soc.,* **123** (1995), 797–799.

{7} P. Borwein and T. Erdélyi, Markov–Bernstein-type inequalities for classes of polynomials with restricted zeros, *Constr. Approx.,* **10** (1994), 411–425.

{8} P. Borwein and T. Erdélyi, *Polynomials and Polynomial Inequalities,* Springer-Verlag (New York, 1995).

{9} P. Borwein and T. Erdélyi, A sharp Bernstein-type inequality for exponential sums, *J. Reine Angew. Math.,* **476** (1996), 127–141.

{10} P. Borwein and T. Erdélyi, Generalizations of Müntz's Theorem via a Remez-type inequality for Müntz spaces, *J. Amer. Math. Soc.,* **10** (1997), 327–329.

{11} P. Borwein and T. Erdélyi, On the zeros of polynomials with restricted coefficients, *Illinois J. Math.,* **41** (1997), 667–675.

{12} P. Borwein, T. Erdélyi, and G. Kós, Littlewood-type problems on $[0,1]$, *Proc. London Math. Soc. (3),* **79** (1999), 22–46.

{13} J. A. Clarkson and P. Erdős, Approximation by polynomials, *Duke Math. J.,* **10** (1943), 5–11.

{14} R. A. DeVore and G. G. Lorentz, *Constructive Approximation,* Springer-Verlag (Berlin, 1993).

{15} R. J. Duffin and A. C. Schaeffer, A refinement of an inequality of the brothers Markoff, *Trans. Amer. Math. Soc.,* **50** (1941), 517–528.

{16} T. Erdélyi and W. B. Johnson, The "Full Müntz Theorem" in $L_p[0,1]$ for $0 < p < \infty$, *J. d'Analyse Math.,* **84** (2001), 145–172.

{17} T. Erdélyi, The phase problem of ultraflat unimodular polynomials: the resolution of the conjecture of Saffari, *Math. Ann.,* **321** (2001), 905–924.

{18} P. Erdős, An extremum-problem concerning trigonometric polynomials, *Acta Sci. Math. Szeged,* **9** (1939), 113–115.

{19} P. Erdős, Note on some elementary properties of polynomials, *Bull. Amer. Math. Soc.,* **46** (1940), 954–958.

{20} P. Erdős, On extremal properties of the derivatives of polynomials, *Ann. of Math. (2),* **41** (1940), 310–313.

{21} P. Erdős, On the uniform distribution of the roots of certain polynomials, *Ann. of Math. (2),* **43** (1942), 59–64.

{22} P. Erdős, On the coefficients of the cyclotomic polynomial, *Bull. Amer. Math. Soc.,* **52** (1946), 179–184.

{23} P. Erdős, Some remarks on polynomials, *Bull. Amer. Math. Soc.,* **53** (1947), 1169–1176.

{24} P. Erdős and H. Fried, On the connection between gaps in power series and the roots of their partial sums, *Trans. Amer. Math. Soc.,* **62** (1947), 53–61.

{25} P. Erdős, On the number of terms of the square of a polynomial, *Nieuw Arch. Wiskunde, Ser. 2,* **23** (1949), 63–65.

{26} P. Erdős, Some unsolved problems, *Michigan Math. J.,* **4** (1957), 291–300.

{27} P. Erdős, An inequality for the maximum of trigonometric polynomials, *Ann. Pol. Math.,* **12** (1962), 151–154.

{28} P. Erdős, On the boundedness and unboundedness of polynomials, *J. d'Analyse Math.,* **19** (1967), 135–148.

{29} P. Erdős, Extremal problems on polynomials, in: *Approximation theory II.,* Proc. Internat. Sympos. Univ. Texas, Austin, Texas, 1976, Academic Press (New York, 1976), pp. 347–355.

{30} P. Erdős and G. Freud, On orthogonal polynomials with regularly distributed zeros, *Proc. London Math. Soc. (3),* **29** (1974), 521–537.

{31} P. Erdős and T. Grünwald, On polynomials with only real roots, *Ann. of Math. (2),* **40** (1939), 537–548.

{32} P. Erdős, F. Herzog, and G. Piranian, Metric properties of polynomials, *Journal d'Analyse Math.,* **6** (1958), 125–148.

{33} P. Erdős and Netanyahu, A remark on polynomials and the transfinite diameter, *Israel J. Math.,* **14** (1973), 23–25.

{34} P. Erdős and G. Szekeres, On the product $\prod_{k=1}^{n}(1-z^{\alpha_k})$, *Publications de L'Institut Math.,* **12** (1952), 29–34.

{35} P. Erdős and P. Turán, On interpolation, II. On the distribution of the fundamental points of Lagrange and Hermite interpolation, *Ann. of Math.*, **39** (1938), 703–724.

{36} P. Erdős and P. Turán, On the uniformly-dense distribution of certain sequences, *Ann. of Math. (2)*, **41** (1940), 162–173.

{37} P. Erdős and P. Turán, On interpolation, III. Interpolatory theory of polynomials, *Ann. of Math. (2)*, **41** (1940), 510–553.

{38} P. Erdős and P. Turán, On the distribution of roots of polynomials, *Ann. of Math. (2)*, **51** (1950), 105–119.

{39} J. Erőd, Bizonyos polinomok maximumának alsó korlátjáról, *Matematikai és Fizikai Lapok*, **XLVI** (1939).

{40} L. Fejér, Az algebrai egyenlet legkisebb abszolutértékű gyökéről, Über die Wurzel vom kleinest absoluten Betrage einer algebraischen Gleichung, *Mat. és Fiz. Lapok*, **17** (1908), 308–324; [Arbeiten] No. 24, I, pp. 334–361.

{41} L. Fejér, Über einen S. Bernsteinschen Satz über die Derivierte eines trigonometrischen Polynoms und über die Szegősche Verschärfung desselben, *Bull. of the Calcutta Math. Soc.*, **20** (1930), 49–54; [Arbeiten] No. 72, II. pp. 281–285.

{42} M. Fekete, Über die Verteilung der Wurzeln bei gewissen algebraischen Gleichungen mit ganzzah ligen Koeffizienten, *Math. Z.*, **17** (1923), 228–249.

{43} M. Fekete, Über den transfinite Durchmesser ebener Punktmengen, *Math. Z.*, **32** (1930), 108–114.

{44} M. Fekete, Approximations par polynomes avec conditions diophantiennes, I at II. *C.R. Acad. Sci. Paris, sér A–B*, **239** (1930), 1337–1339, 1455–1457.

{45} G. Halász, Markov-type inequalities for polynomials with restricted zeros, *J. Approx. Theory (1)*, **101** (1999), 148–155.

{46} J. P. Kahane, Sur les polynomes a coefficient unimodulaires, *Bull. London Math. Soc.*, **12** (1980), 321–342.

{47} T. Kövári, On the distribution of Fekete points II, *Mathematika*, **18** (1971), 40–49.

{48} T. Kövári and Ch. Pommerenke, On the distribution of Fekete points, *Mathematika*, **15** (1968), 70–75.

{49} G. K. Kristiansen, Proof of a polynomial conjecture, *Proc. Amer. Math. Soc.*, **44** (1974), 58–60.

{50} G. K. Kristiansen, Erratum to "Proof of a polynomial conjecture" (*Proc. Amer. Math. Soc.*, **44** (1974), 58–60.), *Proc. Amer. Math. Soc.*, **58** (1976), 377.

{51} G. K. Kristiansen, Some inequalities for algebraic and trigonometric polynomials, *J. London Math. Soc.*, **20** (1979), 300–314.

{52} E. Laguerre, Sur les fonctions du genre zéro et du genre un, *C.R. Acad. Sci.*, **95** (1882), 828–831.

{53} P. Lax, Proof of a conjecture of P. Erdős on the derivative of a polynomial, *Bull. Amer. Math. Soc.*, **50** (1944), 509–513.

{54} A. L. Levin and D. S. Lubinsky, Canonical products and weights $\exp(-|x|^{\alpha})$, $\alpha > 1$, with applications, *J. Approx. Theory (2)*, **49** (1987), 149–169.

{55} A. L. Levin and D. S. Lubinsky, Weights on the real line that admit good relative polynomial approximation with applications, *J. Approx. Theory (2),* **49** (1987), 170–195.

{56} G. G. Lorentz, M. von Golitschek and Y. Makovoz, *Constructive Approximation: Advanced Problems,* Springer-Verlag (New York, NY, 1996).

{57} H. Maier, The coefficients of cyclotomic polynomials, in: *Analytic Number Theory* (Allerton Park IL, 1989), Progr. Math. 85, Birkhäuser Boston (Boston MA, 1990), pp. 349–366.

{58} H. Maier, Cyclotomic polynomials with large coefficients, *Acta Arith.,* **64** (1993), 227–235.

{59} H. Maier, The size of the coefficients of cyclotomic polynomials, in: *Analytic Number Theory,* Vol. II, (Allerton Park IL, 1995), Progr. Math., 139, Birkhäuser Boston (Boston MA, 1996), pp. 633–639.

{60} L. Márki, G. Somorjai, and J. Szabados, Müntz–Jackson type theorems via interpolation, *Acta Math. Hungar. (1–2),* **29** (1977), 185–191.

{61} A. Máté and P. Nevai, Bernstein's inequality in L^p for $0 < p < 1$ and $(C, 1)$ bounds for orthogonal polynomials, *Ann. of Math.,* **111** (1980), 145–154.

{62} A. Máté, P. Nevai, and V. Totik, Szegő's extemum problem on the unit circle, *Ann. of Math. (2),* **134** (1991), 433–453.

{63} C. Müntz, *Über den Approximationsatz von Weierstrass,* H. A. Schwartz Festschrift (Berlin, 1914).

{64} F. Nazarov, Local estimates for exponential polynomials and their applications to inequalities of the uncertainty type, *Algebra i Analiz (4),* **5** (1993), 3–66.

{65} P. Nevai and V. Totik, Weighted polynomial inequalities, *Constr. Approx.,* **2** (86), 113–127.

{66} G. Pólya, Über Annäherung durch Polynome mit lauter reelen Wurlzeln, *Rend. Circ. Mat. Palermo,* **36** (1913), 279–295.

{67} G. Pólya, Beitrag zur Verallgemeinerung des Verzerrungssatzes auf mehrfach Zusammenhängende Gebiete I., *Sitzgsber. Preuss. Akad. Wiss., Berlin, K.L. Math. Phys. Tech.* (1928), 228–232 and 280–282, (1929), 55–62; [Papers] I. pp. 347–362.

{68} G. Pólya, Problem 108, *Jahresber. Deutsch. Math. Verein. II. Abteilung,* **40**, p. 81.

{69} Q. I. Rahman, On a problem of Turán about polynomials with curve majorants, *Trans. Amer. Math. Soc.,* **63** (1972), 447–455.

{70} M. Riesz, Formule d'interpolation pour la dérivée d'un polynôme trigonométrique, *C.R. Acad. Sci. Paris,* **158** (1914), 1152–1154; [Papers] pp. 127–129.

{71} E. B. Saff and T. Sheil-Small, Coefficient and intergal mean estimates for algebraic and trigonometric polynomials with restricted zeros, *J. London Math. Soc.,* **9** (1974), 16–22.

{72} B. Saffari, The phase behavior of ultraflat unimodular polynomials, in: *Probabilistic and Stochastic Methods in Analysis, with Applications,* 555–572, Kluwer Academic Publishers (Dordrecht, 1992).

{73} I. Schur, Untersuchungen über algebraische Gleichungen, *Sitz. Preuss. Akad. Wiss., Phys.-Math. Kl.* (1933), 403–428.

{74} G. Somorjai, A Müntz-type problem for rational approximation, *Acta. Math. Hungar.*, **27** (1976), 197–199.

{75} J. Szabados, *On some extremum problems for polynomials, in Approximation of Function Spaces* (Gdansk, 1979), 731–748, North-Holland (Amsterdam, 1981).

{76} O. Szász, Über die Approximation stetiger Funktionen durch lineare Aggregate von Potenzen, *Math. Ann.*, **77** (1916), 482–496; [Papers] 324–338.

{77} O. Szász, Lösung der Aufgabe 108. Jahresber. Deutsch. Math.-Verein., II., *Abteilung,* **43**, 20–23.

{78} G. Szegő, Ein Grenzwertsatz über die Toeplitzschen Determinanten einer reellen positiven funktion, *Math. Ann.*, **76** (1915), 590–603 [Papers] I, 54–67.

{79} G. Szegő, Über die Entwicklung einer Torplitzschen einer willkürlichen Funktion nach den Polynomen eines Orthogonal system., *Math. Z.*, **12** (1922), 61–94 [Papers] I, 439–472.

{80} G. Szegő, Bemerkungen zu einen Satz von J. H. Grace über die Wurzeln algebraischer Gleighungen, *Math. Z.*, **13** (1922), 28–55; [Papers] I, 506–533.

{81} G. Szegő, Bemerkungen zu einen Arbeit von Herrn Fekete "Über die Verteilung der Wurzeln bei gewissen algebraischen Gleichungen mit ganzzahligen Koeffizienten", *Math. Z.*, **21** (1924), 203–208; [Papers] I, 637–642.

{82} G. Szegő, Über einen Satz von A. Markoff, *Math. Z.*, **23** (1925), 45–61; [Papers] I, 665–681.

{83} G. Szegő, Bemerkungen zu einem Satz von E. Schmidt uber algebraische Gleichungen, *Sitz. Preuss. Akad. Wiss., Phys.-Math. Kl.* (1934), 86–98; [Papers] II, 529–542.

{84} G. Szegő, On some problems of approximation, *Magyar Tud. Akad. Mat. Kut. Int. Közl.*, **2** (1964), 3–9; [Papers] III, 845–851.

{85} V. Totik, Distribution of simple zeros of polynomials, *Acta Math. (1),* **170** (1993), 1–28.

{86} P. Turán, Über die Ableitung von Polynomen, *Compositio Math.*, **7** (1939), 89–95.

{87} P. Turán, Remark on a theorem of Erhard Schmidt, *Mathematica (Cluj), 2(25),* **2** (1960), 373–378.

{88} P. Turán, *On a New Method of Analysis and its Applications,* Wiley (New York, NY, 1984).

{89} G. Wagner, On a problem of Erdős in Diophantine approximation, *Bull. London Math. Soc.*, **12** (1980), 81–88.

Tamás Erdélyi
Department of Mathematics
Texas A&M University
College Station
Texas 77843
U.S.A.

terdelyi@math.tamu.edu

Harmonic Analysis

BOLYAI SOCIETY
MATHEMATICAL STUDIES, 14

A Panorama of Hungarian Mathematics in the Twentieth Century, pp. 159–192.

Commutative Harmonic Analysis

JEAN-PIERRE KAHANE

The present article is organized around four themes: 1. the theorem of Fejér, 2. the theorem of Riesz–Fischer, 3. boundary values of analytic functions, 4. Riesz products and lacunary trigonometric series. This does not cover the whole field of the Hungarian contributions to commutative harmonic analysis. A final section includes a few spots on other beautiful matters. Sometimes references are given in the course of the text, for example at the end of the coming paragraph on Fejér. Other can be found at the end of the article.

1. The theorem of Fejér

On November 19, 1900 the Académie des Sciences in Paris noted that it had received a paper from Leopold FEJEV in Budapest with the title "Proof of the theorem that a bounded and integrable function is analytic in the sense of Euler". On December 10 the Comptes Rendus published the famous note "On bounded and integrable fonctions", in which Fejér sums, Fejér kernel and Fejér summation process appear for the first time, and where the famous Fejér theorem, which asserts that any decent function is the limit of its Fejér sums, is proved. The spelling error of November 19 in Fejér's name (Fejev instead of Fejér) is not reproduced on December 10. It is replaced by an other one: the paper presented by Picard is attributed to Leopold TEJER. This is how Fejér's name enters into history (*C.R. Acad. Sci. Paris,* **131** (1900), 825 and 984–987).

Fejér was a very young man, 20 years old, and unknown. But Fejér's theorem became famous quickly. It was used for proving the completeness of the trigonometric system in Hurwitz's work on the isoperimetric problem, extended by Lebesgue to Lebesgue-integrable fonctions (Fejér–Lebesgue theorem), generalized to other kernels useful in approximation theory (Ch. de la Vallée Poussin), applied in a simple and completely new proof of the Dirichlet–Jordan theorem (Hardy), made more precise by the notion of absolute summability (Hardy and Littlewood), and, above all, its simplicity made it accessible to any mathematics student {1}, {2}, {3}, {4, p. 245}, {5}, {6}. By the 1910's the Fejér theorem had already the status of a classical result, and no mathematician could ignore Fejér's name and its spelling — perhaps except for the place of the accent.

The change can be appreciated in comparing the first and the second edition of Ch. de la Vallée Poussin's Cours d'Analyse {7}. In the first edition (1903–1906) it follows the tradition of the great analysis textbooks of the time: it devotes little place to Fourier series and presents Dirichlet's convergence theorem in the tradition of Dirichlet and Jordan. In the second edition (1912) we can find the Fejér theorem, Hardy's method for deducing the Dirichlet–Jordan theorem from it, and also Fejér's example of a continuous functions whose Fourier series diverges at one point. The difference shows well enough the importance of the revolution which took place.

What was the nature of this revolution? In order to see this, let us go back to 1900.

In the 19th century the theory of analytic functions of one complex variable progressed by giant leaps. It had become "Theory of Functions" par excellence. The theory of functions of several real variables had developed through the investigation of partial differential equations arising in physics. On the other hand, the field of functions of one real variable revealed strange and disquieting creatures: continuous but nowhere differentiable functions (Weierstrass), continuous functions whose Fourier series diverges at one point (du Bois Reymond). Hermite wrote to Stieltjes that he "turned away with fright and horror from that lamentable ulcer: a continuous function with no derivative". Poincaré in l'Enseignement Mathématique complained that examples were not constructed any more in order to illustrate theorems and theories, but just for the purpose of showing that our predecessors were wrong. Gaston Darboux, who wrote an important memoir on discontinuous functions in 1875, turned to geometry quite prudently.

Between 1880 et 1900 there were only a few works on trigonometric series and they did not attract much attention. Fourier series did not appear as a reliable tool; there were too many strange things about them. Maybe there are continuous functions whose Fourier series diverge everywhere, just as there are nowhere differentiable continuous functions (this was still an open question in 1965, before Carleson's theorem {8}). It is even possible that the Fourier series converges but does not represent the function, as it is the case for Taylor series of C^∞-functions? This question seems to have been raised by Minkowski ([40, I, p. 24]).

Emile Picard's Traité d'analyse, which appeared in 1891, is quite instructive in this respect. The problem of finding a harmonic function in a domain when boundary values are given (the Dirichlet problem) is treated for the sphere, in the section of the book devoted to functions of several variables, before being treated for the circle. This is not because things are worse for the circle than for the sphere. But the solution for the circle belongs to the chapter "Fourier series" and this is not a subject to begin with {9}.

However, the pages devoted by Emile Picard to the Dirichlet problem on the circle are quite interesting; in 1900 Fejér knew them well and refers to them in his note. Picard presents the method of Schwarz (1872), which is based on the properties of the "Poisson kernel"

$$\frac{1-r^2}{1-2r\cos t+r^2} = 1+2\sum_1^\infty r^n \cos nt$$

As an application he shows that a continuous function on the circle whose Fourier series is

$$\sum_0^\infty (a_n \cos nt + b_n \sin nt)$$

can be expressed as a uniform limit of trigonometric polynomials of the form

$$\sum_{n\le N(r)} (a_n \cos nt + b_n \sin nt) r^n,$$

providing therefore a new proof of the Weierstrass approximation theorem.

In 1893, Ch. de la Vallée Poussin also used the Poisson kernel in order to establish the Parseval formula, which is essentially equivalent to the totality of the trigonometric system {10}. In 1901 Adolph Hurwitz stated the Parseval formula as a lemma to his solution of the isoperimetric problem by means of Fourier series, saying that he would prove it later {11}.

There were actually a few interesting results on Fourier series, but results and problems were not related to each other. The problem of Minkowski could have been solved easily using the method of Schwarz presented in

Picard's book, but Picard did not know about the problem and Minkowski was apparently unaware of this part of the works of Schwarz and Picard. Schwarz's method also yielded the formula obtained by Ch. de la Vallée Poussin, but de la Vallée Poussin was not aware of it. Certainly Hurwitz did not know the paper of de la Vallée Poussin: this is acknowledged in his article of 1903 {1}. Thus all these were isolated works on a marginal subject.

Fejér wrote at the beginning of his thesis that nothing essentially new appeared on Fourier series between 1885 and 1900. Though this is not completely true, still Fourier series appeared as a stagnant subject, out of fashion.

Fejér's discovery is that the averages of the partial sums

$$\sigma_n = \frac{1}{n}(S_0 + S_1 + \cdots + S_{n-1})$$

approximate the given function f at each point where $f(x+0)$ and $f(x-0)$ exist and $f(x) = \frac{1}{2}\big(f(x+0)+f(x-0)\big)$, and uniformly when f is continuous on the circle. Let us trace the circumstances of that discovery.

The idea of assigning a sum to a divergent series by means of some summation process was familiar to mathematicians. According to Lebesgue, d'Alembert already used the process of taking averages of partial sums for the series

$$\frac{1}{2} + \sum_1^\infty \cos nt \qquad (0 < t < 2\pi).$$

Summation processes became a significant topic of Abel's investigations, then of those of Poisson, Frobenius, Hölder, Cesàro and Borel. Abel proved the famous theorem asserting that

$$\lim_{r\uparrow 1} \sum_1^\infty a_n r^n = \sum_1^\infty a_n$$

whenever the right-hand side exists in the usual sense. Poisson considered the left hand side as the generalized sum of the series, whether or not the series converges. Frobenius generalized Abel's theorem by showing that

$$\lim_{r\uparrow 1} \sum_1^\infty a_n r^n = \lim_{n\to\infty} \frac{1}{n}(S_0 + S_1 + \cdots + S_{n-1})$$

whenever the right side exists. Hölder generalized this theorem of Frobenius by iterating the process of arithmetic means. Cesàro generalized another theorem of Abel, on multiplication of series, and introduced for this purpose the processes that we now denote by (C, k). Emile Borel defined new summation processes and applied them to the analytic continuation of functions defined by a Taylor series. Fejér knew all these works.

In order to solve the Dirichlet problem for the circle in the form

$$f(r\cos t, r\sin t) = \frac{a_0}{2} + \sum_1^\infty (a_n \cos nt + b_n \sin nt) r^n$$

when the function prescribed at the boundary has the Fourier series

$$\frac{a_0}{2} + \sum_1^\infty (a_n \cos nt + b_n \sin nt)$$

one might think of using Abel's theorem. However, the example of P. du Bois Reymond excludes any hope to obtain a solution for an arbitrary continuous function in this way. Could one apply the theorem of Frobenius? This seems to have been the starting point of Fejér.

Fejér learned about this problem during the academic year 1899–1900 which he spent as a student at the University of Berlin. He obtained a solution in Budapest at the end of October. He observed quickly that the method he used was more important that the new solution of the Dirichlet problem. In his Comptes Rendus note, the solution of the Dirichlet problem appears as one of the consequences of his theorem. The proof is based on the kernel

$$K_n(x) = \frac{1 - \cos nx}{n(1 - \cos x)},$$

and it is strongly related to Schwarz's solution of the Dirichlet problem, which makes use of the Poisson kernel.

Fejér indeed did **not** introduce any new summation process for divergent series; on the contrary, he used the most evident of them. It was **not** he who introduced positive kernels in investigating Fourier series: the Poisson kernel was well known and its application to Fourier series could be found in Picard's treatise long before 1900. Fejér did **not** solve a difficult conjecture by sophisticated methods.

What he did is much more than that. He gave a clear, simple and powerful statement in a field where the strange and the bizarre prevailed before. By coupling Fourier series and summation processes he provided a convenient frame for both theories. Since that time, the summation process of Riemann (based on the second differences of the second integral), the method of Schwarz for the Dirichlet problem on the circle (that is, what Fourier already used for computing temperatures inside a heated body) and the newly introduced process of Weierstrass in order to study temperature as a function of time by means of the series

$$\sum (a_n \cos nx + b_n \sin nx) e^{-nt^2}$$

appeared as expressing the same principle, most simply presented in Fejér's note: on one hand, regularization of the function by means of a convenient kernel, on the other, a summation process for Fourier series. The role of positive kernels was emphasized, and developed in many works of Fejér himself later (see section 5). Cesàro processes of different orders appeared for different purposes — for Laplace series, the relevant process is $(C, 2)$ ([40, I no 22, 24, 28]; [40, II no 63]). The two pages long note of Fejér completely changed the position of trigonometric series in mathematics. It also gave impetus to the study of general summation methods.

I shall restrict myself to a very few examples of applications and continuations of Fejér's theorem, that I chose because they involve Hungarian mathematicians. Others can be found in my joint book with Pierre-Gilles Lemarié-Rieusset, Fourier series and wavelets ([83, part I, chapter 7]).

The Fejér kernel can be written as

$$K_n(x) = \sum_{-n}^{n} \left(1 - \frac{|m|}{n}\right) e^{imx}.$$

A linear combination of K_n with positive coefficients yields a positive function whose Fourier coefficients c_m (in the complex form) are even ($c_m = c_{-m}$), positive (≥ 0), convex for $m \geq 0$ and decreasing to 0 as $m \to \infty$. Conversely each sequence (c_m) of this form is a sequence of Fourier coefficients of a positive and integrable function. This is a theorem of W. H. Young (1913), rediscovered many times, and very useful {12}. The analogue for functions $c(t)$ ($t \in \mathbb{R}$) is important in probability theory: if $c(t)$ is even, positive, convex on $\mathbb{R}^+$ and decreasing to 0 at infinity, it is Fourier transform of a positive and integrable function — therefore, a characteristic function if moreover $c(0) = 1$. This was pointed out by György Pólya and such functions $c(t)$ are called Pólya functions {13}.

Another kind of linear combinations of Fejér kernels was introduced by Hardy (1910) {14}, used by Fejér (1913) ([40, I, 715–718]) and later on by Ch. de la Vallée Poussin (1919) {15} in his lectures on the approximation of functions of a real variable, namely

$$(N+1)K_{(N+1)n}(x) - NK_{Nn}(x)$$
$$\Big(= \sum_{|m|\le Nn} e^{imx} + \sum_{Nn<|m|<(N+1)n} \frac{(N+1)n-|m|}{n} e^{imx} \Big),$$

The convolution with a function $f(x)$ reads

$$S_{Nn}(x) + \sum_{Nn<|m|<(N+1)n} \frac{(N+1)n-|m|}{n} c_m e^{imx}$$

and the last sum is controlled when we assume an extra condition on the coefficients. As a consequence, the statement of the Fejér theorem stays valid when Fejér sums are replaced by partial sums, under specific conditions on the coefficients. Hardy's condition is $c_m = O\big(\frac{1}{|m|}\big)(m \to \infty)$. The Fejér condition is

$$(F) \qquad \sum |m|\,|c_m|^2 < \infty.$$

Here is an application: if a function $f(z) = \sum_0^\infty c_m z^m$ is holomorphic in the disc $|z| < 1$ and continuous on $|z| \le 1$ and if the image of the disc has a finite area on the Riemann surface spanned by $f(z)$ (that is, (F) holds), then the Taylor series converges uniformly on the closed disc.

This is the case in particular when $f(x)$ yields a conformal mapping of the disc $|z| < 1$ on the interior of a simple Jordan curve. This theorem was used by Harald Bohr and Gyula Pál and later by Raphaël Salem in order to prove that for every **real** continuous function g on $\mathbb{T}$ there is a homeomorphism h of $\mathbb{T}$ such that the Fourier series of $g\big(f(t)\big)$ $(t \in \mathbb{T})$ converges uniformly {16}, {17}, {18}.

Only in the 1970's was the result extended to complex functions g, by purely real methods {19}, {20}.

Processes of summation raised a number of publications in relation with Dirichlet series and Fourier series between 1910 and 1940. One of the main contributors was Marcel Riesz. In particular, M. Riesz extended the theorem of Fejér by showing that it stays valid when the process of arithmetic means

(Cesàro process of order 1, or $(C, 1)$) is replaced by Cesàro process (C, α) of any positive order ($\alpha > 0$) {21} [158, pp. 62–64].

The effect of Fejér's theorem on the theory of Fourier series was instantaneous. Actually this effect did not decrease along time. Books on harmonic analysis or Fourier series give always Fejér's theorem a special status. In Katznelson's *Introduction to Harmonic Analysis* [86] the first chapter is devoted to Fejér's theorem in the frame of Banach-valued continuous functions when the Banach space under consideration is invariant under translation and the translation is continuous (a "homogeneous Banach space"). In the book of Zygmund *Trigonometric series* [203] only a very few statements are given their author's name: one of them is "The Theorem of Fejér".

2. The theorem of Riesz–Fischer

Another and even more important revolution occurred at the beginning of the century: it was the Lebesgue theory of measure and integral. Though the Lebesgue integral of bounded functions on a bounded interval was already introduced in 1901 {22} and Lebesgue extended it to unbounded functions in 1902 {23}, the first exposition of Lebesgue's integral as we know it now (for unbounded as well as bounded fonctions) appeared first in his 1906 book "Leçons sur les séries trigonométriques" {24}. At the same time, 1906, Fatou defended his thesis on "Séries trigonométriques et séries de Taylor" {25}.

From the very beginning there was a strong linkage between the Lebesgue integral and Fourier series.

However, the intimate relation between the two subjects appeared with the Riesz–Fischer theorem on Fourier coefficients of L^2 functions, that is, the isomorphism between $L^2(\mathbb{T})$ and $\ell^2(\mathbb{Z})$ through the Fourier formulas, called by Frigyes Riesz in a pleasant way "billet aller-retour permanent entre deux espaces à une infinité de dimensions" ([156, I, p. 327]).

The essential tool used by both authors in a more or less explicit way was the completeness of the space L^2. Actually the terms were not yet defined: a rather long sentence was needed in order to express that L^2 is complete. Now the theorem "L^p is complete" is a three-words statement, but the substance of the method elaborated independently by Frederic Riesz and Ernst Fischer is to be found in the definitions of a complete metric space and an L^p space $(1 \leq p < \infty)$. The superiority of Lebesgue's over Riemann's integral and its role to come in functional analysis relies on this three-words statement. The history of the Riesz–Fischer theorem deserves to be related. Here it is.

Frederic Riesz was visiting Göttingen at the beginning of the year 1907, while Ernst Fischer was in Brno (Brünn). Both of them, independently, discovered the following fact: given a sequence $(a_n) \in \ell^2$, there exists a function $f \in L^2$ whose Fourier coefficients are the a_n. Both of them stated the result for general orthonormal systems. Riesz communicated his result in a lecture at Göttingen on February 26, and Fischer in a lecture at Brünn on March 5. Riesz immediately published the theorem in two forms: an article in the Göttingen Nachrichten, presented by Hilbert on March 9, and a note in the Comptes-Rendus de l'Académie des Sciences de Paris, presented by Picard on March 11, published on March 18. The motivation of Riesz was the Hilbert treatment of linear integral equation by means of an orthonormal system. However he proved the theorem first for the trigonometric system, then extended it for an orthonormal system of functions defined on an interval, then to a general orthonormal system. The last step is realized in another note aux Comptes-Rendus, presented on April 2, published on April 8.

As soon as he read Riesz's note of March 18, Ernst Fischer reacted by sending his own contribution to the Comptes-Rendus. In a first note, dated May 13, he recognized the priority of Riesz in publishing the result, but pointed out that he had got it and lectured on it in Brno (Brünn) already on March 5. The result was the same but Fischer's approach was more direct and stated the main point (L^2 is complete) in an explicit (though not so concentrated) form. In a second note, dated May 27, Fischer introduced best approximation in L^2 and announced that he would develop this theory later in the frame of a kind of geometry of functions ("en m'appuyant sur une espèce de géométrie des fonctions").

It was Riesz's turn to react. The Comptes rendus of June 24 published a note of Frédéric Riesz with the title "Sur une espèce de géométrie analytique des systèmes de fonctions sommables". First, he reinforces his priority by mentioning his lecture at Göttingen on February 26. Then he enlarges the question. "After the fundamental result obtained by several geometers during the past last years and based for the main part on those of Mr Fejér, the idea of representing a function by its Fourier constants became very familiar. The set of the summable functions could be represented in this way on a subset of the space with infinitely many dimensions. What is this subspace? Until now we are not able to answer".

In modern notations, what can we say of $A(\mathbb{Z}) = FL^1(\mathbb{T})$? Then Riesz explains what happens when L^1 is replaced by L^2, namely $FL^2(\mathbb{T}) = \ell^2(\mathbb{Z})$. ({26}, {27}, {28}, {29}, {30}).

Frédéric Riesz went back to this subject in a later article when he extended Fischer's theorem (L^2 is complète) in the now classical form known as Riesz's theorem: L^p is complete ($p \geq 1$) ([156, I, C. 10, 441–489]). This is still more fundamental than the theorem of Riesz–Fischer and its goes far beyond harmonic analysis.

Fischer and Riesz worked independently and got the same result about the same time at the beginning of 1907. According to the alphabetical order, it would be justified, as some authors do (Nina Bari in her treatise on Trigonometric series), to say "the theorem of Fischer and Riesz". However most authors and textbooks (and Frédéric Riesz in the first place) call it the theorem of Riesz and Fischer. Anyhow it is the most important theorem relating functions and Fourier coefficients. Let us try to see it in its true perspective.

Given $f \in L^2(T)$ and its Fourier coefficients C_n, the formula

$$\sum |c_n|^2 = \int |f|^2,$$

called the Parseval formula, was known. Fatou had given the appropriate framework, namely the totality of the trigonometric system in L^2, and also credit to Parseval for the formula, which can be found in a memoir, dated 1806, at a time when nobody was able to really prove anything like totality of the trigonometric system. Prior to Fatou, the Parseval formula was called by Hurwitz "Fundamentalsatz der Fourierschen Konstanten" and Fischer had just published "Zwei neue Beweise für der Fundamentalsatz der Fourierschen Konstanten", but the proofs were given only for bounded integrable functions. Fatou was very near to the Fischer–Riesz theorem, but he missed it ({25}, {31}, {32}).

The Parseval formula can be extended to the case of other exponents in the form of two inequalities: when $p \leq 2$ and $\frac{1}{p} + \frac{1}{q} = 1$,

$$\left(\sum |c_n|^q\right)^{1/q} \leq \int (|f|^p)^{1/p}$$

and

$$\int \left(|f|^q\right)^{1/q} \leq \left(\sum |c_n|^p\right)^{1/p}.$$

These are the Hausdorff–Young inequalities (Young 1912 when q is an even integer, Hausdorff 1923 in the general case). Therefore

$$FL^p(\mathbb{T}) \subset \ell^q(\mathbb{Z}), \quad F\ell^p(\mathbb{Z}) \subset L^q(\mathbb{T}).$$

Frédéric Riesz generalized these inegalities by considering a bounded orthonormal system instead of the trigonometric system (1923). After a conversation with his collegue Alfréd Haar in Szeged he observed that his result can be expressed in a purely algebraic form, namely

$$\left(\sum_1^N |y_n|^q\right)^{\frac{1}{q}} \leq M^{(2-p)/p}\left(\sum_1^N |x_n|^p\right)^{\frac{1}{p}}$$

when $(y_1, y_2, \ldots, y_N)$ is obtained from $(x_1, x_2, \ldots, x_N)$ through an orthogonal matrix with entries majorized by M in absolute value. These inequalities are easy consequences of the convexity theorem of Marcel Riesz (1924), generalized with a simplified proof by his student G. O. Thorin in 1948, which now forms a part of the large theory of interpolation of linear operators ([158, C13], and other references in [203, chapter XII]).

The above inclusions, involving FL^p and $F\ell^p$, are strict except when $q = p = 2$. Frédéric Riesz pointed out in 1907 that no characterisation was known of $FL^1(\mathbb{T})$, in terms of usual sequence spaces. We know that all spaces $FL^p(\mathbb{T})$ $(1 \leq q \leq \infty)$ as well as $FC(\mathbb{T})$ (Fourier coefficients of continuous functions) are intrisically intricate except for the case $q = 2$. The same is true for the spaces $F\ell^p(\mathbb{Z})$. Actually I wrote a whole book on $F\ell^1(\mathbb{Z})$ [82] and it is an inexhaustible subject.

Let me relate a personal experience. When I got interested in the subject it was not clear whether or not $c \in FL^1(\mathbb{T})$ implies $|c| \in FL^1(\mathbb{T})$, and $f \in F\ell^1(\mathbb{Z})$ implies $|f| \in F\ell^1(\mathbb{Z})$. Now we know, by Katznelson's theorem {33}, that only analytic function operate, and consequently the answer is negative. The first proof of this was given by means of explicit constructions, namely a positive function on $(0, \pi)$ such that its expansion as a sine series is absolutely convergent and its expansion as a cosine series is not absolutely convergent, and a positive sequence (a_n) such that $\sum a_n \sin nx$ is a Fourier–Lebesgue series and $\sum a_n \cos nx$ is not. And I was led to such constructions by an article of Béla Sz. Nagy in which he proved the following theorem: given a positive and decreasing function g on $(0,\pi)$, and

$$a_n = \frac{2}{\pi}\int_0^\pi g(x)\cos nx\,dx, \quad b_n = \frac{2}{\pi}\int_0^\pi g(x)\sin nx\,dx$$

(assuming $g \in L^1$ for the computation of the a_n, and $xg \in L^1$ for the b_n), then, given $0<\gamma<1$,

$$x^{\gamma-1}g(x) \in L^1(0,\pi) \Longleftrightarrow \sum_1^\infty n^{-\gamma}|a_n| < \infty \Longleftrightarrow \sum_1^\infty n^{-\gamma}|b_n| < \infty,$$

while, for $\gamma = 1$,

$$g \in L^1(0,\pi) \Longleftrightarrow \sum_1^\infty |a_n| < \infty,$$

$$g(x)\log x \in L^1(0,\pi) \Longleftrightarrow \sum_1^\infty |b_n| < \infty \qquad \{34\}.$$

This nice theorem extends results of Young, Pólya and Zygmund concerning the case $\gamma = 1$ (Young and Pólya are mentioned above), and actually I got aware of these previous results through the article of B. Sz.-Nagy.

I shall not dwell on $FL^1(\mathbb{T})$ (the question raised by F. Riesz) nor on $FC(\mathbb{T})$. However, I have to explain their relation with the Riesz–Fischer theorem.

If $\sum|c_n|^2 < \infty$ $(n \in \mathbb{Z})$, then $\sum c_n e^{inx}$ is a Fourier–Lebesgue series. Is there a better condition on the modulus $|c_n|$ with the same conclusion? The answer is negative, and the clearest proof is given by a theorem of Paley and Zygmund on random trigonometric series: if $\sum|c_n|^2 = \infty$, almost surely $\sum \pm c_n e^{inx}$ is not a Fourier–Lebesgue series; on the other hand, if $\sum|c_n|^2 < \infty$, the random series represents almost surely a function belonging to all $L^p(\mathbb{T})$ $(1 \le p < \infty$; but **not** $L^\infty)$.

If $\sum c_n e^{inx}$ $(n \in \mathbb{Z})$ represents a continuous function, then $\sum|c_n|^2 < \infty$ (Parseval). Is there a better conclusion? The answer is negative if the following sense: given any $(d_n) \in \ell^2(\mathbb{Z})$ there exist (c_n) such that $|c_n| \ge |d_n|$ and $\sum c_n e^{inx}$ represents a continuous function. The original proof by De Leeuw, Kahane and Katznelson used randomization; another proof by Nazarov avoids it {35}, {36}.

The question raised by F. Riesz in 1907 has a rather disappointing answer not only for $L^1(\mathbb{Z})$, but for $C(\mathbb{T})$ and all $L^p(\mathbb{T})$ except $L^2(\mathbb{T})$: the spaces of Fourier coefficients are not easily described and in particular they are not stable by the operation of taking absolute values. Simple characterizations of Fourier coefficients exist only for very few classes of functions: analytic, C^∞, Sobolev spaces, Schwartz distributions.

The success of the modern theory of wavelets comes from the fact that many properties of functions can be recognized easily on the coefficients of their wavelet expansion. There is a simple description of many function spaces in terms of their wavelet transforms. Let me recall that a wavelet system is an orthonormal system in $L^2(\mathbb{R})$ of the form

$$2^{\frac{j}{2}}\Psi(2^j x - k) \qquad (j \in \mathbb{Z}, k \in \mathbb{Z})$$

where Ψ is a "good" function, for example $\Psi \in S(\mathbb{R})$, the Schwartz space of C^∞ rapidly decreasing functions.

The wavelet theory is recent and explosive since 1985. However the paradigm of wavelets is rather old and it deserves attention. It is the Haar system, corresponding to $\Psi(x) = 1$ on $\left(0, \frac{1}{2}\right), \Psi(x) = -1$ on $\left(\frac{1}{2}, 0\right)$ and $\Psi(x) = 0$ elsewhere. Haar considered only the Haar wavelets on $(0,1)$ and had to add the constant 1 in order to have a complete orthonormal system in $L^2(0,1)$. The article of A. Haar was published by Mathematische Annalen in 1910, and his modern impact on harmonic analysis can not be overestimated ([62, 331–371]).

The history of wavelets is very interesting example of interactions between physicists, engineers, and mathematicians [83]. The Hungarian physicist Dénes Gábor, who was awarded the Nobel Prize, is one of the important figures of this history; his fundamental contribution can be found in his article "Theory of communication" (1946) {37}.

3. Boundary values of analytic functions

The Taylor series of an analytic function in the unit disc of the complex plane,

$$\sum_0^\infty c_n z^n$$

can be considered as a one-parameter family of trigonometric series by writing $z = re^{it}$ $(0 < r < 1,\ 0 \leq t \leq 2\pi)$. The real and imaginary parts of the Taylor series are conjugate trigonometric series.

Conversely, given two conjugate trigonometric series S and $\widetilde{S}$, $S + i\widetilde{S}$ can be written formally as

$$\sum_0^\infty c_n e^{int},$$

There is therefore an intimate relation between conjugate trigonometric series and boundary values of analytic functions inside the unit disc.

The subject was explored and developed after 1906 by Fatou, Young and Hardy. In the following decades three Hungarian names emerge: Frédéric and Marcel Riesz, and Gábor Szegő. The book of Henry Helson, Harmonic analysis [70] is in a large part devoted to their works and their consequences.

"Über (die) Randwerte einer analytischen Funktion" is the common title of three important papers of F. and M. Riesz (1916), G. Szegő (1921) and F. Riesz (1923) and also the matter of a joint article of F. Riesz and G. Szegő (1920) ([156, D 4]). The first paper is the only one that Frédéric and Marcel Riesz signed together ([156, D 1]). The second and the third come from an exchange of letters between F. Riesz and G. Szegő, exploited separately by both of them after their joint article ([173, 21–6], [156, D 1]). The notation H^p for Hardy spaces, that is, spaces of analytic functions in the unit disc such that

$$\int_0^{2\pi} \left| f(re^{it}) \right|^p dt = O(1) \qquad (r \to 1)$$

was introduced by F. Riesz in the third paper. I use $H^p(\mathbb{T})$ for the boundary value.

Helson explains the meaning of the F. and M. Riesz theorem in the following way:

1. "Theorem of F. and M. Riesz". If μ is a mesure of analytic type, meaning that its Fourier coefficients of order n vanish when $n < 0$, then μ is absolutely continuous. Equivalently, if F is analytic in the unit disc and $\int \left| F(re^{it}) \right| dt = O(1), F$ is the Poisson then integral of a function in $H^1(\mathbb{T})$ (here $H^1(\mathbb{T})$ denotes the subspace of $L^1(\mathbb{T})$ consisting of functions with vanishing coefficients of negative order).

2. If $f \in H^1(\mathbb{T})$ and f vanishes on a set of positive measure, then $f = 0$. What F. and M. Riesz stated and proved combines the two statements.

Let g be an analytic function in the disc, of bounded variation on $\mathbb{T}$. It was already known that g must be continuous. Thus g maps $\mathbb{T}$ into a continuous and rectifiable curve in the plane. The arc length along the curve determines a measure on the curve. The original theorem asserts that the image under g of a Lebesgue null set on $\mathbb{T}$ is a null set on the curve and vice-versa. This is the equivalent to statements 1 and 2.

The starting point of F. and M. Riesz was Fatou's theorem on boundary values of bounded analytic functions — each bounded analytic function in the unit disc has non-tangential limits almost everywhere on the circle.

The starting point of the article of G. Szegő and F. Riesz was the theorem of Fejér (1916) which states that each positive trigonometric polynomial on the circle is of the form $|a_0 + a_1 e^{it} + \cdots + a_m e^{imt}|^2$. Szegő showed that the square moduli of the functions in $H^2(\mathbb{T})$ are characterized by the property that they are positive and their logarithm is Lebesgue integrable (except for the zero function). Riesz extended this to all $H^p(\mathbb{T})$ by means the following decomposition theorem: every function in $H^p (p > 0)$ can be written in the form of a product gh, where h is bounded, g has no zero in the open unit disc, and $g \in H^p$.

This decomposition plays an essential role in the study of invariant subspaces of $H^2(\mathbb{T})$. According to Arne Beurling, g is an outer and h an inner function, if moreover the boundary values of h have modulus 1 almost everywhere. Outer functions satisfy

$$\int \log |g| = \log \left| \int g \right| > -\infty$$

and this characterizes them. Inner functions can, in turn, be represented as the products of a Blaschke factor and a singular factor:

$$h(z) = \prod \left(\frac{z - z_j}{1 - z\bar{z}_j} \right) \exp \left(- \int \frac{e^{it} + z}{e^{it} - z} d\mu(t) \right)$$

where the z_j are inside the unit disc and $\Pi |z_j| > 0$, and $d\mu$ is a positive measure on $\mathbb{T}$. Beurling's theorem, generalized by Helson, is as follows. Let M be an invariant subspace of $L^2(\mathbb{T})$ ($e^{it} M \subset M$). Either M consists precisely of all functions of $L^2(\mathbb{T})$ that are supported on some measurable subset of $\mathbb{T}$, or it is $qH^2(\mathbb{T})$ for some inner function q. (The first kind of subspace is a Wiener subspace, the second a Beurling subspace). It is worth mentioning that already in 1907 Carathéodory and Fejér introduced the so-called Blaschke factors, in the case of a finite number of zeros z_j, as solutions of an extremal problem ([40, I, 19]).

An earlier theorem of Szegő (1920–21) answers an extremal problem, namely to compute the infimum of $\int |1+P|^2\omega$ when ω is a non negative integrable function on $\mathbb{T}$, and P ranges over all trigonometric polynomials of the form $P(x) = a_1 e^{ix} + \cdots + a_n e^{inx}$. The Szegő theorem is

$$\inf_P \int |1+P|^2\omega = \exp \int \log\omega.$$

The interpretation when $\log\omega$ is not integrable, that is, the right hand side is zero, is obvious but very important, and it is of constant use in prediction theory ([173, 20–3, 21–1]).

The subject of H^p spaces was renewed in the 60s and 70s from the probabilistic point of view by Austin, Gundy, Burkholder and Silverstein. The H^p martingales associated with Brownian motion give an elegant way to recover the classical result on H^p spaces of analytic functions. Boundary values of analytic functions is both an old and a modern subject {38}, {39}, {40}, {41}, {42}, {43}, {44}.

Together with H^p spaces of analytic functions, the subject of conjugate fonctions was in the air in the 1920s. Is $L^1(\mathbb{T})$ stable under conjugacy? The answer is easily seen to be negative (consider a cosine series with convex coefficients) and Kolmogorov in 1924 {45} and Zygmund in 1929 {46} gave substitutes involving either "weak L^1" or the space $L\log L$. Is $L^p(\mathbb{T})$ stable under conjugacy when $1 < p < \infty$? Now the answer is positive and it is due to Marcel Riesz (a Comptes rendus note in 1924 and a developed article in 1927) ([158, 29, 33]). The theorem is also expressed by inequalities

$$\int |\widetilde{f}|^p \leq A_p \int |f|^p.$$

It is a deep result, and many proofs of it were given (a pretty proof was published in 1990 in the Comptes rendus by S. Pichorides {47}). Marcel Riesz himself gave two proofs. One is long and direct, the other applies the convexity theorem that he found exactly at that time (1927) ([158, 32]) as a tool for interpolating the inequality between $p = 2$ (obvious case) and $p =$ even integer (amenable case).

The convexity theorem, in the form given by Thorin (1948) is still more important than the Hausdorff–Young and Marcel Riesz inequalities. But undoubtebly Marcel Riesz had these applications in mind when he discovered and proved his convexity theorem.

4. Riesz product and Sidon sets

As a consequence of the F. and M. Riesz theorem the following holds: let f be a function of bounded variation on the circle, together with $\widetilde{f}$, the conjugate function; then the Fourier coefficients of f are $o\left(\frac{1}{n}\right)(n \to \infty)$.

In 1918, two years after the communication of F. and M. Riesz at the fourth Congress of Scandinavian Mathematicians, Frédéric Riesz published a paper on the following question: given a **continuous** function of bounded variation in the circle, are its coefficients necessarily $o\left(\frac{1}{n}\right)$ $(n \to \infty)$?

He provided a negative answer with the example

$$f(x) = -x + \lim_{m\to\infty} \int_0^x (1+\cos x)(1+\cos 4x)\dots(1+\cos 4^{m-1}x)\,dx.$$

Nowadays we are more familiar with the language of measures and we can say that the derivative of $f(x)+x$ is the first example of a positive continuous measure whose Fourier coefficients do not tend to zero. It can be written as a "Riesz product"

$$\mu = \Pi_1^\infty (1+\cos 4^n x).$$

The partial products are positive and can be written in the form

$$1 + \sum \cos 4^n x + R,$$

R containing only frequencies of the form $\pm 4^{n_1} \pm 4^{n_2} \cdots \pm 4^{n_j}$ $(1 \le n_1 \le n_2 \le \cdots \le n_j,\ j \ge 2)$. Their L^1-norm is bounded (actually, constant). Taking a weak limit of a subsequence defines μ as a positive measure, with a Fourier expansion of the type just described: $\widehat{\mu}(4^n) = \frac{1}{2}$ $(n = 1, 2, \dots)$.

There are other ways to answer F. Riesz's question and they appeared in the 1920's in the investigations of A. Rajchman on sets of uniqueness. Maybe the simplest is to consider the probability measure carried by the standard triadic Cantor set, whose Fourier transform is

$$\widehat{v}(u) = \Pi_1^\infty \cos(2\pi 3^{-n} u);$$

it is easy to see that $\widehat{v}(3^m)$ tends to a non-zero limit as $m \to \infty$.

The construction of F. Riesz is versatile and proved very useful in harmonic analysis. Given a positive sequence of integers λ_n such that $\lambda_{n+1}/\lambda_n \ge q \ge 3$ $(n = 1, 2, \dots)$ a sequence (a_n) such that $0 < a_n \le 1$ and a real sequence φ_n, the Riesz product corresponding to these data is

$$R\big((\lambda_n), (a_n), (\varphi_n)\big) = \Pi_1^\infty \big(1 + a_n \cos(\lambda_n x + \varphi_n)\big).$$

It defines a positive and continuous measure whose Fourier expansion reads

$$1+\sum_{1}^{\infty} a_n \cos(\lambda_n x+\varphi_n)+R,$$

R containing only frequencies different from 0 and from the λ_n. Such a measure is either absolutely continuous (when $(a_n)\in \ell^2$) or purely singular (when $(a_n)\notin \ell^2$). When the sequence λ_n is fixed, the measure depends only on the sequence $c=(c_n)$ with $c_n=a_n e^{i\varphi_n}$ and it can be written as

$$\mu_c=\Pi_1^{\infty}\big(1+Re(c_n e^{i\lambda_n x})\big).$$

It is known that, given c and c', μ_c and $\mu_{c'}$ are either equivalent (same null-sets) or orthogonal (carried by disjoint Borel sets). When all c_n are in a compact subset of the open unit disc, the equivalence condition is $c-c'\in\ell^2$ (Peyrière, Brown and Moran). In the general case it is still unknown and it is an active field of research {48}, {49}, {50}, {51}.

The lacunary condition on the λ_n implies that the factors of the Riesz product are pretty independent. The study of Riesz products anticipated the theory of multiplicative processes and the random measures that they generate. Actually random Riesz products ((λ_n) and (a_n) fixed, φ_n random with the usual distribution) are parts of both theories {52}.

In 1924 Simon Sidon went back to the original question: given a continuous function of bounded variation on the circle, what can we say about its Fourier coefficients? Sidon proved that they satisfy the equivalent conditions $\lim_{n\to\infty}\frac{1}{n}\sum_1^n |ka_k|^2=0$ and $\lim_{n\to\infty}\frac{1}{n}\sum_1^n |ka_k|=0$.

Using an expression introduced by Mihály Fekete in 1916, the sequence (ka_k) "quasi-converges" to 0. Actually Sidon's condition are necessary and sufficient for the continuity of the function whose Fourier coefficients are a_k, when bounded variation is assumed; this is Wiener's condition for the continuity of measures. Though Wiener found it independently, part of the result belongs to Sidon.

Riesz products in the modern acceptation were introduced by Sidon in his study of lacunary trigonometric series. In the 1920's Hadamard lacunary trigonometric series, meaning series of the form

$$(*)\qquad \sum_{1}^{\infty} a_n \cos(\lambda_n x+\varphi_n),\quad \lambda_{n+1}/\lambda_n\ge q>1\quad (n=1,2,\dots),$$

became a popular subject, with contributions of Kolmogorov, Banach and Zygmund in particular. The contribution of Sidon proved crucial. First he proved that if $(*)$ is the Fourier series of a function bounded from above (or from below), then $\sum_1^\infty |a_n| < \infty$ (1927). Then he identified the sequences $(\hat{f}(\lambda_n))\,(f \in L^1(\mathbb{T}))$ with the sequences tending to zero; in brief,

$$FL^1 \mid \wedge = c_0(\wedge), \quad \wedge = \{\lambda_n\} \quad (1932).$$

The two results are strongly related to each other. Nowadays Sidon sets are defined as subsets $\wedge$ of $\mathbb{Z}$ satisfying one of the equivalent relations:

$$\sup_x \sum_{\lambda \in \wedge} Re(a_\lambda e^{i\lambda x}) \geq c \sum_{\lambda \in \wedge} |a_\lambda|$$

for some $c > 0$ and all finite sequences $(a_\lambda)(\lambda \in \wedge)$,

$$FL^1 \mid \wedge = c_0(\wedge)$$

already mentioned, and

$$FM \mid \wedge = \ell^\infty(\wedge),$$

where M is the space of complex measures on the circle. The relation of the last condition with the Riesz product is clear: if $\lambda = (\pm\lambda_n)$ with $\lambda_{n+1}/\lambda_n \geq q > 3$, given any sequence (c_n) in the unit disc, the measure that I denoted by μ_c satisfies $\widehat{\mu}_c(\lambda_n) = \frac{1}{2}c_n$. From this fact it is easy to deduce Sidon's theorems of 1927 and 1932.

Sidon sets were investigated vigorously since the end of the 1950's. A crucial step was accomplished in 1970 by S. Drury when he proved that the union of the two Sidon sets is a Sidon set. New characterizations were given by G. Pisier and J. Bourgain in the 1980's. Meaningfully the 1985 paper of Bourgain is entitled: "Sidon sets and Riesz products". The subject is a crossing point between harmonic analysis, combinatorics and probability theory {53}, {54}, {55}, {56}.

There are many other properties of Hadamard lacunary trigonometric series. Let me mention two of them, involving Hungarian mathematicians.

If the coefficients of a Hadamard lacunary trigonometric series are real and tend to zero, the series converges at some point. This is theorem of Zygmund. What can we say of the sequence (λ_n) $(1 \leq \lambda_1 < \lambda_2 \ldots)$ such that every series

$$\sum_1^\infty a_n \cos(\lambda_n x + \varphi_n), \quad \lim_{1 \to \infty} a_n = 0,$$

converges at some point x? This question was considered by Erdős in 1966, and the subject seems difficult. There are necessary conditions, and sufficient conditions that look like the necessary conditions, and sufficient conditions for $\wedge$ to be a Sidon set, but it is not known whether the Sidon implies the Zygmund properties, nor the reverse {57}.

Suppose now that $f(t)$ is a continuous function and

$$f(t) = \sum_{1}^{\infty} a_n \cos(\lambda_n x + \varphi_n), \quad (t \in \mathbb{R}),$$

(λ_n) being a Hadamard sequence. Then there is an intimate relation between the local properties of the function and the order of magnitude of the coefficients. Very precise results were obtained by Géza Freud in 1962 and 1966. Here are two of them {58}, {59}.

1. Assume $a_n = 1/\lambda_n$. Then

a) $f(t+h) - f(t) = O\left(|h| \log \frac{1}{|h|}\right)$ $(h \to 0)$ everywhere

b) $\overline{\lim\limits_{h \to 0}} \left(f(t+h) - f(t)\right) / \left(|h| \log \frac{1}{|h|}\right) > 0$ quasi everywhere

c) $\overline{\lim\limits_{h \to 0}} \left(f(t+h) - f(t)\right) / \left(|h| \sqrt{\log \frac{1}{|h|} \log\log \frac{1}{|h|}}\right) < \infty$ almost everywhere

d) $f(t+h) - f(t) = O(|h|)$ $(h \to 0)$ on a dense t-set

e) f is nowhere differentiable.

This last statement, e), is due to Hardy, and the function under consideration is called a Hardy–Weierstrass function. The theorem expresses that the run of the function is as fast as possible (taking a) into account) when t belongs to some set of the second category of Baire (it is the meaning of "quasi everywhere"), and that is as slow as possible (taking e) into account) on some dense set, with an average behaviour ("almost everywhere") inbetween. Such a behaviour was discovered later for the Brownian motion, with $|h|^{\frac{1}{2}}$ instead of $|h|$, and it is very much in the spirit of the modern investigations on multifractal analysis.

2. Assume

$$\int_{-\pi}^{\pi} \left| \frac{f(t+h) - f(t)}{h} \right| dh < \infty$$

for **some** t. Then $\sum |a_n| < \infty$.

Assume $f(t+h) - f(t) = O(|h|^\alpha)$ $(h \to 0)$ for **some** t $(0 < \alpha < 1)$. Then $a_n = O(\lambda_n^{-\alpha})$ $(n \to \infty)$ and $f \in \Lambda_\alpha$, the Lipschitz–Hölder class of order α.

The last statement was discovered independently by Masako Sato–Izumi and there are several variations about in a paper of Izumi–Izumi–Kahane {60}. It is one the very rare properties of Hadamard lacunary trigonometric series where the Hadamard condition proves necessary and sufficient.

5. Miscellaneous*

I insisted on the importance of the theorem that Fejér published in 1900. Fejér's works and personality go far beyong this first and essential result. On the other hand, Fourier analysis in Hungary involved many first-class mathematicians and many more topic than what I considered in the preceding sections. Let me try to fill some of the large gaps of this review.

Let me begin with a theorem of Ferenc Lukács (1919).

Suppose $f \in L^1(\mathbb{T})$ and $\int_0^h \left| f(x+t) - f(x-t) - D_x \right| dt = O(h)$. Then

$$D_x = -\lim_{n\to\infty} \pi \frac{\widetilde{S}_n(x)}{\log n}$$

where $\widetilde{S}_n(.)$ denotes the n-th partial sum of the conjugate Fourier series of f.

This is theorem B in the original paper of Lukács {61}. Theorem A is a particular case of special interest, when it is assumed that both $f(x+0)$ and $f(x-0)$ exist, and $D_x = f(x+0) - f(x-0)$. Theorem C is a consequence, namely that $\widetilde{S}_n(x) = o(\log n)$ $(n \to \infty)$. almost everywhere.

*Most of the references of this section are given in the text. A few others are listed at the References section of the paper.

The paper was published in 1920, but the actual date is 1918. F. Lukács was a young man, who died on November 30, 1918, and this is his legacy. It is worth seeing the context.

Theorem A answers in a complete way a question that Fejér asked in 1913: to determine the jump of a function at a point through its Fourier expansion. Fejér himself indicated different ways and gave a complete answer in the case of a function satisfying the Dirichlet conditions. Both Fejér's and Lukács papers are entitled: "Über die Bestimmung des Sprunges einer Funktion aus ihrer Fourierreihe" ([40, I p. 718], {61}). In between several articles appeared in Hungarian journals and Hungarian language, by Fejér again (1913), Pál Csillag (1918) and S. Sidon (1918) ([40, I p. 744], {62}, {63}). The formula of Lukács was discovered by Fejér when Dirichlet's conditions are satisfied, then proved by Csillag for functions of bounded variation, and reproved by Sidon in a simpler way for the same functions. Both Csillag and Sidon were stimulated by the questions F. Riesz asked at the end of this paper of 1918 when he introduced the Riesz product:

"Gibt es eine stetige, nach 2π periodische Funktion von beschränkter Schwankung, für welche die Folgen na_n, nb_n konvergieren und wenigstens einer der beiden Grenzwerte von Null verschieden ist? Gibt es eine unstetige Funktion von beschränkter Schwankung, für welche $a_n = o\big(\frac{1}{n}\big)$ und $b_n = o\big(\frac{1}{n}\big)$?" ($a_n$ et b_n are the cosine and sine coefficients). The formula answers both questions in a negative way.

Though quite different, Lukács's theorem originated from the same area as Riesz products and Sidon sets. Theorem A is well known, theorems B and C deserve to be known also.

The thesis of Marcel Riesz was published in 1910, in Hungarian, in the *Mathematikai és Physikai Lapok.* The main result was already published in the Comptes-rendus in October 1907 {64}, when M. Riesz was not yet 21. It answers a question asked by Fejér, namely, to give a condition on the coefficients of a trigonometric series $\sum(a_n \cos nx + b_n \sin nx)$ such that, if this condition is satisfied and if the arithmetic means $\sigma_n(x)$ of the partial sums converge to 0 everywhere, it is necessarily the null series. The condition given by M. Riesz is

$$\sum_1^\infty \frac{|a_n| + |b_n|}{n^2} < \infty.$$

Moreover, the assumptions $|a_n| + |b_n| = o(1)$ and $\sigma_n(x) = o(1)$ except on a countable compact (= reductible) set imply the same conclusion. As a test of the sharpness of the result one can consider the series

$$\sum_{2}^{\infty} \frac{n \sin nx}{\log n}, \quad \frac{1}{2} + \sum_{1}^{\infty} \cos nx$$

the first is $(C, 1)$ summable to 0 everywhere (last example in the thesis) and the second everywhere except 0 (mod 2π) (example given by Fejér). There are very good results but the subject was not explored later anymore neither by Riesz nor by other Hungarian mathematicians (see a comment on {64} in the references).

However, the subject belongs to an important stream of real and harmonic analysis, the Riemann theory of trigonometric series and the Cantor theory of uniqueness, and it was cultived later by Russian and Polish mathematicians, Menšov, Bari, Rajchman and Zygmund. Moreover, the thesis as a whole is a masterpiece of exposition of old and new ideas, not unlike the historical models, Riemann's and Cantor's theses on trigonometric series. The thesis of M. Riesz was translated into English by J. Horváth and published in English in the Collected papers of M. Riesz in 1988. It is available now to a large international public and it is a delightful piece of reading. It connects the Riemann process of summation, the sets of uniqueness introduced by Cantor, the process of summation of Fejér, the Lebesgue integral, the summability processes used in Taylor series and a critical examination of Hadamard's statement on analytic continuation, with also a series of examples among which I have mentioned just one [158].

A permanent question of interest in Fourier analysis is the relation between properties of functions and behaviour of Fourier coefficients. For instance, Marcel Riesz in his thesis asked for a characterisation of Fourier coefficients of continuous functions (in a simple form which involves only the coefficients and not a variable x); the specific question is still unsolved and very likely unsolvable. We already discussed the question after the Riesz–Fischer theorem. Let us write again c_m $(m \in \mathbb{Z})$ for the Fourier coefficients of a function f. A theorem of S. Bernstein (1914) asserts that the assumption $f \in \operatorname{Lip} \alpha$, $\alpha > \frac{1}{2}$ implies $\sum |c_n| < \infty$. Here $f \in \operatorname{Lip} \alpha$ means

$$\sup_{x,h} \left(|h|^{-\alpha} \big| f(x+h) - f(x) \big| \right) < \infty.$$

The subject was renewed by Ottó Szász (see [172]), then, using Szász's results, by Hardy and Littlewood (1928). Here is the first result of Szász

(1922): the assumption

$$(L_2) \qquad \int_0^{2\pi} \left| f(x+2t) + f(x-2t) - 2f(x) \right|^2 dx = O(t^{2\alpha}) \quad (t \downarrow 0)$$

with $0 < \alpha < 1$, implies $\sum |c_m|^k < \infty$ when $k > \frac{2}{2\alpha+1}$, but not when $k = \frac{2}{2\alpha+1}$. The best way to see it is to observe that

$$(L_2) \Longleftrightarrow \sum_{|m| \le n} m^4 |c_m|^2 = O(n^{4-2\alpha}) \quad (n \to \infty)$$

as Szász does at the end of this 1928 paper. The latter is devoted to the consequences of a more general assumption, namely

$$(L_p) \qquad \int_0^{2\pi} \left| f(x+2t) + f(x-2t) - 2f(x) \right|^p dx = O(t^{p\alpha}) \quad (t \downarrow 0)$$

with $p > 1$ and $-\frac{1}{p} < \alpha < 1$. In particular, when $1 < p \le 2$ and $0 < \alpha < 1$, L_p implies $\sum |c_m|^k < \infty$ when $k > \frac{p}{p(1+\alpha)-1}$, but not necessarily when $k = \frac{p}{p(1+\alpha)-1}$. The same statement appears as Theorem 8 in the article of Hardy and Littlewood devoted to the Lip (α, p) classes (essentially, classes of functions that satisfy condition (L_p)), with reference to the results and observations of Szász (see collected papers of G. H. Hardy, vol III, pp. 631–632).

Late in life (1948) L. Fejér gave an expository talk on singular and positive kernels, and he made use of a pleasant image: at a first look, "elles se ressemblent comme les pingouins", they look all the same, like penguins ([40, II, pp. 750]). Of course, Fejér then shows how different these creatures can be. It is worth visiting some pieces of the Fejér collection of positive trigonometric polynomials, the Fejér personal zoo. The favorite, born in 1900, is the Fejér kernel

$$K_n(x) = 1 + 2\left(1 - \frac{1}{n}\right)\cos x + \cdots + \frac{2}{n}\cos(n-1)x.$$

Then comes the integrated Dirichlet kernel (divided by 2 as usual)

$$F_n(x) = \sin x + \frac{1}{2}\sin 2x + \cdots + \frac{1}{n}\sin nx,$$

positive on $]0,\pi[$ and uniformly bounded with respect to x and n, that serves as a basic cell in order to construct continuous functions whose Fourier series diverge at 0 ([40, I p. 258], 1909). Not far from those we meet the Lukács polynomials

$$L_n(x) = \sin x + \left(1 - \frac{1}{n}\right)\sin 2x + \cdots + \frac{1}{n}\sin nx,$$

also positive on $]0,\pi[$ ([40, II. p. 230]), which a little more should be said. As we saw, Lukács died in 1918, and he never published that result, but he mentioned it to Fejér, with a direct proof. Fejér included this "Satz von Lukács" in an extended article in 1928, with a new and interesting proof, using generating functions of the form

$$\sum_{n=1}^{\infty} \frac{nL_n(x)}{\sin x} r^{n-1} \left(= \frac{1}{(1-r)^2}\frac{1}{1-2r\cos x + r^2}\right).$$

The method of generating functions was used again in order to derive properties of the Cesàro means of order > 1 of the Dirichlet kernels, as presented in the 1948 survey ([40, II p. 742–4]). The most striking result, going back to 1932, is that the Cesàro means of order 3

$$K_n^{(3)}(x) = \binom{n+3}{3}^{-1}\left(\binom{n+3}{3} + 2\binom{n+2}{3}\cos x + 2\binom{n+1}{3}\cos 2x + \cdots + 2\binom{3}{3}\cos nx\right)$$

are not only positive on $]0,\pi[$, but also decreasing on that interval. The zoo contains other interesting creatures, but I'll stop the visit here.

The properties of $K_n^{(3)}(x)$ were used immediately by Fejér in order to compare the graph of a function and those of the Cesàro means of order 3 of its Fourier series, that is, the convolutions

$$S_n^{(3)} = f * K_n^{(3)}.$$

For example, if f is a continuous, odd, and concave on $[0,\pi]$, the same holds for $S_n^{(3)}$ ([40, II p. 480], 1933). After Fejér, the comparison between the graph of a function and those of approximating polynomials became a Hungarian speciality, with important contribution of Pál Turán and Pólya in particular. Here are some of them.

Though different shapes of functions were considered, let me stick to functions f of the above type (odd, $f(0) = f(\pi) = 0$, concave on $[0,\pi]$). In

1935, Turán proved a spectacular result for the Cesàro means $S_n^{(k)}$ of these functions, namely

$$f(x) \geq S_n^{(1)}(x) \geq S_n^{(2)}(x) \geq S_n^{(3)}(x) \ldots (0 \leq x \leq \pi),$$

implying that the best approximation is provided by the $S_n^{(1)}$, the Fejér sums (*Journal of the London Mathematical Society,* **10** (1935), pp. 277–280). For such functions again, Turán proved in 1938 that the pointwise approximation by the $S_n^{(k)}$ improves when n increases:

$$f(x) \geq S_{n+1}^{(k)}(x) \geq S_n^{(k)}(x) \quad (0 \leq x \leq \pi;\ k \geq 2,\ n \geq 0),$$

(Theorem 3, *Proceedings of the Cambridge Philosophical Society,* **34** (1938), pp. 134–143); the same holds when $k = 1$ and f satisfies the symmetry condition.

$$f(\pi - x) = f(x) \quad (0 \leq x \leq \pi)$$

(Theorem 4, ibidem). The situation is different for ordinary partial sums $S_n^{(0)}$ which are positive on $]0, \pi[$ (L. Koschmieder, *Monatshefte für Mathematik und Physik,* **39** (1932), pp. 321–344). What Turán proved is

$$0 \leq S_n^{(0)}(x) \leq 2f(x) \quad (0 \leq x \leq \pi)$$

and 2 is the best constant (*Journal of the London Mathematical Society,* **13** (1938), pp. 278–282).

In 1958, G. Pólya and I. I. Schoenberg returned to the general topic introduced by Fejér: comparing the shapes of approximating trigonometric polynomials and the shape of the function. They considered the de la Vallée Poussin kernels

$$V_n(x) = c_n \left(\cos \frac{x}{2}\right)^{2n} \quad \left(\int_{-\pi}^{\pi} V_n(x) \frac{dx}{2\pi} = 1\right)$$

and the convolutions $f * V_n$. They observed that, compared to the Cesàro means $S_n^{(k)}$, "the de la Vallée Poussin means possess such shape-preserving properties to a much higher degree, thanks to their variation diminishing character" (*Pacific Journal of Mathematics,* **8** (1958), pp. 295–334). Among the numerous results of Pólya and Schoenberg, one is related to the simple situation mentioned above (odd functions, concave on $[0, \pi]$); then the de la Vallée Poussin sums enjoy the property that Fejér pointed out for $S_n^{(3)}$, namely

$$0 \leq (f * V_n)(x) \leq f(x) \quad (0 \leq x \leq \pi)$$

and $f * V_n$ is concave on $[0, \pi]$. For general oscillating shapes, they were able to prove results of the same nature.

At the end of the thirties multiple Fourier series became a topic of interest, with new techniques developed by Marcinkiewicz and Zygmund {65}. The history involves Fejér again. Going back to 1913, Hardy and Littlewood proved that

$$(*) \qquad \lim_{n\to\infty} \frac{1}{n+1}\left(S_0^2(x) + S_1^2(x) + \cdots + S_n^2(x)\right) = f^2(x)$$

when $f \in L^1(T)$ and f is continuous at x, $S_n (= S_n^{(0)})$ denoting the n-th partial sum of the Fourier series. As a consequence

$$\lim_{n\to\infty} \frac{1}{n+1}\left(S_0(x) - f(x)\right)^2 + \cdots + \left(S_n(x) - f(x)\right)^2 = 0$$

and

$$\lim_{n\to\infty} \frac{1}{n+1}\left(\left| S_0(x) - f(x)\right| + \cdots + \left| S_n(x) - f(x)\right|\right) = 0$$

the "strong summability" theorem. The proof of $(*)$ is not easy. Fejér had the idea of deriving it from a statement on double Fourier series ([40, II p. 725], 1938): if $f \in L^1(\mathbb{T})^2$ and f is continuous at (x, y) the "square partial sums" $S_{n,n}(x, y)$ converge to $f(x, y)$ through the Cesàro process of order 3 $(C, 3)$. This relies on the positivity of the corresponding kernel, one more creature of the Fejér zoo. Actually Géza Grünwald extended the result by proving $(C, 1)$ summability almost everywhere, without the continuity condition (*Proceedings of the Cambridge Philosophical Society,* **35** (1939), pp. 343–350) and it was extended later to (C, α) summability $(\alpha > 0)$ by J. Herriot (*Transactions of the American Mathematical Society,* **53** (1952)).

The strong summability theorem of Hardy and Littlewood can be applied in order to give conditions for a cosine series to be a Fourier–Lebesgue series, as the Fejér theorem is applied in order to get the Young convexity condition mentionel in section 1. This was done by S. Sidon (Hinreichende Bedingungen für den Fourier Charakter einer trigonometrische Reihe, *Journal London Math. Soc.,* **14** (1939), 158–160) and generalized by S. A. Teljakowskii (On a sufficient condition of Sidon for integrability of trigonometric series (in Russian), *Mat. Zametki,* **14** (1973), 317–328) and more recently by F. Móricz (Sidon type inequalities, *Publ. Inst. Math. Szeged,* **48** (1990), 101–109).

Summability processes can be considered from a quantative point of view: given a process and a function space, how far does the process converge to functions that belong to the space? Given a process, is there an optimal speed of convergence, and a maximal function space for which the process converges that fast? The last question was asked by Jean Favard in 1947 (Colloque d' Analyse Harmonique, CNRS, Nancy); the maximum function space is called the "saturation class" of the process. The first question goes back to the beginning of the century and the works of S. Bernstein, Dunham Jackson and others.

Both questions were investigated in Hungary. The first main contribution is due to Béla Sz. Nagy (Sur une classe générale de procédés de sommation pour les séries de Fourier, *Hungarica Acta Mathematica,* **1**, 3 (1948), 14–52). For quite general processes of summation Sz. Nagy obtained very precise estimates on the order of approximation on one hand, on "Lebesgue constants" on the other. The particular processes given by fractional integration were investigated by M. Mikolás and related to some one-sided generalized limits, denoted by $f\langle x \pm 0\rangle$ (Application d'une nouvelle méthode de sommation aux séries trigonométriques et de Dirichlet, *Acta. Math. Acad. Sci. Hungar.*, **11** (1960), 317–344). For the strong summability process $h(p)$ $(p \leq 1)$ which associates with a given function f and its Fourier sums $S_k(f,.)$ the pointwise approximations

$$h_n(f,x,p) = \frac{1}{n+1}\left(\sum_{k=0}^{n} |\, f(x) - s_k(f,x)|^p\right)^{\frac{1}{p}},$$

the pioniering work was that of György Alexits (*Acta. Sci. Math. Szeged,* **26** (1965), 211–224): if f belongs to the Hölder class $\wedge_\alpha$ and $\alpha < \frac{1}{p}$, then $h_n(f,x,p) = O(n^{-\alpha})$ $(n \to \infty)$ uniformly with respect to x, and that is false for $\alpha = \frac{1}{p}$. In 1969, Géza Freud investigated the boundary case and the saturation problem, and obtained very precise results: given $h(p)$, the best speed of convergence is $n^{-1/p}$, that is, $h_n(f,x,p) = O(n^{-1/p})$ uniformly in x when $f \in X_p$, the saturation class (containing non-constant functions), while $h_n(f,x,p) = o(n^{-1/p})$ implies that f is a constant. The space X_p is included in $\wedge_{\frac{1}{p}}$ and consists of functions f satisfying $f(x+h) - f(x) = o\big(|h|^{1/p}\big)$ a.e. The space X_2 is defined by

$$X_2 = \left\{ f \, : \, \int \left| \frac{f(x+t) + f(x-t) - 2f(x)}{t} \right|^2 dt < \infty \right\}$$

(*Acta Math. Acad. Sci. Hungar.*, **20** (1969), pp. 275–279). More information about the topic can be found in the article of L. Leindler in *Journal of Approximation Theory,* **46** (1986) pp. 58–64.

Let us turn back to the favorite animals of Fejér, the positive (≥ 0) trigonometric polynomials. The fundamental theorem expresses that they are nothing but the square moduli of polynomials in e^{ix}:

$$f(x) = \left|P(e^{ix})\right|^2.$$

It was guessed by Fejér, proved by F. Riesz, proved again in several ways by J. Egerváry and Fejér himself (see [40, I. p. 843], 1915). From this representation Fejér derived a number of inequalities. Here are the simplest. If

$$f(x) = 1 + \lambda_1 \cos x + \mu_1 \sin x + \cdots + \lambda_n \cos nx + \mu_n \sin nx \geq 0,$$

then $f(0) \leq n+1$, with equality if and only if f is the Fejér kernel of order n. If moreover f is even ($\mu_1 = \cdots = \mu_n = 0$), then

$$|\lambda_1| \leq 2\cos\frac{\pi}{n+2}$$

([40, I. p. 869]). This last inequality plays a role in the modern theory of operators in Hilbert spaces. It can be translated into the following statement (V. Haagerup, P. de la Harpe, Procedings of the *American Mathematical Society,* **115** (1992), pp. 371–379). If T is a linear operator in a Hilbert space H such that $\|T\| \leq 1$ and $T^n = 0$, its "numerical radius"

$$\omega_2(T) = \sup\left\{\left|\langle Tx, x\rangle\right. : x \in H, \|x\| = 1\right\}$$

satisfies $\omega_2(T) \leq \cos\frac{\pi}{n-1}$. Inequalities of this type are a topic of current interest (see for example C. Badea and G. Cassier, Constrained von Neumann Inequalities, *Adv. Math.*, **166** (2002), 260–297).

After such a random walk through the legacy of Fejér a natural question is: why stop here? We visited only a few spots, leaving aside many more places of interest. Instead of filling the gaps between the first sections of this survey, I created more and more gaps — as it is natural for a plane Brownian motion. The best I can do is to invite the reader to a promenade inside journals and books, to discover by himself the Hungarian jewels in harmonic analysis, of which I uncovered only a small part.

References

[40] Fejér, Lipót, *Összegyűjtött Munkái = Gesammelte Arbeiten,* ed. Pál Turán, Akadémiai Kiadó (Budapest, 1970).

[62] Haar, Alfréd, *Összegyűjtött Munkái = Gesammelte Arbeiten,* ed. Béla Sz.-Nagy, Akadémiai Kiadó (Budapest, 1959).

[70] Helson, Henry, *Harmonic Analysis,* Addison-Wesley (Reading, MA, 1983).

[82] Kahane, Jean-Pierre, *Séries de Fourier absolument convergentes,* Ergebnisse der Mathematik und Ihrer Grenzgebiete, Band 50, Springer-Verlag (Berlin–Heidelberg–New York, 1970).

[83] Kahane, Jean-Pierre and Lemarié-Rieusset, Pierre Gilles, *Séries de Fourier et Ondelettes,* Cassini (Paris, 1998). *Fourier Series and Wavelets,* Gordon and Breach (New York, 1995).

[86] Katznelson, Yitzhak, *An Introduction to Harmonic Analysis,* Dover (New York, 1976).

[156] Riesz, Frigyes, *Összegyűjtött Munkái = Œuvres complètes = Gesammelte Arbeiten,* ed. Á. Császár, Akadémiai Kiadó (Budapest, 1960).

[158] Riesz, Marcel, *Collected Papers,* ed. L. Gårding and L. Hörmander, Springer-Verlag (Berlin–Heidelberg–New York, 1988).

[172] Szász, Ottó, *Collected Mathematical Papers,* ed. H. D. Lipsich, University of Cincinnati (1955).

[173] Szegő, Gábor, *Collected Papers,* ed. R. Askey, Birkhäuser (Boston–Basel–Stuttgart, 1982).

[203] Zygmund, Antoni, *Trigonometric Series,* Cambridge University Press (London–New York, 1959).

References, Section 1

{1} A. Hurwitz, Über die Fouriershen Konstanten integrierbarer Funktionen, *Math. Ann.,* **57** (1903), 425–446 = Math. Werke, Band I, S. 555–576, Basel (Birkäuser), 1932.

{2} H. Lebesgue, Sur une condition de convergence des séries de Fourier, *Comptes Rendus Acad. Sci. Paris,* **140** (1905), 1378–1381 = Oeuvres sci., vol III, p. 95–98, Genève (Enseign. Math.), 1972.

{3} H. Lebesgue, Recherche sur la convergence des séries de Fourier, *Math. Ann.,* **61** (1905), 251–280 = Oeuvres sci., vol III, p. 181–210, Genève (Enseign. Math.), 1972.

{4} C. de la Vallée Poussin, Sur l'approximation des fonctions d'une variable réelle et de leurs dérivées par des polynômes et des suites limitées de Fourier, *Acad. Royale Belgique, Bull. Classe Sci.,* 1908, 193–254.

{5} G. H. Hardy, *On the Summability of Fourier's series*, Proceed. London Math. Soc., (2), 12 (1913), 365-372 = Collected papers, vol. III p. 118-125, Oxford (Clarendon Press), 1969.

{6} G. H. Hardy, Sur la série de Fourier d'une fonction à carré sommable, *Comptes Rendus Acad. Sci. Paris,* **156** (1913) 1307–1309 = G. H. Hardy, Collected papers, vol. III, p. 118–125, Oxford (Clarendon Press), 1969.

{7} C. de la Vallée Poussin, *Cours d'analyse infinitésimale*, Louvain (Oystruyst–Dieudonné)

{8} L. Carleson, On Convergence and Growth of Partial Sums of Fourier Series, *Acta. Math.,* **116** (1966), 135–157.

{9} E. Picard, *Traité d'analyse,* t. I, Paris (Gauthier-Villars), 1891.

{10} C. de la Vallée Poussin, Sur quelques applications de l'intégrale de Poisson, *Ann. Soc. Sci. Bruxelles,* **17** (1893), B, 18–34.

{11} A. Hurwitz, Sur les séries de Fourier, *Comptes Rendus Acad. Sci. Paris,* **132** (1901), 1473–1475 = Math. Werke, Band I, p. 506–508, Basel (Birkhäuser), 1932.

{12} W. H. Young, On the Fourier series of bounded functions, *Proc. London Math. Soc.,* **12** (1913), 41–70.

{13} G. Pólya, Remarks on characteristic functions, *Proc. Berkeley Symp. Math. Statistics and Probability,* Berkeley, 1949, 115–123.

{14} G. H. Hardy, Summability and convergence of slowly oscillating series, *Proc. London Math. Soc. (2),* **8** (1910), 301–320.

{15} C. de la Vallée Poussin, *Leçons sur l'approximation des fonctions d'une variable réelle,* Paris (Gauthier-Villars), 1919.

{16} J. Pál, Sur des transformations de fonctions qui font converger leurs séries de Fourier, *Comptes Rendus Acad. Sci. Paris,* **158** (1914), 101–103.

{17} H. Bohr, Über einen Satz von J. Pál, *Acta Univ. Szeged, Sect. Sci. Math.,* **7** (1934–1935), 129–135 = Collected Math. Works, vol. III, E. 13, Kobenhavn (Dansk Mat. Forening), 1952.

{18} R. Salem, On a theorem of Bohr and Pál, *Bull. Amer. Math. Soc.,* **50** (1944), 579–580) = Oeuvres Math., p. 334–335, Références et errata, p. 5, Paris (Hermann), 1967.

{19} J.-P. Kahane et Y. Katznelson, *Séries de Fourier des fonctions bornées, Study in Pure Mathematics, to the memory of Paul Turan,* Basel, Académie des Sciences de Hongrie (1983), 395–410.

{20} A. A. Saakian, Integral modulus of continuity and Fourier coefficients under superposition of functions, *Mat. Sbornik,* **110** (152) (1979), 597–608.

{21} M. Riesz, Sur les séries de Dirichlet et les séries entières, *Comptes Rendus,* **149** (1909), 309–312.

References, Section 2

{22} H. Lebesgue, Sur une généralisation de l'intégrale définie, *C.R. Acad. Sc. Paris* (1901).

{23} H. Lebesgue, *Intégrale, longueur, aire,* Thèse (1902).

{24} H. Lebesgue, *Leçons sur les séries trigonométriques,* Paris, Gauthier-Villars (1906).

{25} P. Fatou, Séries trigonométriques et séries de Taylor, *Acta Math.,* **30** (1906), 335–400.

{26} F. Riesz, Sur les systèmes orthogonaux de fonctions, *C.R.A.S. Paris,* **144** (1907), 615–619 (Oeuvres C2).

{27} E. Fischer, Sur la convergence en moyenne, *C.R.A.S. Paris,* **144** (1907), 1022–1024.

{28} F. Riesz, Sur les systèmes orthogonaux de fonctions et l'équation de Fredholm, *C.R.A.S. Paris,* **144** (1907), 734–736, (oeuvres C3).

{29} E. Fischer, Application d'un théoreme sur la convergence en moyenne, *C.R.A.S. Paris,* **144** (1907), 1148–1151.

{30} F. Riesz, Sur une espèce de géométrie analytique des systèmes de fonctions sommables, *C.R.A.S. Paris,* **144** (1907), 1409–1411 (Oeuvres C4).

{31} A. Hurwitz, Über die Fourierschen Konstanten integrierbarer Funktionen, *Mathematische Annalen* **57** (1903), 425–446.

{32} E. Fischer, Zwei neue Beweise für der Fundamentalsatz der Fourierschen Konstanten, *Monatshefte für Mathematik und Physik,* **15** (1904), 69–92.

{33} Y. Katznelson, Sur les fonctions opérant sur l'algèbre des séries de Fourier absolument convergentes, *C.R.A.S. Paris,* **247** (1958), 404–406.

{34} B. Sz.-Nagy, Séries et intégrales des fonctions monotones non bornées, *Acta. Sci. Math. Szeged,* **13** (1949), 118–135.

{35} K. de Leeuw, J.-P. Kahane et Y. Katznelson, Sur les coefficients de Fourier des fonctions continues, *C.R.A.S. Paris,* **285** (1978), 1001–1003.

{36} F. Nazarov, The Bang solution of the coefficient problem, *Algebra i Analisis* (in Russian), **9** (1997), 2, *St. Petersburg Math. J.,* **9** (1998), 2, 407–419 (In English).

{37} D. Gabor, Theory of communication, *J. Inst. Elect. Engng.* **93** (1946), 249–457.

References, Section 3

{38} D. G. Austin, A sample function property of martingales, *Ann. Math. Statis.,* **37** (1966), 1396–1397.

{39} D. L. Burkholder, Martingales Transforms, *Ann. Math. Statis.,* **37** (1966), 1494–1504.

{40} D. L. Burkholder and R. F. Gundy, Extrapolation and interpolation of quasi-linear operators on martingales, *Acta. Math.*, **124** (1970), 249–304.

{41} D. L. Burkholder, R. F. Gundy and M. L. Silverstein, A maximal characterization of the class H^p, *Trans. Amer. Math. Soc.*, **157** (1971), 137–153.

{42} D. L. Burkholder, Brownian motion and classical analysis, *Proc. Symp. Pure Math.*, **31** (1977), 5–14.

{43} R. F. Gundy, *Inégalités pour martingales à un ou deux indices; l'espace H^p*, Ecole d'été Saint-Flour 1978, Lect. Notes in Math 774, Springer-Verlag (1980).

{44} F. Weisz, *Martingale Hardy Spaces and their application in Fourier Analysis,* Lecture Notes in Mathematica 1568, Springer (1994).

{45} A. Kolmogorov, Sur les fonctions harmoniques conjugées et les séries de Fourier, *Fondam. Math.*, **7** (1925), 23–28.

{46} A. Zygmund, Sur les fonctions conjuguées, *Fondam. Math.*, **13** (1929), 284–302; **18** (1932), 312.

{47} S. Pichorides, Une remarque sur le théorème de M. Riesz concernant la norme L^p de sommes partielles de séries de Taylor, *C.R.A.S. Paris I* **310** (1990), 549–551.

References, Section 4

{48} F. Parreau, Ergodicité et pureté des produits de Riesz, *Annales de l'Institut Fourier,* **40**, 2 (1990), pp. 391–405.

{49} J. Peyrière, Almost everywhere convergence of lacunary trigonometric series with respect to Riesz products, *Journal Australien Math. Society A,* **48** (1990), pp. 376–383.

{50} A. H. Fan, Quelques propriétés des produits de Riesz, *Bulletin Sciences Mathématiques 2e série,* **117** (1993), pp. 421–439.

{51} S. J. Kilmer and S. Saeki, On Riesz product measures: mutual absolute continuity and singularity, *Annales de l'Institut Fourier,* **38**, 2 (1988), pp. 63–93.

{52} A. H. Fan, Equivalence et orthogonalité de mesures aléatoires engendrées par martingales positives homogènes, *Studia Mathematica* (1991), pp. 249–266.

{53} S. Drury, Sur les ensembles de Sidon, *C.R.A.S. Paris,* **271** (1970), pp. 162–163.

{54} D. Rider, Randomly continuous fonctions and Sidon sets, *Duke Mathematical Journal,* **42** (1975), pp. 759–764.

{55} G. Pisier, Arithmetic characterizations of Sidon sets, *Bulletin American Mathematical Society,* **8** (1983), pp. 87–89.

{56} J. Bourgain, Sidon sets and Riesz products, *Annales de l'Institut Fourier,* **35**, 1 (1988), pp. 136–148.

{57} J.-P. Kahane, *Almost surely everywhere*, Paul Erdős and His Mathematics I, Budapest, 2003, pp. 333-343.

{58} G. Freud, Über trigonometrische Aproximation und Fouriersche Reichen, *Mathematischer Zeitschrift,* **78** (1962), pp. 252–262.

{59} G. Freud, On Fourier series with Hadamard gaps, *Studia Sc. Math. Hungarian I* (1966), pp. 87–96.

{60} M. Izumi, S. Izumi and J.-P. Kahane, Théorèmes élémentaires sur séries de Fourier lacunaires, *Journal d'Analyse Mathématique* **14** (1965), pp. 235–246.

References, Section 5

{61} F. Lukács, Über die Bestimmung des Sprunges einer Funktion aus ihrer Fourier-Reihe, *J.f. reine und angew. Math,* **150** (1920), 107–112.

{62} P. Csillag, Über die Fourierschen Konstanten der Funktionen von beschränker Schwankung (Hungarian), *Math. és Phys. Lapok,* **27** (1918), 301–308.

{63} S. Sidon, *Bestimmung des Sprunges der Funktion aus der Fourier-Reihe* (Hungarian), ibidem, 309–311.

{64} M. Riesz, Sur les séries trigonométriques, *Comptes rendus Acad. Sci. Paris,* **145** (1907), 583–586.

{65} J. Marcinkiewicz and A. Zygmund, On the summability of double trigonometric series, *Fund. Math.,* **32** (1939), 122–132.

Comment on {64}. The note {64} is worth reading by those who know French. It is beautifully written and states the main results of the thesis. Surprisingly it is not reprinted in the Collected Works of M. Riesz. As a sign of bad luck it cannot be found in the 1907 issue of the Jahrbuch über die Fortschritte der Mathematik under the name of M. Riesz; it is attributed to F. Riesz.

Jean-Pierre Kahane
11 Rue du Val de Grace
75005 Paris
France

BOLYAI SOCIETY
MATHEMATICAL STUDIES, 14

A Panorama of Hungarian Mathematics in the Twentieth Century, pp. 193–209.

NON-COMMUTATIVE HARMONIC ANALYSIS

JONATHAN ROSENBERG

For present purposes, we shall define *non-commutative harmonic analysis* to mean the decomposition of functions on a locally compact G-space X,[1] where G is some (locally compact) group, into functions well-behaved with respect to the action of G. The classical cases are of course Fourier series, when $G = X = \mathbb{T}$, the circle group, and the Fourier transform, when $G = X = \mathbb{R}$, but we will mostly be concerned with the case when G is non-commutative. Since this subject is inextricably linked with the subject of representations of G (unitary representations, if we specialize to the case of L^2-functions), we will also consider the general theory of representations of locally compact groups and of various related structures, such as Lie algebras and Jordan algebras.

The subject of group representations was created by Georg Frobenius {9} in a remarkable series of papers in the 1890's, and continued in the first decade of the twentieth century in the work of his student Issai Schur {17}. However, Frobenius worked exclusively with finite groups, and his treatment was purely algebraic. It took a while before it was realized that Frobenius' theory had important implications for harmonic analysis. The generalization of the theory to compact groups was largely carried out by Hermann Weyl, and applications to harmonic analysis on compact groups did not come until the Peter–Weyl Theorem ({16}; reprinted in {28}, pp. 387–404).

It is against this background that we shall consider the contributions of a few great Hungarian mathematicians: Alfred Haar, John von Neumann, and Eugene Wigner in the 1920's, 1930's, and 1940's; and in somewhat later generations, Béla Szőkefalvi-Nagy and Lajos Pukánszky. As there is room here to discuss only a few of their contributions, we refer the reader to the

[1]This means that X is a locally compact space and we are given a continuous map $G \times X \to X$, $(g,x) \mapsto gx$, such that $(gh)x = g(hx)$ and $ex = x$ for all $g, h \in G$ and $x \in X$, where e is the identity element of G.

scientific obituaries {18}, {23}, {15}, {10}, {14}, {29}, {24}, {6}, and {5} for more details.

1. Haar, von Neumann, and Wigner

Alfred (Alfréd) Haar, Eugene Paul (Jenő Pál) Wigner, and John (János) von Neumann were all born in Budapest: Haar in 1885, Wigner in 1902, von Neumann in 1903. All three had the good fortune to have as their secondary school mathematics teacher László Rátz of the Evangelical Lutheran High School in Budapest. Rátz seems to have done a remarkable job in encouraging mathematical talent, and was also the founder of the *Mathematical Journal for Secondary Schools,* or *Középiskolai Matematikai Lapok.* (See {23} and "Eugene Paul Wigner: A Biographical Sketch" in [199], pp. 3–14.) These three students of Rátz were among the most important contributors to the development of non-commutative harmonic analysis.

1.1. Hilbert's Fifth Problem

Hilbert's Fifth Problem {11} asked "how far Lie's concept of continuous groups of transformations is approachable in our investigations without the assumption of the differentiability of the functions." Hilbert's Fifth Problem can be said to mark the beginning of the subject of non-commutative harmonic analysis.

Among the very first results in the direction of a solution was a paper of von Neumann, "Zur Theorie der Darstellungen kontinuierlichen Gruppen" (*Sitzungber. der Preuss. Akad.* (1927), 76–90; reprinted in [118], vol. 1, pp. 134–148). This paper basically proves that any continuous finite-dimensional representation of a Lie group is automatically differentiable, in fact analytic.

1.2. Invariant Measures and Analysis on Locally Compact Groups

It was soon realized that a reasonable attack on more substantial cases of Hilbert's Fifth Problem requires a means of doing analysis on a locally compact group, comparable to the sort of analysis one does with functions in Euclidean space. Since analysis on Euclidean space is based in large part on Lebesgue integration, the "search was on" for a means of invariant integration on a general locally compact group. In the case of an n-dimensional Lie group G, since G is an orientable manifold, G always admits left-invariant smooth measures, which can be identified with non-zero left-invariant differential n-forms, or in other words with non-zero elements of the one-dimensional vector space $\bigwedge^n \mathfrak{g}^*$, where $\mathfrak{g}$ is the Lie algebra of G, that is, the vector space of left-invariant vector fields. Note that one-dimensionality of $\bigwedge^n \mathfrak{g}^*$ implies that left-invariant measures on G are unique up to a scalar multiple.

One of Haar's greatest mathematical contributions was his proof, in the paper "Der Massbegriff in der Theorie der kontinuierlichen Gruppen," (*Ann. of Math.* (2) **34** (1933), 147–169; reprinted in [62], 600–622), that *every* locally compact group G admits a left-invariant measure. Haar's paper also appeared slightly earlier in Hungarian (*Mat. Term. Ért.* **49** (1932), 287–307; reprinted in [62], 579–599). While various improved reformulations of Haar's method have been given, notably the elegant ones due to Weil {27} and Cartan {3}, to this author's knowledge, no one has ever improved on his main idea, which is to compare the relative size of two compact sets A and B with dense interiors, by letting $h(A; B)$ be the minimal number of translates of B required to cover A. From this data Haar constructs his measure m by letting

$$m(A) = \lim_n \frac{h(A; B_n)}{h(C; B_n)},$$

where C is fixed once and for all and where the sets B_n run over a compact neighborhood base of the identity. Haar was well aware that his theorem made possible harmonic analysis on non-Lie topological groups, and he remarks at the end of his paper that it immediately follows that one can prove the Peter–Weyl Theorem ({16}; reprinted in {28}, pp. 387–404), giving a decomposition of $L^2(G)$ into an orthogonal direct sum of matrix coefficients of irreducible representations, for any compact group G, not just a Lie group.

Haar and von Neumann were in close contact at the time of this work, and a paper of von Neumann on Hilbert's Fifth Problem, "Die Einführung analytischer Parameter in topologischen Gruppen" ([118], vol. 2, pp. 366–386) was submitted to the *Annals* the same day as Haar's paper and published right next to it. In it, von Neumann proves that every compact group which is *topologically* locally Euclidean is a Lie group, i.e., admits an analytic structure. From this and the Peter–Weyl Theorem, it follows that every compact group is an inverse limit of Lie groups.

Two other important papers of von Neumann follow up on the theme of Haar's work. In "Zum Haarschen Maß in topologischen Gruppen" (*Compositio Math.* **1** (1934), 106–114; also [118], vol. 2, pp. 445–453), von Neumann gives an easier proof of the existence of Haar measures on a compact group G, by proving that if f is a continuous function on G, then the closed convex hull of the translates of f contains a unique constant function (the value of the constant being of course $\int f(g)\,dg$). This remains the easiest proof of existence of Haar measure for compact groups. Then in "The uniqueness of Haar's measure" (*Mat. Sb.* **1** (1936), 721–734; also [118], vol. 4, pp. 91–104), von Neumann gives a proof that a left (or right) Haar measure is unique up to scalar multiples, just as in the case of invariant smooth measure on a Lie group. This important result was proved independently, using different methods, by André Weil {27} and Henri Cartan {3}.

1.3. Representation Theory and Quantum Physics

Among the most important motivations for the development of non-commutative harmonic analysis in the years between the two World Wars was the development of quantum mechanics. Indeed, Weyl, von Neumann, and Wigner all approached the subject of non-commutative harmonic analysis with quantum mechanics in mind, and Wigner always considered himself more of a physicist than a mathematician.

As early as his paper "Über nicht kombinierende Terme in der neueren Quantentheorie" of 1926 (*Z. für Physik* **40** (1926–27), 492–500 and 883–892; reprinted in [199], pp. 34–52), Wigner realized that Frobenius' theory of representations of the symmetric group was relevant to the study of wave functions of multi-particle systems. This can be regarded as an example of non-commutative harmonic analysis in the sense of this article, with $G = S_n$. In his paper, Wigner thanks von Neumann for telling him about the work of Frobenius and Schur.

The following year, 1927, Wigner spent at Göttingen as Hilbert's assistant. There he met several mathematicians and physicists and began to collaborate with Pascual Jordan. In their paper "Über das Paulische Äquivalenzverbot" of 1928 (*Z. für Physik* **47** (1928), 631–651; reprinted in [199], pp. 109–129), Jordan and Wigner first reformulated the Pauli exclusion principle in terms of representations of the "canonical anticommutation relations" (CAR). The Clifford algebras defined by the CAR of course play a pivotal role in the Dirac equation of the electron, and in fact the connection between Dirac and Wigner was more than scientific: Dirac later married Wigner's sister.

The work of Wigner and Jordan was the precursor of the famous paper "On an algebraic generalization of the quantum mechanical formalism" by Jordan, von Neumann, and Wigner (*Ann. of Math.* (2) **35** (1934), 29–64; reprinted in [199], pp. 298–333 and in [118], vol. 2, pp. 408–444) that founded the study of what are now called Jordan algebras. In their paper, Jordan, von Neumann, and Wigner obtain the classification of the finite-dimensional simple Jordan algebras over $\mathbb{R}$, including the exceptional ones coming from the Cayley octonians. We now know that one can trace the existence of the exceptional compact Lie groups G_2, F_4, E_6, E_7, and E_8 to these exceptional Jordan algebras.

We have mentioned the CAR, the anticommutation relations that govern the behavior of fermions. Of equal importance are the "canonical commutation relations" (CCR) for bosons, that the position operators Q_k and momentum operators P_j should satisfy

$$Q_k P_j - P_j Q_k = i\hbar\delta_{kj}, \qquad Q_k^* = Q_k, \qquad P_j^* = P_j, \tag{1}$$

where $\hbar$ is Planck's constant. These simple relations, familiar to every physics student, hide a serious mathematical difficulty: the equations (1) have no finite-dimensional solutions,[2] in fact no solutions in bounded operators! And for unbounded operators which are not everywhere defined, what is the meaning of the commutator? The problem of making rigorous sense of (1), and of showing that there is essentially only one irreducible solution (satisfying certain nice regularity properties), was solved by Hermann Weyl, Marshall Stone, and von Neumann. The final result, in von Neumann's important paper "Die Eindeutigkeit der Schrödingerschen Operatoren" (*Math. Ann.* **104** (1931), 570–578; also [118], vol. 2, pp. 221–229), turned out to be important not only for theoretical physics but also for the future development of unitary representation theory.

The idea is to note that (1) amounts to looking for a representation through skew-adjoint operators of what is now called the *Heisenberg Lie algebra* $\mathfrak{g}$ of dimension $2n+1$, with a basis $X_1, \dots, X_n, Y_1, \dots, Y_n, Z$ satisfying

$$[X_j, Y_k] = \delta_{kj} Z, \qquad [Z, X_j] = 0, \qquad [Z, Y_k] = 0. \tag{2}$$

[2]The reason is that for any finite-size matrices Q and P, $\mathrm{Tr}\,(QP - PQ) = 0$, so $QP - PQ$ cannot be a non-zero multiple of the identity.

Now to the Lie algebra $\mathfrak{g}$ we can attach a simply connected nilpotent Lie group G. Topologically it looks like $\mathbb{R}^{2n+1}$, but the multiplication is slightly twisted. We can impose a regularity condition on our Lie algebra representations by requiring that they come from unitary representations π of G, that is, strongly continuous homomorphisms from G to the unitary group of some Hilbert space, for which the corresponding "infinitesimal representation" is $d\pi(Z) = i\hbar$, $d\pi(X_j) = iP_j$, $d\pi(Y_k) = iQ_k$. Von Neumann's Theorem says that up to unitary equivalence, there is one and only one irreducible such representation, and every unitary representation π of G with $d\pi(Z) = i\hbar$ is a multiple of this irreducible representation.

1.4. Invariant Means and Almost Periodic Functions and Groups

In von Neumann's paper "Zum Haarschen Maß in topologischen Gruppen," cited above, there appears an interesting "Zusatz während der Korrektur," in which von Neumann mentions that he had noticed that his argument for constructing the Haar measure on a compact group can be extended to some non-compact groups as well, but that in this case it gives rise not to Haar measure but to an invariant "mean" $f \mapsto \int f$ for which the constant function 1 has "mean value" 1. (On the other hand, Haar measure on a non-compact group is never a finite measure, so on a non-compact group, constant functions are never integrable with respect to Haar measure.) The study of such means was to lead to a whole other direction in non-commutative harmonic analysis. Von Neumann's Zusatz asserts that "die Ausführung erscheint demnächst in den *Annals of Mathematics*." Evidently he misspoke; the paper von Neumann refers to, "Almost periodic functions in a group, I" was to come out in the *Transactions* of the *Amer. Math. Soc.* (**36** (1934), 445–492; also [118], vol. 2, 454–501), not the *Annals*.

To explain von Neumann's discovery, we have to back up a bit and review Harald Bohr's theory of almost periodic functions {2}. In its simplest version, this refers to bounded uniformly continuous functions f on the line with a "Fourier series" expansion $f(x) \sim \sum_j c_j e^{i\lambda_j x}$ (convergent in a suitable sense), where the frequencies λ_j are not necessarily rationally related to one another, and thus f is not necessarily periodic. Of course we cannot expect expect the Fourier series of f to converge uniformly to f, since this is not always true even when f is literally periodic. Instead, Bohr found that a natural notion of convergence in this context is convergence in mean over bigger and bigger intervals, i.e., that

$$\lim_{N\to\infty} \lim_{T\to\infty} \frac{1}{2T} \int_{-T}^{T} \left| f(x) - \sum_{j=1}^{N} c_j e^{i\lambda_j x} \right|^2 dx = 0,$$

and that there is a well-defined notion of *mean* for such functions, namely

$$\overline{f} = \lim_{T\to\infty} \frac{1}{2T} \int_{-T}^{T} f(x)\, dx.$$

Existence of the limit here is Bohr's "Mittelwertsatz" in {2}. What von Neumann noticed is that for almost periodic functions f on $\mathbb{R}$, the closed convex hull of the translates of f contains a unique constant function, and the value of this constant is the Bohr mean value of f.

Now in the earlier paper "Zur allgemeinen Theorie des Masses" (*Fund. Math.* **13** (1929), 76–116 and 333; reprinted in [118], vol. 1, pp. 599–643), von Neumann had studied the similar subject of means on *discrete* groups, and had defined a group G to be "messbar" (literally *measurable*, but we will instead use the now-standard terminology *amenable*) if it admits a left-invariant mean. Such a mean can be viewed as an "integration" process $f \mapsto \int f$ on bounded functions, for which the constant function 1 has "mean value" 1, and for which $\int f = \int \lambda(x) f$, if $\lambda(x) f$ denotes the left translate of f by $x \in G$, i.e., $\big(\lambda(x)f\big)(y) = f(x^{-1}y)$. Von Neumann proved that finite and abelian (discrete) groups are amenable, and that the class of amenable groups is closed under extensions and direct limits. It follows that there is a fairly large class of "obviously" amenable groups, what are now called *elementary amenable* groups: the smallest class containing the solvable and finite groups and closed under extensions and direct limits. On the other hand, von Neumann exhibited countable groups that are *not* amenable, for example, the free group on two generators. It is important to note that on infinite amenable groups, invariant means are not at all unique. For example, a bounded function f on the group $G = \mathbb{Z}$ is simply a two-sided bounded infinite sequence $f = \{f_n\}_{n\in\mathbb{Z}}$, and the possible means of f are all the various limit points of the sequence of approximating averages

$$\frac{1}{2N+1}\big(f_{-N} + \cdots + f_0 + \cdots + f_N\big)$$

as $N \to \infty$.

Let us come back to von Neumann's work on almost periodic functions, continued in two later papers (the first with Salomon Bochner, "Almost periodic functions in a group, II," *Trans. Amer. Math. Soc.* (**37** (1935), no. 1, 21–50; also [118], vol. 2, 528–557; and the second with E. Wigner, "Minimally almost periodic groups," *Ann. of Math.* (2) **41** (1940), 746–750; in [118], vol. 4, 220–224 and in [199], pp. 390–394). To make a long story short, von Neumann defines two classes of groups, which are in some sense opposite extremes. *Minimally almost periodic groups* admit no non-constant almost periodic functions (in a natural sense extending Bohr's). *Maximally almost periodic groups* admit enough almost periodic functions to separate points. André Weil {27} eventually cleaned up the theory and showed that, for every maximally almost periodic group G, there exists a compact group G^* which contains G as a dense subgroup and such that every continuous real-valued almost periodic function on G can be uniquely extended to a continuous (and hence almost periodic) function on G^*. In fact, a locally compact group is maximally almost periodic if and only if it has a continuous embedding into a compact group. Thus as von Neumann pointed out, the Bohr mean on almost periodic functions really comes from the construction of Haar measure on compact groups, and in the case of an abelian locally compact group, it coincides with the restriction of an invariant mean on all bounded uniformly continuous functions. However, residually finite discrete groups[3] are always maximally almost periodic, but not always amenable, so even in the case of discrete groups, the Bohr mean on almost periodic functions does not always extend to a mean on all bounded functions.

1.5. Von Neumann Algebras

Among von Neumann's greatest contributions was the development of the theory of what he called *rings of operators,* and what are now called *von Neumann algebras.* A von Neumann algebra is simply a subalgebra of $\mathcal{B}(\mathcal{H})$, the algebra of all bounded operators on a Hilbert space $\mathcal{H}$, which is stable under the involution $T \mapsto T^*$ and closed under the strong (or equivalently, weak) operator topology. The massive papers of von Neumann on this subject, almost all of them joint with Francis Joseph Murray, fill all of volume III of [118]. It would be impossible to do justice to them here, so we refer the reader to {10} and to the article on functional analysis by Á. Császár and D. Petz in this volume for more information, but we

[3]This means groups like $SL(n, \mathbb{Z})$, the $n \times n$ matrices with integer entries and determinant 1, with enough homomorphisms to finite groups to separate points.

just briefly point out what this huge body of work has to do with non-commutative harmonic analysis.

Suppose G is a locally compact group. A *unitary representation* π of G on a Hilbert space $\mathcal{H}$ means a homomorphism π from G to the group of unitary operators on $\mathcal{H}$, which is continuous with respect to the strong operator topology (or the weak operator topology—it gives exactly the same notion). Note that we do *not* require continuity in the norm topology for operators, since this fails for the standard example of a unitary representation, namely the left regular representation λ of G on $L^2(G)$ (L^2 being defined with respect to Haar measure). Then the possible decompositions of π into subrepresentations are governed by the structure of the *commutant* of the representation,

$$\pi(G)' = \{T \in \mathcal{B}(\mathcal{H}) : T\pi(g) = \pi(g)T \text{ for all } g \in G\}.$$

For example, Schur's Lemma says that π is irreducible if and only if $\pi(G)' = \mathbb{C}$. But $\pi(G)'$ is a von Neumann algebra, so the classification theory of von Neumann algebras comes into play at this point. For example, we call π *multiplicity-free* if $\pi(G)'$ is abelian and a *factor representation* if $\pi(G)'$ is a factor, that is, a von Neumann algebra with one-dimensional center. The Murray-von Neumann papers classify factors into three types. If $\pi(G)'$ is a type I factor, then π is simply a multiple of a single irreducible representation. But if $\pi(G)'$ is a type II or type III factor, then there is no canonical way to decompose π into irreducible representations, so that type II or type III factor representations should themselves be regarded as basic building blocks of representation theory. Some groups (for example, abelian or compact groups, or the Heisenberg group defined by (2)) are *type I*, in the sense that the commutants of their unitary representations are *always* type I. For such groups, at least if G is second countable, von Neumann's theory of direct integral decompositions ("On rings of operators. Reduction theory," *Ann. of Math.* (2) **50** (1949), 401–485; [118], vol. 3, pp. 400–484) provides a canonical way of decomposing all unitary representations (on separable Hilbert spaces) into irreducible pieces, and there is a hope to copy many features of the Frobenius–Schur theory for finite groups. But for non-type I groups, one is forced to contend with non-type I factor representations. For example, Murray-von Neumann proved that the commutant of the regular representation of a discrete group G is a finite type II factor if and only if G is an ICC-group,[4] that is, if the identity of G is the only element whose conjugacy class is finite.

[4]This stands for "infinite conjugacy classes".

One of the other great contributions of Murray and von Neumann was the theory of the trace on a type II factor. If π is a finite-dimensional unitary representation, then one can define its character χ_π, a class function on G, by $\chi_\pi(g) = \operatorname{Tr}\pi(g)$. (This definition was introduced by Frobenius.) Furthermore, all one needs to know about π can be recovered from the character χ_π. It would be nice to do something similar for certain other factor representations. If π is a *finite* type II factor representation, then there is a continuous linear functional Tr on the von Neumann algebra generated[5] by $\pi(G)$, satisfying $\operatorname{Tr}(1) = 1$ and the usual trace property $\operatorname{Tr}(ab) = \operatorname{Tr}(ba)$, and so it is again possible to develop a theory of group characters similar to the one for finite groups. If π is an infinite-dimensional irreducible representation or a II_∞ factor representation, then $\pi(G)''$ admits a trace, but it is only partially defined, and in particular $\operatorname{Tr}(u)$ is undefined for u unitary. So in this case, while it is possible that $\chi_\pi(g) = \operatorname{Tr}\pi(g)$ might make sense as an equality of *distributions*, provided there are enough functions ϕ on G for which $\pi(\phi) = \int\phi(g)\pi(g)\,dg$ is trace-class, $\chi_\pi = \operatorname{Tr}\circ\pi$ does not make sense directly as a function on the group G.[6]

1.6. Development of Unitary Representation Theory of Non-Compact Lie Groups

The modern development of the unitary representation theory of non-compact Lie groups, which today is now a large subject, grew out of the work of Gelfand–Naimark, Bargmann, Mackey, and Harish-Chandra in the late 1940's and the 1950's. One of the key papers that prompted this development was Wigner's paper "On unitary representations of the inhomogeneous Lorentz group" (*Ann. of Math.* (2) **40** (1939), 149–204; reprinted in [199], pp. 334–389). Curiously, this paper was first submitted to the *American Journal of Mathematics*, usually regarded as being somewhat less prestigious than the *Annals*, and was rejected there with the remark that "this work is not interesting for mathematics" ([199], p. 9). In any event, this

[5]By von Neumann's Double Commutant Theorem, found in §II of "Zur Algebra der Funktionaloperationen und Theorie der normalen Operatoren" (*Math. Ann.* **102** (1929), 370–427; [118], vol. 2, pp. 86–143), this is just the commutant $\pi(G)''$ of $\pi(G)'$. The Double Commutant Theorem implies that a $*$-subalgebra A of $\mathcal{B}(\mathcal{H})$ is a von Neumann algebra if and only if it is equal to its double commutant A''.

[6]Note that $\pi(\phi)$ as we have just defined it can be viewed as a sort of operator-valued Fourier coefficient of ϕ. Its trace $\chi_\pi(\phi)$, when defined, is a sort of scalar-valued Fourier coefficient.

paper is important for two reasons: it promoted interest in the unitary representations of the actual Lorentz groups $SO_0(2,1)$ and $SO_0(3,1)$, classified soon afterwards by Bargmann {1}, and it amounted to the working out of an important special case of what was later formulated as the Mackey Imprimitivity Theorem {13}, and thus motivated the modern point of view on the decomposition of representations of group extensions.[7] More precisely, Wigner's paper studies unitary representations of the semidirect product $G = V \rtimes H$, where $V = \mathbb{R}^{3,1}$ is Minkowski space and $H = SO_0(3,1)$ is the Lorentz group acting on V the usual way. Wigner proves that each irreducible unitary representation of G is supported on a single H-orbit in $\widehat{V} \cong V$. Wigner only studies the cases where this orbit is either one of the light cones or else half of a two-sheeted hyperboloid. (The representations supported on the trivial H-orbit $\{0\}$ factor through H, and were only classified later in {1}.) In either case, if one thinks of an elementary particle corresponding to such an irreducible representation, and views its wave function ϕ as a vector-valued L^2-function on Minkowski space V, then this analysis shows that the Fourier transform of ϕ is supported on one of the two components of the variety where $\hat{x}_4^2 - \hat{x}_1^2 - \hat{x}_2^2 - \hat{x}_3^2 = m^2$, where m is a constant corresponding to the mass of the particle and the $\hat{x}_j$ are the Fourier transform variables, and thus ϕ itself satisfies the Klein–Gordon equation

$$\Box\phi = m^2\phi, \quad \text{where} \quad \Box = \frac{\partial^2}{\partial x_1^2} + \frac{\partial^2}{\partial x_2^2} + \frac{\partial^2}{\partial x_3^2} - \frac{\partial^2}{\partial x_4^2}.$$

We have already discussed (in section 1.5) the work of von Neumann on unitary representations of discrete groups, and the application of von Neumann's theory of direct integral decompositions to the decomposition of unitary representations of arbitrary second countable locally compact groups. However, Haar also worked on representation theory. A little-known paper of Haar, "Über die Gruppencharaktere gewisser unendlichen Gruppen" (*Acta Sci. Math.* (Szeged) **5** (1932), 172–186; reprinted in [62], pp. 172–186) extended the Frobenius theory of group characters from finite groups to what are now usually called *FC-groups*,[8] groups in which every conjugacy class is finite. What Haar called characters in this case turn out to be the same thing as characters in the more modern sense of traces of

[7] The other key case of the Imprimitivity Theorem that was known before was von Neumann's theorem on uniqueness of representations of the Heisenberg commutation relations, discussed above in Section 1.3.

[8] "FC" stands for "finite conjugacy classes."

finite factor representations, as later studied by Elmar Thoma {25}. An interesting fact about FC-groups is that if they are finitely generated, then they are virtually abelian groups, that is, have an abelian subgroup of finite index. It was eventually shown by Thoma {26} that the virtually abelian groups are precisely the class of type I groups, discrete groups whose unitary representations always generate type I von Neumann algebras, and in fact the dimensions of their irreducible representations are bounded. Hence for finitely generated FC-groups, non-commutative harmonic analysis in the sense of Haar is precisely the decomposition of functions into matrix coefficients of irreducible representations, as in the situation of the Peter–Weyl Theorem.

2. Sz.-Nagy and Pukánszky

2.1. Béla Sz.-Nagy

Béla Sz.-Nagy, one of the great operator theorists of the twentieth century, made a few interesting contributions to non-commutative harmonic analysis, even though this was not his primary mathematical interest.[9] Here we will just mention four of them. The first is a strengthening of von Neumann's automatic analyticity theorem for homomorphisms of Lie groups (see Section 1.1): Nagy {19} proves that measurability is enough to guarantee analyticity; one does not need to assume continuity from the start.

The second contribution, on its face, only deals with *commutative* harmonic analysis. In {20}, Nagy proved that a one-parameter group $\{T^n\}_{n\in\mathbb{Z}}$ or $\{T_s\}_{s\in\mathbb{R}}$ of invertible linear operators on a Hilbert space $\mathcal{H}$ is similar to a unitary representation (of $\mathbb{Z}$ or $\mathbb{R}$) if and only if it is *uniformly bounded*, i.e., there is a constant $K > 0$ such that $\|T^n\| \leq K$ for all $n \in \mathbb{Z}$ or $\|T_s\| \leq K$ for all $s \in \mathbb{R}$. The connection with unitary representation theory is that essentially the same theorem, with almost the same proof, holds for uniformly bounded representations π of arbitrary amenable locally compact groups G, as was pointed out by Jacques Dixmier {4}. In other words, if π is a strongly continuous homomorphism from G to the invertible linear

[9]The bulk of Sz.-Nagy's work is discussed in the companion article in this volume by Á. Császár and D. Petz.

operators on $\mathcal{H}$, and if $\|\pi(g)\| \leq K$ for all $g \in G$, then

$$\langle \xi, \eta \rangle' = \int_G \langle \pi(g)\xi, \pi(g)\eta \rangle \, dg,$$

with $\int_G$ denoting an invariant mean on G, defines a new inner product on the Hilbert space, equivalent to the original one, with respect to which the representation π is unitary. Incidentally, the requirement of amenability is essential here; the group $SL(2, \mathbb{R})$ was shown in {7} to have uniformly bounded representations that are not unitarizable.

The third concerns the following problem. Suppose $\mathcal{H}$ is a Hilbert space and $\sigma : G \to \mathcal{B}(\mathcal{H})$ is a map from a discrete group to bounded operators on $\mathcal{H}$. What is the necessary and sufficient condition for σ to be the compression of a unitary representation, or in other words, for there to be a Hilbert space $\mathcal{H}' \supseteq \mathcal{H}$ and a unitary representation π of G on $\mathcal{H}'$ such that if $P : \mathcal{H}' \to \mathcal{H}$ is the orthogonal projection, then $\sigma(g) = P\pi(g)P$ for all $g \in G$? Nagy solved this problem in {21}. This can be viewed as an operator-valued analogue of the characterization of matrix coefficients $g \mapsto \langle \pi(g)\xi, \xi \rangle$ of unitary representations as the positive-definite functions, which follows from the Gelfand–Naimark–Segal construction.

The fourth problem studied by Nagy (in joint work with Ciprian Foiaş and László Gehér {8} and in {22}) can be viewed as a postscript to von Neumann's work (cited above in Section 1.3) on uniqueness of representations of the canonical commutation relations (1). Recall that the method of Weyl and von Neumann was to study unitary representations of the Heisenberg Lie group, not representations by unbounded operators of the original relations, which are more numerous and which pose difficult analytic questions. However, Nagy was able to give necessary and sufficient conditions for a representation of the CCR to come from a representation of the group: an irreducible pair of closed symmetric operators Q and P on a Hilbert space $\mathcal{H}$ is a "Schrödinger couple" if and only if $QP - PQ = iI$ holds on a subspace $\mathcal{D} \subset \mathcal{D}QP - PQ$ which is large enough so that the restrictions of Q and P to $\mathcal{D}$ are essentially selfadjoint and that at least one of the eight sets $(Q \pm iI)(P \pm iI)\mathcal{D}$, $(P \pm iI)(Q \pm -iI)\mathcal{D}$ is dense in $\mathcal{H}$.

2.2. Lajos Pukánszky

Lajos Pukánszky, who was born in Budapest in 1928, studied with Sz.-Nagy in Szeged. His earliest papers (dating from 1951 through 1960) deal with von Neumann algebras; all his subsequent publications were on the unitary representation theory of Lie groups. While it was von Neumann who first anticipated the importance of the theory of rings of operators (i.e., von Neumann algebras) to non-commutative harmonic analysis, it was Pukánszky who finally achieved a deep synthesis of these two subjects. Again, we have no room here to go into details, some of which may be found in {6} and {5}. The reader interested in Pukánszky's work could also see his posthumous monograph [139]. The subject of this book, indeed of much of Pukánszky's work, concerns the following question. For a type I Lie group G, there is a bijective correspondence between $\widehat{G}$, the set of unitary equivalence classes of irreducible unitary representations π, and the set of "characters" of G, the generalized functions $g \mapsto \operatorname{Tr} \pi(g)$. What is the substitute for this bijection in the case of non-type I Lie groups?

The answer, which is quite remarkable, is that for non-type I Lie groups, one needs to replace $\widehat{G}$ by $\overset{\cap}{G}$, the set of quasi-equivalence classes[10] of "normal" representations. These are factor representations π of G of types I or II, for which the (usually unbounded) trace Tr on the factor $\pi(G)''$ is finite and non-zero on some ideal in $C^*(G)$, the C^*-completion of the convolution algebra $L^1(G)$. On a type I group, there is no difference between $\widehat{G}$ and $\overset{\cap}{G}$. But on non-type I connected Lie groups, Pukánszky showed that $\overset{\cap}{G}$ is big enough to support the canonical (i.e., central) direct integral decomposition of the left regular representation of G on $L^2(G)$. Thus *harmonic analysis of L^2 functions on the group sometimes forces one to consider normal representations that are not irreducible*, and the left regular representation of G on $L^2(G)$ generates a von Neumann algebra with no type III summand. (This last fact was proved through the joint efforts of Pukánszky and Dixmier.) Not only this, but every primitive ideal of $C^*(G)$ (that is, the kernel of an irreducible representation of this algebra) is the kernel of a unique quasi-equivalence class of normal representations, so that there is a natural bijection between $\overset{\cap}{G}$ and $\operatorname{Prim} C^*(G)$. Finally, in a stunning generalization of Alexander Kirillov's character formula {12}, Pukánszky [139] was able to give a geometric parametrization of the normal representations and a formula for their characters, at least for connected solvable Lie groups.

[10] Two factor representations π_1 and π_2 of G are called *quasi-equivalent* if there is an isomorphism $\Phi : \pi_1(G)'' \to \pi_2(G)''$ such that $\pi_2 = \Phi \circ \pi_1$. This is the natural equivalence relation on factor representations. The difference between this and unitary equivalence is that Φ need not be given by conjugation by a unitary operator.

REFERENCES

[62] Haar, Alfréd, *Összegyűjtött Munkái = Gesammelte Arbeiten,* ed. Béla Sz.-Nagy, Akadémiai Kiadó (Budapest, 1959).

[118] Neumann, János (John von Neumann), *Collected Works,* 6 volumes, ed. A. H. Taub, Pergamon Press (New York, 1961). Of interest to this survey are Vol. I: Logic, theory of sets and quantum mechanics; Vol. II: Operators, ergodic theory and almost periodic functions in a group; Vol. III: Rings of operators; and Vol. IV: Continuous geometry and other topics.

[139] Pukánszky, Lajos, *Characters of Connected Lie Groups,* With a preface by J. Dixmier and M. Duflo, Mathematical Surveys and Monographs, Volume 71, American Mathematical Society (Providence, RI, 1999).

[199] Wigner, Jenő (E.P.), *Collected Works,* ed. A. Wightman, Springer-Verlag (Berlin–New York, 1993).

{1} V. Bargmann, Irreducible unitary representations of the Lorentz group, *Ann. of Math.* (2), **48** (1947), 568–640.

{2} H. Bohr, Zur Theorie der Fastperiodischen Funktionen, I, *Acta Math.,* **45** (1925), 29–127; II, *ibid.,* **46** (1925), 101–214.

{3} H. Cartan, Sur la mesure de Haar, *C.R. Acad. Sci. Paris,* **211** (1940), 759–762.

{4} J. Dixmier, Les moyennes invariantes dans les semi-groups et leurs applications, *Acta Sci. Math. Szeged,* **12** (1950), "Leopoldo Fejer et Frederico Riesz LXX annos natis dedicatus," Pars A, 213–227.

{5} J. Dixmier, M. Duflo, A. Hajnal and A. Koranyi, Pukánszky Lajos (1928–1996): About his mathematical work: a personal commemoration (Hungarian), *Mat. Lapok* (N.S.) **4**, no. 2–3 (1994), 71–81 (1998).

{6} J. Dixmier, M. Duflo, A. Hajnal, R. Kadison, A. Koranyi, J. Rosenberg and M. Vergne, Lajos Pukánszky (1928–1996), *Notices Amer. Math. Soc.,* **45**, no. 4 (1998), 492–499.

{7} L. Ehrenpreis and F. I. Mautner, Uniformly bounded representations of groups, *Proc. Nat. Acad. Sci. U.S.A.,* **41** (1955), 231–233.

{8} C. Foiaş, L. Gehér and B. Sz.-Nagy, On the permutability condition of quantum mechanics, *Acta Sci. Math. (Szeged),* **21** (1960), 78–89.

{9} F. G. Frobenius, *Gesammelte Abhandlungen,* J.-P. Serre, ed., 3 vols., Springer-Verlag, Berlin and New York, 1968.

{10} J. Glimm, J. Impagliazzo and I. Singer, editors, *The legacy of John von Neumann,* Proc. of Symposia in Pure Math., 50., Amer. Math. Soc. (Providence, RI, 1990).

{11} D. Hilbert, Mathematische Probleme: Vortrag, gehalten auf dem internationalen Mathematiker-Congress zu Paris 1900, *Gött. Nachr.* (1900), 253–297, translated and reprinted as Mathematical Problems, *Bull. Amer. Math. Soc.,* **37**, no. 4 (2000), 407–436.

{12} A. A. Kirillov, Unitary representations of nilpotent Lie groups, *Uspekhi Mat. Nauk,* **17**, no. 4 (1962) (106), 57–110 (Russian); English translation in *Russian Math. Surveys,* **17**, no. 4 (1962), 52–104.

{13} G. W. Mackey, Imprimitivity for representations of locally compact groups, I, *Proc. Nat. Acad. Sci. U.S.A.,* **35** (1949), 537–545.

{14} G. W. Mackey, Introduction to the mathematical papers of E. P. Wigner, in: [199], pp. 241–290.

{15} J. C. Oxtoby, B. J. Pettis and G. B. Price (editors), *John von Neumann, 1903–1957, Bull. Amer. Math. Soc.,* **64**, no. 3, Part 2 (1958); reprinted as a monograph, Amer. Math. Soc. (Providence, RI, 1988).

{16} F. Peter and H. Weyl, Der Vollstandingkeit der primitiven Darstellungen einer geschlossen kontinuierlichen Gruppe, *Math. Ann.,* **97** (1927), 737–755.

{17} I. Schur, *Gesammelte Abhandlungen,* Alfred Brauer and Hans Rohrbach, eds., 3 vols., Springer-Verlag (Berlin and New York, 1973).

{18} J. Szabados and K. Tandori, eds., *A. Haar Memorial Conference,* 2 vols., János Bolyai Math. Soc. and Elsevier (Amsterdam and New York, 1987).

{19} B. Szőkefalvi-Nagy, Über messbare Darstellungen Liescher Gruppen, *Math. Ann.,* **112** (1936), 286–296.

{20} B. Szőkefalvi-Nagy, On uniformly bounded linear transformations in Hilbert space, *Acta Univ. SzegedṠect. Sci. Math.,* **11** (1947), 152–157.

{21} B. Szőkefalvi-Nagy, Transformations de l'espace de Hilbert, fonctions de type positif sur un groupe, *Acta Sci. Math. (Szeged),* **15** (1954), 104–114.

{22} B. Szőkefalvi-Nagy, The commutation relation of quantum mechanics, in: *Functional analysis and its applications* (Nice, 1986), ICPAM Lecture Notes, World Sci. Publishing (Singapore, 1988), pp. 323–337.

{23} B. Szőkefalvi-Nagy, Alfred Haar (1885–1933): invariant measure of mathematical excellence, in: {18}, vol. 1, pp. 17–24.

{24} B. Szőkefalvi-Nagy (1913–1998), *Acta Sci. Math. (Szeged),* **65**, no. 1–2 (1999), 3–18.

{25} E. Thoma, Zur harmonischen Analyse klassenfiniter Gruppen, *Invent. Math.,* **3** (1967), 20–42.

{26} E. Thoma, Über unitäre Darstellungen abzählbarer, diskreter Gruppen, *Math. Ann.,* **153** (1964), 111–138; Eine Charakterisierung diskreter Gruppen vom Typ I, *Invent. Math.,* **6** (1968), 190–196.

{27} A. Weil, *L'intégration dans les groupes topologiques et ses applications,* Hermann (Paris, 1940).

{28} H. Weyl, *Selecta,* Birkhäuser (Basel, 1956).

{29} D. P. Zhelobenko, S. M. Nikol'skii and S. A. Telyakovskii, Béla Szőkefalvi-Nagy (obituary), *Uspekhi Mat. Nauk,* **54**, no. 4 (1999), 143–146 (Russian); English translation in *Russian Math. Surveys,* **54**, no. 4 (1999), 819–822.

Jonathan Rosenberg
Department of Mathematics
University of Maryland
College Park, MD 20742-4015
U.S.A.
`jmr@math.umd.edu`
URL: `http://www.math.umd.edu/~jmr`

BOLYAI SOCIETY
MATHEMATICAL STUDIES, 14

A Panorama of Hungarian Mathematics in the Twentieth Century, pp. 211–244.

A Panorama of the Hungarian Real and Functional Analysis in the 20th Century

ÁKOS CSÁSZÁR and DÉNES PETZ[1]

The theory of real functions is a relatively young chapter of mathematical analysis. In fact, in the 19th century, the expression "function theory" was applied to mean the theory of complex-valued analytic functions of one or more complex variables. The first investigations that initiated the part of analysis called today "theory of real functions" or briefly "real analysis" were constructions of various real-valued functions of a real variable whose characteristic properties are very far from those of analytic functions; as a typical example, we can mention the construction, due to *Karl Weierstrass,* of a real-valued function continuous in an interval but not differentiable at any point of this interval.

The investigations in this direction obtained a very powerful instrument in the theory of sets discovered and developed by *Georg Cantor.* As the first important results of the theory, we can mention the investigations of *Camille Jordan* on properties of the functions of bounded variation, or the theory of the area of plane point sets due to the same author. However, the last discovery that finished the acceptance of real analysis as a well-adopted chapter of analysis was the concept and theory of a very general kind of *integral,* due to *Henri Lebesgue,* in the first years of the 20th century. This acceptance was perhaps due to the fact that Lebesgue not only developed the theory of the integral but also reached amazing applications of his theory so that the importance of the new, general theory of the integral convinced everybody interested in mathematical analysis. On the other hand, Lebesgue's integration theory catalyzed the cristallization of the ideas of *functional analysis* in the setting of infinite dimensional function spaces.

[1]The authors are grateful to János Horváth, József Szűcs and László Zsidó for many helpful suggestions.

There were Hungarian mathematicians who joined the investigations on real analysis in the earliest period of its existence, i.e., in the last decades of the 19th century. We must first mention **Gyula [Julius] Kőnig** (1849–1913) who, in his courses on analysis given at the Technical University Budapest, presented a definition of the *integral* that included not only the classical definition of the Riemannian integral but also the Stieltjes integral. Despite the fact that he did it in about 1890, he formulated these ideas not earlier than 1897 in a paper in Hungarian language {27} so that the priority evidently belongs to *Thomas Jean Stieltjes* who published his discovery in 1894.

However, Kőnig was very probably the first researcher who constructed a real-valued function continuous in an interval and having an extremum in every subinterval {26}. This result certainly influenced the disciple of Kőnig, **Zoárd Geőcze** (1873–1916), who constructed a continuous function that is not rectifiable in any subinterval {10}. Many years later, it turned out that Geőcze's construction yields essentially more; in fact it is one of the simplest constructions furnishing a continuous nowhere differentiable function i.e., it presents the *Weierstrass singularity* (see work {21} of *Sándor Kántor*). On the other hand, the same method of construction, under another choice of the parameters, furnishes a continuous function that is increasing and singular (i.e., its derivative is equal to 0 almost everywhere, see {8}). In the last sentence, the expression "almost everywhere" means, of course, *"with the exception of a set of points of Lebesgue measure zero"*. It is an imperishable merit of the history of mathematics in Hungary that we had a researcher who, a few years after its birth, not only made himself a master of Lebesgue's theory of measure and integral, but also added to this theory essential contributions. This researcher is **Frigyes [Frédéric] Riesz** (1880–1956).

Frédéric Riesz was definitely the first mathematician in Hungary who understood the great importance of the new theory of integration. He was born in Győr, a town in the midway between Budapest and Vienna. After a two-year-study at the polytechnic in Zürich, he continued at the science university in Budapest. From 1904 to 1912 he was a high-school teacher and wrote fundamental papers already in this period. Although he published very good works also in Hungarian all his life, he was clever enough to understand that Paris was not only the capital of France but the capital of modern analysis as well. His publications in *Comptes Rendus*, the journal of the French Academy of Sciences, earned a world fame for him very early and he became a professor of the University of Kolozsvár (called Cluj-Napoca now in Roumania) in 1912.

Frédéric Riesz was one of the fathers of functional analysis. Although functional analysis in the sense of nowadays had several roots in the 19th century, such as Fourier expansion of functions and spectral theory of some differential equations, its genesis could be put at the beginning of 20th century. Even at the end of 1800's linear algebra was very finite dimensional and dealt with n-tuples of real numbers, and Fréchet's thesis on metric spaces appeared in 1906. The cristallization of the ideas was catalyzed by Lebesgue's integration theory. Many of the basic concepts of functional analysis were born in the setting of infinite dimensional function spaces. The intimate relation of the Lebesgue integral and infinite dimensional functional analysis is very transparent in the work of Frédéric Riesz.

The space L^2 of square integrable functions on an interval of the real line was the first infinite dimensional space on which functional analysis in the modern sense was studied. The so-called *Riesz–Fischer theorem* (1907) {59} claims that the space L^2 is complete, that is, all Cauchy sequences are convergent in the L^2-sense. That time it was shown that if $f_n \in L^2$ and $\int \big| f_n(x) - f_m(x) \big|^2 dx$ is arbitrarily small when n and m are large enough, then there exists a function $f \in L^2$ such that

$$\int \big| f_n(x) - f(x) \big|^2 dx \to 0 \quad \text{as} \quad n \to \infty.$$

This is Fischer's version of the Riesz–Fischer theorem but Riesz was aware of the equivalence. Riesz actually showed that given an orthogonal sequence $f_1, f_2, \ldots$ of functions of unit length and a sequence $c_1, c_2, \ldots$ of scalars such that $\sum_i |c_i|^2$ is finite, there exists a function f in the L^2 space such that $\langle f_i, f \rangle = c_i$.

Another early result of Riesz, discovered independently by Fréchet relying on an earlier paper of Riesz, tells us that any bounded linear functional A of L^2 is induced by an element g of L^2 in the form of integration of the product {55}:

$$A(f) = \int f(x) g(x)\, dx \qquad (f \in L^2) \tag{1}$$

for some $g \in L^2$ if there exists a constant M_A such that

$$\int \big| f(x) \big|^2 dx \le 1 \quad \text{implies} \quad \big| A(f) \big| \le M_A \qquad (f \in L^2).$$

It is still a pleasure to read the original works of Riesz. In 1910 he published a paper in Hungarian (Integrálható függvények sorozatai {57}), in which he explains his understanding of L^2 and the above mentioned two results. It is remarkable that abstract functional analysis did not exist at that time, nevertheless he understood his own (as well as Lebesgue's, Fréchet's and Fischer's) result in a very modern and intrinsic way. For example, he wrote that *"The extension of the concept of integral due to Lebesgue is an indispensable condition for my theorem, similarly to the fact that validity of certain theorems of algebra or arithmetics require the appropriate extension of the concept of numbers"*. In this paper he defined the *weak topology* on the space L^2 and shows that a bounded sequence contains a weakly convergent subsequence.

In our present language, (1) describes the dual of the L^2 space. The *dual space* was certainly a concept that should be attributed to Riesz. He defined the dual of L^2 in 1907 and in 1909 he dealt with the dual of the space of continuous functions, *Sur les opérations fonctionnelles linéaires*. Let $C[a,b]$ denote the set of all continuous real-valued functions on the interval $[a,b]$. In 1903 Hadamard wanted to describe all linear functionals $U : C[a,b] \to \mathbb{R}$ such that $Uf_n \to Uf$ whenever $f_n \to f$ uniformly. He took a function F such that

$$f(x) = \lim_{n\to\infty} n \int_a^b f(t)F\big(n(t-x)\big)\,dt$$

uniformly in $x \in [a,b]$, for example $F(x) = \exp(-x^2)$ would do, and he showed that

$$U(f) = \lim_{n\to\infty} \int_a^b f(t)\Phi_n(t)\,dt,$$

where $\Phi_n(t)$ is the value of the functional U at the function $x \mapsto nF\big(n(t-x)\big)$. Riesz described the continuous linear functionals of $C[a,b]$ by means of the Stieljes integral and removed the arbitrariness of the function F in Hadamard's theorem {56}. He proved that there exists a function α of bounded variation such that

$$U(f) = \int_a^b f(x)\,d\alpha(x), \tag{2}$$

moreover α is unique if $\alpha(a) = 0$ and the left continuity of α are required. For any $a < t < b$ he considered the function

$$f_t(x) = \begin{cases} x-a, & \text{if } a \le x \le t, \\ t-a, & \text{if } t \le x \le b. \end{cases}$$

He showed that the function $A : t \mapsto U(f_t)$ is Lipschitz and took $-\alpha(t)$ as one of the derived numbers of A at the point t. Then it was a standard procedure to modify α to fulfil the additional requirements and to keep (2).

Extending his work on the space L^2, Riesz devoted a fundamental paper to L^p spaces in 1910, *Untersuchungen über Systeme integrierbarer Funktionen* {58}. L^p is the set of all complex valued measurable functions such that $|f|^p$ is integrable. He restricted himself to the case $p > 1$ and extended the Hölder and Minkowski inequalities

$$\left|\sum_{k=1}^{n} a_k b_k\right| \le \left(\sum_{k=1}^{n} |a_k|^p\right)^{1/p} \left(\sum_{k=1}^{n} |b_k|^q\right)^{1/q}, \quad \text{where} \quad \frac{1}{p} + \frac{1}{q} = 1,$$

$$\left(\sum_{k=1}^{n} |a_k + b_k|^p\right)^{1/p} \le \left(\sum_{k=1}^{n} |a_k|^p\right)^{1/p} + \left(\sum_{k=1}^{n} |b_k|^p\right)^{1/p}$$

to measurable functions. If $f \in L^p$ and $g \in L^q$, then fg is integrable and

$$\left|\int f(x)g(x)\,dx\right| \le \left(\int |f(x)|^p\,dx\right)^{1/p} \left(\int |g(x)|^q\,dx\right)^{1/q}.$$

Moreover, if $f, g \in L^p$, then $f + g \in L^p$ and

$$\left(\int |f(x) + g(x)|^p\,dx\right)^{1/p} \le \left(\int |f(x)|^p\,dx\right)^{1/p} + \left(\int |g(x)|^p\,dx\right)^{1/p}.$$

He extended several definitions and results from the theory of L^2 spaces. He defined *strong convergence* in L^p as $f_n \to f$ if and only if $\int |f_n(x) - f(x)|^p\,dx \to 0$. His first definition of *weak convergence* was different from today's usual one. He said that $f_n \to f$ weakly if

$$\int_a^t f_n(x)\,dx \to \int_a^t f(x)\,dx$$

for all t in the interval on which the functions are defined. He showed that this is equivalent to

$$\int (f(x) - f_n(x))\,g(x)\,dx \to 0 \quad \text{for all} \quad g \in L^q.$$

He proved the *weak compactness* of the unit ball of L^p and he was particularly interested in the solution of the infinite system of linear equations

$$\int_a^b f_i(x)\xi(x)\,dx = c_i, \tag{3}$$

where $\xi(x)$ is the unknown and the $f_i(x)$'s belong to L^q. (The subscript i can run over an arbitrary set, countable, or not.) One cannot give an easy condition for the existence of the solution. He claimed that the condition is the existence of a constant M such that

$$\left|\sum_{i\in I}\mu_i c_i\right|^q \le M^q \int_a^b \left|\sum_{i\in I}\mu_i f_i(x)\right|^q dx \tag{4}$$

holds for all finite subsets I of the index set and for all complex numbers μ_i. Riesz was aware of the importance of the case where the f_i's are all the functions in L^q. Then (4) is exactly the roundedness of the functional L defined as $L(f_i) = c_i$ and he discovered that the dual of L^q can be identified with L^p. He achieved the first example of what we call today *reflexive Banach space.*

The system of equations (3) is related to the moment problem. This means that given the continuous functions f_i on an interval $[a, b]$ and a sequence c_i of real numbers $(i = 1, 2, \dots)$, an increasing function α should be found such that

$$c_i = \int_a^b f_i(x)\, d\alpha(x) \qquad (i = 1, 2, \dots) \tag{5}$$

(cf. (2)). In the original moment problem $f_i(t) = t^i$. A trivial necessary condition for the existence of α is the property that $\sum_{j=1}^n \lambda_j f_j \ge 0$ should imply $\sum_{j=1}^n \lambda_j c_j \ge 0$. If this is fulfilled then

$$U_0\left(\sum_{j=1}^n \lambda_j f_j\right) = \sum_{j=1}^n \lambda_j c_j$$

defines a positive functional on the linear span of the functions f_i and this functional should be extended to all continuous functions. The representation theorem (2) could be used. The moment problem belonged to the circle of ideas Frédéric Riesz worked on. His brother, **Marcell [Marcel] Riesz** (1886–1969), considered the moment problem as the question of extension of a positive functional. His method works in a very general setting, where a linear functional is defined on a subspace and the positivity is determined by a convex cone. His beautiful method is applicable not only to the power moment problem but many related problems in function theory (see Section II.6 in {1}).

In 1920 Frédéric Riesz published a detailed paper dedicated to an elementary presentation of Lebesgue's integral (B4 in [156]). The method is based on a completely elementary particular case of Lebesgue measure, namely on the definition and simplest properties of the sets of measure zero (briefly null sets): A set $A \subset R$ is a null set if it can be covered, for an arbitrary $\varepsilon > 0$, with a sequence of intervals $[a_n, b_n]$ such that $\sum_{n=1}^{\infty}(b_n - a_n) < \varepsilon$. A statement is true a.e. if it is true everywhere with the exception of the points of a null set.

The starting point is the, still quite elementary, definition of the integral of *simple functions,* i.e., functions in $[a, b]$ such that there is a decomposition of $[a, b]$ into finitely many pairwise disjoint subintervals in each of which the function is constant. For simple functions, the (Riemann) integral can be given by a finite sum.

Now a function f, bounded in $[a, b]$, is said to be *integrable* if there exists a bounded sequence of simple functions f_n (i.e., $|f_n| \leq M$ for some M) such that $f_n \to f$ a.e. in $[a, b]$. It can be shown that, under these conditions, the integrals $\int_a^b f_n(x)\,dx$ converge to a limit depending only on f (i.e., independent of the sequence); this limit defines $\int_a^b f(x)\,dx$.

In order to define the integral of an unbounded function, let us first say that a (bounded or unbounded) function is *measurable* in $[a, b]$ if it is the pointwise limit of an a.e. convergent sequence of simple functions. The integral $\int_a^b f(x)\,dx$ of a measurable function f is defined as the limit of the sequence $\int_a^b f_{c_n,d_n}(x)\,dx$, where (c_n) is an arbitrary sequence tending to $-\infty$ and (d_n) is one tending to $+\infty$, while f_{c_n,d_n} is the truncation of f: it is equal to $f(x)$ if $c_n \leq f(x) \leq d_n$, to c_n if $f(x) < c_n$, and to d_n if $f(x) > d_n$; the function f is said to be *integrable* if the above limit exists, is finite and independent of the choice of the sequences (c_n) and (d_n).

Based on these definitions, it is not difficult to deduce the usual properties of the integral (linearity, theorems on the integration of sequences of functions, etc.) It can be easily shown that, for a function integrable in the sense of Riemann, the new integral exists and is equal to the Riemann integral.

Riesz presented his exposition of the theory of the Lebesgue integral in his courses on analysis. A detailed exposition can be found in the monograph [157] which was later translated into several languages.

In two papers (B5 and B6 in [156]) Riesz analyzes the role of *Egoroff's theorem,* which states that a convergent sequence of measurable functions is uniformly convergent eliminating a subset of arbitrarily small measure

{9}, in the theory of the Lebesgue integral. In particular, he indicates the modifications necessary for extending the theorem for applications in the theory of the *Lebesgue–Stieltjes integral.* The Stieltjes integral apparently captured Riesz's attention because it played a decisive role in his result concerning the integral representation of bounded linear operations on the function space $C(a, b)$ of continuous functions (see (2)).

In three short papers (B7, B8 and B9 of [156]) and in his letters to *G. H. Hardy,* Riesz gives simple proofs for some integral inequalities in particular, the celebrated maximal inequality of Hardy and Littlewood; in general, arguments are based on the use of the *distribution function* $m(y) = m\big(\{x \in [a, b] : f(x) < y\}\big)$ associated with a function f measurable in $[a, b]$, where m denotes Lebesgue measure. In B9, he uses the so called *Riesz lemma* to furnish an elementary proof of Lebesgue's theorem: every monotone function is almost everywhere differentiable (see B10, B11 and B12 in [156]). In its simplest form, i.e., for continuous functions, the Riesz lemma is so elementary that its proof can be included here.

Riesz lemma. *If f is continuous in the interval $[a, b]$, then the set H of points $x \in [a, b]$ for which there exists some point $x < x' \le b$ such that $f(x) < f(x')$ is open: $H = \bigcup_k (a_k, b_k)$ and $f(a_k) \le f(b_k)$ for each k.*

The set H can be empty; in this case we have nothing to prove. If $H \neq \emptyset$, it is evidently open by the continuity of f so that the representation $H = \bigcup (a_k, b_k)$ is clearly possible. Fix a k and consider $a_k < x < b_k$. Let x_0 be one of the points in the interval $[x, b]$ where the value of f is maximal. Then $x \le x_0 < b_k$ is impossible since it would imply $x_0 \in H$ and the existence of an $x_0 < x' \le b$ satisfying $f(x_0) < f(x')$. Thus $b_k \le x_0 \le b$ and then $f(b_k) \ge f(x_0)$ as $b_k \notin H$. On the other hand, $f(x) \le f(x_0)$ by the choice of x_0, hence $f(x) \le f(b_k)$ and, from the continuity of f, $x \to a_k$ yields $f(a_k) \le f(b_k)$.

Using the Riesz lemma, the proof of Lebesgue's theorem becomes almost completely elementary. The Riesz lemma quite automatically provides coverings of the exceptional set by systems of intervals of arbitrarily small total length in the proof of Lebesgue's theorem on the a.e. differentiability of a monotone function.

Riesz himself was aware that his lemma can be used for proving further interesting theorems of measure theory (B9 and B13 of [156]) and ergodic theory (G5, G7 and G8 in [156]). Much later, the lemma was generalized to several variables (cf. {5} and [166]).

The fact that Lebesgue's theorem on the differentiability of monotone functions obtained an elementary proof through the Riesz lemma, suggested to Riesz a new approach to Lebesgue's integral theory, based on the differentability of monotone functions. He presents his ideas in two papers (B14 and B15 of [156]) The starting point is the following observation: Let $f \geq 0$ in the interval $[a, b]$ and suppose that there exists a function F, increasing in $[a, b]$ and satisfying $F'(x) = f(x)$ a.e. in (a, b). Then there exists, among these F, one for which the difference $F(b) - F(a)$ is the smallest possible.

After having proved this, we say that $f \geq 0$ is *integrable* in $[a, b]$ if there is an F as above, and define $\int_a^b f$ as the minimum of $F(b) - F(a)$. A function f of arbitrary sign is said to be integrable if $f^+ = \max(f, 0)$ and $f^- = -\min(f, 0)$ are integrable and then we define $\int_a^b f = \int_a^b f^+ - \int_a^b f^-$. From these definitions, one can deduce without any difficulty the usual properties of the integral, e.g., the theorems on the integration of sequences of functions.

In the years after World War II, Riesz wrote some big expository papers on the evolution of the concept of the integral (B16, B17 in [156]) and on the role of null sets in real analysis (B18 in [156]). It is natural that his own ideas played a central role in all these summaries.

The original proof of the Riesz lemma due to Frédéric Riesz, was slightly more complicated; the idea of applying the above point x_0 is due to his brother **Marcell [Marcel] Riesz** (1886–1969). Marcel Riesz was also an outstanding mathematician, he lived most of his life in Sweden and had a wide scientific interest, including functional analysis, partial differential equations and algebra. Assume that a linear operator A is defined on a set of measurable functions and its values are also measurable functions on a different space. Assume that A has a finite norm $C(p, q)$ when it is regarded as a map from L^p to L^q ($1 \leq p \leq \infty$, $1 \leq q \leq \infty$). The *Riesz convexity theorem* of Marcel Riesz tells us that $\log C(p, q)$ is a convex function of the variables $(p^{-1}, q^{-1}) \in [0, 1] \times [0, 1]$. The convexity theorem became a starting point of abstract interpolation theorems. The spaces L^p and $L^{p'}$ are connected by a path of Banach spaces (namely the L^q spaces, when q is between p and p'). Under some conditions a construction works for any two Banach spaces, this is a very concise description of the *interpolation theory* due Calderón, Lions and Peetrewhich has its roots in the work of Marcel Riesz.

A considerable part of the work of several Hungarian mathematicians in the first part of the 20th century was devoted to an important application of Lebesgue integral, namely to the calculation of the area of surfaces. Surface area is only seemingly an easy two-dimensional analogue of arc length. Since the work of *Hermann Amandus Schwarz,* we know that the theory of surface area is essentially more complicated. Recall that if, say,

$$x = \varphi(t), \quad y = \psi(t), \quad z = \chi(t) \quad (a \le t \le b)$$

is the parametric representation of a continuous curve in R^3 (i.e., φ, ψ, χ are continuous in $[a, b]$) then the *length* of the curve can be defined as the least upper bound of the lengths of polygons inscribed in the curve, namely obtained with the help of a subdivision of $[a, b]$ by points $a = t_0 < t_1 < t_2 < \cdots < t_n = b$ and taking as vertices of the polygon the points of the curve with parameters t_i; the curve is *rectifiable* iff this least upper bound is finite.

Schwarz discovered that the area of a surface *cannot* be defined in a similar way. Even for very simple surfaces, e.g., for a circular cylinder, it can happen that the areas of all inscribed polyhedra are unbounded from above, and the surface can be uniformly approximated by inscribed polyhedra so that their area tends to an arbitrary limit not less than the usual (elementary) area of the surface.

Consider, for the sake of simplicity, a surface having an equation $z = f(x, y)$ where f is continuous in a rectangle $R = [a, b] \times [c, d]$. In this case, an idea due to Lebesgue again produces a suitable definition of the area of the surface. Consider a subdivision of R into pairwise non-overlapping triangles $T_1, \ldots, T_n$ and a function g continuous in R and linear in each of the triangles T_i. The equation $z = g(x, y)$ corresponding to the *piecewise linear* function g can be considered as representing a polyhedron P having an elementary area $a(P)$. Let us consider a sequence of subdivisions having the property that the functions g_n converge uniformly to f and the areas $a(P_n)$ have a limit l; this limit may depend on the sequence (g_n) and then the smallest possible limit can be considered as the area of the surface; we shall call it the *Lebesgue area* $L(f)$ of $z = f(x, y)$.

In the case of a good function f (e.g. if f has continuous partial derivatives f_x and f_y in R) it is not difficult to show that the Lebesgue area can be computed with the help of the classical formula

$$L(f) = \iint \sqrt{f_x^2 + f_y^2 + 1}\, dx\, dy; \tag{6}$$

however, in the general case of a continuous f, the definition of $L(f)$ does not directly involve any method for calculating it.

This was the motivation for **Zoárd Geőcze,** in one of his first papers on the theory of surfaces, presented as a Thesis to the Sorbonne in Paris in 1908 (Quadrature des surfaces courbes, *Ungar. Ber.*, **26** (1910), 1–88.), to introduce the following expressions:

$$G_1(f,I) = \int_\alpha^\beta \left| f(x,\delta) - f(x,\gamma) \right| dx,$$

$$G_2(f,I) = \int_\gamma^\delta \left| f(\beta,y) - f(\alpha,y) \right| dy,$$

$$G(f,I) = \left(G_1(f,I)^2 + G_2(f,I)^2 + |I| \right)^{1/2},$$

where $I = [\alpha,\beta] \times [\gamma,\delta]$ is a subinterval of R and $|I|$ denotes the area $(\beta-\alpha)(\delta-\gamma)$ of I. He considered further the limit that we call nowadays the *Burkill integral* of the interval function G; in order to define it, let us consider a subdivision $\mathcal{I} = \{I_1,\dots,I_n\}$ of R into pairwise non-overlapping subintervals I_i, then take the sum

$$s(\mathcal{I}) = \sum_1^n G(f,I_i)$$

and the (always existing) limit $H(f,R)$ of $s(\mathcal{I})$, i.e., a (finite or infinite) number to which $\left(s(\mathcal{I}_n)\right)$ converges whenever the subdivision $\mathcal{I}_n$ is *infinitely refining* (i.e., varies such a manner that the maximum of the diameters of the intervals belonging to $\mathcal{I}_n$ tends to 0).

Now Geőcze proposes to consider the value $H(f,R)$ as the area of the surface $z = f(x,y)$. This is motivated by the result that $H(f,R) = L(f)$ whenever the function f satisfies a *Lipschitz condition* (i.e., there is a constant M such that $\left| f(x',y') - f(x,y) \right| < M\left(|x'-x| + |y'-y| \right)$ whenever $(x,y),(x',y') \in R$). This proposal is well-motivated because **Tibor Radó** (1895–1967) proved later that the equality $H(f,R) = L(f)$ is valid for any continuous function f {48}. Thus Geőcze found in fact a method for calculating the Lebesgue area of an arbitrary continuous surface defined by an equation $z = f(x,y)$ $\left((x,y) \in R \right)$.

Geőcze found also a necessary and sufficient condition for the value $H(f,R)$ to be finite, i.e., by Radó's theorem, for the continuous surface $z = f(x,y)$ to have a finite Lebesgue area. This is the following: let the function f be of bounded variation as a function of x in the interval $[a, b]$ for almost every fixed $y \in [c, d]$ and as a function of y in the interval $[c, d]$ for almost every fixed $x \in [a, b]$; let us denote by $V_1(y)$ the total variation of $f(x, y)$ as a function of x over the interval $[a, b]$ and by $V_2(x)$ the total variation of $f(x, y)$ as a function of y over the interval $[c, d]$; the condition postulates that V_1 should be (Lebesgue) integrable in $[c, d]$ and V_2 be integrable in $[a, b]$.

This condition due to Geőcze was rediscovered by *Leonida Tonelli* {66}; a function satisfying this condition is said to be of *bounded variation in the Tonelli sense.* Tonelli found also a necessary and sufficient condition for the classical formula (6) to give the Lebesgue area of the continuous surface $z = f(x, y)$. A function f satisfying this condition is said to be *absolutely continuous in the Tonelli sense.* This theory fills Chapter V of the brilliant monograph {63} of Stanislaw Saks, where works due to Geőcze and Radó are often quoted.

The problem of calculation of the area of surfaces is essentially more complicated if we consider continuous surfaces having a parametric representation; suppose the surface S is given in the form

$$x = f(u, v), \quad y = g(u, v), \quad z = h(u, v),$$

where f, g, h are continuous in a rectangle $R = [a, b] \times [c, d]$ of the uv-plane. Geőcze made a few first steps in this direction in {12} and in the works "A rectifiabilis felületről" {14} "A felület területének Peano-féle definitiójáról" {15} written in Hungarian. However, the thorough discussion of this problem was mainly accomplished by Radó who not only published a long series of papers on this subject but is also the author of a great monograph [144] containing a deep analysis of the serious difficulties of the problem.

Besides the Lebesgue area $L(S)$ of the surface S, defined with the help of sequences of polyhedra quite similarly as in the case of surfaces represented in the form $z = f(x, y)$, it is convenient to introduce another kind of area $a(S)$ playing a role similar to the expression $H(f, R)$ in the theory of surfaces $z = f(x, y)$. This is done, based on ideas of Geőcze {12} by Radó in {49, 50, 51} and in [144]. The concept of $a(S)$ can be used in examining the properties of the surfaces of zero area {13, 53}.

The role of the surface area $a(S)$ in calculating the Lebesgue area $L(S)$ is emphasized by the fact that, in many cases, we have $a(S) = L(S)$ and, at the same time, $a(S)$ is often equal to the value of the classical integral formula

$$\iint_R W(u,v)\,dudv, \tag{7}$$

where

$$W(u,v) = \left(J_1(u,v)^2 + J_2(u,v)^2 + J_3(u,v)^2\right)^{1/2}$$

and

$$J_1(u,v) = \frac{\partial(g,h)}{\partial(u,v)}, \quad J_2(u,v) = \frac{\partial(h,f)}{\partial(u,v)}, \quad J_3(u,v) = \frac{\partial(f,g)}{\partial(u,v)}$$

are Jacobians. Radó {52} has shown that the value of (7) is always $\le L(S)$, whenever the partial derivatives f_x, f_y, g_x, g_y, h_x, h_y exist a.e. in R.

In the general case, Radó has shown in {54} that, instead of the concept of functions of bounded variation and absolutely continuous in the Tonelli sense, it is possible to introduce the concept of a surface S of *essential bounded variation* and *essentially absolutely continuous,* respectively, further instead of the ordinary Jacobians J_i, *essential generalized Jacobians* and, with the help of them, a generalized function $W_e(u,v)$. Now if $L(S)$ is finite, then S is of essential bounded variation, $W_e(u,v)$ exists a.e. on R and we have the inequality

$$\iint_R W_e(u,v)\,du\,dv \le L(S).$$

The sign of equality holds if S is essentially absolutely continuous. Moreover, if the partial derivatives $f_x, \ldots, h_y$ exist a.e. in R, then $W_e(u,v)$ can be replaced here by $W(u,v)$. As to the equality $a(S) = L(S)$, it holds whenever $L(S)$ is finite and also if $a(S) = 0$.

Radó's results in the theory of surface area play, of course, an important role in his monograph on a famous question in differential geometry [142]. He also published a monograph together with Reichelderfer [145], where the methods developed in the theory of surface area play an essential role. In his last papers, he combines the methods of this theory with methods of general measure theory {52} and of three papers written in collaboration with *E. J. Mickle* {31, 32, 33}.

Geőcze and Radó were decisive personalities in the theory of surface area and their works are quoted everywhere in the literature of this important chapter of Analysis.

György [George] Alexits (1899–1978) became later a famous researcher in the theory of orthogonal series; however, one of his early papers {2} is an essential contribution to an important chapter of real analysis, namely to the theory of Baire functions. A paper of **Pál [Paul] Veress** (1893–1945) {67} is concerned with the same theory. Both papers are quoted in the monograph of *Hans Hahn* (Reelle Funktionen, Leipzig, 1932). Veress was the author of the first textbook on real analysis in Hungarian.

At the beginnings of functional analysis *integral equations* enjoyed a lot of attention. They are of the form

$$\varphi(s) = f(s) + \lambda \int_a^b K(s,t) f(t)\, dt,$$

where φ is a continuous function on $[a,b]$, λ is some complex parameter, $K(s,t)$ is continuous on $[a,b] \times [a,b]$ and $f(t)$ is the unknown function. For example, the Dirichlet problem could be reduced to such an integral equation. *David Hilbert,* in his very famous and fundamental paper of 1906, replaced the integral equation by an older concept of an infinite system of linear equations. Let u_n be a complete orthonormal sequence of continuous functions on $[a,b]$. If f is a solution of the equation, then we can consider the generalized Fourier coefficients

$$k_{ij} = \int_a^b \int_a^b k(s,t) u_i(s) u_j(t)\, ds\, dt,$$

$$x_i = \int_a^b \varphi(s) u_i(s)\, ds \quad \text{and} \quad y_i = \int_a^b f(s) u_i(s)\, ds.$$

In this way we arrive at the infinite system of linear equations

$$y_i + \lambda \sum_{j=1}^{\infty} k_{ij} y_j = x_i,$$

where the sequences x_i and y_i are square summable and k_{ij} is an infinite matrix (of a certain bilinear form). Hilbert himself worked with square integrable sequences and introduced the important concepts of continuity and complete continuity, mostly for symmetric bilinear forms. It is not our aim to give more details about Hilbert's work on integral operators, we want to turn to the work of Riesz on *completely continuous operators.*

His lecture delivered in a session of the Hungarian Academy of Sciences in 1916 appeared in the journal Mathematikai és Természettudományi

Értesítő with the title *Lineáris függvényegyenletekről* in Hungarian {60} in 1917, and the German translation *Über lineare Funktionalgleichungen* {61} was published in 1918. The subject of this paper is the invertibility of certain transformations and Riesz gave the definition and spectral theory of completely continuous transformations. He works on the space of all continuous functions on an interval, but he notes that similar methods work on other function spaces, i.e. on L^2, where they are even simpler. He uses the *norm* of a function f: $\|f\|$, which is the maximal value of the function $|f(x)|$. The same concept and the same notation is standard today. He carried over Hilbert's definition of a completely continuous bilinear form (based on the weak topology) to the new situation. He defined a linear mapping as completely continuous if the image of a bounded sequence is compact. (In today's language one would replace "compact" by "precompact" or by "relative compact".) The novelty of his paper is that he realized that Fréchet's concept of compactness is the proper tool to deal with completely continuous operators and he uses only the axiomatic definition of norm years before it was introduced under the name of "Banach space". He gives in an entirely geometric language what is known nowadays as the *Riesz–Fredholm theory of compact operators.*

In 1918 Riesz left Kolozsvár, after the World War I the town became part of Roumania. For two years Riesz lived in Budapest and in 1922 he became professor of the newly founded university at Szeged.

A partially ordered real linear space L has an order structure which is compatible with the linear structure. This means, that for any pair of elements f and g in L satisfying $f \leq g$ it follows that $f + h \leq g + h$ holds for all $h \in L$ and $af \leq ag$ holds for all real numbers $a \geq 0$. If, in addition, the order structure in L is a lattice structure, then L is called a *Riesz space.* In the present language Frédéric Riesz was interested in the ordered dual of an ordered vector space and the basic example was the space of continuous functions. His lecture at the International Mathematical Congress at Bologna in 1928 was devoted to this subject and he returned to it in an 1940 Annals of Mathematics paper (which was the translation of his 1937 inaugural lecture at the Hungarian Academy of Science). Riesz put emphasis on the following decomposition property: If $f_1 + f_2 = g_1 + g_2$, then there are elements f_{11}, f_{12}, f_{21} and f_{22} such that $f_1 = f_{11} + f_{12}$, $f_2 = f_{21} + f_{22}$, $g_1 = f_{11} + f_{21}$ and $g_2 = f_{12} + f_{22}$. In the space of continuous functions one can choose $f_{11} := \min(f_1, g_1)$ and $f_{22} = f_1 + f_2 - \max(f_1, g_1)$.

Between 1938 and 1948 Riesz dealt in eight papers with *ergodic theorems.* The ergodic and quasi-ergodic hypothesis were born in statistical mechanics and von Neumann gave the following mathematical formulation. Let $\mathcal{H}$ be a Hilbert space and T be a bounded linear transformation on $\mathcal{H}$. According to *von Neumann's mean ergodic theorem* the averages

$$s_n(f) := \frac{1}{n}(f + Tf + \cdots + T^{n-1}f)$$

converge to a T-invariant vector for every vector $f \in \mathcal{H}$, when T is a unitary operator. Riesz gave a very elegant proof for von Neumann's result. Riesz showed that the orthogonal complement of the set

$$\{f - s_n(f) : n \in \mathbb{N},\ f \in \mathcal{H}\}$$

is the fixed point set of T. From this fact one can prove the convergence of the averages and the proof requires only the hypothesis $\|Tf\| \le \|f\|$ for every $f \in \mathcal{H}$, that is T is a contraction. The Hilbert space version of the mean ergodic theorem corresponds to L^2 spaces and Riesz considered other L^p spaces as well. Much later ergodic theory appeared again in Hungarian functional analysis in the context of operator algebras: **István Kovács** and **József Szűcs** obtained the first mean ergodic theorem in von Neumann algebras {25}.

Their result implies that if α is an automorphism of a von Neumann algebra admitting a faithful normal invariant state, then the averages

$$s_n := \frac{1}{n}(I + \alpha + \cdots + \alpha^{n-1})$$

converge to a conditional expectation onto the fixed point algebra, pointwise in the strong operator topology.

John von Neumann was born Neumann János in 1903 in Budapest. He was a child prodigy, a prodigious student and he left his mark not only on pure mathematics but on theoretical physics, on meteorology, on economics, on digital computers and on more. He was the mathematician admired by most scholars outside of his own discipline. In its December 24 issue in 1999, The Financial Times has declared John von Neumann to be *"Man of the Century"*.

In the years 1914–21 von Neumann studied in Budapest's Lutheran Gymnasium. In 1921 he went to become a chemical engineer first to Berlin University and then in 1923 he took the entrance examination for the prestigious course in the chemical engineering department of the famous

Eidgenössische Technische Hochschule in Zürich. When Hermann Weyl was absent from Zürich, the undergraduate chemist von Neumann took over the teaching of some of his classes. During his university years at the ETH, von Neumann was passing courses in Budapest University (which he never attended) from where received his Ph.D. with highest honors in 1925.

In early autumn of 1926 von Neumann arrived in Göttingen. He immediately learnt quantum theory from Heisenberg's lectures. Von Neumann became an axiomatizer of quantum mechanics on behalf of the so-called Copenhagen school (which did not include Schrödinger.) To Hilbert's delight, von Neumann's mathematical exposition made much use of Hilbert's own concept of Hilbert space. However, it is not sure that axiomatization of the Hilbert space and its linear operators (as a substitute for infinite matrices) by the twenty-three-year-old von Neumann was to Hilbert's delight. Our present concept of Hilbert space, infinite dimensional complex vector space endowed with an inner product whose metric is complete and separable, was formulated by von Neumann. The rigorous quantum mechanics required the use of unbounded operators defined only on a subspace of a Hilbert space. Von Neumann developped several technicalities concerning such operators. The role of the graph, the difference between symmetric and selfadjoint operators, the spectral decomposition of unbounded selfadjoint operators were discovered by him. In his excellent textbook {29} **Peter Lax** makes the following historical comment: *In the 1960s Friedrichs met Heisenberg, and used the occasion to express to him the deep gratitude of the community of mathematicians for having created quantum mechanics, which gave birth to the beautiful theory of operators in a Hilbert space. Heisenberg allowed that this was so; Friedrichs then added that the mathematicians have, in some measure, returned the favor. Heisenberg looked noncommittal, so Friedrichs pointed out that it was a mathematician, von Neumann, who clarified the difference between a selfadjoint operator and one is merely symmetric. "What's the difference," said Heisenberg.*

After some earlier work on single operators, von Neumann turned to families of operators. He initiated the study of rings of operators, which are commonly called *von Neumann algebras* today. The papers which constitute the series *"Rings of operators"* opened a new field in mathematics and influenced research for half a century (or even longer). In the standard theory of modern operator algebras, many concepts and ideas have their origin in von Neumann's work.

A von Neumann algebra consists of bounded linear Hilbert space operators. The characteristic feature of the concept of von Neumann algebra is its very rich structure. A von Neumann algebra contains the spectral projections of all selfadjoint operators belonging to the algebra. In particular, there are many orthogonal projections in the algebra itself. Roughly speaking, the point in the concept of von Neumann algebra is that formation of product and spectral diagonalization of selfadjoint elements are posssible within the algebra. It turns out that the projections of a von Neumann algebra form a lattice in the sense that any two of them have a least upper bound and a greatest lower bound with respect to an appropriate and natural ordering. The lattice of projections is the starting point in the classification of von Neumann algebras and a ground for quantum logic. Von Neumann algebras are classified in terms of the range of a dimension function defined on the lattice of projections. The dimension function is the extension of the simple concept of rank (for matrices) and the peculiarity of the subject begins with the observation that in nontrivial cases this "rank" can be noninteger. Below the classification of von Neumann algebras is described. Also, the influence of measure theory on early operator algebra theory is demonstrated by a comparison of a measure-theoretic construction of **Alfréd Haar** with the dimension function of *Murray* and von Neumann. This example shows that the connection with measure theory and ergodic theory has been very important for operator algebras since the very beginning.

We denote by $B(\mathcal{H})$ the set of all bounded operators acting on the Hilbert space $\mathcal{H}$. For a subset $\mathcal{S} \subseteq B(\mathcal{H})$, its commutant $\mathcal{S}'$ is defined as the set of operators commuting with $\mathcal{S}$:

$$\mathcal{S}' = \{ K \in B(\mathcal{H}) \, : \, KS = SK \text{ for all } S \in \mathcal{S} \}.$$

Note that $\mathcal{S} \subseteq (\mathcal{S}')'$ holds obviously for any $\mathcal{S} \subseteq B(\mathcal{H})$. A family of operators acting on a Hilbert space and containing the identity operator is called a von Neumann algebra if it contains the adjoint, the linear combinations and the products of its elements and forms a closed subspace of the space of all bounded operators with respect to the topology of pointwise convergence. A von Neumann algebra is linearly spanned by its selfadjoint elements and the spectral resolution of the latter ones lies conveniently in the algebra. One of the first results of von Neumann, the *von Neumann's double commutant theorem,* was an equivalent algebraic definition of von Neumman algebras. Von Neumann's double commutant theorem asserts that a family of operators is a von Neumann algebra if and only if it contains the adjoint of its elements and coincides with its second commutant (that is, the commutant of its commutant). The remarkable point in the double commutant theorem is the lack of any topological requirement. In the concept of von Neumann algebra, topology and pure algebra are in great harmony.

The selfadjoint idempotents, called (orthogonal) projections, of a von Neumann algebra form an orthomodular, complete lattice with respect to the lattice operations $\wedge$, $\vee$, $\perp$ and the partial ordering $\leq$. Below we describe how these operations are defined in terms of the algebaric operations. The projections are in natural correspondence with the closed subspaces of the underlying Hilbert space and the set theoretical inclusion of subspaces induces a partial ordering on the projections. This ordering is equivalently defined as

$$p \leq q \quad \text{if} \quad pq = p. \tag{8}$$

The smallest projection with respect to this ordering is 0 and the largest one is the identity. For projections p and q, their meet (that is, greatest lower bound) $p \wedge q$ is the orthogonal projection onto the intersection of the range spaces of p and q. The projection $p \wedge q$ may be obtained as the strong limit of $(pq)^n$ as $n \to \infty$. The projection $p \vee q$ is defined as the smallest upper bound in the lattice of projections. ($p \vee q$ projects onto the closed subspace spanned by the range spaces of p and q.)

The orthocomplementation $\perp$ is defined as $p^\perp = I - p$. The orthomodularity of the lattice of projections means that the following so-called orthomodularity condition is fulfilled in the lattice:

$$q = p \vee (p^\perp \wedge q) \quad \text{for} \quad p \leq q. \tag{9}$$

This relation is a weakening of the distributivity condition and is an essential property of the lattice of projections.

Let p and q be two projections in a von Neumann algebra $\mathcal{M}$. The projections p and q are called equivalent (with respect to $\mathcal{M}$), $p \sim q$ in notation, if there is an operator x in $\mathcal{M}$ such that $p = x^*x$ and $q = xx^*$. In terms of the underlying Hilbert space, the equivalence of p and q means that there exists a partial isometry x in the given von Neumann algebra which sends the range space of p isometrically onto the range of q. An extended positive-valued function $D : \mathcal{P}(\mathcal{M}) \to [0, \infty]$ on the set $\mathcal{P}(\mathcal{M})$ of all projections of $\mathcal{M}$ is called a dimension function if it satisfies the following requirements:

(a) $D(p) > 0$ if $p \neq 0$ and $D(0) = 0$.

(b) $D(p) = D(q)$ if p and q are equivalent projections.

(c) $D(\sum_i p_i) = \sum_i D(p_i)$ if $p_i p_j = 0$ whenever $i \neq j$.

It is fundamental in the theory of von Neumann algebras that the dimension function is determined up to a positive multiple if the center of the algebra is trivial, that is, the algebra is a *factor.*

We sketch how the dimension function was obtained in {34}. A nonzero projection is called finite if it is not equivalent to a smaller projection. "Smaller" is understood here in the sense of the partial ordering (8). Murray and von Neumann proved in {34} that if f is a finite and e is an arbitrary projection in a factor then there exists a unique integer k such that

$$f = q_1 + q_2 + \cdots + q_k + p,$$

where $q_1, q_2, \ldots, q_k$ are pairwise orthogonal projections equivalent to e, p is a projection orthogonal to all q_i and equivalent to a subprojection of f. This integer k is denoted by

$$\left[\frac{f}{e}\right] \tag{10}$$

and this is the number of projections equivalent to e which may be packed into f in a pairwise orthogonal way. (10) is an integer and is only an approximate measure of the ratio of the subspaces corresponding to f and e. Now we fix a finite non-zero projection e_0 and a sequence e_n of non-zero finite projections converging to 0. The limit

$$\lim_{n\to\infty} \frac{\left[\frac{f}{e_n}\right]}{\left[\frac{e_0}{e_n}\right]} = \left(\frac{f}{e_0}\right) \tag{11}$$

forms a quantitative ratio of relative dimensionality, when the sequence e_n converges to 0 strongly. (Heuristically, the projection e_0 will have dimension 1, first we estimate the dimension e_n by comparison with e_0 and then the dimension of f is estimated by comparison with f_n.)

The relative dimension was defined in {34} as

$$D(e) = \begin{cases} 0 & \text{if } e = 0, \\ \left(\dfrac{e}{e_0}\right) & \text{if } e \text{ is finite}, \\ +\infty & \text{if } e \text{ is not finite}. \end{cases}$$

The use of the relative dimension in the classification of factors will be discussed below. Now we make a detour and compare the construction of the dimension function with that of the Haar measure on a locally compact topological group. The existence of a measure on an abstract locally compact group which is invariant under right translations was proven in 1932 by the Hungarian mathematician **Alfréd Haar** {17}. Von Neumann was in contact with Haar and knew his celebrated result before it was published. It is instructive to trace back the dimension function of a ring of operators to Haar's beautiful idea for the construction of the invariant measure.

Let G be a locally compact topological group and for a precompact $B \subset G$ and an open $U \subset G$ denote by $h(B;U)$ the number which gives at least how many right-translates of the set U are needed to cover the set B. This is an integer showing the size of B compared to U. $h(B;U)$ is translation invariant by construction. Of course, the smaller the U, the larger the $h(B;U)$. The latter one may increase to infinity when U runs over the neighbourhoods of a point. We need a normalization of $h(B;U)$. A compact set S of nonempty interior is chosen to normalize the measure. (S will be a set of unit measure.)

$$\lim_n \frac{h(B;R_n)}{h(S;R_n)} = \mu(B) \tag{12}$$

gives the measure of a compact set B if (R_n) is the filter of neighbourhoods of a point. The set function μ is right-translation invariant and additive on disjoint compact sets. After the measure μ of compact sets is obtained, measure-theoretic arguments are used to extend μ to a larger class of sets.

It is difficult to refrain from comparing Haar's idea with the construction of dimension function of projections in a von Neumann algebra: the similarity between the formulas (12) and (11) is striking. (12) yields the right-translation invariant size of subsets of a group G and (11) defines an invariant under partial isometries for projections in a von Neumann algebra. This example demonstrates how measure-theoretic arguments can survive in the apparently different discipline of operator algebras. Von Neumann devoted two papers to Haar measure. In {39}, he gave another proof for the existence and uniqueness in the compact case and in {40} he obtained uniqueness in the general locally compact case. Von Neumann presented several courses on measure theory and invariant measures at the *Institute for Advanced Study.* His lecture notes were published in 1999 by the American

Mathematical Society {43}. For him operator algebra theory was a *noncommutative outgrowth* of measure theory. Rosenberg's article (in this volume) is a complementary reading about noncommutative harmonic analysis {62}.

Now we continue the comparison of the relative dimension and Haar measure. The objective of integration theory is to construct a linear functional, called integral, from a certain measure. Murray and von Neumann extended the relative dimension functional to arbitrary selfadjoint elements of the given von Neumann algebra $\mathcal{M}$. Let $A = A^* \in \mathcal{M}$ and let $\int \lambda \, dE(\lambda)$ be its spectral resolution with a projection-valued measure E on the real line. Then by property (c) of the relative dimension, $D(E)$ is an ordinary measure and

$$\mathrm{Tr}_{\mathcal{M}}(A) = \int \lambda \, dD(E)(\lambda) \tag{13}$$

determines a real number when the integral on the right-hand side exists. The inconveniency of definition (13) is in the fact that for noncommuting self-adjoint operators A and B one cannot say much about the spectral resolution of $A+B$ in terms of the spectral resolutions of A and B. Murray and von Neumann expected that

$$\mathrm{Tr}_{\mathcal{M}}(A+B) = \mathrm{Tr}_{\mathcal{M}}(A) + \mathrm{Tr}_{\mathcal{M}}(B)$$

but this was proven in {34} only for commuting A and B. The general case was postponed to the subsequent paper {35}. It was established there that the abstract trace functional $\mathrm{Tr}_{\mathcal{M}}$ is linear. $\mathrm{Tr}_{\mathcal{M}}$ yields an analogue of an integral. (This analogy has developed into an operator-algebraic integration theory, including L^p spaces, measurable operators and so on. For this Segal proposed the term *"noncommutative integration"* in {64} since a commutative von Neumann algebra admits representations by functions.)

In {37} von Neumann established the structure of commutative von Neumann algebras: The selfadjoint part of a commutative von Neumann algebra consists of all bounded measurable functions of a certain selfadjoint operator. The classification of nonabelian algebras was carried out in {34}. Murray and von Neumann recognized that the center of the algebra plays an important role in the structure problem. The center of a von Neumann algebra $\mathcal{M}$ is a von Neumann algebra again and if it contains a projection z, then $\mathcal{M}$ becomes the direct sum of $z\mathcal{M}$ and $(I-z)\mathcal{M}$. Hence to decrease the complexity of an algebra, one may assume that its center does not contain a nontrivial projection. A von Neumann algebra is called a *factor* if its center is trivial, that is, if it contains the multiples of the identity operator only.

On a von Neumann factor, the dimension function is unique up to a scalar multiple. Murray and von Neumann proved that there are the following possibilities for the range of the dimension function of projections:

(I_n) $\{0, 1, \ldots, n\}$, where n is a natural number.

(I_∞) $\{0, 1, \ldots, n, \ldots, \infty\}$.

(II_1) The interval $[0, 1]$.

(II_∞) The interval $[0, +\infty]$.

(III) The two-element set $\{0, +\infty\}$.

In this classification all von Neumann factors were found to belong to the classes type *I*, type *II* or type *III*. (However, it is worth mentioning that at the time of the discovery of the classification it was not known whether type *III* factors exist.)

Factors are the building blocks of von Neumann algebras, hence the understanding of their structure has primary interest. According to the range of the dimension function of projections, a factor might be "trivial", "regular" or "singular". The trivial or type *I* is characterized by integer dimension, in the regular or type *II* case the dimension function has a continuous range and the singular or type *III* case is free of finite nonzero projections. To investigate the type *I* and type *II* cases Murray and von Neumann could utilize the dimension function; however, that tool was insufficient for type *III* factors. To have a feeling about the "singularity" of type *III* factors, one can think of a measure space in which all nonempty measurable sets have infinite measure. The full understanding of the type *III* case needed half a century. Ergodic theory was the first source of factors. Classification of von Neumann algebras is strongly related to conjugacy classes of transformations of measure spaces. The *Tomita–Takesaki theory* provided the new tools and revolutionized operator algebras in the 1970's. (The book {65} by Serban Strătilă and **László Zsidó** is a suggested introductory reading about von Neumann algebras.)

Factors of type *I* are characterized by the existence of minimal projections. If a maximal pairwise orthogonal family of minimal projections has cardinality n, then the factor is isomorphic to $B(\mathcal{H})$, where $\mathcal{H}$ is a Hilbert space of dimension n. In particular, for every $s \in \mathbb{N} \cup \{+\infty\}$, there exists only one factor of type I_s up to isomorphism. The existence of factors of type *II* and type *III* is not at all apparent, however. Murray and von Neumann constructed factors of type II_1 and type II_∞ by means of ergodic theory in {34}. Below we describe a method called *"group measure space construction"*. This construction yields factors of different type.

Let $(X, \mathcal{B}, \mu)$ be a measure space and let G be a countable group of measure-preserving transformations of X. The group measure space construction yields a von Neumann algebra acting on the Hilbert space $L^2(\mu) \otimes l^2(G)$, which is regarded as a set of functions ξ defined on G and with values in $L^2(\mu)$. (In this identification $\delta_g \otimes f$ corresponds to $\delta_g \times f$ for $g \in G$ and $f \in L^2(\mu)$.) For every $f \in L^\infty(\mu)$ define a bounded operator M_f acting on $L^2(\mu) \otimes l^2(G)$ as

$$(14) \quad \big((M_f\xi)(g)\big)(x) = f(g^{-1}x)\big(\xi(g)(x)\big) \quad \big(\xi \in L^2(\mu) \otimes l^2(G),\ g \in G\big)$$

and for every $g \in G$ we define a unitary V_g by the formula

$$(15) \qquad V_g(\xi)(g')(x) = \xi(g^{-1}g')(x) \qquad \big(\xi \in L^2(\mu) \otimes l^2(G),\ g' \in G\big).$$

Let $\mathcal{M}(\mu, G)$ be the von Neumann algebra generated by the operators

$$\big\{M_f : f \in L^\infty(\mu)\big\} \cup \{V_g : g \in G\}.$$

Then the choice of the unit circle with Lebesgue measure and (the powers of) an irrational rotation yields a factor of type II_1. The real line with Lebesgue measure and the rational translations give a factor of type II_∞. A factor of type *III* was constructed only in the third paper of the "Rings of Operators" series {42}. Von Neumann modified the above measure theoretic procedure by allowing measurable transformations preserving measure 0, nowadays they are called nonsingular transformations. In this way he produced a factor of type *III* from the Lebesgue measure of the real line and the group of all rational linear transformations. (Although Murray and von Neumann used the group measure space construction for the production of factors, which are called *Krieger factors* nowadays, the difficult question of isomorphism of factors that arose from different actions was clarified only 40 years later {28}. Krieger proved that two ergodic nonsingular transformations of a Lebesgue space give rise to isomorphic factors if and only if the two transformations are orbit equivalent.)

Von Neumann believed that among all factors the case II_1 has the strongest interest and expected that not all factors of type II_1 are isomorphic to each other. Von Neumann preferred the type II_1 case for two main reasons. One of these is the nice behavior of the unbounded operators affiliated with a type II_1 factor. It is well-known that addition and multiplication of such operators are particularly troublesome. The crux of the difficulty lies in the unrelatedness of the domain and range of such an operator with the domain of another one. Much of the difficulties evaporates, however, if one considers selfadjoint operators with spectral resolution in a factor of type II_1. The other reason why von Neumann attributed great importance to continuous finite factors is that he interpreted this lattice as the proper logic of a quantum system. The lattice of projections of such a factor is modular, that is, in addition to the orthomodularity property (9), the stronger condition

$$p \vee (p' \wedge q) = (p \vee p') \wedge q \quad \text{for} \quad p \leq q$$

holds for every p' (and not only $p' = p^{\perp}$). (Non-modularity of the projection lattice of an infinite dimensional factor of type I was considered by von Neumann as a pathology of the usual Hilbert space quantum mechanics as a non commutative probability theory.)

The paper "Rings of Operators IV" {36} has two important achievements concerning type II_1 factors. It is proved that there exist nonisomorphic type II_1 factors, and that there is only one hyperfinite type II_1 factor. A von Neumann factor is called *hyperfinite* if it is generated by an increasing sequence of finite dimensional subalgebras. (Nowadays such algebras are preferably called approximately finite dimensional, or AFD for short.) The hyperfinite type II_1 factor $\mathcal{R}$ may be produced in many different ways; for example, the above group measure space construction yields $\mathcal{R}$. The uniqueness of $\mathcal{R}$ reminds us of the uniqueness of a finite, atomless separable measure space. Factors of type II_1 did not play much role in the theory of von Neumann algebras until recent years. After *Jones* founded his index theory {19}, the study of subfactors of type II_1 factors has received much interest. Even a concise review of index would require a lot of space (cf. {23}) but its flavour is given below.

Let $\mathcal{N}$ be a von Neumann algebra acting on a Hilbert space $\mathcal{H}$ and having commutant $\mathcal{N}'$. Assume that both $\mathcal{N}$ and $\mathcal{N}'$ are type II_1 factors and let $\mathrm{Tr}_{\mathcal{N}}$ and $\mathrm{Tr}_{\mathcal{N}'}$ be the canonical normalized traces. For any vector $\xi \in \mathcal{H}$ the projection $[\mathcal{N}\xi]$ onto the closure of $\mathcal{N}\xi$ belongs to $\mathcal{N}'$ and similarly $[\mathcal{N}'\xi] \in \mathcal{N}$. The quotient

$$\dim_{\mathcal{N}}(\mathcal{H}) \equiv \frac{\mathrm{Tr}_{\mathcal{N}'}\left([\mathcal{N}\xi]\right)}{\mathrm{Tr}_{\mathcal{N}}\left([\mathcal{N}'\xi]\right)} \tag{16}$$

is known to be independent of the vector ξ and is called the coupling constant since the work of Murray and von Neumann. In a certain sense the coupling constant is the dimension of the Hilbert space $\mathcal{H}$ with respect to the von Neumann algebra $\mathcal{N}$. (When $\mathcal{N} \equiv \mathbb{C}I$, the coupling constant is the usual dimension of $\mathcal{H}$, hence the notation $\dim_{\mathcal{N}}(\mathcal{H})$.) V. Jones used the coupling constant to define the size of a subfactor of a finite factor. He was inspired by the notion of the index of a subgroup of a group, he therefore called this the relative size index.

Let $\mathcal{N}$ be a subfactor of a type II_1 von Neumann factor $\mathcal{M}$ possessing a unique canonical normalized trace $\mathrm{Tr}_{\mathcal{M}}$. The index is obtained as the quotient

$$[\mathcal{M} : \mathcal{N}] = \frac{\dim_{\mathcal{N}}(\mathcal{H})}{\dim_{\mathcal{M}}(\mathcal{H})}. \tag{17}$$

The number $[\mathcal{M} : \mathcal{N}]$ is not always an integer, and the possible values of the index form the following set:

$$\{t \in \mathbb{R} : t \geq 4\} \cup \left\{4\cos^2(\pi/p) : p \in \mathbb{N},\ p \geq 3\right\}. \tag{18}$$

This is the fundamental result of Jones which influenced a huge amount of subsequent research and renewed the almost forgotten coupling constant. *Vaughan F. R. Jones* was awarded the Fields Medal in 1992 for discovering a surprising relationship between von Neumann algebras and geometric topology (see {4} for a review). The index theorem was the first step towards his discovery.

Construction of factors was the main activity in the field of operator algebras after the papers "Rings of operators" for many years. It is out of the scope of this survey to summarize the constructions that were used to get more and more factors. Instead, we turn to the very end of the story. By the time the paper "Rings of Operators IV" was published (year 1943) it was known that the classes of type I_n, II_1 contain a unique (up to algebraic isomorphism) hyperfinite von Neumann factor. However, the types II_∞ and *III* remained unclear for many years. In 1956 **Lajos Pukánszky** constructed two different factors of type *III* {47, 24}. After his breakthrough infinitely many factors were constructed but the final classification was not achieved until the discovery of new invariants. Operator algebras achieved a revolutionary development in the late 60's after a relative isolation of 30 years.

Type *III* factors may be produced by means of infinite tensor product. Let $M_2(\mathbb{C})$ be the algebra of 2-by-2 matrices. Fixing $0 < \lambda < 1$ we can define a state φ on this algebra as follows.

$$\varphi(A) = \mathrm{Tr}\,(AD), \quad \text{where} \quad D = \begin{pmatrix} \frac{1}{\lambda+1} & 0 \\ 0 & \frac{\lambda}{\lambda+1} \end{pmatrix}.$$

(The matrix D is called the density matrix inducing φ.) A representation of the inductive limit of the n-fold tensor product of copies of $M_2(\mathbb{C})$ can be constructed by means of tensor product states of copies of φ. (The so-called *Gelfand–Naimark–Segal* construction is involved here, but we do not give more details.) The generated von Neumann algebra is a hyperfinite factor. For $\lambda = 1$, the type II_1 factor shows up, for $\lambda = 0$ we obtain a type I_∞ factor and for $0 < \lambda < 1$ a type III_λ factor $\mathcal{R}_\lambda$ appears. In fact, $\mathcal{R}_\lambda$ is the only hyperfinite type III_λ factor. Confined to hyperfinite type III_λ factors with $0 < \lambda < 1$ the Connes spectrum is a complete invariant due to the results of *Alain Connes.* He received the Fields Medal in 1983 for his work on von Neumann algebras including the classification of type *III* factors, approximately finite dimensional factors and automorphisms of the hyperfinite type II_1 factor {3}. After the work of Connes, the uniqueness of the hyperfinite type III_1 factor remained undecided. This question was answered positively somewhat later by *Uffe Haagerup* {16}. (In the case of type III_0, there are infinitely many nonisomorphic hyperfinite factors.)

Quantum mechanics influenced von Neumann to develop several ideas. He was the first person who summarized quantum theory in a comprehensive and mathematical form, his monograph [119] has been a standard reference in mathematical physics. Operator algebras consist of bounded operators but quantum mechanics needs unbounded ones. Von Neumann understood the importance of maximal symmetric operators on Hilbert spaces and introduced the entropy of statistical operators ([119] and {46}. The von Neumann entropy got a new information theoretic interpretation recently.

Béla Szőkefalvi-Nagy was born in Kolozsvár in Transylvania on July 29, 1913. His father was also a mathematician and his mother was a high school teacher. In his scientific papers he did not use his full name but the abbreviation B. Sz.-Nagy. Native Hungarians have always been surprised about the strange pronounciation of his name by foreigners. After World War I, the family moved to Szeged (Hungary).

During his university studies Sz.-Nagy was deeply influenced by *Frigyes Riesz, Béla Kerékjártó and Alfréd Haar.* Von Neumann's book on the foundations of quantum mechanics [119] and van der Waerden's book on group theory and quantum mechanics were his favorite readings. This was the time when quantum theory revolutionized both physics and mathematics. Between 1937 and 1939 Sz.-Nagy spent some time in Leipzig, Grenoble and Paris. From 1939 he worked for the University of Szeged; he became full professor in 1948.

Sz.-Nagy had rather wide mathematical interests. In one of his first papers he gave a new proof for Stone's theorem about the spectral representation of a strongly continuous one-parameter semigroup of unitary Hilbert space operators. Such a semigroup $U(t)$ is obtained in the form

$$U(t) = \int_{-\infty}^{\infty} e^{i\lambda t}\, dE(\lambda),$$

by means of a projection-valued measure E on the real line. Later he extended this result to semigroups of normal operators. He also wrote a very concise book on spectral theory. Generations learnt the spectral theorem from [177] published in 1942 by Springer Verlag.

Although Sz.-Nagy contributed to the theory of Fourier series and to approximation theory, the center of his interest was the Hilbert space and its linear operators. The basic example of a Hilbert space is the L^2 space, the space of square integrable functions over a measure space. On top of the standard Hilbertian structure L^2 has an order structure which is determined by the cone of positive functions. In an early paper Sz.-Nagy gave an abstract characterization of the positive cone. In other words, he listed the requirements of a cone of an abstract Hilbert space under which an isomorphism of the space with an L^2 space exists, such that the positive cones correspond to each other. He also proved that an invertible Hilbert space operator whose positive and negative powers are uniformly bounded is similar to a unitary operator.

The analysis of Hilbert space operators mostly concerns some particular classes of operators such as self-adjoint, unitary etc. The highlight of the scientific activity of Sz.-Nagy was the theory of contractions. It started with the unitary dilation theorem obtained in 1953. Let $\mathcal{H}$ be a Hilbert space and let T be a general bounded linear operator. Hence $\|T\|$ is finite and multiplying T by some constant we can achieve that $\|T\| \le 1$. (Such a T is called a contraction.) The dilation theorem says that there exist a Hilbert space $\mathcal{K} \supset \mathcal{H}$ and a unitary operator U on $\mathcal{K}$ such that

$$T^n f = PU^n f \quad \text{and} \quad (T^*)^n f = PU^{-n} f \qquad (f \in \mathcal{H})$$

for any $n \in \mathbb{N}$, where P denotes the orthogonal projection from $\mathcal{K}$ onto $\mathcal{H}$. The space $\mathcal{K}$ could be the direct sum of infinitely many copies of $\mathcal{H}$ and U can be written in the form of an infinite matrix (with operator entries) as follows:

$$U = \begin{bmatrix} & \vdots & \vdots & \vdots & \vdots & \\ \cdots & I & 0 & 0 & 0 & \cdots \\ \cdots & 0 & D_* & T & 0 & \cdots \\ \cdots & 0 & -T^* & D & 0 & \cdots \\ \cdots & 0 & 0 & 0 & I & \cdots \\ & \vdots & \vdots & \vdots & \vdots & \end{bmatrix},$$

where $D = (I - T^*T)^{1/2}$ and $D_* = (I - TT^*)^{1/2}$. Since the structure of a unitary operator is rather well-understood, the contractions could be investigated through the dilation.

In the study of contractions Sz.-Nagy had a longstanding cooperation with *Ciprian Foiaş* from 1956 to the end of his life. They wrote together 50 papers and the monograph [44]. An interesting class of operators is formed by the completely non-unitary contractions, they do not act unitarily on any subspace. The class C_0 is formed by the completely non-unitary contractions T for which there exists a function $0 \neq w \in H^\infty$ such that $w(T) = 0$. An operator $T \in C_0$ has the following remarkable properties:

(1) For every vector f, $T^n f \to 0$ and $(T^*)^n \to 0$ as $n \to \infty$.

(2) T has a nontrivial invariant subspace.

Recall that a linear operator T of a finite dimensional space always admits a polynomial p such that $p(T) = 0$. The definition and several properties of the class C_0 resemble the finite dimensional scenario. Sz.-Nagy and Foiaş found a quasisimilarity model for the C_0-contractions and a unitary equivalence model for arbitrary completely non-unitary contractions. Their lifting theorem is connected with the minimal isometric dilation of a contraction T. Let T_i be a contraction acting on a Hilbert space $\mathcal{H}_i$ and let V_i be the minimal isometric dilation of T_i acting on the space $\mathcal{K}_i$, $i = 1, 2$. If a bounded linear operator X from the space $\mathcal{H}_1$ to $\mathcal{H}_2$ has the property $T_2X = XT_1$, then there exist an operator Y from the space $\mathcal{K}_1$ to $\mathcal{K}_2$ such that $V_2Y = YV_1$. The lifting theorem of Sz.-Nagy and Foiaş extends earlier results of *T. Ando* and *D. Sarason.* Many applications are known, in particular to interpolation problems.

References

[44] Foiaş, Ciprian – Sz.-Nagy, Béla, *Harmonic Analysis of Operators on Hilbert Space,* North-Holland Publishing Co. (Amsterdam, 1970).

[62] Haar, Alfréd, *Összegyűjtött Munkái = Gesammelte Arbeiten,* ed. Béla Sz.-Nagy, Akadémiai Kiadó (Budapest, 1959).

[118] Neumann, János (John von Neumann), *Collected Works,* 6 volumes, ed. A. H. Taub, Pergamon Press (New York, 1961).

[119] Neumann, Johann von, *Mathematische Grundlagen der Quantenmechanik,* Die Grundlehren der Mathematischen Wissenschaften in Einzeldarstellungen, Band XXXVIII, J. Springer (Berlin, 1932).

[142] Radó, Tibor, *On the Problem of Plateau,* Ergebnisse der Mathematik und ihrer Grenzgebiete, Band 2, No. 2, Springer-Verlag (Berlin, 1933).

[144] Radó, Tibor, *Length and Area,* Amercian Mathematical Society Colloquium Publications, Vol. XXX. (1948).

[145] Radó, Tibor – Reichelderfer, P. V., *Continuous Transformations in Analysis with an introduction to algebraic topology,* Die Grundlehren der Mathematischen Wissenschaften in Einzedarstellungen, Band LXXV, Springer-Verlag (Berlin–Göttingen–Heidelberg, 1955).

[156] Riesz, Frigyes, *Összegyűjtött Munkái = Œuvres complètes = Gesammelte Arbeiten,* ed. Á. Császár, Akadémiai Kiadó (Budapest, 1960).

[157] Riesz, Frigyes – Sz.-Nagy, Béla, *Leçons d'Analyse Fonctionnelle,* Akadémiai Kiadó (Budapest, 1952). English translation: *Functional Analysis,* Ungar (New York, 1955).

[166] Stein, Elias M., *Harmonic Analysis: Real Variable Methods, Orthogonality and Oscillatory Integrals,* Princeton Mathematical Series, 43, Princeton Unversity Press (Princeton, NJ, 1993).

[177] Sz.-Nagy, Béla, *Spektraldarstellung linearen Transformationen des Hilbertschen Raumes,* Ergebnisse der Mathematik und ihrer Grenzgebiete, Band 5, No. 5. Springer-Verlag (Berlin, 1942).

{1} N. I. Akhiezer, *The classical moment problem,* Hafner Publishing Company, 1965.

{2} G. Alexits, Über die Erweiterung einer Baireschen Funktion, *Fund. Math.,* **15** (1930), 51–56.

{3} H. Araki, The work of Alain Connes, *Proceedings of the Int. Congress of Mathematicians, Warszawa 1983* (ed.: Z. Ciesielski, C. Olech), PWN - Polish Scientific Publisher, 1984.

{4} J. S. Birman, The work of Vaughan F. R. Jones, *Proceedings of the Int. Congress of Mathematicians, Kyoto 1992* (ed.: I. Satake), Springer, 1992, 9–18.

{5} A. P. Calderón and A. Zygmund, On the existence of certain singular integrals, *Acta Math.,* **88** (1952), 85–139.

{6} A. Connes, Une classification des facteurs de type III, *Ann. Sci. École Norm. Sup., (4)* **6** (1973), 133–252.

{7} A. Connes, Classification of injective factors, *Ann. Math.*, **104** (1976), 73–115.

{8} Császár Á., Megjegyzés Geőcze Zoárd függvényéhez [Remark to Z. Geőcze's function], *Mat. Lapok,* **8** (1957), 268–271.

{9} D. Th. Egoroff, Sur les suites de fonctions mesurables, *C.R. Acad. Sci. Paris,* **152** (1911), 244–246.

{10} Geőcze Z., *Folytonos rendszert képező síkgörbék ívhosszáról [On the arc length of plane curves forming a continuous system],* Az Ungvári Reáliskola 1904/05. évi értesítője, 32 pp.

{11} Z. Geőcze, Recherches générales sur la quadrature des surfaces courbes, *Ungar. Ber.*, **27** (1911), 1–21 and 131–163, **30** (1914), 1–29.

{12} Z. Geőcze, Sur la quadrature des surfaces courbes, *C.R. Acad. Sci. Paris,* **154** (1912), 1211–1213.

{13} Geőcze Z., A zérus területű felületről [On the surface of zero area], *Mathematikai és Természettudományi Értesítő,* **33** (1915), 730–748.

{14} Geőcze Z., A rectifiabilis felületről [On the rectifiable surface], *Mathematikai és Természettudományi Értesítő,* **34** (1916), 337–354, 587.

{15} Geőcze Z., A felület területének Peano-féle definitiójáról [On Peano's definition for the area of a surface], *Mathematikai és Természettudományi Értesítő,* **35** (1917), 325–358.

{16} U. Haagerup, Connes' bicentralizer problem and the uniqueness of the injective factor of type III_1, *Acta Math.,* **158** (1987), 95–147.

{17} A. Haar, Der Maßbegriff in der Theorie der kontinuierlichen Gruppen, *Annals of Math.,* **34** (1933), 147–169; [62] H2, pp. 600–622.

{18} H. Hahn, *Reelle Funktionen,* Akademische Verlagsgesellschaft, Leipzig, 1932.

{19} V. Jones, Index for subfactors, *Invent. Math.,* **72** (1983), 1–25.

{20} P. Jordan, E. Wigner and J. von Neumann, On an algebraic generalization of the quantum mechanical formalism, *Ann. of Math.,* **35** (1934), 29–64; [118].

{21} Kántor S., Geőcze Zoárd függvénye mindenütt folytonos, de sehol sem differenciálható [Z. Geőcze's function is everywhere continuous but nowhere differentiable], *Mat. Lapok,* **8** (1957), 264–267.

{22} L. Kérchy and H. Langer, Béla Szőkefalvi-Nagy, in *Recent Advances in Operator Theory and Related Topics. The Béla Szőkefalvi-Nagy Memorial Volume,* 11–38, eds. L. Kérchy et al, Birkhäuser, 2001.

{23} H. Kosaki, Index theory for operator algebras, *Sugaku Expositions,* **4** (1991), 177–197.

{24} Kovács I., Pukánszky Lajos és a faktorok elmélete [Lajos Pukánszky and the theory of factors], *Mat. Lapok.,* **6** (1996), 28–39.

{25} I. Kovács and J. Szűcs, Ergodic type theorems in von Neumann algebras, *Acta Sci. Math. Szeged,* **27** (1966), 233–246.

{26} Gy. Kőnig, Über stetige Funktionen, die innerhalb jedes Intervalles extreme Werte besitzen, *Monatshefte f. Math.*, **1** (1890), 7–12.

{27} Kőnig Gy., A határozott integrálok elméletéhez [On the theory of definite integrals], *Mathematikai és Természettudományi Értesítő,* **15** (1897), 380–384.

{28} W. Krieger, On ergodic flows and the isomorphism of factors, *Math. Ann.,* **223** (1976), 19–70.

{29} P. Lax, *Functional analysis,* John Wiley, 2002.

{30} N. Macrae, *John von Neumann. The Scientific Genius Who Pioneered the Modern Computer, Game Theory, Nuclear Deterrence and Much More,* American Mathematical Society, 1992.

{31} E. J. Mickle and T. Radó, Density theorems for outer measures in n-space, *Proc. Amer. Math. Soc.,* **9** (1958), 433–439.

{32} E. J. Mickle and T. Radó, On reduced Carathéodory outer measures, *Rend. Circ. Mat. Palermo (2),* **7** (1958), 5–33.

{33} E. J. Mickle and T. Radó, A uniqueness theorem for Haar measure, *Trans. Amer. Math. Soc.,* **93** (1959), 492–508.

{34} F. J. Murray and J. von Neumann, On rings of operators, *Ann. of Math.,* **37** (1936), 116–229; [118] Vol. III., No. 2, 6–119.

{35} F. J. Murray and J. von Neumann, On rings of operators II, *Trans. Amer. Math. Soc.,* **41** (1937), 208–248; [118] Vol. III., No. 3, 120–160.

{36} F. J. Murray and J. von Neumann, On rings of operators IV, *Ann. of Math.,* **44** (1943), 716–808; [118] Vol. III., No. 5., 229–321.

{37} J. von Neumann, Zur Algebra der Funktionaloperatoren und Theorie der normalen Operatoren, *Math. Ann.,* **102** (1929), 370–427; [118] Vol. II., No. 2, 86–143.

{38} J. von Neumann, Die Eindeutigkeit des Schrödingerschen Operatoren, *Math. Ann.,* **104** (1931), 570–578; [118] Vol. II., No. 7, 221–229.

{39} J. von Neumann, Zum Haarschen Maß in Topologischen Gruppen, *Compos. Math.,* **1** (1934), 101–114; [118] Vol. II., No. 22, 220–229.

{40} J. von Neumann, The uniqueness of Haar's measure, *Matem. Sbor.,* **1** (1936), 721–734; [118] Vol. IV., No. 6, 91–104.

{41} J. von Neumann, On algebraic generalization of the quantum mechanical formalism (Part I), *Matem. Sbor.,* **1** (1936), 415–484; [118] Vol. III., No. 9, 409–444.

{42} J. von Neumann, On rings of operators III, *Ann. Math.,* **41** (1940), 94–161; [118] Vol. III., No. 4, 161–228.

{43} J. von Neumann, *Invariant measures,* American Mathematical Society (Providence, 1999).

{44} M. Ohya and D. Petz, *Quantum Entropy and Its Use,* Springer-Verlag, Heidelberg, 1993.

{45} D. Petz and M. Rédei, John von Neumann and the theory of operator algebras, in: *The Neumann Compendum,* eds. F. Bródi, T. Vámos, World Scientific Series in 20th Century Math. vol. 1, World Scientific, Singapore, 1995, 163–185.

{46} D. Petz, Entropy, von Neumann and the von Neumann entropy, in: *John von Neumann and the Foundations of Quantum Physics,* eds. M. Rédei and M. Stöltzner, Kluwer, 2001.

{47} L. Pukánszky, Some examples of factors, Publ. Math. Debrecen, **4**, 135–156 (1956).

{48} T. Radó, Sur le calcul de l'aire des surfaces courbes, *Fund. Math.,* **10** (1927), 197–210.

{49} Radó T., A felszínmérés elméletéhez [On the theory of measuring surface area], *Mathematikai és Természettudományi Értesítő,* **45** (1928), 225–244.

{50} T. Radó, Sur l' aire des surfaces continues, *Atti. Congr. Internaz. Bologna,* **6** (1928), 355–360.

{51} T. Radó, Über das Flächenmass rektifizierbarer Flächen, *Math. Ann.,* **100** (1928), 445–479.

{52} T. Radó, On the derivative of the Lebesgue area of continuous surfaces, *Fund. Math.,* **30** (1938), 34–39.

{53} T. Radó, On continuous path-surfaces of zero area, *Ann. of Math.,* **44** (1943), 173–191.

{54} T. Radó, On surface area, *Proc. Nat. Ac. Sci. U.S.A.,* **31** (1945), 102–106.

{55} F. Riesz, Sur une espèce de géométrie analytique des systèmes de fonctions sommables, Comptes Rendus, *Acad. Sci. Paris,* **144** (1907), 1409–1411; [156] C4, pp. 386–388.

{56} F. Riesz, Sur les opérations fonctionnelles linéaires, Comptes Rendus, *Acad. Sci. Paris,* **149** (1909), 947–977; [156] C7, pp. 400–402.

{57} Riesz F., Integrálható függvények sorozatai [Sequences of integrable functions], *Math. és Phys. Lapok,* **19** (1910), 165–182; [156] C9, pp. 407–440.

{58} F. Riesz, Untersuchungen über Systeme integrierbarer Funktionen, *Math. Annalen,* **69** (1910), 449–497; [156] C10, pp. 441–489.

{59} F. Riesz, Sur le systèmes orthogonaux de fonctions, Comptes Rendus, *Acad. Sci. Paris,* **144** (1907), 615–619; [156] C2, pp. 378–381.

{60} Riesz F., Lineáris függvényegyenletekről, *Mathematikai és Természettudományi Értesítő,* **35** (1917), 544–579; [156] F5, pp. 1017–1052.

{61} F. Riesz, Über lineare Funktionalgleichungen, *Acta. Math.,* **41** (1918), 71–98; [156] F6, pp. 1053–1080.

{62} J. Rosenberg, *A Panorama of Hungarian Mathematics in the Twentieth Century: Non-Commutative Harmonic Analysis,* this volume.

{63} S. Saks, *Theory of the integral,* Nakl. Pols. Towarzystwa Matemarycznego, Warszawa, 1937.

{64} I. Segal, On non-commutative extension of abstract integration, *Ann. of Math.,* **57** (1953), 401–457.

{65} Ş. Strătilă and L. Zsidó, *Lectures on von Neumann algebras,* Abacuss Press, Tunbridge Wells, 1979.

{66} L. Tonelli, Sulla quadratura delle superficie, *Atti Accad. Naz. Lincei (6)* **3**, 357–363; (1926), 445–450 and 633–658.

{67} P. Veress, Über kompakte Funktionenmengen und Bairesche Klassen, *Fund. Math.*, **7** (1925), 244–249.

Ákos Császár

Alfréd Rényi Institute of Mathematics
P.O.B. 127
Budapest H–1364
Hungary

csaszar@renyi.hu

Dénes Petz

Budapest University of Technology and Economics
Department of Analysis
Budapest H–1111
Egry J. u. 1.
Hungary

petz@math.bme.hu

BOLYAI SOCIETY
MATHEMATICAL STUDIES, 14

A Panorama of Hungarian Mathematics in the Twentieth Century, pp. 245–294.

Differential Equations: Hungary, the Extended First Half of the 20th Century

ÁRPÁD ELBERT and BARNABÁS M. GARAY

Introduction. It is well known that the theory of differential equations does not belong to the most important chapters in the history of Hungarian mathematics. Yet, when making preparations for this paper, both of us were astonished to realize how many prominent Hungarian scholars had been concerned with the theory of differential equations, even if marginally, and how much they had been aware of the relations of their primary fields to our topic.

Summary. We give a detailed account on

- the relation between Fejér's summation theorem and Dirichlet's problem on the unit disc;
- Fejér's work in stability theory (in connection to his habilitation lecture);
- F. Riesz' subharmonic functions;
- Haar's inequality for partial differential equations of the first order;
- the Haar–Radó results in the calculus of variations (with a particular emphasis on the minimal surface problem)
- what is called 'von Neumann's stability analysis' and the underlying Lax equivalence "consistency & stability $\Leftrightarrow$ convergence";
- Lax's contribution to the theory of conservation laws (a field of research he entered under the influence of Neumann's interest in shock waves);
- M. Riesz' theory of fractional potentials;
- the work of Pólya and Szegő on isoperimetric inequalities.

The concluding pages are devoted to differential equations in Hungary after the second world war. As before, the emphasis is placed on

- results providing a significant contribution to classical problems (like Bihari's 1956 inequality, the first nonlinear version of Gronwall's Lemma);
- results which pave the way to modern theories (like the 1950–60 contributions by Rényi and Barna to the emerging theory of interval maps).

In connection to some statistical data from the

- first decade of the twentieth century;
- years 1928 and 1953 (representing the period before and after the second world war)

a couple of general remarks are also made.

At the turn of the century. At the beginning of the twentieth century the general theory of differential equations was closer to physics, and within this, to mechanics than it is nowadays. Let us mention here that even the famous paper on linear inequalities by Gyula Farkas had its roots in the one-sided constraints of mechanics. In the volumes of Mathematische Annalen between 1900 and 1910 34 articles (among approx. 450) had Hungarian authors. Out of them 16 are concerned with differential equations in a wider sense, including 8 papers on physical applications, namely: 4 papers by Mór Réthy discuss the variational principles of mechanics, 1 paper by Győző Zemplén treats hydrodynamics, another is on electrodynamics, 1 paper by Gyula Farkas is on shock waves and, finally, 1 paper by Lipót Fejér is concerned with the variational principles of mechanics. The last one will be discussed in detail later. Three papers on linear ordinary differential equations by Lajos Schlesinger come under pure mathematics, definitely. One paper by Károly Goldziher and 4 papers by József Kürschák, who is known mainly as an algebraist, are on partial differential equations. E.g. Kürschák provides a new proof to a theorem of Lie according to which, under certain compatibility conditions, Monge–Ampère equations are transferred into Monge–Ampère equations by contact transformations {44}.

Although the authors received their most important professional stimulations from abroad, even, while staying abroad, these results could not have been achieved without certain Hungarian scholarly antecedents. Gyula König taught courses on differential equations regularly at the Budapest University of Technology, so did Gyula Vályi at the University of Kolozsvár. Their most important results obtained in the field of partial differential equations were highly appreciated by the international scientific community. The studies in partial equations of second order by Vályi, König and József Kürschák, a follower of König, belong to the theory of formal integrability. Their main aims were to set up compatibility criteria on the reduction of equations (i.e. to a system of partial equations of the first order or that of ordinary ones). At the Budapest University of Technology Gusztáv Kondor, who belonged to an earlier generation and had much more modest abilities than Vályi and König, also read lectures on differential equations. Among the predecessors, the name of József Petzval, a pioneer of photography, who left the chair of the University of Budapest for that of Vienna University in 1837, should also be mentioned. His two-volume work on differential equations was highly appreciated all over the Austro–Hungarian Monarchy.

The papers mentioned, as well as the preliminaries discussed, represent demonstratively the state of the theory of differential equations in the early twentieth century: it had a very close relation to various branches of physics and there was a great effort to solve equations in a closed form. Even in our days it seems to be surprising how many of them could be solved by integral representations and series expansions in terms of elementary and special (higher transcendental) functions. By 1900 the development of the general theory of linear ordinary differential equations was already at a fairly high level. This can be exemplified, primarily, by the generalization of Galois theory, widely known in the field of algebraic equations, to linear ordinary differential equations (whose coefficients are meromorphic functions). On the other hand, at that time only the germs of a general theory of nonlinear ordinary differential equations, namely, the existence theorems of Peano and Picard, and elements of the qualitative theory began to emerge. But the number of known nonlinear types of equation which could be solved by quadrature became several dozens. As far as partial differential equations are concerned, we cannot speak about a general theory at all, except for the equations of first order. Here emphasis was laid on the concrete solutions of initial-boundary value problems for linear and, to a much lesser extent, for nonlinear equations closely connected with physical applications.

Lajos Schlesinger and his work. In the history of differential equations in Hungary Lajos Schlesinger holds a special and distinguished place for two reasons: he was the first mathematician in Hungary whose prime field of activity was the study of abstract differential equations, namely that of the linear ordinary differential equations, during most of his life and who gained an international fame and reputation for this. His major work, a two-volume monograph on linear ordinary differential equations {73} was published again by the Johnson Reprint Corporation in 1968. While the activities and lives of Gyula König and Gyula Vályi are discussed by Barna Szénássy in his excellent book on the history of Hungarian mathematics of early times, (i.e. before the twentieth century) in detail, the life and work of Lajos Schlesinger are not treated at all since most of his activity took place after 1900.

Lajos Schlesinger was born in Nagyszombat in 1864. He started studying mathematics at Heidelberg, finished his studies in Berlin, and took his doctor's degree with Lazarus Fuchs whose daughter he married (and, thus, he confirmed a peculiar law of genetics according to which the genes carrying mathematical abilities are passed from a father–in–law to a son–in–law. Another example of this is Aurél Wintner born and educated in Hungary, who married Otto Hölder's daughter). Lajos Schlesinger spent most of his life in Germany and died as Professor of the University of Giessen in 1933. His activities are attached closely to his father–in–law's. Similar to the latter, he used methods of the theory of analytic functions to a greater extent and those of group theory to a lesser extent in the theory of ordinary differential equations. Further on, he wrote a great number of papers on Fuchsian equations and Fuchsian groups and published the correspondence of Fuchs and Weierstrass. By the evidence of his references, based on studying and processing almost 2000 professional articles, he wrote his survey {74} on the development of the theory of linear ordinary differential equations from 1865 on, the publication year of his father-in-law's famous paper. Together with Abraham Plessner he wrote a remarkable book on Lebesgue integrals and Fourier series and took part in processing Gauss's unpublished manuscripts. Undoubtedly, he had a very rich career in mathematics. In a late paper related to the solution of the differential equation $\dot{z} = C(t)z$ where $z \in \mathbb{C}^N$ and C is a complex-valued matrix function, he became the forerunner of the theory of product integrals (i.e. of generalizations of Lie's matrix formula $e^{A+B} = \lim_{n\to\infty} \left(e^{A/n}e^{B/n}\right)^n$). This activity of Schlesinger is mentioned with admiration by Felix Browder, too, in the preface to the Mathematical Encyclopedia volume 'J. D. Dollard & C. N. Friedman, *Product Integration with Applications to Differential Equations*, Addison–Wesley, Reading, Mass., 1979'.

Lajos Schlesinger was Professor at the University of Kolozsvár between 1897 and 1911. Among Hungarian mathematicians he stimulated Manó Beke and his impact can be felt on a paper by Lipót Fejér. Beke investigated the irreducibility of homogeneous linear ordinary differential equations the coefficients of which are rational functions. Also, for the equations with such coefficients the concept of irreducibility itself was introduced by him {3}. (An equation is irreducible if it has no solution common with a homogeneous linear ordinary differential equation which is of a lower order and has coefficients of the same class.) With a highly effective application of Cauchy's majorant method, Fejér {21} gave a new proof to Fuchs's theorem on the singularities of the solutions of homogeneous linear differential equations.

For Hungarians, Schlesinger's reputation is indicated by the fact that he became the subject of the well-known anecdote attached to Henri Poincaré's visit to Budapest. In 1905 Poincaré came to Budapest to receive the Bolyai Prize. At his arrival his reply to the greetings of the notables who were meeting him was as follows: 'Thank you! But where is Fejér?' Similar to the legends about the lives of medieval saints, this story has become extant in another version in which the name of Schlesinger replaces that of Fejér's. Indeed, it is conceivable that Poincaré wanted to meet Schlesinger since he himself had been concerned with homogeneous linear ordinary differential equations with rational coefficients. It was he who gave the name of Fuchsian functions, in honour of Immanuel Lazarus Fuchs, to a certain class of automorphic functions which played an important role in the integration of the aforementioned equations.

Fejér summation theorem and the Dirichlet problem on the unit disc. In the first decade of the twentieth century Lipót Fejér was concerned with the theory of ordinary differential equations in a wider sense in several papers. Pál Turán mentions in the Introduction to Lipót Fejér's Collected Works that the discovery of the famous Fejér's summation theorem is closely related to the Dirichlet problem on the unit disc $B = \{ (x, y) \in \mathbb{R}^2 \mid x^2+y^2 \le 1\}$. Given a continuous function $g : \partial B \to \mathbb{R}$, does there exist a continuous function $u : B \to \mathbb{R}$ with the properties that u is harmonic on $B \setminus \partial B$ (i.e. u is twice continuously differentiable on $B \setminus \partial B$ and satisfies $\Delta u = 0$ on $B \setminus \partial B$) and $u|_{\partial B} = g$? The positive answer, together with the form of the solution function

$$(1) \quad u(r\cos\varphi, r\sin\varphi) = \begin{cases} a_0/2 + \sum_{k=1}^{\infty} (a_k \cos k\varphi + b_k \sin k\varphi) r^k & \text{if } r < 1 \\ g(\varphi) & \text{if } r = 1 \end{cases}$$

(where $a_0, a_1, b_1, a_2, b_2, \ldots$ are the corresponding Fourier coefficients) had already been guessed earlier but only H. A. Schwarz succeeded in proving it with the help of the so-called Poisson integral representation of the series expansion

$$a_0/2+\sum_{k=1}^{\infty}(a_k \cos k\varphi+b_k \sin k\varphi)r^k = \frac{1}{2\pi}\int_0^{2\pi} \frac{1-r^2}{1-2r\cos(\varphi-\psi)+r^2} g(\psi)\, d\psi$$

valid for $r < 1$.

Fejér spent the academic year 1899/1900 in Berlin where he was influenced greatly by the lectures of H. A. Schwarz. When discussing the Dirichlet problem on the unit disc Schwarz stated that it would be expedient to give an existence proof by the theory of series exclusively, moreover he spoke about the unsuccessful attempts made in that direction. The problem was that due to the possible divergence of the Fourier series of function g the Abel summation theorem could not be applied. The way out of the situation was that — at the points of ∂B — continuity of the solution function defined by formula (1) followed from a new summation procedure. Fejér was given the decisive impetus to frame a new summation procedure by the theorem of Frobenius according to which Abel's convergence assumption on $\sum c_k$ for the existence of $\lim_{r\to 1-}\sum c_k r^k$ can be weakened to assuming the convergence of $\sum s_n$ where $s_n = (c_0+c_1+\cdots+c_n)/(n+1)$, $n \in \mathbb{N}$. All these considerations led Fejér to prove that the arithmetic means of the partial sums of the Fourier series of continuous 2π-periodic functions are uniformly convergent. Compared to his revolutionary discovery the original question posed, i.e. to prove the solvability of the Dirichlet problem on the unit disc by the theory of series exclusively, remained entirely in the background.

In Fejér's famous Comptes Rendus note {13} there is only one sentence which indicates that his summation theorem is also applicable to the theory of Poisson integrals. He worked out the details in two separate papers published in Hungary. The first one {14} discusses the Dirichlet problem on the unit disc. The second one {15} treats the heat equation $u_t = u_{xx}$ equipped with the initial condition $u(0,x) = g(x)$ where $g : \mathbb{R} \to \mathbb{R}$ is a 2π-periodic continuous function. The results of these two short papers are included in Section 3/a of an extensive paper on his summation theorem he published in Mathematische Annalen in 1904 {16}.

The work of Fejér in mechanics and his habilitation lecture. Between 1905 and 1911 Lipót Fejér worked at the Department of Mathematics and Physics of the University of Kolozsvár. This must explain the fact that he chose the topic of his "habilitation" lecture not from the theory of Fourier series but from the stability theory of ordinary differential equations. First, he gave an outline of the definitions of stability used in those days (strangely enough, including the recurrence property as well): 'The concept of stability carries highly different contents even within the framework of mass point systems. It is no use arguing which one of them is the best since, except for some inherent features of it, stability as a popular concept is so indefinite and so relative that, owing to the variety of existing relations, stability definitions highly differing from one another may be formulated without getting into contradiction with the popular concept'. Among the definitions of stability listed by him we can find the one accepted in general nowadays but it is considered too narrow by Fejér, joining Felix Klein's opinion. Then, he discusses some simpler aspects of the three-body problem and finishes his habilitation lecture with the discussion of the Lagrange–Dirichlet theorem.

In connection with his habilitation lecture {19} Fejér published a finding of his own in the problem of equilibrium instability {18}, {20}. (Ref. {20} is a word–for–word German translation of the Hungarian original {18}.) Slightly changing notations, Fejér investigated equation

$$\ddot{\underline{r}} = \operatorname{grad} \pi(\underline{r}) - \dot{\underline{r}} f\left(|\dot{\underline{r}}|\right) / |\dot{\underline{r}}|, \quad \underline{r} = (x, y, z) \in \mathbb{R}^3 \tag{2}$$

under the conditions below: The potential function $\pi : \mathbb{R}^3 \to \mathbb{R}$ is analytic in a small vicinity of the origin $\underline{0}$ and its Taylor series about $\underline{0}$ begins with a negative definite 3-variable homogeneous polynomial of order $2n$ for some integer $n \geq 1$ (in particular, the potential function π has an isolated maximum at $\underline{0}$), $f : \mathbb{R} \to \mathbb{R}$ is a continuous increasing function, $f(0) = 0$, $f(v) > 0$ for $v > 0$ and

$$\limsup_{v \to 0^+} f(v)/v < \infty. \tag{3}$$

Starting from Jacobi's identity

$$\frac{\mathrm{d}^2}{\mathrm{d}t^2}\left(\frac{1}{2}|\underline{r}|^2\right) = |\dot{\underline{r}}|^2 + \langle \operatorname{grad} \pi(\underline{r}), \underline{r} \rangle - \langle \dot{\underline{r}}, \underline{r} \rangle f\left(|\dot{\underline{r}}|\right) / |\dot{\underline{r}}|$$

Fejér proved via elementary ad hoc inequalities that the equilibrium state $\underline{0}$ is unstable. He raised the question of whether the result of instability will remain true without the condition (3), e.g. for the function $f(v) = v^\alpha$, $\alpha \in (0, 1)$. The positive answer is a consequence of La Salle's invariance principle elaborated more than 50 years later: A 1968 theorem of Luigi Salvadori (see Thm. III. 5.8 as well as the accompanying discussion) in 'N. Rouche, P. Habets & M. Laloy, *Stability Theory by Liapunov's Second Method*, Springer, Berlin, 1977' can be directly applied.

Another paper of Fejér's {17} published in 1906 is related to the so-called Ostwald principle. In his work entitled 'Lehrbuch der allgemeinen Chemie' published in 1893 Ostwald stated that 'von allen möglichen Energieumwandlungen wird diejenige eintreten, welche in gegebener Zeit den grösstmöglichen Umsatz ergibt' (from among all possible conversions of energy the one which produces the greatest increment in the given time will happen). For a given planar potential field $U(\underline{r})$ and starting point $\underline{r}_0$ Fejér studied such equations of motion

$$\ddot{\underline{r}} = -\operatorname{grad} U(\underline{r})$$

for which <u>*Case 1*</u> $U(\underline{r}_0) - U\big(\underline{r}(t)\big)$ is maximal for a single and fixed $t = T_0 > 0$ or, alternatively, <u>*Case 2*</u> $U(\underline{r}_0) - U\big(\underline{r}(t)\big)$ is maximal for all $t \in [0, T_0]$. He stated that in <u>*Case 1*</u> the initial velocity $\dot{\underline{r}}(0)$ has to meet some conditions of compatibility and the motion itself will be brachistochronal, and in <u>*Case 2*</u> the orthogonal semitrajectory through $\underline{r}_0$ should be a half-line and $\dot{\underline{r}}(0)$ should be parallel to the direction vector of this half-line. Otherwise, Ostwald wanted to demonstrate the general validity of his principle by the motions along half-lines obtained in <u>*Case 2*</u>. In spite of the great prestige and merits of Ostwald — he was awarded a Nobel Prize in chemistry in 1909 — this energy principle had already been highly criticized. All parties in the debate emphasized that the original phrase 'von allen...' provided several opportunities for different mathematical interpretations. Most of them, including Mór Réthy {66}, investigated which modifications and improved versions of the Ostwald principle were in harmony with the general laws of mechanics and other physical sciences.

On partial differential equations, more precisely on the equation of the form

$$\sum_{i,k=1}^{n} a_{ik}(x)u_{x_i x_k} + \sum_{r=1}^{n} b_r(x)u_{x_r} + c(x)u = 0, \quad a_{ik} = a_{ki} \tag{4}$$

Lipót Fejér wrote only a single article {22}. He assumed that the coefficients were analytic functions and $c(\underline{0}) < 0$. By using the Cauchy–Kowalevskaya theorem he proved that (4) had a solution having a positive maximum at

$\underline{0}$ if and only if matrix $\{a_{ik}(\underline{0})\}_{i,k=1}^{n}$ is not positive semidefinite. This is a contribution to the theory of maximum principles.

F. Riesz' subharmonic functions. Out of the papers of Frigyes Riesz those on subharmonic functions are related to the theory of differential equations directly. The concept of subharmonic functions was introduced by Riesz in a lecture in Stockholm in 1924 {68}. The definition is as follows: Let $\Omega \subset \mathbb{R}^n$ be open and connected, and $f : \Omega \to [-\infty, \infty)$ an upper semicontinuous function, $f(x) \neq -\infty$ for at least one $x \in \Omega$. The function f is subharmonic if for any pair (Ω', u), the inequality $f|_{\partial\Omega'} \leq u|_{\partial\Omega'}$ implies that $f|_{\Omega'} \leq u$. Here Ω' is a bounded domain, $\overline{\Omega}' = \mathrm{cl}(\Omega') \subset \Omega$, $u : \overline{\Omega}' \to \mathbb{R}$ is a continuous function which is harmonic on Ω'. The definition of subharmonic functions was inspired, partly, by some characteristics of analytic functions discovered by Hardy and Landau and, partly, by Perron's method, which was elaborated for proving the existence of classical $C(\overline{\Omega}) \cap C^2(\Omega)$ solutions to the Dirichlet boundary value problem $\Delta u = 0$, $u|_{\partial\Omega} = g$ (where $g : \partial\Omega \to \mathbb{R}$ is a given continuous function and $\Omega \subset \mathbb{R}^n$ is a bounded domain having a 'nice' boundary.) Prior to his lecture in Stockholm Riesz had generalized {67} the results of Hardy and Landau, and in a joint paper with Tibor Radó, he gave a simplification of Perron's method {72}. The core of Perron's method as conceived by Riesz and Radó is the following assertion: If $\Omega \subset \mathbb{R}^n$ is a bounded domain and $g : \partial\Omega \to \mathbb{R}$ is a continuous function, then

$$u(x) = \sup\left\{\, v(x) \mid v \in C(\overline{\Omega}), v|_{\partial\Omega} \leq g \text{ and } v|_{\Omega} \text{ is subharmonic} \right\}$$

defines a harmonic function on Ω. (If $\partial\Omega$ is nice e.g. if it is of class C^2 or satisfies the outer sphere condition, then $u \in C(\overline{\Omega})$ and $u|_{\partial\Omega} = g$.) Connections to the celebrated Fejér–Riesz proof of Riemann's conformal mapping theorem (published as part of the Introduction in a paper by Radó' (*Acta Sci. Math. (Szeged),* **1** (1922/23), 240–251. (reprinted as pp. 841–843 of the Appendix to Fejér's 'Gesammelte Arbeiten')) are transparent.

The basic characteristics of subharmonic functions and their relation to potential theory was clarified by Frigyes Riesz in two consecutive papers {69}, {70}. A function $f \in C^2(\Omega)$ is subharmonic, if and only if $\Delta f \geq 0$ on Ω. The main result is that any subharmonic function can be represented as a potential plus a harmonic function. If $f : \Omega \to [-\infty, \infty)$ is a subharmonic function and, as before, Ω' is a bounded domain with $\overline{\Omega}' \subset \Omega$, then

$$f(x) = \int_{\Omega'} \Gamma(x,y)\, d\mu_y + h(x), \quad x \in \Omega' \tag{5}$$

where Γ is the fundamental solution of Laplace's equation $\Delta u = 0$ on $\mathbb{R}^n$, μ_y is a (uniquely defined) Borel measure and $h : \Omega' \to \mathbb{R}$ is a harmonic function. It is worth comparing (5) with Green's classical representation formula

$$w(x) = \int_{\Omega'} \Gamma(x,y)\Delta w(y)\, dy + \int_{\partial\Omega'} \left(w(y)\frac{\partial \Gamma(x,y)}{\partial \nu_y} - \Gamma(x,y)\frac{\partial w(y)}{\partial \nu} \right) dS_y,$$

$x \in \Omega'$ for $C^2(\Omega)$ functions. (Note also that the function defined by the boundary integral is harmonic.) If $f = w \in C^2(\Omega)$ then $d\mu_y = \Delta w(y)\, dy$. In the book written by Tibor Radó [143] several variations of formula (5) can be found. Some of them originate from Frigyes Riesz. (Radó reviewed the development of the theory of subharmonic functions till 1937 but he was not concerned with its applications to partial differential equations.)

The late period of Frigyes Riesz had little relevance to differential equations. Techniques applied in his 1938–48 series of papers on ergodic theory belong to general functional analysis and the theory of measure and integration. However, he considers ergodic theory as part of a dynamical system theory for measure–preserving transformations and interprets, occasionally, his own results from this view–point. It is amazing how much the old Riesz felt the importance of functional iterations! He asked (himself and) each reader of Matematikai Lapok (**3** (1952), Problem No. 54) to "determine the set of those complex initial points z_1 for which, given a complex parameter a, the iteration $z_{n+1} = (z_n^2 + a)/2$ is converging." The Editor of the Problem Session had waited two years for the solution in vain when he stated (**5** (1954), p. 283) that Problem No. 54 was entirely open and promised to publish the solution any time it arrives. (Actually, Problem No. 54 concerns the structure (of the complement) of Julia sets for the quadratic family, a basic problem in fractal geometry... Unfortunately, Mandelbrot was apparently unaware of the existence of Matematikai Lapok and published his results somewhere else ...)

The works of F. Riesz and Haar on linear integral equations. It is important to mention that integral equations played a fundamental role in the formulation of the papers on compact linear operators by Frigyes

Riesz, as well as in the emergence of the whole functional analysis. Alfréd Haar, a colleague of Frigyes Riesz at Kolozsvár and, later, at Szeged was also concerned with integral equations. As a student in Göttingen, while applying Hilbert's method elaborated for the problem $\Delta u = 0$, $u|_{\partial\Omega} = g$ in his first paper, Haar {30} discusses the reducibility of the boundary value problem

$$\Delta^2 u = \frac{\partial^4 u}{\partial x^4} + 2\frac{\partial^4 u}{\partial x^2 \partial y^2} + \frac{\partial^4 u}{\partial y^4} = 0, \; u|_{\partial\Omega} = g_0, \; \left.\frac{\partial u}{\partial \nu}\right|_{\partial\Omega} = g_1$$

to integral equations and proves Fredholm type alternative theorems.

Haar's inequality for partial equations of the first order. The most important results in differential equations of Hungarian mathematics in the first half of the twentieth century are attached to the name of Alfréd Haar. He made essential contributions to both the theory of general equations of the first order and of quasilinear elliptic equations of the second order. His results in the latter topic are related to his pioneering work in the field of the calculus of variations, to put it more precisely, in the study of the minimal surface problem and of the minimal surface equation

$$(1 + u_y^2)u_{xx} - 2u_x u_y u_{xy} + (1 + u_x^2)u_{yy} = 0.$$

Here we sum up Alfréd Haar's papers on partial differential equations of the first order based on his lecture delivered in Bologna in 1928 {39}.

Consider a finite interval $[x_1, x_2] \subset \mathbb{R}$. Given a parameter $a > 0$, define the triangle

$$\mathcal{T} = \left\{ (x, y) \in \mathbb{R}^2 \mid 0 \le y \le (2a)^{-1}(x_2 - x_1) \text{ and } x_1 + ay \le x \le x_2 - ay \right\}.$$

Finally, let $z : \mathcal{T} \to \mathbb{R}$ be a C^1 function and set $M = \max\left\{ \left| z(x, 0) \right| \mid x_1 \geq x \geq x_2 \right\}$. Assume that, for some constants $b > 0$ and $c > 0$,

$$\left| z_y(x, y) \right| \le a \left| z_x(x, y) \right| + b \left| z(x, y) \right| + c \quad \text{whenever} \quad (x, y) \in \mathcal{T}.$$

Then

$$\left| z(x, y) \right| \le M e^{by} + cb^{-1}(e^{by} - 1) \quad \text{whenever} \quad (x, y) \in \mathcal{T}. \tag{6}$$

Inequality (6) — just like its one-variable counterpart, the famous Gronwall lemma in the theory of ordinary differential equations — has far-reaching consequences. As a preliminary, we may state that the problem

$\mathcal{F}(u_y, u_x, u, x, y) = 0$, $u|_S = h$ ($h : S \to \mathbb{R}$ is a continuous function, $S \subset \partial\Omega$, $\Omega \subset \mathbb{R}^2$) can be transformed to the form (7) under very general conditions.

Let $g : [x_1, x_2] \to \mathbb{R}$ be a continuous function and let $F : \mathbb{R}^2 \times \mathcal{T} \to \mathbb{R}$ be a continuous function with the property that

$$\left| F(p, z, x, y) - F(\tilde{p}, \tilde{z}, x, y) \right| \leq a|p - \tilde{p}| + b|z - \tilde{z}|$$

whenever $(p, z, x, y), (\tilde{p}, \tilde{z}, x, y) \in \mathbb{R}^2 \times \mathcal{T}$. Applying inequality (6) to the difference of two possible solutions, it is immediate that the first order problem

$$u_y = F(u_x, u, x, y) \quad \text{and} \quad u|_{[x_1,x_2]\times\{0\}} = g \tag{7}$$

has at most one C^1 solution on $\mathcal{T}$ {36}. Formerly, uniqueness results were known only for C^2 solutions and proved within the framework of the theory of characteristics. Inequality (6) has some consequences to characteristics in return {37}. As has been mentioned by Hadamard in his additional comment published jointly with Haar's paper {36}, Haar's inequality leads not only to a result on uniqueness but also to the assertion that the C^1 solutions of equation (7) on the triangle $\mathcal{T}$ depend continuously on function g.

Haar finished his lecture delivered at Bologna with the extension of inequality (6) to systems of partial differential equations with a simple structure. His promise to devote another paper to the topic could not be fulfilled due to his early death. Incidentally, Alfréd Haar was already concerned with systems of first order partial differential equations in one of his early articles {41} the coauthor of which was Tódor Kármán, one of his fellow students in Göttingen.

Haar's existence and uniqueness theorem in the calculus of variations. We are going to sum up Alfréd Haar's work on calculus of variations based on his own lecture held in Hamburg in 1930 {40} as well as on the monograph of Tibor Radó [142] who himself achieved fundamental results in this field. The focus of the discussion will be placed on Alfréd Haar's existence and uniqueness theorem {35}. In the proof a lemma originating from Tibor Radó {63} plays an important part which, by its external form, is a geometric statement on saddle surfaces but in essence is an a priori estimate for the gradient of solutions to quasilinear elliptic equations which will be reviewed separately here. Naturally, the abstract existence and uniqueness theorem has important consequences for the classical minimal surface problem {34} which motivates it. Finally, we are going to discuss these consequences. Emphasis will be laid on the regularity properties of the solution, more exactly, on the analytic feature of the minimal surface.

Consider a bounded domain $\Omega \subset \mathbb{R}^2$ with a convex Jordan curve $\Gamma = \partial\Omega$ as its boundary, a C^2 function $F : \mathbb{R}^2 \to \mathbb{R}$, and a continuous function $\varphi : \Gamma \to \mathbb{R}$. The variational problem

$$I(u) = \iint_\Omega F(p,q)\,dx\,dy \longrightarrow \min_{u|_\Gamma=\varphi} \tag{8}$$

is called regular if the Hessian matrix of F is positive definite for all $p, q \in \mathbb{R}^2$. Of course, $p = u_x$ and $q = u_y$. The Euler–Lagrange equation of problem (8) is

$$F_{pp}u_{xx} + 2F_{pq}u_{xy} + F_{qq}u_{yy} = 0. \tag{9}$$

In the most important special case, i.e. in the minimal surface problem we have $F(p,q) = (1+p^2+q^2)^{1/2}$, and thus equation (9) simplifies to

$$(1+q^2)u_{xx} - 2pqu_{xy} + (1+p^2)u_{yy} = 0. \tag{10}$$

Using geometrical terms equation (10) expresses that the mean curvature of a minimal surface equals zero.

As a preliminary, observe that the functional I can be defined for all elements of the function class

$$\mathcal{L} = \{u : \overline{\Omega} \to \mathbb{R} \mid u \text{ is a Lipschitz function}\}$$

and the double integral in (8) is understood in the sense of Lebesgue. The validity of the formula

$$\operatorname{Area}(u,\Omega) = \iint_\Omega (1+u_x^2+u_y^2)^{1/2}\,dx\,dy, \quad u \in \mathcal{L}$$

was proven first by Zoárd Geőcze {25} where $\operatorname{Area}(u,\Omega)$ stands for the area of surface $S = \{(x,y,u(x,y)) \in \mathbb{R}^3 \mid (x,y) \in \overline{\Omega}\}$ as defined by Lebesgue for continuous surfaces in his famous doctoral thesis. Also, observe that both (9) and (10) are quasilinear elliptic equations. The eigenvalues of the coefficient matrix of equation (10) are $\lambda_1(p,q) = 1+p^2+q^2$ and $\lambda_2(p,q) = 1$. Since the ratio λ_1/λ_2 is unbounded, (10) is nonuniformly elliptic. It is well-known that the solvability of a Dirichlet boundary value problem for nonuniformly elliptic quasilinear equations depends crucially on the geometric assumptions on the pair (Γ, φ). The crucial assumption of Haar's existence and uniqueness theorem is that the pair (Γ, φ) satisfies Hilbert's so-called three–point condition. In other words, it is assumed that, for some constant $K > 0$, every set of three distinct points on the curve $\{(x, y, \varphi(x, y)) \in \mathbb{R}^3 \mid (x, y) \in \Gamma\}$ lies in a plane of slope $\leq K$. The three–point condition is the assumption which implies the property $\mathcal{L} \neq \emptyset$. It implies also the strict convexity of Γ.

After such preparations we are in a position to formulate Haar's existence and uniqueness theorem {35}. Consider the variational problem (8). Assume that (8) is regular and that the three–point condition with constant K is satisfied. Then (8) has a unique solution in the function class $\mathcal{L}$ and the Lipschitz constant of this solution is $\leq K$.

T. Radó's regularity Lemma. Uniqueness is a rather elementary consequence of the regularity assumption on (8). The starting point of the existence proof is the observation that the functional $I : \mathcal{L} \to \mathbb{R}$ is lower semicontinuous (with respect to uniform convergence). The difficulty is that minimizing sequences are not precompact. A nice geometric property of saddle functions, which was conjectured by Haar and proved by Radó {63}, helps, instead. A continuous function $u : \overline{\Omega} \to \mathbb{R}$ is a saddle function if, given arbitrarily three constants $\alpha, \beta, \gamma \in \mathbb{R}$ and an open set $\emptyset \neq \Omega' \subset \Omega$, the maximum-minimum principle

$$\min_{(x,y)\in\overline{\Omega'}} \big(u(x,y) - (\alpha x + \beta y + \gamma)\big) = \min_{(x,y)\in\partial\Omega'} \big(u(x,y) - (\alpha x + \beta y + \gamma)\big),$$

$$\max_{(x,y)\in\overline{\Omega'}} \big(u(x,y) - (\alpha x + \beta y + \gamma)\big) = \max_{(x,y)\in\partial\Omega'} \big(u(x,y) - (\alpha x + \beta y + \gamma)\big)$$

is satisfied. Radó's lemma concerns saddle functions $u : \overline{\Omega} \to \mathbb{R}$ for which the pair $(\Gamma, u|_\Gamma)$ is subject to the three–point condition with constant K and states that such functions are Lipschitz continuous on $\overline{\Omega}$ and the Lipschitz constant is $\leq K$.

Radó's lemma was given a new and a much simpler proof by János Neumann {51}. Both the original method of proving and the one given by Neumann lead, automatically, to Lemma 12.6 in 'D. Gilbarg & N. S. Trudinger, *Elliptic Partial Differential Equations of the Second Order*, Springer, Berlin, 1983': Let Ω be a bounded domain in $\mathbb{R}^2$, and let $\varphi : \partial\Omega \to \mathbb{R}$ be a continuous function satisfying the three-point condition with constant K. Suppose

$u \in C(\overline{\Omega}) \cap C^2(\Omega)$ with $u|_{\partial\Omega} = \varphi$ satisfies a quasilinear elliptic equation of the form $au_{xx} + 2bu_{xy} + cu_{yy} = 0$ where a, b, c are continuous functions in the variables (x, y, u, p, q). Then $|u_x|, |u_y| \leq K$ on Ω. (Besides the Lax–Milgram lemma, which is considered 'half-Hungarian' due to Péter Lax, this is the only result in the well-known book of Gilbarg and Trudinger which originates from a Hungarian mathematician. It is worth mentioning that Lemma 12.6 and, in relation to this, the name of Tibor Radó are mentioned in the preface of this monograph.

Haar's Lemma on the variation of double integrals. Applying the existence and uniqueness theorem for the variational problem (8) in the special case $F(p, q) = (1 + p^2 + q^2)^{1/2}$ we obtain that there exists a $u^* \in \mathcal{L}$ for which

$$\text{Area}\,(u^*, \Omega) < \text{Area}\,(u, \Omega) \quad \text{for all} \quad u^* \neq u \in \mathcal{L}.$$

Actually, $u^*|_\Omega$ is analytic. The argumentation leading to this will be outlined below. The starting point is one of Alfréd Haar's earlier results {32}, namely the two-variable counterpart of Du Bois–Reymond's fundamental lemma of the calculus of variations. In order that the novelty of Haar's result should be emphasized, we would like to present Du Bois–Reymond's lemma: If $f : [a, b] \to \mathbb{R}$ is a continuous function such that

$$\int_a^b f(x)\varphi'(x)\,dx = 0 \;\text{ for each }\; \varphi \in C[a, b] \cap C^1(a, b) \;\text{ with } \varphi(a) = \varphi(b) = 0,$$

then f is a constant function.

Using the previously introduced notation, let $v, w : \overline{\Omega} \to \mathbb{R}$ be continuous functions and assume that

$$\iint_\Omega (v\xi_x + w\xi_y)\,dx\,dy = 0 \tag{11}$$

for each continuous function $\xi : \overline{\Omega} \to \mathbb{R}$ satisfying $\xi|_\Gamma = 0$ and for which the partial derivatives derivatives $\xi_x, \xi_y : \Omega \to \mathbb{R}$ are continuous and bounded. Then there exists a C^1 function $\omega : \Omega \to \mathbb{R}$ with the properties that

$$v = \omega_y \quad \text{and} \quad w = -\omega_x \quad \text{on} \quad \Omega. \tag{12}$$

If we assume merely that that functions v and w are measurable and bounded, then function ω satisfies only the Lipschitz condition and (12) is valid almost everywhere. Now let u denote the solution of the variational problem (8). Since

$$\delta I(u;\xi) = \iint_\Omega (F_p(u_x,u_y)\xi_x + F_q(u_x,u_y)\xi_y)\,dx\,dy = 0$$

for all ξ (which have the above mentioned properties), (12) goes over into the system of equations

$$F_p(u_x,u_y) = \omega_y \quad \text{and} \quad F_q(u_x,u_y) = -\omega_x. \tag{13}$$

We arrived at a system of first order partial equations which replaces the Euler–Lagrange equation and can be used excellently for clarifying the differentiability properties of the solutions of (8). Under adequate differentiability conditions, of course, (9) can be derived from (13) easily. The hard fact is that we cannot foresee whether these differentiability conditions will be met by u, moreover, examples are also known when the unique solution to (8) is not twice differentiable.

Fortunately, Lipschitz solutions of the system

$$\frac{u_x}{(1+u_x^2+u_y^2)^{1/2}} = \omega_x \quad \text{and} \quad \frac{u_y}{(1+u_x^2+u_y^2)^{1/2}} = -\omega_y \tag{14}$$

— this is to what (13) is simplified in the special case $F(p,q) = (1+p^2+q^2)^{1/2}$ — are analytic functions. Eventually, the latter assertion depends on the fact proven by Rademacher according to which Lipschitz solutions of the Cauchy–Riemann system are analytic functions. Analyticity of C^1 solutions of the system (14) was shown by Tibor Radó {62} while Haar had been working on the proof to the existence and uniqueness theorem. (Analyticity of C^2 solutions of equation (10) had been known much earlier.) Radó's argumentation works for Lipschitz solutions as well. Thus, the proof for the existence and uniqueness theorem was also a justification for the analytic nature of function $u^*|_\Omega$, at the same time. As has been shown by Tibor Radó, Area $(u^*,\Omega) \leq$ Area (u,Ω) for all $u \in \mathcal{C}$, where

$$\mathcal{C} = \{u : \overline{\Omega} \to \mathbb{R} \mid u \text{ is a continuous function and } u|_\Gamma = \varphi\}.$$

From a later result of McShane it also follows that

$$\text{Area}\,(u^*,\Omega) < \text{Area}\,(u,\Omega) \quad \text{for all } \; u^* \neq u \in \mathcal{C}.$$

The implication (11) ⇒ (12) is called Haar's Lemma in the literature. Incidentally, this was Haar's first result in the field of the calculus of variations which he published in Hungarian in 1917 {31} and in German in 1919 {32}. The converse of Haar's Lemma, the implication (12) $\Rightarrow$ (11) is also true. The latter was proven by Schauder (*Acta Sci. Math (Szeged)*, **4** (1929), 38–50). For the fulfilment of the equivalence between (11) and (12) the convexity assumption on Ω is too restrictive. Even Haar and Schauder themselves worked with Jordan domains. Definitely, Haar was aware of the fact that his lemma could be generalized for multiple integrals on n-dimensional domains and it was in close relation to the n-dimensional Stokes theorem but in those directions he worked out very few or rather no details at all. His work had been continued by others, amongst them, by Jenő Gergely {26}, Antal Sólyi {76} and Adolf Szücs {80} in Hungary. It is important to mention that the concept of the adjoint variational problem was introduced also by Alfréd Haar {38}. The pattern to that was provided by adjoint minimal surfaces as well as by an earlier result {33} of his generalizing an observation of Zermelo on the Du Bois–Reymond lemma in one dimension.

In the development of the minimal surface problem it turned out that the three–point condition for the pair (Γ, φ) is superfluous. Keeping only the convexity assumption on $\partial\Omega = \Gamma$, Tibor Radó [142] proved that equation (10) has a solution in the function class $\mathcal{C} \cap C^2(\Omega)$. (The proof depends, essentially, upon the existence theorem for conformal mappings of approximating polyhedra and it does not indicate at all that the three–point condition can be omitted from the general conditions of Haar's existence and uniqueness theorem.) This $\mathcal{C} \cap C^2(\Omega)$ function is analytic on Ω and constitutes the unique solution of the variational problem $J : \mathcal{C} \to \mathbb{R}$, $J(u) = \text{Area}\,(u, \Omega) \longrightarrow min$ as well. Extensions of the minimal surface problem allow surfaces which cannot be parametrized globally in the form $\big\{\big(x, y, u(x, y)\big) \in \mathbb{R}^3 \mid (x, y) \in \overline{\Omega}\big\}$, whose boundary is disconnected or may be just a knotted Jordan curve, etc. If the problem is studied in this generalized sense, great difficulties arise both for existence and uniqueness and, eventually, neither of them holds true. However, the relevant methods (in the elaboration of which Tibor Radó himself participated according to his monograph [142]) are quite far from those in differential equations.

An early paper on billiards. There is a paper of Adolf Szücs, jointly with Dénes König {43} which is worth to be discussed in detail. They considered the motion of a single, dimensionless particle included in an immobile cube. The impacts on the walls follow the laws of elastic reflection. By using elementary geometry and elementary number theory, they classified the orbits as closed, dense in a polyhedral surface, and dense within the whole cube. The three possibilities do not depend on the initial state but only on the three components of the initial velocity. This is one of the earliest results in the theory of rational billiards.

Neumann's method of stability analysis. In connection with the argumentation leading to Haar's existence and uniqueness theorem we have already mentioned the name of János Neumann who gave a new proof for Radó's lemma on saddle surfaces which was the most important part of the aforementioned argumentation from the aspect of the general theory of partial differential equations. Neumann need not be introduced here. He was one of the 'Martians' who came out from the leading secondary schools of Budapest at the turn of the XIX and XX centuries. Once Jenő Wigner was asked how it happened that so many geniuses had been born in Budapest a century earlier. Wigner replied that he did not understand the question because in those days, as he had said it before, just one genius was born and that person was called János Neumann. In the early phase of planning this volume of studies the idea cropped up that a special chapter should be written about Neumann, while keeping the division by the broad mathematical themes. Although this proposal was not realized, it demonstrates well what a great personality Neumann was and what a prominent role he played in the history of mathematics.

Neumann exerted a great influence on the general development of differential equations. This was not achieved by his articles but by his consulting activities with which he kept track of the operation of the first digital computer. The first genuine tasks solved by eletronic computers were as follows: initial-boundary value problems of thermonuclear reactions, neutron diffusion and transport, radiation flows, and fluid dynamics. The respective partial differential equations were solved by numerical stepsize integration procedures, namely, by the so-called method of finite differences. At each point of the integration net, partial derivatives, i.e. differential quotients were replaced by (finite) difference quotients. That procedure led to large-scale systems of algebraic equations. The number of unknowns was the same as that of the points of the integration net. If the original partial differential equation was linear, then the system of approximating algebraic equations became also linear. The computer was used for the solution of that system of algebraic equations. Actually, in most of the cases the computing task to be performed was to solve a system of linear equations, or to put it in another way, the inversion of a large-scale matrix. Needless to say, the original initial-boundary value problems emerged during the development of wartime technology. All related research results were considered top military secrets and their publication was quite out of question. What Neumann was allowed to publish as a 'by-product' of the numerical solution of the initial-boundary value problems was only an expository article in 1947 on the numerical inversion of matrices of high order with the co–authorship of H. H. Goldstine {27} (in today's terminology, they established the stability of the LU and the Choleski factorizations) and a six-page paper with R. D. Richtmyer {53} on the numerical calculation of hydrodynamic shocks in 1950. (Some of) his earlier reports to various divisions of the National Defence Research Committee on shocks — actually, the generals were interested in detonation/blast waves — remained unpublished for nearly two decades, see pp. 178–347 in Volume 6 of the Collected Works of John von Neumann (and others may remain unpublished still). The fact that the discretization of partial differential equations might lead to matrices of high order could be found out by any reader of that time, therefore it was mentioned in the introduction of the first paper briefly. However, it was not mentioned at all whether partial differential equations and, mainly which ones, had been discretized concretely, and whether their solutions had been computed. Moreover, no mention was made of the capacity and character of the computing devices used. (With the pride of a father and the precision of a book-keeper, however, he lists technical and finantial data of the first high-speed 'fully automatic electronic computing machines' in his confidential reports.) Yet, the aim was just to compute the concrete solutions of the concrete initial-boundary value problems with the aid of the first electronic/digital computers which had been built by the mid-forties. The second paper with Goldstine {28}, a 1951 continuation of {27} contains a probabilistic analysis of rounding errors.

Neumann kept track of the numerical solution of those concrete initial-boundary value problems closely and, in particular, he was concerned with the computational stability of the discretization methods applied. The main difficulty was that truncation errors (which emerged from replacing partial derivatives by the respective finite differences) could be amplified considerably in the course of the numerical-computational procedure. Thus, the establishment of stability criteria, the fulfilment of which would prevent elementary errors (truncation errors, rounding errors, or (local) errors of any kind) from becoming so amplified as to make gibberish the whole calculation, was of vital importance. Neumann acquainted his colleagues with his ideas on the concept of computational stability in his lectures held at the Los Alamos laboratories and other military research institutes. The 'von Neumann's method of stability analysis' has become known since the early l950s, primarily, owing to the paper written by G. C. O'Brien, M. A. Hyman & S. Kaplan (*J. Math. Phys.*, **29** (1950), 223–239) who incorporated and presented Neumann's heuristic techniques of stability with his consent. (The 1947 interior report had been remained unknown for a long time.) Now let us quote A. H. Taub's words from his Editorial Note on page 664 of Volume 5 of the Collected Works: '... his procedure was, as he always emphasized, at all times heuristic. It was used in various "practical" situations for practical purposes but, as he once wrote (letter of November 10, 1950 to Werner Leutert, "as far as any evidence that is known to me (J. v. N.) goes, ... it has not led to false results so far".' Robert Richtmyer (*Difference Methods for Initial-Value Problems,* Interscience, New York, 1957), a further close witness of the emergence of 'von Neumann's method of stability analysis' declares his opinion in the very same sense: '... the new development has been based more on empiricism and intuition and less on a mathematical basis than the classical development. One should not blame the new development for this, if we were to wait for convergence proofs and error estimates for the new methods, most of the computers now in use in technology and industry would come grinding to a halt.'

For Neumann the pattern was provided by R. Courant, K. Friedrichs H. Lewy (*Math. Ann.,* **100** (1928), 32–74), who studied the convergence of the solutions of the approximating system of difference equations and, through this, proved existence theorems for partial differential equations of various types. Courant, Friedrichs and Lewy observed for the first time that restrictions on the permissible size of Δt in terms of the size of the other increment (e.g. inequality $\Delta t \leq \Delta x/c$ for the wave equation $u_{tt} = c^2 u_{xx}$ expressing that the triangle-shaped domain of dependence on local data cannot become larger under discretization) are necessary for the convergence of the finite difference approximations.

However, as it is shown by the quotations above, Neumann's interest in analysing difference approximations was focused on error amplification and not on convergence. The essence of his approach can be easily extracted from his 1947/48 interim "First/Second Report on the Numerical Calculation of Flow Problems" reprinted on pages 652–750 of Volume 5 of the Collected Works: Consider, for simplicity, the one-dimensional heat equation $u_t = u_{xx}$, $t \geq 0$, $u(0,x) = f(x)$, $x \in \mathbb{R}$. The simplest discretization procedure leads to a system of linear equations of the form

$$\frac{u_{i+1,j} - u_{i,j}}{\Delta t} = \frac{u_{i,j+1} - 2u_{i,j} + u_{i,j-1}}{(\Delta x)^2} \quad i \in \mathbb{N},\ j \in \mathbb{Z}. \tag{15}$$

Neumann's basic idea was to investigate how errors (existing at some initial or intermediate time level) evolve in the course of solving system (15). Substituting $\varepsilon(t,x) = a(t)\sin kx$ into the error formula

$$\varepsilon(t+\Delta t, x) - \varepsilon(t,x) = \mu\big[\varepsilon(t, x+\Delta x) - 2\varepsilon(t,x) + \varepsilon(t, x-\Delta x)\big]$$

where $\mu = \Delta t/(\Delta x)^2$ (cf. (15) but note that $\varepsilon(t,x)$ is different from $e(t,x) = u(t,x) - \overline{u}(t,x)$, the difference between the computed and the exact values of the solution), it is readily checked that the coefficient $a(t)$ is amplified at each integration step by a factor of $1 + 2\mu(\cos k\Delta x - 1)$, see p. 653. If the interval of variability of x is from 0 to π and $u_t = u_{xx}$ is equipped with Dirichlet boundary conditions, then $\varepsilon_{i,j}$ can be represented as $\varepsilon_{i,j} = \sum_{k=1}^{N-1} a_{i,k} \sin k(j\Delta x)$, and the amplification factor is the same, i.e.

$$a_{i,k+1} = a_{i,k}\big(1 + 2\mu(\cos k\Delta x - 1)\big) \quad \text{for} \quad k = 1, 2, \ldots, N-1 \text{ with } N = \pi/\Delta x,$$

see p. 691. Observe that $\big\{1 + 2\mu(\cos k\Delta x - 1) \mid k = 1, 2, \ldots, N-1\big\}$ is the set of eigenvalues of the tridiagonal error amplification matrix (cf. (15))

$$A = \{\alpha_{i,j}\}_{i,j=1}^{N-1}, \quad \alpha_{i,j} = 1 - 2\mu \text{ if } i = j \text{ and } \mu \text{ if } |i-j| = 1 \text{ (and 0 otherwise)}.$$

Hence, for two different reasons,

$$\big|1 + 2\mu(\cos\beta - 1)\big| \le 1 \quad \text{for all} \quad \beta \in (0, \pi) \;\Leftrightarrow\; \mu \le 1/2$$

is a necessary condition for preventing error amplification. He also proved that an implicit version of (15) is computationally stable, independently of the fact how much the concrete value of the mesh ratio $\Delta t/(\Delta x)^2$ is, see p. 707. (It is worth mentioning here that the inequality $\mu \le 1/2 \Leftrightarrow \Delta t \le (\Delta x)^2/2$ plays no role in the Courant, Friedrichs & Lewy paper.) As for the linear system (15) equipped with initial conditions $u_{0j} = f(j\Delta x)$ ($j \in \mathbb{Z}$, f sufficiently nice), they only prove that, in the special case $\Delta t/(\Delta x)^2 = 1/2$, the solutions of (15) converge to the exact solution $\overline{u}$ of the original initial value problem. In fact, they point out by a direct computation that

$$u_{i,j} = \sum_{k=0}^{i} \frac{1}{2^i}\binom{i}{k} f\big((j+i-2k)\Delta x\big) \quad \longrightarrow$$

$$\longrightarrow \quad \overline{u}(t,x) = \frac{1}{(4\pi t)^{1/2}} \int_{-\infty}^{\infty} \exp\left(-\frac{(\xi - x)^2}{4t}\right) f(\xi)\, d\xi$$

whenever $i,j \to \infty$ with $i\Delta t \to t$ and $j\Delta x \to x$ for any $t > 0$ and $x \in \mathbb{R}$ fixed. (Actually, they consider equation $2u_t = u_{xx}$ and — in order to make the combinatorics for $u_{i,j}$ simpler — they take $\Delta t = (\Delta x)^2$ (and mention that similar convergence results hold true for general parabolic equations as well).) In the Courant, Friedrichs & Lewy paper the emphasis is put on hyperbolic equations. On the other hand, though the affirmative answer is implicitely contained in his considerations, Neumann does not seem to pay any attention to the question if, at least in some technical sense and for f and $\overline{u}$ sufficiently nice, inequality $\mu \leq 1/2$ is enough to imply that $e(t,x) \to 0$ as $\Delta t \to 0$.)

The model results collected in the previous paragraph illustrate how Neumann applied his Fourier series and eigenvalue techniques to studying the computational stability of finite difference approximations. Real-life examples can be found in his weather forecast paper {8} and in the aforementioned 1947/48 confidential reports intended for military and industrial use. Thus, Neumann can be regarded, rightly, as the founder of the stability theory of the numerical methods of differential equations even if he always used the concept of stability in an empirical sense. Several applications of the Fourier approach and the eigenvalue one initiated by him can be found in Arieh Iserles' remarkable textbook. With good reason, 'A. Iserles, *A First Course in the Numerical Analysis of Differential Equations,* Cambridge University Press, Cambridge, 1996' criticizes that the bulk of the literature (in particular, the older literature) terms each of these approaches as 'von Neumann's method of stability analysis', without any distinction and in a rather confusing way. A more descriptive and less ambiguous terminology is preferred.

Lax equivalence theorem, the theoretical result behind. Neumann's 'naive' argumentations concerning the computational stability of difference approximations were raised to the level of an abstract theory in linear functional analysis by Péter Lax in the early fifties. Together with his parents, the 15 year old Lax left Hungary for the US in 1941. Letters of recommendation from Dénes König and Rózsa Péter to Neumann accompanied him. As a student and young researcher, he was very much influenced by Neumann, Friedrichs (his PhD adviser), and Courant. Thirty years after the death of Neumann, he remembered his mentor in an interview by saying: 'Von Neumann, who was the central figure of the mid-century, firmly believed that computing was central not only to the numerical side of applied mathematics but also to progress in theory. That is why he invented computers and pushed for their development. He foresaw that computations are essential to discover basic phenomena in nonlinear systems.'

Drafted for the war, Lax ended up at Los Alamos and remained there until May 1946. His Los Alamos stay shaped his general attitude to mathematics as well the choice of his research subjects considerably. He belongs to a minority of mathematicians who consider themselves both pure and applied.

For Banach space problems $u_t = Au$, $u(0) = u_0$ generating C^0 operator semigroups, e.g. for well-posed linear evolutionary partial differential equations, Lax gave a precise definition of the stability of approximating linear procedures. The core of the definition is, on each time interval $[0, T]$, the uniform boundedness of all products of the approximating small-stepsize linear operators. Simultaneously, Lax opened up the series of equivalence theorems

$$\text{consistency} \quad \& \quad \text{stability} \;\Leftrightarrow\; \text{convergence}$$

which played a vital role in the theory of discretizations. For details, see e.g. 'R. D. Richtmyer & K. W. Morton, *Difference Methods for Initial-Value Problems,* Wiley, New York, 1967'. In the background of the later equivalence theorems, too, there is the Neumann idea according to which the convergence of numerical procedures can be proven through an argumentation of the following type: 'small local errors' plus 'no (significant) error amplification' imply 'small global error'.

The work of Lax on a single conservation law. In what follows we give a brief description of Lax's work on shock waves which had grown out of his Los Alamos experiences. 'The existing literature on this question is unsatisfactory' — summarized Neumann his opinion in 1943. Thanks to the development that followed, in particular to Lax's contribution {45}, {46}, Neumann's statement had lost much of its validity by the time of his death in 1957. We follow the respective chapters in 'L. C. Evans *Partial Differential Equations,* AMS, Providence, R.I., 1998' and 'J. Smoller, *Shock Waves and Reaction-Diffusion Equations*, Springer, Berlin, 1982' very closely.

Consider the scalar conservation law in a single space variable

$$u_t + \big(f(u)\big)_x = 0, \quad x \in \mathbb{R},\ t \geq 0 \tag{16}$$

with initial data $u(x, 0) = u_0(x)$, $x \in \mathbb{R}$. Characteristics are straight lines of the form $\big\{\big(f'\big(u_0(x_0)\big)\,t + x_0, t\big) \mid t \geq 0\big\}$, $x_0 \in \mathbb{R}$. If $x_0 < \tilde{x}_0$ and $f'\big(u_0(x_0)\big) < f'\big(u_0(\tilde{x}_0)\big)$ for some $x_0, \tilde{x}_0 \in \mathbb{R}$, the characteristic lines through $(x_0, 0)$ and $(\tilde{x}_0, 0)$ meet at some point in $t > 0$. Since classical solutions are constant along characteristics, one has either to accept that solutions cannot be defined for all time or, alternatively, to look for a more general concept of a solution that allows the appearance of discontinuities (even in solutions with continuous initial data). A function $u \in L_\infty(\mathbb{R}\times\mathbb{R}^+, \mathbb{R})$ is called a weak solution of (16) with initial data $u_0 \in L_\infty(\mathbb{R}, \mathbb{R})$ if

$$\int_0^\infty \int_{-\infty}^\infty \big(\varphi_t u + \varphi_x f(u)\big)\, dx\, dt + \int_{-\infty}^\infty \varphi(\cdot, 0) u_0\, dx = 0$$

for any C^∞ function $\varphi : \mathbb{R}^2 \to \mathbb{R}$ with compact support in $\{(x,t) \mid x \in \mathbb{R}, t \geq 0\}$. In general, unfortunately, as it is demonstrated by the example of Burger's equation $u_t + uu_x = 0$ (i.e. $f(u) = u^2/2$) equipped with Riemann initial data $u_0(x) = u_\ell \in \mathbb{R}$ if $x < 0$ and $u_r \in \mathbb{R}$ if $x > 0$, there exists an abundance of weak solutions defined for all $t \geq 0$. Now equations of the form (16) arise in the physical sciences and so one must have some mechanism to pick out the 'physically relavant' one. Mathematically, the basic question is to impose an a priori condition on weak solutions that ensures existence and uniqueness. This a priori assumption is O. Oleinik's entropy condition (we present as inequality (18) below) found in 1957 (*Usp. Mat. Nauk.*, **12** (1957), 3–73. (in Russian) (*AMS Transl. Ser. 2*, **26** (1957), 95–172.). Until very recently, no such results were known for systems of conservation laws of general type.) In the very same year, Lax {46} also proved an existence and uniqueness theorem. He considered a subclass of systems of conservation laws and proved existence and uniqueness within a class of piecewise continuous functions with a finite number of certain shock and contact discontinuities. His abstract result applies to Riemann's classical tube problem in gas dynamics and gives a rigorous proof for the earlier results.

The key is to look at the Hamilton–Jacobi equation

$$(17) \qquad w_t + f(w_x) = 0 \quad \text{with initial data} \quad w_0(x) = \int_0^x u_0(y)\, dy.$$

(Formal differentiation shows that u can be taken for w_x.) From now on, assume that $f \in C^2$, $f(0) = 0$, $\inf\{f''(u) \mid u \in \mathbb{R}\} > 0$ and let L denote the Legendre transform of f. The uniform convexity assumption on f implies that f' is a C^1 self-diffeomorphism of $\mathbb{R}$. For later purposes, set $g = (f')^{-1}$, the inverse of f'. The Hopf–Lax formula

$$w(x,t) = \inf\left\{ \int_0^t L\big(\dot{q}(s)\big)\,ds + w_0(y) \,\Big|\, q \in C^1\big([0,t],\mathbb{R}\big),\right.$$

$$\left. q(0) = y,\ q(t) = x \right\}$$

$$= \min\left\{ tL\left(\frac{x-y}{t}\right) + w_0(y) \,\Big|\, y \in \mathbb{R} \right\},$$

a basic result in the calculus of variation defines a weak solution to (17) in the sense that $w : \mathbb{R} \times \mathbb{R}^+ \to \mathbb{R}$ is Lipschitz continuous, (and hence, by the Rademacher theorem, w is almost everywhere differentiable), it satisfies $w_t(x,t) + f\big(w_x(x,t)\big) = 0$ for almost every $(x,t) \in \mathbb{R} \times \mathbb{R}^+$, and $w(x,0) = w_0(x)$ for each $x \in \mathbb{R}$. Actually, given $t > 0$ arbitrarily, the mapping $x \to w(t,x)$ is differentiable for almost every $x \in \mathbb{R}$. In addition, there exists for all but at most countably many values of $x \in \mathbb{R}$ a unique $y(x,t) \in \mathbb{R}$ such that

$$w(x,t) = tL\left(\frac{x - y(x,t)}{t}\right) + w_0\big(y(x,t)\big),$$

the mapping $x \to y(x,t)$ is nondecreasing, and $\frac{\partial}{\partial x} w(x,t) = g\big(\frac{x-y(x,t)}{t}\big)$ holds true for almost every $x \in \mathbb{R}$. The final result is that the Lax–Oleinik formula

$$u(x,t) = g\left(\frac{x - y(x,t)}{t}\right)$$

defines a weak solution for (16) satisfying, with some absolute constant $C = C(u_0)$, for each $t > 0$, the one-sided Lipschitz estimate

$$u(x+z,t) - u(x,t) \le Cz/t \tag{18}$$

for almost every $(x,z) \in \mathbb{R} \times \mathbb{R}^+$. As a reformulation of (18), the function $x \to u(x,t) - Cx/t$ is nondecreasing for each $t > 0$. Thus, even though the initial data u_0 is merely an $L_\infty(\mathbb{R},\mathbb{R})$ function, the Lax–Oleinik solution u immediately becomes fairly regular in $t > 0$. It is a crucial result of Oleinik that (in some technical sense) u depends continuously on u_0 and — up to a set of measure zero — no other weak solution of (16) with initial data u_0 satisfies (18).

Consider a C^1 curve $\Gamma = \{(x(t), t)\}$ of discontinuties in a weak solution u and assume that u extends continuously from either side of Γ to Γ. By choosing the test function to concentrate at the discontinuity, one arrives easily at the Rankine–Hugoniot jump identity

$$s(u_+ - u_-) = f(u_+) - f(u_-). \tag{19}$$

Here $s = \dot{x}(t)$ is the speed of the discontinuity in $(x(t), t)$ and $u_\pm = u(x(t) \pm 0, t)$ are the states at the jump. The motion along the discontinuity curve is called a shock wave. In a rough analogy to the thermodynamic principle that physical entropy cannot decrease as time goes forward, Lax introduced the entropy condition

$$f'(u_+) < s < f'(u_-) \tag{20}$$

required at each point of Γ. A more direct analogy for requiring (20) is that information may vanish at the shock but may not be created at a shock — geometrically, (20) means that characteristic lines may enter a shock but may not leave it. Armoured with (19) and (20), it is not hard to single out the 'physically relevant' solution in a great number of cases. For conservation laws with uniformly convex f, (20) is an easy consequence of (18).

The work of Lax on systems of conservation laws. The modern theory of systems of conservation laws $\mathbf{u}_t + (\mathbf{f}(\mathbf{u}))_x = 0$, $(x \in \mathbb{R},\ t \geq 0)$ started with Lax's fundamental paper {46}. It is there where one first encounters the basic ideas in the subject: the shock inequalities (that replace Lax's entropy condition (20) for systems), the notion of genuine nonlinearity, the one-parameter families of shock- and rarefaction-wave curves, as well as the solution to the general Riemann problem. We do not enter the details here but, indicating the complexity of {46}, describe the solution of Riemann's classical tube problem in gas dynamics instead. Consider a long, thin, cylindrical tube containing gas separated at $x = 0$ by a thin membrane. It is assumed that the gas is at rest on both sides of the membrane, but it is of different constant pressures and densities on each side. At time $t = 0$, the membrane is broken, and the problem is to determine the ensuing motion of the gas. This leads to a system of conservation laws with dependent variable $\mathbf{u} = (v, \rho, p) = (\text{velocity}, \text{density}, \text{pressure})$ and initial data $(v_\ell, \rho_\ell, p_\ell) \in \mathbb{R}^3$ for $x < 0$ and $(v_r, \rho_r, p_r) \in \mathbb{R}^3$ for $x > 0$. Note that $v_\ell = v_r = 0$ and consider the case $\rho_\ell > \rho_r$, $p_\ell > p_r$. By symmetry,

$v(x,t) = \alpha(s)$, $\rho(x_t) = \beta(s)$, $p(x,t) = \gamma(s)$ for some real functions α, β, γ where $s = x/t$, $x \in \mathbb{R}$, $t > 0$. The solution $\mathbf{u}(x,t)$ can be described as follows. The initial discontinuity breaks up into two discontinuities, the shock wave and a contact discontinuity with constant speed $s_4 > 0$ and $s_3 \in (0, s_4)$, respectively. In addition, there exist constants $s_1 < 0$ and $s_2 \in (s_1, 0)$ such that

$$u(s) = \begin{cases} 0 \\ mif(s) \\ u_0 \\ u_0 \\ 0 \end{cases}, \quad \rho(s) = \begin{cases} \rho_\ell \\ mdf(s) \\ \rho_1 \\ \rho_2 \\ \rho_r \end{cases}, \quad p(s) = \begin{cases} p_\ell & \text{if } s < s_1 \\ mdf(s) & \text{if } s_1 < s < s_2 \\ p_0 & \text{if } s_2 < s < s_3 \\ p_0 & \text{if } s_3 < s < s_4 \\ p_r & \text{if } s_4 < s \end{cases}$$

(i.e. gas in original high pressure state, rarefaction wave, rarified gas, compressed gas, gas in original low pressure state) where $u_0 > 0$, $p_0 \in (p_r, p_\ell)$, $\rho_1 \in (0, \rho_\ell)$, $\rho_2 > \min\{\rho_1, \rho_r\}$ are constants and mdf and mif stand for certain decreasing and increasing functions, respectively.

The Lax–Milgram Lemma. It is a must to discuss the Lax–Milgram Lemma. Consider, for simplicity, the Dirichlet problem for the Laplacian

$$\Delta u = f \in L_2(\Omega), \quad u|_{\partial\Omega} = 0$$

on a bounded domain $\Omega \subset \mathbb{R}^n$ with boundary $\partial\Omega$ nice. Function $u \in H_0^1(\Omega)$ is called a weak solution if

$$\int_\Omega \nabla u \cdot \nabla v \, dx + \int_\Omega fv \, dx = 0$$

for each C^∞ function v with compact support in Ω. The Lax–Milgram Lemma is an abstract result in linear functional analysis. (Actually, a simple consequence of Riesz' theorem on the dual space of a Hilbert space: Given a bounded, coercive bilinear form b on a Hilbert space H with scalar product $\langle .,. \rangle$, there exists a uniquely defined linear self-homeomorphism S of H such that $b(Sf, v) + \langle f, v \rangle = 0$ whenever $f, v \in H$. Coercivity of the not necessarily symmetric bilinear form b means that $b(v,v) \geq \beta\langle v, v \rangle$ for some $\beta > 0$ and all $v \in H$.) The Lax–Milgram Lemma guarantees existence and uniqueness for the weak solution. It also works in the case when the Laplacian is replaced by a more general elliptic operator. Thus the Lax–Milgram Lemma reduces the existence proof for a solution in $H^2(\Omega) \cap H_0^1(\Omega)$

to a regularity lemma for the weak solution. Here $H^2(\Omega)$ and $H_0^1(\Omega)$ are Sobolev spaces. As it is suggested by the observation that follows (18), note that the natural space for weak solutions of a conservation law is some bounded variation space.

In line with the editorial principles, we do not pursue the scientific career of Lax any further but restrict ourselves to call the attention of the reader to [107], a survey on systems of conservation laws. The *MathSciNet Full Search* option gives evidence that his name — in expressions like Lax difference operator, Lax equations, Lax–Friedrichs scheme, Lax integrability, Lax monoids, Lax pairs, Lax–Phillips scattering theory, Lax representation, Lax–Wendroff scheme etc. — appears in the respective titles of more than 600 mathematical research papers.

A cross-section in 1928. The state of art. Neumann must have known the R. Courant, K. Friedrichs & H. Lewy (*Math. Ann.,* **100** (1928), 32–74) paper very early because an article of his own, the one in which he proved the minimax theorem of game theory, was published in the same volume of Mathematische Annalen. If we have started our study with mentioning how many papers of Hungarian mathematicians were published in Mathematische Annalen between 1900 and 1910, let us cite similar statistics here, based on Volumes 98–99–100 which were issued in 1928. Out of about 100 papers 17 were written by Hungarian authors or coauthors. Since Gyula Szőkefalvi-Nagy, János Neumann and Gábor Szegő are represented by several papers, the number of Hungarian authors is 13. Out of them Pólya lived in Zürich, Switzerland, Neumann, Szász and Szegő lived in Germany, and Szőkefalvi-Nagy was a resident in Romania. Later Szőkefalvi-Nagy moved to Szeged and the other four, threatened by the worsening of their working conditions in a continental Europe under German influence and foreseeing the dimensions of a racial persecution that culminated in the *Endlösung*, emigrated to the USA. Tibor Radó, after the failure of his 1929 application for a professorship at Debrecen University, joined them in the self-chosen/from-outside-enforced exile. Though supported by a commission of the four most respected Hungarian professors who recommended him *primo et unico loco,* Radó was surpassed by a protègè of a clique of local potentates with some political support in Budapest. (Also Riesz' and Haar's applications were refused in their days. They applied for a professorship at Budapest University but neither of them was appointed.)

Returning to the Hungarian contribution to Volumes 98–99–100 of the Mathematische Annalen, we note that only two of the 17 papers treat differential equations. These are Alfréd Haar's aforementioned paper on adjoint problems of the calculus of variations {38} and an additional one by Aurél Wintner {82}. During his U.S. years Wintner did not consider himself a Hungarian mathematician, therefore in this report, in compliance with the editorial principles (which exclude discussing the oeuvre of Arthur Erdélyi and that of Paul Halmos, for example) we are going to discuss only his earliest career. The cited paper is concerned with analytic solutions of differential equations in Hilbert spaces: Wintner provides an infinite–dimensional generalization of the Cauchy–Kowalewskaya theorem. By the way, in one of his papers, the coauthor of which was S. Bochner, Neumann treats ordinary differential equations in Hilbert spaces, most precisely, with almost periodic solutions of one type of these equations {6}. It is worth mentioning another paper of his, written jointly with G. W. Brown {7}, in which they give a new proof using Liapunov functions of game dynamics for the existence of good strategies for zero–sum two-person games. Similarly to his works on partial differential equations, here also, practical aspects are emphasized: "The proof is 'constructive' in a sense that lends itself to utilization when, actually, computing the solutions of specific games." Our report on the relation of Neumann's oeuvre to the theory of differential equations will be complete if we mention his contribution of great impact to ergodic theory — his 1940/41 Princeton lectures on invariant measures were published quite recently {52} — as we did in the case of Frigyes Riesz.

Methods of differential equations appeared occasionally in the works of Károly Jordan. We restrict ourselves to quoting his monograph on difference calculus (actually, on basic combinatorial enumeration from a probabilist's view but including a long chapter on linear difference equations and a short one on linear equations of partial differences) {42}. On the other hand, differential equation methods infiltrated the works of István Grynaeus, whose illness and untimely death in 1936 deprived the circle of Hungarian differential geometers of its most talented member, to a much deeper extent, e.g. in {29} which is an application of the Ricci calculus to a Pfaffian system.

Pólya and Szegő on isoperimetric inequalities. Several works of György Pólya and Gábor Szegő can be considered to be about differential equations; primarily the ones in which they proved certain isoperimetric inequalities with the aid of the methods of potential theory and calculus of variations. Pólya and Szegő were led to isoperimetric inequalities, partly, by their notorious problem-solving attitude and, partly, by their profound knowledge of complex function theory (and, within complex function theory, by the fact that potential theory in dimension two is essentially equivalent to the theory of conformal mapping). Among the antecedents it should be mentioned that Gábor Szegő translated Webster's partial differential equations textbook in the late 1920s. Since Szegő added some mathematical details to the original text of the physicist Webster, who neglected, or elaborated only roughly, certain parts of it, the translation became a revision and, thus, the translator became a co-author. The German edition of the work was published under both of their names [195]. Also, it is worth mentioning that in the encyclopedic work of 'P. Frank and R. von Mises *Die Differential- und Integralgleichungen der Mechanik und Physik*, Vieweg, Braunschweig, 1925' the chapter on potential theory was written by Szegő whose own first result in that field was on a relationship between Green functions and the transfinite diameter of plane curves. This latter concept was introduced by Mihály Fekete in his famous work on generalizing Chebishev polynomials (which arise in the case of a line segment) {23}. Later, Pólya joined Szegő's research of this type. By isoperimetric inequalities we mean statements on extremal properties of set functions which have obvious geometric or physical interpretations. The model statement (which was due, originally, to Pólya in 1920) can be taken from Pólya and Szegő [129, Problem IX. I. 2]: Consider a corn hill the base of which is a unit disc in a horizontal plane. Then $V/S \geq \pi/3$ where V and S stand for the volume and the maximal slope, respectively. Equality is attained for circular cones. Based on the machinery necessary to their formulation and proof, isoperimetric inequalities can be classified as belonging to the relevant branches of mathematics.

In what follows let $V_1 \subset V \subset V_0$ denote a nested triplet of closed solids in $\mathbb{R}^3$ with closed regular surface boundaries. It is assumed that $\partial V_1 \subset V \setminus \partial V$ and $\partial V \subset V_0 \setminus \partial V_0$. Let u denote the uniquely defined solution to the Dirichlet problem

$$(21) \qquad \Delta u = 0 \quad \text{on} \quad V_0 \setminus (V_1 \cup \partial V_0), \quad \text{and} \quad u|_{\partial V_1} = 1, \quad u|_{\partial V_0} = 0.$$

The capacity of the nested pair $(\partial V_1, \partial V_0)$ is defined as

$$C = -\frac{1}{4\pi} \int_{\partial V} \frac{\partial u}{\partial \nu}\, dS \qquad \text{(the normal vector } \nu \text{ points outwards)}$$

— the integral does not depend on the particular choice of V. The nested pair $(\partial V_1, \partial V_0)$ itself is termed a condenser. The terms 'capacity' and 'condenser' refer to the meaning of the Dirichlet problem (21) in electrostatics. (Of course, the function u can be interpreted as an equilibrium solution of the heat equation also.) The capacity of ∂V_1 is defined via the Dirichlet problem

$$\Delta u = 0 \quad \text{or} \quad \mathbb{R}^3 \setminus V_1, \quad \text{and} \quad u|_{\partial V_1} = 1, \quad u(\infty) = 0$$

(that corresponds to the limiting process $\partial V_0 \to \infty$).

Szegő's main result {78} is as follows: Among all nested pairs $(\partial V_1, \partial V_0)$ with volume (V_1) and volume (V_0) given, the capacity is minimal if and only if V_1 and V_0 are concentric balls. The limiting process $\partial V_0 \to \infty$ leads to the proof of a conjecture due to Poincaré: Of all surfaces ∂V_1 with volume (V_1) given, the sphere has the smallest capacity. Similarly {78}, of all surfaces $\partial(V_1)$ with area (V_1) given, the sphere has the largest capacity. Naturally, the abovementioned statements, the planar versions of which had been known before, could be reformulated in the form of inequalities as well. In his latter work {79} Szegő verified Maxwell's conjecture $C \leq d/2$, too. Here d stands for the usual diameter and the equality holds only for spheres. It should be mentioned that exact indications to a satisfactory proof of Poincaré's conjecture can be found in an earlier paper by G. Faber (*Sitzungsber. Bayr. Acad. Wiss.* (**1923**), 169–172).

The main finding of Georg Faber's aforesaid paper is the proof of one of Rayleigh's important conjectures: of all vibrating membranes, the closed disc emits the gravest fundamental tone. The mathematical task is to minimize $\lambda_1(D)$ where D is a closed regular domain in $\mathbb{R}^2$ with area (D) given, say π, and $\lambda_1(D)$ stands for the principal eigenvalue of the negative Laplacian equipped with the Dirichlet boundary condition $u|_{\partial D} = 0$. Recall that

$$\lambda_1(D) = \min\left\{\frac{\iint_D (u_x^2 + u_y^2)\,dx\,dy}{\iint_D u^2\,dx\,dy} \,\middle|\, u \in C^1(D \setminus \partial D) \cap C(D) \text{ and } u|_{\partial D} = 0\right\},$$

the minimum is attained for the principal eigenfunction e_1, the level sets of e_1 are (except for one point) simple closed curves, and $e_1(x,y) > 0$ for $(x,y) \in D \setminus \partial D$. In essence, the major observation of Faber and of Edgar Krahn (*Math. Ann.*, **94** (1924), 97–100) (the latter obtained the same result nearly simultaneously but independently of the former) is that

$$\iint_D \left((e_1)_x^2 + (e_1)_y^2\right) dx\,dy \geq \iint_{B_1} (v_x^2 + v_y^2)\,dx\,dy$$

where B_1 is the closed unit disc and the function $v = v(e_1, D) : B_1 \to \mathbb{R}^+$ is (uniquely) defined by the property as follows: For any $\kappa \in \big[0, \max\big\{ e_1(x,y) \mid (x,y) \in D\big\}\big]$,

(22)
$$v(x,y) = \kappa \text{ if and only if } x^2 + y^2 = \pi^{-1} \cdot \text{area}\big(\{ (x,y) \in D \mid e_1(x,y) \geq \kappa\}\big).$$

From this the proof of Rayleigh's conjecture can be derived easily. The name of the procedure applied in formula (22) is symmetrization with respect to a point. Faber remarks that the very same method leads to the proof of Poincaré's conjecture on minimal capacities and, that is indeed the case.

Szegő {78} followed a totally different, simpler and ad hoc way, but the family of symmetrization methods some elements of which had already been known by Jacob Steiner and Hermann Amandus Schwarz in the nineteenth century proved to be much more successful in the long run. At least, a dozen quantities in geometry and physics increase or decrease under a certain symmetrization procedure. Pólya and Szegő, jointly and individually, proved several assertions of this type and, through them, isoperimetric inequalities {61}. With the help of the symmetrization methods Pólya {57} proved de Saint-Venant's conjecture of 1856 (which de Saint-Venant supported by convincing physical considerations and several particular cases, but did not prove in a mathematical sense): Of all cross-sections with a given area, the circular cross-section has the largest torsional rigidity. The torsional rigidity or stiffness $P(D)$ of a cross-section D (i.e. of an infinite beam with a given plane domain D as cross-section) can be defined as

$$P(D) = 4 \cdot \max\left\{ \frac{\left(\iint_D u\,dx\,dy\right)^2}{\iint_D (u_x^2 + u_y^2)\,dx\,dy} \,\middle|\, u \in C^1(D \setminus \partial D) \cap C(D) \text{ and } u|_{\partial D} = 0 \right\}.$$

Note that the maximum is attained if and only if $u = cv$ where $c \neq 0$ is a real constant and v solves the boundary value problem

$$v_{xx} + v_{yy} + 2 = 0 \quad \text{on} \quad D \setminus \partial D, \quad \text{and} \quad v|_{\partial D} = 0.$$

In their 1951 book Pólya and Szegő [130] presented the 'state of the art' of the questions concerning isoperimetric inequalities of that age. The influence of this so-called 'smaller Pólya–Szegő' can be felt even nowadays and this work of theirs continues to be the source of inspirations. At least half the book discusses Pólya and Szegő's own results. They formulated, improved and optimized in it the inequalities about various set functions. They treated the cases of nearly circular and nearly spherical domains as well as several techniques for handling parameters. After the publication of their book the study of the topic was continued by both of them. Among their co-authors Menahem Schiffer's name should be mentioned: With him, Pólya proved {59} an old conjecture of his according to which the transfinite diameter of a convex plane curve is no less than one-eighth of the perimeter.

A special mention should be made of Pólya and Szegő's joint paper {60} on qualitative properties of the one-dimensional heat equation. Applying Descartes's generalized rule of signs and Sturm's oscillation theorem they state that the number of roots and/or the extrema of each individual solution is a decreasing function of time. In one of his papers {57} Pólya treats similar questions again but the longer study intended has never been written. If it had been written, it might have accelerated the recognition how important a role is played by the number of sign changes in the qualitative theory of linear and nonlinear parabolic equations of one dimension. The 1952 paper of Pólya {57} on combining finite differences with the Rayleigh–Ritz method is frequently interpreted as a preparatory step towards the discovery of finite element methods.

M. Riesz' fractional potentials. Now we are going to discuss the contribution to differential equations of the younger Riesz brother. Marcel Riesz was concerned with differential equations only in a rather late period of his career, from the early 1930's on. His most important results were in the field of potential theory and wave propagation. His interest was motivated, partly, by the application to the theory of relativity. All his work on partial differential equations until that time was summarized by Marcel Riesz himself in a book-size paper written in a book style, published in 1949 {71}. We are going to discuss this monumental work below.

Marcel Riesz worked out several basic techniques in multidimensional fractional integration and generalized the concept of the classical Riemann–Liouville integral

$$(I^\alpha f)(x) = \frac{1}{\Gamma(\alpha)} \int_a^x f(t)(x-t)^{\alpha-1}\,dt, \qquad \alpha > 0$$

in different directions. The one associated with the m-dimensional Laplacian Δ is

$$(23) \qquad (I_\Delta^\alpha f)(x) = \frac{1}{H_{\Delta,m}(\alpha)} \int_{\mathbb{R}^m} |x-y|^{\alpha-m} f(y)\, dy$$

where $H_{\Delta,m}(\alpha) = \pi^{m/2} 2^\alpha \Gamma(\alpha/2)/\Gamma\big((m-\alpha)/2\big)$. Case $m = 3$, $\alpha = 2$ simplifies to the standard Newtonian potential. Patterned on the simple identity

$$(I^\alpha f)(x) = \frac{f(a)}{\Gamma(\alpha+1)}(x-a)^\alpha + (I^{\alpha+1} f')(x) = (I^{\alpha+1} f')(x)$$

valid for $f \in C^1(\mathbb{R})$ with $f(a) = 0$ and $\alpha > -1$, Green's formula applies for $f : \mathbb{R}^m \to \mathbb{R}$ sufficiently nice, extends the operator $\alpha \to (I_\Delta^\alpha f)$ by analytic continuation and leads to the properties $\Delta(I_\Delta^{\alpha+2} f) = -I_\Delta^\alpha f$ and $\Delta(I_\Delta^2 f) = -f$. If $f\,dy$ is a mass distribution with a finite total mass in $\mathbb{R}^m$, then the integral in (23) makes sense for $0 < \alpha < m$ and $I_\Delta^\alpha f$ is called the fractional potential of order α of $f\,dy$. By passing to the limit, $I_\Delta^0 f = f$. A further fundamental fact established by Riesz is that $I_\Delta^{\alpha+\beta} f = I_\Delta^\alpha (I_\Delta^\beta f)$ whenever $\alpha > 0$, $\beta > 0$ and $\alpha + \beta < m$. The very same semigroup properties hold true if $f\,dy$ is replaced by $d\mu(y)$ where μ is a general mass distribution in $\mathbb{R}^m$. In this setting $(I_\Delta^\alpha f)$ is the fractional potential of the mass while the energy of μ with respect to the fractional potential is defined by $\int (I_\Delta^\alpha \mu)(x)\, d\mu(x)$. Existence, uniqueness and basic properties of the equilibrium distribution in a compact set $F \subset \mathbb{R}^m$ (i.e. of a distribution having minimal energy in the class of mass distributions supported by F and having a given total mass) were proven rigorously in the 1935 PhD thesis of Otto Frostman, a famous disciple of Riesz. In fact, Frostman's very general approach and method of proving the existence of the equilibrium measure is considered the foundation of modern general potential theory. Riesz' functional potentials thus generated a far reaching development including weighted potentials as well as the Wiener theory of Brownian motion.

The fractional integral associated with the D'Alembertian operator $\square = \partial_1^2 - \partial_2^2 - \cdots - \partial_m^2$, $m \geq 2$ is

$$(I_\square^\alpha f)(x) = \frac{1}{H_{\square,m}(\alpha)} \int_{x-C} \big(r(x-y)\big)^{\alpha-m} f(y)\, dy.$$

Here $1/H_{\square,m}(\alpha)$ is a suitable 'Γ–factor' — suitable to imply $\square(I_\square^{\alpha+2} f) = I_\square^\alpha f$ and $\square(I_\square^2 f) = f$ for $f : \mathbb{R}^m \to \mathbb{R}$ sufficiently nice —, $r^2(x-y) =$

$(x_1 - y_1)^2 - (x_2 - y_2)^2 - \cdots - (x_m - y_m)^2$ is the square of the Lorentzian distance and

$$x - C = \{x - y \in \mathbb{R}^m \mid r^2(y) \geq 0 \text{ and } y_1 > 0\},$$

the retrograde light-cone with its vertex at x. Semigroup properties for $\alpha, \beta \geq 0$ are also established. The last chapter of {71} contains a similar theory for the wave operator in arbitrary Riemannian spaces.

A major part of {71} is devoted to the Cauchy problem for the wave equation $\Box u = f$ with initial data on a codimension one surface of the form $S = \{x \in \mathbb{R}^m \mid x_1 = s(x_2, \ldots, x_m)\}$. Riesz establishes integral representations for the solution involving certain divergent integrals which obtain a meaning by analytic continuation methods. This is more elegant than the parallel theory of Hadamard on 'finite parts' of divergent integrals because it does not distinguish between even and odd numbers of dimensions. On the basis of his formula, Riesz clarifies that Huygens phenomenon is a consequence of the fact that, for $m > 2$ even, function $H_{\Box,m}$ has a simple pole at $\alpha = 2$. He gives a purely geometric interpretation of the solution for the physically most important case, namely $m = 4$, with a disussion of certain line congruences and caustics. The entire discussion is important with respect to the Lorentz group. Then he applies his method to the Maxwell and Dirac equations and analyses the Liénard-Wiechert potential of a moving electron, too. Similar to Hadamard, Riesz also extends his solution representation formula for the wave equation with variable coefficients and initial data on S.

Marcel Riesz's work {71} reflects the state of differential equations which preceded the introduction of Schwartz distributions and Sobolev spaces. Since that time one of his main goals, the proper interpretation of divergent integrals, has been attained in a much larger framework through the theory of distributions. Although he must have been rather distant from defining the appropriate function spaces, his results in potential theory pointed towards the introduction of fractional powers of the Laplacian. In the later development of linear partial differential equations from among his disciples two of them, Lars Gårding and Lars Hörmander played a basically important role. Apart from his work on spinors and Clifford algebras in the late period of his life Riesz himself contributed relatively little to his earlier differential equation results.

The work of Egerváry. The first result obtained in the post–war period in Hungary we present is due to Jenő Egerváry and Pál Turán {11} and devoted to the memory of D. König and A. Szücs who did not survive the tragic days of 1944/45. Combined with hard analytic tools which go back to H. Weyl, Egerváry and Turán used the geometric ideas of D. König and A. Szücs {43} in proving a weak, somewhat artificial form of the Boltzmannian Hypothesis in the kinetic theory of gases. They considered an oversimplified differential equation model (which is very carefully chosen but not a differential equation model any more — nevertheless, we feel that the differential equation chapter is the right place to discuss it) of n particles: the n particles are included in an immobile cube $C = \big\{(x_1, x_2, x_3) \mid 0 \le x_1, x_2, x_3 \le \pi\big\}$, they are dimensionless, of equal mass, no attractive or exterior forces acting, the impacts on the walls according to the laws of elastic reflection, collisions between three or more particles excluded, collisions between two particles according to the law of elastic impact, the initial conditions of the n particles at time $t_0 = 0$ are arbitrary and, with $\vartheta_1 = 1$, $\vartheta_2 = 2^{1/2}$, $\vartheta_3 = 3^{1/2}$, the initial velocities satisfy

$$v_k^i \in n^{2/5}\left(1 + \frac{k}{n^{101/100}}\right) \cdot \left(\vartheta_i - \frac{1}{n^{10}}, \vartheta_i + \frac{1}{n^{10}}\right)$$

$$i = 1, 2, 3 \quad \text{and} \quad k = 1, 2, \ldots, n.$$

For simplicity, Egerváry and Turán assumed that the n particles are *equidistributed* at time t if for any rectangular body R in C, the number of particles $N(R, t)$ in R at t satisfies

$$\left| \frac{N(R,t)}{n} - \frac{\mathrm{vol}\,(R)}{\pi^3} \right| \le \frac{1}{n^{1/10}}.$$

They prove that the particles are equidistributed for the time interval $0 \le t \le n^{1/4}$ except time intervals whose total length does not exceed $c_0 n^{-1/10} \log^4 n$ where c_0 stands for a moderate numerical constant. If n is of the order 10^{23}, then $n^{1/4}$ is on the order of several days, and $c_0 n^{-1/10} \log^4 n$ is on the order of several seconds long. Estimates which are slightly better and work for more realistic initial velocities can be found in {12} which is a technically improved version of {11}. In both papers, the intention of the authors is to support the opinion that (some reasonable variant of) the Boltzmannian hypothesis can be derived as a consequence of the basic laws of mechanics.

Jenő Egerváry, a professor at the Budapest University of Technology, is one of the very few Hungarian mathematicians whose entire career is closely related to applied mathematics. Starting from his 1913 PhD Thesis (dedicated to a single linear Fredholm integral equation {9}) to his latest results (including his 1956 paper on a large system of fourth–order linear differential equations modelling suspension bridges {10}) he wrote several articles on the convergence of the method of finite differences. He had papers on the three–body problem, on heat conduction, and on the motion of the electron as well.

A cross-section in 1953. The state of art. As far as the application of mathematics is concerned, the decade preceding 1956 played a role of special importance in Hungary. Obviously, in the age of reconstruction of war–time damages (with the priority of rebuilding the bridges destroyed on the Danube) and during a period of an unprecedented development of heavy industry most of the applications were closely related to differential equations. (Motivations of mathematicians to take part in this work were diverse: with genuine enthusiasm some supported the efforts to establish the new society which was called people's democracy officially; others did the same out of fear of the Communist Party under external and internal pressures; there were still others who just wanted to earn money.) In the meantime, central industrial research institutes were set up and even the Research Institute of Applied Mathematics organized by Egerváry and Rényi, which was the legal predecessor of today's Rényi Institute (Research Institute of Mathematics of the Hungarian Academy of Sciences), there was a Department of Chemical Industry, a Department of Mechanics and Statics as well as a group on Electrotechnic (precisely, an Independent Group on Electronics and Function Approximation). In compliance with the above mentioned administrative structure of mathematical research dozens of papers of practical importance were born in the field of the application of differential equations. From the mid- and late fifties researchers of mathematical analysis in a broader sense turned to more abstract research topics.

From 1960 to 1970 the Department of Differential Equations of the Research Institute of Mathematics — the attribute 'Applied' was taken away after the fifties — was led by Károly Szilárd, the brother of Leó Szilárd. Károly Szilárd left Hungary in 1919 and returned in 1960. He spent 14 years in Germany (PhD in Göttingen, 1925) and 27 years in the USSR (Stalin Prize in 1953, after several years in a 'prison-research-institute'). A further emblematic figure of applied mathematical analysis was Samu Borbély. He worked in a research laboratory of the German aviation industry in the thirties, then returned to Hungary for reasons of conscience, and fled the Gestapo in 1944. While in a USSR 'prison-research-institute' after the war, he could conceale his expertize in aviation matters and worked for the artillary. Being a member of the Hungarian Academy of Sciences from 1949 onward, he taught mathematics in Miskolc and, later, at the Budapest University of Technology. Like Szilárd, he has a very limited number of publications (in the literature with general availability).

Having discussed the data between 1900 and 1910 as well as those in 1928, let us have a look at the state of differential equations in light of the statistical figures found in the 1953 volumes of the old Acta Scientiarum Mathematicarum (Szeged) **1/32** and the recently founded new journals Acta Mathematica Hungarica (Budapest) **1/23**, Publicationes Mathematicae (Debrecen) **2/33** — papers only in English, French, German, and Russian; MTA III. Osztály Közleményei (Proceedings of the Third Branch (Mathematics and Physics) of the Hungarian Academy of Sciences) **1/21**, MTA Alkalmazott Matematika Intézetének Közleményei (Proceedings of the Research Institute of Applied Mathematics of the Hungarian Academy of Sciences) **10/36** — papers only in Hungarian. The name of each journal is followed by a fraction. The denominator is the number of papers in the journal written by Hungarian authors whereas the numerator is the number of papers that may be ranked among differential equations in a broader sense. Since in 1953 mathematicians in Hungary could hardly think to publish their work abroad, actually, the number of their papers in the five periodicals mentioned were almost identical with the total number of their publications of that year.

Bihari inequality. The 1956 paper of Imre Bihari {4} is probably the most frequently cited ordinary differential equation paper ever written by a Hungarian mathematician. It contains what we call today the Bihari inequality, the first nonlinear version of the classical Gronwall lemma. Let $u, v : [a, b) \to \mathbb{R}^+$, $\omega : \mathbb{R}^+ \to \mathbb{R}^+$ be continuous functions. Assume that ω is increasing and $\omega(u) > 0$ whenever $u > 0$. In addition, let K be a nonnegative constant and assume that

$$u(t) \le K + \int_a^t v(s)\omega\big(u(s)\big)\,ds \quad \text{whenever} \quad t \in [a, b).$$

Then

(24)

$$u(t) \le \begin{cases} \Omega^{-1}\left(\Omega(K) + \displaystyle\int_a^t v(s)\,ds\right) & \text{if } \Omega(K) > -\infty \\ 0 & \text{if } \Omega(K) = -\infty \end{cases} \qquad \text{whenever } t \in [a, c)$$

where, with some fixed positive u_0,

$$\Omega(u) = \int_{u_0}^{u} \frac{1}{\omega(t)}\, dt, \quad u \geq 0,$$

$$c = \begin{cases} \min\left\{ b, \sup\left\{ t \geq a \,\middle|\, \Omega(K) + \int_a^t v(s)\, ds < \lim_{u\to\infty} \Omega(u) \right\}\right\} & \\ & \text{if } \Omega(K) > -\infty \\ b & \text{if } \Omega(K) = -\infty \end{cases}$$

and Ω^{-1} stands for the inverse function of Ω. Note that

$$\Omega(K) = -\infty \quad \text{if and only if} \quad K = 0 \text{ and } \int_{u_0}^{0} \frac{1}{\omega(t)}\, dt = -\infty.$$

(The result in the degenerate case follows from the inequality in the nondegenerate case simply by choosing $K = k^{-1}$, $k = 1, 2, \dots$ and letting $k \to \infty$. Bihari did not specify the domains of his functions.) Neither c (the case $c = b = \infty$ is not excluded) nor the right–hand side of inequality (24) depends on the particular choice of u_0. If $\omega(u) = u$ for each $u \geq 0$, then (24) simplifies to

$$u(t) \leq K \exp\left(\int_a^t v(s)\, ds \right) \qquad \text{whenever} \quad t \in [a, b),$$

i.e. to Bellman's version of the classical Gronwall lemma. Bihari's inequality (24) has direct implications on questions of uniqueness and continuous dependence. The relationship between (24) and the Alexeev–Gröbner nonlinear variation-of-constants formula is more or less the same as the relationship between the classical Gronwall lemma and the standard, linear variation-of-constants formula. Bihari {4} himself discusses the uniqueness criteria of Osgood, Perron, and Nagumo as well as the nonuniqueness criterion of Tamarkine in the light of his inequality and presents an application to continuous dependence on initial conditions. In an accompanying paper {5}, he applies inequality (24) to problems of stability and boundedness. Inequality (24) has been generalized in various directions, by a great number of authors.

In the sixties the interest of Bihari was focused on establishing a Sturm–Liouville theory for certain types of second-order nonlinear ordinary differential equations he called half-linear. The one-dimensional p-Laplacian

$$\big(q(x)\Phi(y')\big)' + r(x)\Phi(y) = 0$$

(where $q, r > 0$ and $\Phi(s) = |s|^{p-2}s$ ($s \in \mathbb{R}$) with some $p \in (1, \infty)$) provides an example. (We have to admit Bihari's terminology was not always consistent. One can venture to state the more an equation is subject to Sturm–Liouville theory the more this equation is half-linear.)

The contributions of Makai. Three years after Bihari's inequality, a paper by Endre Makai {47} attracted international interest, too. He proved that the principal eigenvalue $\lambda_1(D)$ in Reyleigh's conjecture and the torsional rigidity $P(D)$ in Saint-Vernant's conjecture (we discussed in connection with the work of Pólya and Szegő on isoperimetric inequalities) satisfy

$$\lambda_1(D)\frac{\text{area}^2(D)}{\text{length}^2(\partial D)} \le 3 \quad \text{and} \quad P(D)\frac{\text{area}^3(D)}{\text{length}^2(\partial D)} \le 1, \tag{25}$$

respectively. An important ingredient of Makai's proof is the observation that, with $D(\varepsilon)$ denoting the Euclidean ε-neighborhood of D in $\mathbb{R}^2$, $\text{length}(\partial D(\varepsilon))$ is an increasing function in ε. Makai proved this observation in a generality which suited his purposes — the difficulty is of course related to the existence of the length (the exceptional ε-set in $\mathbb{R}^+$ where $\text{length}(\partial D(\varepsilon))$ does not make sense is countable) — nevertheless, in the last version of his paper finally published he refers to the more general geometric inequalities of Béla Szőkefalvi-Nagy {50} obtained in the meantime. The 'method of interior parallels' of Makai and Szőkefalvi-Nagy helped Pólya {58} to find the sharp upper bounds $\pi^2/4$ and $3/4$ in (25) later — the equalities are approached as D approaches an infinite strip.

A further interesting result of Makai {49} concerns the eigenfunctions of the Laplacian for the Dirichlet and the Neuman problem on the m-dimensional symplex

$$S_m = \{(x_1, x_2, \ldots, x_m) \mid 0 \le x_1 \le x_2 \le \cdots \le x_m \le \pi\}.$$

The eigenfunctions are

$$\text{determinant}\,[\sin n_i x_j]_{i,j=1}^m \quad \text{whenever} \quad 0 < n_1 < \cdots < n_m, \ \text{integers}$$

and

$$\text{permanent}\,[\cos n_i x_j]_{i,j=1}^m \quad \text{whenever} \quad 0 \le n_1 \le \cdots \le n_m, \ \text{integers}$$

with eigenvalues $\sum n_i^2$, respectively. Related results for the isosceles rectangular triangle S_2 as well as for the equilateral triangle were obtained by Makai {48} a couple of years earlier.

The papers of Rényi and Barna on interval maps. In a 1957 paper {65}, Alfréd Rényi elaborated a method for proving that certain interval maps admit absolutely continuous ergodic measures. His examples include what is called today Rényi transformation

$$R_\beta : [0,1] \to [0,1], \quad x \to \beta x \pmod 1$$

where $\beta > 1$ is a real parameter and the absolutely continuous ergodic measure ν_β is equivalent to the standard Lebesgue measure λ on $[0,1]$. (Note that an absolutely continuous ergodic measure is necessarily unique.) Rényi's interest comes from number theory: If $\beta = 10$, then $\nu_\beta = \lambda$ and his result simplifies to Borel's Normal Number Theorem stating that, for almost every $x \in [0,1]$, the frequency of any digit in the decimal expansion of x is $1/10$. He reproves the corresponding result for continued fraction expansions and has also a similar application to an 1832 algorithm of Farkas Bolyai.

A further class of transformations with absolutely continuous ergodic measures Rényi investigates are mappings of the form

$$S : [0,1] \to [0,1], \quad x \to \tau(x) \pmod 1$$

where $\tau : [0,1] \to \mathbb{R}^+$ is a C^1 function with $\tau(0) = 0$, $\tau(1) \in \{2, 3, \dots\}$, and satisfying the expanding condition $\tau'(x) > 1$ for each $x \in [0,1]$ as well as a technical condition (C). While keeping condition (C), Rényi points out that the remaining set of assumptions can be replaced by three alternative sets of conditions under which the existence of an absolutely continuous ergodic measure can be established. Condition (C) itself is a so-called distortion inequality, a uniform bound for the build-up of nonlinearities under the iterates of S. Though condition (C) involves an infinite number of iterates of S, it can be checked in a number of various circumstances. As it was observed by Adler in the afterword to a posthumous paper by R. Bowen (*Comm. Math. Phys.*, **69** (1979), 1–17), condition (C) is automatically satisfied if $\tau : [0,1] \to \mathbb{R}^+$ is a C^2 function with $\tau(0) = 0$, $\tau(1) \in \{2, 3, \dots\}$, and the expanding condition $\tau'(x) > 1$ for each $x \in [0,1]$. Condition (C) and other distortion inequalities have remained extremely useful in the later development of the subject. The number of contributors in the sixties and the seventies became so large that, following Adler, the collection of Rényi-type results on the existence of absolutely continuous ergodic measures for general Markov maps of the interval is termed usually as The Folklore Theorem.

The theory of invariant measures for interval maps began with the 1947 result of Stanislaw Ulam & János Neumann {81} who pointed out that

$d\lambda/(\pi(x(1-x))^{1/2})$ defines an absolutely continuous ergodic measure for the logistic map $[0,1] \to [0,1]$, $x \to 4x(1-x)$.

The most interesting Hungarian contribution to the early theory of interval maps is the work of Béla Barna on divergence properties of Newton's method when applied for approximating real roots of real polynomials. His results remained unnoticed for about two decades. In an 1985 survey paper, however, S. Smale (*Bull. Amer. Math. Soc.*, **13** (1985), 87–121) mentions his name, together with those of Fatou and Julia, as one of the pioneers of the iteration theory of rational functions.

The work of Barna originates in two questions of Rényi posed at the end of his 1950 half-scientific, half-educational paper {64} on Newton's method. Rényi's interest is mainly qualitative. He describes in detail the results of Cauchy and Fourier on convergence criteria but does not mention that the order of convergence is, in fact, quadratic. He turns his attention to "bad initial points" instead and gives a sufficient condition for a particularly strong form of divergence.

Let $f : \mathbb{R} \to \mathbb{R}$ be a C^1 function. For $x \in \{y \in \mathbb{R} \mid f'(y) \neq 0\}$, set $N_f(x) = x - f(x)/f'(x)$. A point $x_0 \in \mathbb{R}$ is *convergent* if the infinite orbit sequence $x_0, x_1 = N_f(x_0), x_2 = N_f(x_1), \dots$ is (defined and) convergent (and then, necessarily, $\lim_{n\to\infty} x_n$ is a zero of f). Otherwise x_0 is *divergent*. For an arbitrary C^2 function with the properties that f'' is strictly increasing and f has exactly three simple roots say A_1, A_2, A_3, Rényi {64} proves that the set of divergent points is countable, there exists a unique period–two orbit $x_0^*, x_1^*, x_0^*, \dots$ and, last but not least, for $i = 1, 2, 3$, any neighborhood of x_0^* contains a point whose orbit converges to A_i, a strikingly sensible dependence on initial values near x_0^*. Rényi asks 1.) if for real polynomials without complex roots the set of divergent points is always countable and 2.) if there is a real polynomial with the properties that not all roots are complex and the set of divergent points contains an interval.

Answering the first question of Rényi in the negative, Barna {1} shows that, given a fourth–degree real polynomial with four simple real roots, the set of divergent points is a compact set of the form $C \cup F$ where C is a Cantor set, $C \cap F = \emptyset$, and

$$F = \{x \in \mathbb{R} \mid \text{the iteration } x_0, x_1, \dots \text{ breaks up in a finite number of steps}\}$$

is a countable set of isolated points. Moreover, the set

$$S = \{x \in C \mid \text{ the infinite orbit } x_0, x_1, \dots \text{ is eventually periodic}\}$$

is also countable and, for each $k \geq 2$, contains periodic orbits with minimal period k. The number of such periodic orbits is $2k^{-1}(2^{k-1} - 1)$ if $k \neq 2$ is a prime number and 3 if $k = 2$. (If k is not a prime number, Barna asserts that the number of periodic orbits with minimal period k can be computed via a complicated recursion but gives no details at all.) Given $x_0 \in C \setminus S$ arbitrarily, Barna — in today's terminology — shows that the omega–limit set $\omega(x_0) = \cap_{k=0}^{\infty} \mathrm{cl}\left(\{x_k, x_{k+1}, \dots\}\right)$ is not finite.

The consecutive four papers of Barna {2a}, {2b}, {2c}, {2d} are devoted to real polynomials of degree m, $m \geq 4$. He proves that all the $m = 4$ cardinality and topological results on C, F, S, and the structure of the set of divergent points remain valid under the condition that the roots of the polynomial are real and simple say $A_1, A_2, \dots, A_m$. In addition, he proves that, given $x_0 \in C$ and $i \in \{1, 2, \dots, m\}$ arbitrarily, any neighborhood of x_0 in $\mathbb{R}$ contains a point whose orbit converges to A_i. He provides two different proofs to this latter result. The first one {2a} is based on the general, complex–variable theory of Fatou and Julia on iterating rational functions whereas the second one {2c}, like the whole approach of Barna, is completely elementary. No $m \geq 5$ version of the "$2k^{-1}(2^{k-1} - 1)$ if $k \neq 2$" combinatorial result is given. In the last paper of the series {2d}, Barna proves that his Cantor set C is a Lebesgue null set.

The answer to Rényi's second question is, in contrast to the conjecture in {64}, affirmative. Barna's example in {2b} is $f(x) = 11x^6 - 34x^4 + 39x^2$ for which $N_f(1) = -1$, $N_f(-1) = 1$ and $N_f'(1) = N_f'(-1) = 0$. Thus $1, -1, 1, \dots$ is an asymptotically stable period–two orbit of N_f and sufficiently small intervals about $x_0 = 1$ consist entirely of divergent points (attracted by the period–two orbit $1, -1, 1, \dots$).

A further early contribution to the modern theory of dynamical systems is due to György Szekeres, a childhood friend of Pál Erdős. He presents a detailed study of the one-dimensional conjugacy equation $\mathcal{H}\big(f(x)\big) = \mu\mathcal{H}(x)$, $\mu \neq 0, 1$ in 1958 {77}. Here the real function f is strictly increasing, defined on some finite or infinite interval $[0, c) \subset \mathbb{R}^n$, and satisfies $f(0) = 0$, $f(x) < x$ for $x \neq x_0 = 0$. He looks for strictly monotone solutions $\mathcal{H}$ representable as limits of iterations like $\mathcal{H}(x) = \eta \lim_{n\to\infty} \mu^{-n} f^n(x)$, $x \in [0, c)$ in the regular case $f'(0) = \mu \in (0, 1)$, ($f \in C^r$, $r \geq 1$; $\eta \neq 0$ is a real parameter) on some interval $[0, b)$ or $(0, b)$. In the singular cases $\mu \neq f'(0) = 0$ or $\mu \neq f'(0) = 1$, the existence of such solutions is pointed out under certain asymptotic conditions on f at the fixed point $x_0 = 0$. Similar results are proved for Abel's functional equation $\alpha\big(f(x)\big) = \alpha(x) + c$ ($c \neq 0$) as well as for the embeddability of $f = \Phi(1, \cdot)$ in a local iteration group (i.e. in a continuous-time local dynamical system satisfying) $\Phi(t+\tau,\cdot) = \Phi\big(t,\Phi(\tau,\cdot)\big)$, $t,\tau \in \mathbb{R}$. Szekeres' results generalize, complete, and unify those of Koenigs and Schröder who worked with f analytic, and can be considered as 'variations and fugue' on the 1959/60 Grobman–Hartman Lemma in one dimension.

Further workers, further works. Finally, we mention the names of several mathematicians who started their careers in the late fourties and whose research topics were, to a lesser or greater extent, connected to the theory of differential equations. These are György Alexits (actually, he belonged to an earlier generation but his scientific activities could not freely develop in the pre-war period — he worked mainly in approximation theory but his 1924 PhD Thesis was devoted to the Laplace equation), István Fenyő (his main area was, as it is demonstrated by the title of his major work with H. W. Stolle {24}, the theory and praxis of linear integral equations), Géza Freud (he is well known as an expert on orthogonal polynomials — but published several papers on partial differential equations in the early years of his career), Miklós Mikolás (his fractional calculus papers contain several applications to ordinary differential equations with fractional derivatives), and György Targonski (who combined the theory of iterations with those of functional equations). As for representatives of the ten years younger generation, the names of Tamás Fényes and of his blind friend, Pál Kosik, are mentioned (a great part of their joint papers is devoted to the Mikusinski operational calculus).

Epilogue. And this ends our report on the history of differential equations: Hungary, the extended first half of the 20th century, a terrific place and time.

Acknowledgement. The authors received help from a great number of their colleagues during the preparation of this paper. We are greatly indebted to Miklós Farkas, László Hatvani, Tibor Krisztin, and Antal Varga (who have seen an almost final version of the manuscript) for their suggestions and remarks. Special thanks to Róbert Kersner and Szilárd Révész who helped us explaining several aspects of the oeuvres of M. Riesz and P. Lax. We obtained valuable hints from Aurél Galántai, József Kolumbán, Endre Makai Jr. as well as from János Tóth — the latter (informed by József Salánki) called our attention to Problem No. 54 in the 1952 issue of Matematikai Lapok. Sincere thanks to Mrs. Éva Németh who helped a lot by patiently translating some parts of the manuscript to English — the strictly–mathematical sections were written directly in English — as well as to the librarians in the major mathematical libraries in Budapest and Szeged. Last but not least, we would like to thank Professor János Horváth for his advices, constant care and encouragement.

Addendum. *Árpád Elbert,* a devoted researcher in the fields of half-linear and delay equations as well as in the theory of special functions, died during the preparation of this paper. He intended this report to be a tribute to the memory of his masters, *Imre Bihari* and *Kató Rényi.*

References

The *number* in parentheses at the end of an entry is the number assigned to the work in the respective 'Collected Papers', 'Oeuvres Complètes', 'Gesammelte Arbeiten' etc.

[107] Lax, Péter, *Hyperbolic Systems of Conservation Laws and the Mathematical Theory of Shock Waves,* CBMS-NSF Regional Conference Series in Applied Mathematics, No. 11, Society for Industrial and Applied Mathematics (Philadelphia, PA, 1973), Second printing, 1984.

[129] Pólya, György – Szegő, Gábor, *Aufgaben und Lehrsätze aus der Analysis,* Grundlehren der Mathematischen Wissenschaften in Einzeldarstellungen, Band 19–20, Springer-Verlag (Berlin, 1925), Dover (New York, 1945), 2nd edition, Springer, 1954, 3rd edition, Springer 1964, 4th edition, Heidelberger Taschenbücher, Band 73–74, Springer, 1970. English translation: *Problems and Theorems in Analysis,* Die Grundlehren der Mathematischen Wissenschaften, Band 193, 216, Springer (Berlin–New York–Heidelberg, 1972, 1976).

[130] Pólya, György – Szegő, Gábor, *Isoperimetric Inequalities in Mathematical Physics,* Annals of Mathematics Studies, No. 27, Princeton University Press (Princeton, NJ, 1951).

[142] Radó, Tibor, *On the Problem of Plateau,* Ergebnisse der Mathematik und ihrer Grenzgebiete, Band 2, No. 2, Springer-Verlag (Berlin, 1933).

[143] Radó, Tibor, *Subharmonic Functions,* Ergebnisse der Mathematik und ihrer Grenzgebiete. Band 5, No. 1, Springer-Verlag (Berlin, 1937)/Chelsea (New York, 1949).

[195] Webster, A. G. – Szegő, Gábor, *Partielle Differentialgleichungen der mathematischen Physik,* B. G. Teubner (Leipzig, 1930).

{1} B. Barna, Über das Newtonsche Verfahren zur Annäherung von Wurzeln algebraischer Gleichungen, *Publ. Math. (Debrecen),* **2** (1951), 50–63.

{2} B. Barna, Über die Divergenzpunkte des Newtonschen Verfahrens zur Bestimmung von Wurzeln algebraischer Gleichungen, I.–IV, *Publ. Math. Debrecen,* **3** (1953), 109–118; **4** (1956), 384–397; **8** (1961), 193–207; **14** (1967), 91–97.

{3} E. Beke, Die Irreduzibilität der homogenen linearen Differentialgleichungen, *Math. Ann.,* **45** (1894), 278–294.

{4} I. Bihari, A generalization of a lemma of Bellman and its application to uniqueness problems of differential equations, *Acta Math. Hungar.,* **7** (1956), 71–94.

{5} I. Bihari, Researches of the boundedness and stability of the solutions on non–linear differential equations, *Acta Math. Hungar.,* **8** (1957), 261–278.

{6} S. Bochner and J. v. Neumann, On compact solutions of operational–differential equations, *Ann. Math.,* **36** (1935), 255–291. *(54–IV.1)*

{7} G. W. Brown and J. v. Neumann, Solutions of games by differential equations, In: *Contributions to the Theory of Games* (Annals of Math. Studies 24, Princeton University Press), pp. 73–79. *(121–VI.14)*

{8} J. G. Charney, R. Fjortoft and J.v. Neumann, Numerical integration of barotropic vorticity equation, *Tellus,* **2** (1950), 237–254. *(123–VI.29)*

{9} E. Egerváry, On a class of integral equations, *Math. Phys. Lapok,* **23** (1913), 301–355. (in Hungarian)

{10} E. Egerváry, Begründung und Darstellung einer allgemeinen Theorie der Hängebrücken mit Hilfe der Matrizenrechnung, *Abhandlungen der Internationalen Vereinigung für Brückenbau und Hochbau,* **16** (1956), 149–184.

{11} E. Egerváry and P. Turán, On a certain point of the kinetic theory of gases, *Studia Math.,* **12** (1951), 170–180. *(53)*

{12} E. Egerváry and P. Turán, On some problems in the kinetic theory of gases, *MTA Mat Fiz. Oszt. Közl.,* **1** (1951), 303–314. (in Hungarian) *(56; English translation)*

{13} L. Fejér, Sur le fonctions bornées et intégrables, *C.R. Acad. Sci. Paris,* **131** (1900), 984–987. *(2)*

{14} L. Fejér, Zur Theorie des Poissonsches Integrals, *Mat. Term. Ért.,* **19** (1901), 322–325. (in Hungarian) *(3; German translation)*

{15} L. Fejér, Über zwei Randwertaufgaben, *Math. u. Naturwissenschaftliche Berichte aus Ungarn,* **19** (1903), 329–331. *(4)*

{16} L. Fejér, Untersuchungen über Fouriersche Reihen, *Math. Ann.,* **58** (1904), 51–69. *(9)*

{17} L. Fejér, Das Ostwaldsche Prinzip in der Mechanik, *Math. Ann.,* **59** (1906), 422–436. *(11)*

{18} L. Fejér, Tömegpont egyensúlya ellenálló közegben, *Mat. Term. Ért.,* **24** (1906), 109–116. (in Hungarian) *(13)*

{19} L. Fejér, Untersuchungen über Stabilität und Labilität in der Mechanik der Massenpunktsysteme, *Mat. és Fiz. Lapok,* **15** (1906), 152–172. (in Hungarian) *(14; German translation)*

{20} L. Fejér, Über Stabilität und Labilität eines materiellen Punktes im widerstrebenden Mittel, *J. reine angew. Math.*, **131** (1906), 216–223. *(16; German version of {18})*

{21} L. Fejér, Sur le calcul des limites, *C.R. Acad. Sci. Paris,* **143** (1906), 957–959. *(18)*

{22} L. Fejér, Über die Eindeutigkeit der Lösung der linearen partiellen Differentialgleichungen zweiter Ordnung, *Math. Zeitschr.*, **1** (1918), 70–79. *(57)*

{23} M. Fekete, Über die Verteilung der Wurzeln bei gewissen algebraischen Gleichungen mit ganzzahligen Koeffizienten, *Math. Zeitschr.*, **17** (1923), 228–249.

{24} M. Fenyő and H. W. Stolle, *Theorie und Praxis der Linearen Integralgleichungen I–IV,* VEB Deutscher Verlag der Wissenschaften, Berlin, 1982–84.

{25} Z. Geőcze, Quadrature des surfaces courbes, *Math. naturwiss. Ber. Ungarn,* **26** (1910), 1–88.

{26} E. Gergely, Über die Variation von Doppelintegralen mit einer variierender Begrenzung, *Acta Sci. Math. (Szeged),* **2** (1924–26), 139–146.

{27} H. H. Goldstine and J.v. Neumann, Numerical inverting of matrices of high order, *Amer. Math. Soc. Bull.*, **53** (1947), 1021–1099.

{28} H. H. Goldstine and J.v. Neumann, Numerical inverting of matrices of high order II., *Amer. Math. Soc. Proc.*, **2** (1951), 188–202.

{29} E. Grynaeus, Sur les systèmes de Pfaff, *Bull. Soc. Math. France,* **56** (1927), 74–97.

{30} A. Haar, Die Randwertaufgabe der Differentialgleichung $\Delta\Delta U = 0$, *Göttinger Nachr.*, **1907**, 280–287. *(1)*

{31} A. Haar, On the variation of double integrals, *Mat. Term. Ért.*, **35** (1917), 1–19. (in Hungarian) *(11)*

{32} A. Haar, Über die Variation der Doppelintegrale, *J. reine angew. Math.*, **149** (1919), 1–18. *(15; German version of {31})*

{33} A. Haar, Über eine Verallgemeinerung des Du Bois–Reymond'schen Lemmas, *Acta Sci. Math. (Szeged),* **1** (1923), 167–179. *(16)*

{34} A. Haar, Über das Plateausche problem, *Math. Ann.*, **97** (1926), 124–158. *(20)*

{35} A. Haar, Über reguläre Variationsprobleme, *Acta Sci. Math. (Szeged)* **3** (1927), 224–234. *(21)*

{36} A. Haar, Sur l'unicité des solutions des équations aux derivées partielles, *C.R. Acad. Sci. Paris,* **187** (1928), 23–25. *(22)*

{37} A. Haar, Zur Charakteristikentheorie, *Acta Sci. Math. (Szeged),* **4** (1928), 103–114. *(23)*

{38} A. Haar, Über adjungierte Variationsprobleme und adjungierte Extremalflächen, *Math. Ann.*, **100** (1928), 481–502. *(24)*

{39} A. Haar, Über Eindeutigkeit und Analyzität der Lösungen partiellen Differentialgleichungen, *Atti del Congresso Internationale dei Matematici, Bologna 3–10 Settembre 1928,* **3** (1928), 5–10. *(26)*

{40} A. Haar, Zur Variationsrechnung. Drei Vorträge gehalten am Mathematischen Seminar der Hamburgischen Universität (23–25 Juli 1929), *Abhandlungen aus dem Mathematischen Seminar der Hamburgischen Universität,* **8** (1930), 1–27. *(27)*

{41} A. Haar and T.v. Kármán, Zur Theorie der Spannungszustände in plastischen und sandartigen Medien, *Göttinger Nachr.*, **1909**, 204–218. *(3)*

{42} Ch. Jordan, *Calculus of Finite Differences,* Eggenberger, Budapest, 1939.

{43} D. König and A. Szücs, Mouvement d'un point abandonné à l'intérieur d'un cube, *Palermo Rend.,* **36** (1913), 79–90.

{44} J. Kürschák, Zur Theorie der Monge–Ampèreschen Differentialgleichungen, *Math. Ann.,* **61** (1905), 109–116.

{45} P. D. Lax, Weak solutions of nonlinear hyperbolic equations and their numerical computation, *Commun. Pure Appl. Math.,* **7** (1954), 159–193.

{46} P. D. Lax, Hyperbolic systems of conservation laws II, *Commun. Pure Appl. Math.,* **10** (1957), 537–566.

{47} E. Makai, Bounds for the principal frequency of a membrane and the torsional rigidity of a beam, *Acta Sci. Math. (Szeged),* **20** (1959), 33–35.

{48} E. Makai, Complete orthogonal systems of eigenfunctions of three triangular membranes, *Studia Sci. Math. Hungar.,* **5** (1970), 51–62.

{49} E. Makai, Complete systems of eigenfunctions of the wave equation in some special cases, *Studia Sci. Math. Hungar.,* **11** (1976), 139–144.

{50} B. Sz.-Nagy, Über Parallelmengen nichtkonvexer ebener Bereiche, *Acta Sci. Math. (Szeged),* **20** (1959), 36–47.

{51} J.v. Neumann, Über einen Hilfsatz der Variationsrechnung, *Hamburger Abh.,* **8** (1930), 28–31. *(31–II.4)*

{52} J.v. Neumann, *Invariant Measures,* AMS, Providence, R.I., 1999.

{53} J.v. Neumann and R. D. Richtmyer, A method for the numerical calculation of hydrodynamic schocks, *J. Appl. Phys.,* **21** (1950), 232–237. *(120–VI.27)*

{54} J.v. Neumann and R. D. Richtmyer, On the numerical solution of partial differential equations of parabolic type, In: *John von Neumann Collected Works,* Vol. V. (ed. A. H. Taub), Pergamon Press, Oxford, 1963; pp. 652–663. *(111–V.18)*

{55} G. Pólya, Qualitatives über Wärmeausgleich, *Z. angew. Math.,* **13** (1933), 125–128. *(135)*

{56} G. Pólya, Torsional rigidity, principal frequency, electrostatic capacity and symmetrization, *Quarterly of Applied Math.,* **6** (1948), 267–277. *(177)*

{57} G. Pólya, Sur une interprétation de la méthode des différences finies qui peut fournir des bornes supérieures ou inférieures, *C.R. Acad. Sci. Paris,* **235** (1952), 995–997. *(195)*

{58} G. Pólya, Two more inequalities between physical and geometrical quantities, *J. Indian Math. Soc.,* **24** (1960), 413–419. *(219)*

{59} G. Pólya and M. Schiffer, Sur la représentation conforme de l'extérieur d'une courbe fermée convexe, *C.R. Acad. Sci. Paris,* **248** (1959), 2837–2839. *(211)*

{60} G. Pólya and G. Szegő, Sur quelques propriétés qualitatives de la propagation de la chaleur, *C.R. Acad. Sci. Paris,* **192** (1931), 340–341. *(124)*

{61} G. Pólya and G. Szegő, Über den transfiniten Durchmesser (Kapazitätskonstante) von ebenen und räumlichen Punktmengen, *J. reine angew. Math.,* **165** (1931), 4–49. *(125)*

{62} T. Radó, Über den analytischen Charakter der Minimalflächen, *Math. Zeitschr.,* **24** (1925), 321–327.

{63} T. Radó, Geometrische Betrachtungen über zweidimensionale reguläre Variationsprobleme, *Acta Sci. Math.,* **2** (1926), 228–253.

{64} A. Rényi, On Newton's method of approximation, *Mat. Łapok,* **1** (1950), 278–293. (in Hungarian) *(33; English translation)*

{65} A. Rényi, Representations for real numbers and their ergodic properties, *Acta Math. Hungar.,* **8** (1957), 477–493. *(139)*

{66} M. Réthy, Das Ostwaldsche Prinzip von Energieumsatz in der Mechanik, *Math. Ann.,* **59** (1904), 554–571.

{67} F. Riesz, Sur le valeurs moyennes du module des functions harmoniques et des fonctions analytiques, *Acta Sci. Math. (Szeged),* **1** (1922–23), 27–32. *(45)*

{68} F. Riesz, Über subharmonische Funktionen und ihre Rolle in der Funktionentheorie und in der Potentialtheorie, *Acta Sci. Math. (Szeged),* **2** (1924–26), 87–100. *(49)*

{69} F. Riesz, Sur les functions subharmoniques et leur rapport à la théorie du potentiel I., *Acta Math.,* **48** (1926), 329–343. *(54)*

{70} F. Riesz, Sur les functions subharmoniques et leur rapport à la théorie du potentiel II., *Acta Math.,* **54** (1930), 321–360. *(66)*

{71} M. Riesz, L'intégrale de Riemann–Liouville et le problème de Cauchy, *Acta Math.,* **81** (1949), 1–223.

{72} F. Riesz and T. Radó, Über die erste Randwertaufgabe für $\Delta u = 0$, *Math. Zeitschr.,* **22** (1925), 41–44. *(52)*

{73} L. Schlesinger, *Handbuch der Theorie der linearen Differentialgleichungen I–II,* Teubner, Leipzig, 1895–1898.

{74} L. Schlesinger, *Bericht über die Entwickelung der Theorie der linearen Differentialgleichungen seit 1865,* Teubner, Leipzig, 1909.

{75} L. Schlesinger, Neue Grundlagen für einen Infinitesimalkalkül der Matrizen, *Math. Zeitschr.,* **33** (1931), 33–61.

{76} A. Sólyi, Über das Haarsche Lemma in der Variationsrechnung und seine Anwendungen, *Acta Sci. Math. (Szeged),* **11** (1946), 1–16.

{77} G. Szekeres, Regular iteration of real and complex functions, *Acta Math.,* **100** (1958), 203–258.

{78} G. Szegő, Über einige Extremalaufgaben der Potentialtheorie, *Math. Zeitschr.,* **31** (1930), 583–593.

{79} G. Szegő, Über einige neue Extremaleigenschaften der Kugel, *Math. Zeitschr.,* **33** (1931), 419–425.

{80} A. Szücs, Sur la variation des intégrales triples et le théorème de Stokes, *Acta Sci. Math. (Szeged),* **3** (1927), 81–95.

{81} S. M. Ulam and J. von Neumann, On combination of stochastic and deterministic processes (preliminary report), *Bull. Amer. Math. Soc.,* **53** (1947), 1120.

{82} A. Wintner, Zur Lösung von Differentialsystemen mit unendlich vielen Veränderlichen, *Math. Ann.,* **98** (1928), 273–280.

Barnabás M. Garay
Budapest University of Technology and Economics
Department of Differential Equations
Budapest H–1111
Egry J. u. 1.
Hungary

garay@math.bme.hu

BOLYAI SOCIETY
MATHEMATICAL STUDIES, 14

A Panorama of Hungarian Mathematics in the Twentieth Century, pp. 295–371.

Holomorphic Functions

JÁNOS HORVÁTH

1. Prologue

It is quite well known that, as Jean-Pierre Kahane tells us elsewhere in this volume, Hungarian mathematics started the twentieth century with a bang when in October 1900 a twenty year old student, who had just returned from Berlin after a year there, proved that the Fourier series of a continuous function is uniformly Cesàro-summable to the function. It is, however, less well known that Lipót Fejér has already previously published an article containing some simple theorems concerning power series (Mat. Fiz. Lapok **9** (1900), 405–410; [40], No. 1, p. 29). A typical result is the following: if g is a positive integer and f an entire function of genus $\leq g-1$, then the radius of convergence of $\sum_{n=0}^{\infty} c_n f(n) x^{n^g}$ is not smaller than the the radius of convergence of $\sum_{n=0}^{\infty} c_n x^{n^g}$. An application is the fact that a power series and the series obtained from it by termwise differentiation have the same radius of convergence.

2. The Jensen formula

Fejér got to know Constantin Carathéodory during the academic year he spent in Berlin. At that time Carathéodory was an engineer for the Suez Canal Company. His family, as many Greeks, was in the diplomatic service of the Ottoman Empire: his father was ambassador in Brussels, his uncle, who was also his father-in-law, ambassador in Berlin. Carathéodory was always interested in mathematics, so while visiting his uncle he went to see what is going on at the seminar of Hermann Amandus Schwarz. According to the legend, he found a man seven years his junior who presented a proof of the characterization of the triangle with shortest perimeter inscribed into a triangle ([40], Appendix III, vol. II, p. 847; [140]). Carathéodory was so impressed that then and there he decided to abandon his career as an engineer and to become a mathematician.

The Comptes Rendus note of Carathéodory and Fejér from 1907 (**145**, 163–165; [40], No. 19) is based on the geometric theorem according to which if P is a point inside a circle with center O and radius R, and Q is its polar, i.e. OPQ are collinear and $OP \cdot OQ = R^2$, then the ratio $PM : QM$ of the distances from P and from Q to a point M on the circumference equals the constant OP/R.

Consider now a circle in the complex plane $\mathbb{C}$ with radius R and center at the origin. If $a \in \mathbb{C}$ is such that $|a| < 1$, then its polar with respect to the circle is

$$\hat{a} = \frac{R^2}{\bar{a}} = \frac{R^2 a}{|a|^2}$$

and so

$$\left| \frac{z-a}{z-\hat{a}} \right| = \frac{|a|}{R}$$

for all $z \in \mathbb{C}$ with $|z| = R$. Carathéodory and Fejér consider all functions f holomorphic in $\{z \in \mathbb{C} : |z| < R\}$ with $f(0) = A$, which vanish at points $a_1, \dots, a_n$ inside the circle, and ask for which such function is $M = \max_{|z|<1} |f(z)|$ the smallest. They divide f by the product

$$Q(z) = \frac{z-a_1}{z-\hat{a}_1} \cdot \frac{z-a_2}{z-\hat{a}_2} \cdots \frac{z-a_n}{z-\hat{a}_n}$$

and obtain from the fact that $|Q(z)|$ is constant on $|z| = R$ that

$$M \geq \frac{|A| R^n}{|a_1|\,|a_2| \dots |a_n|},$$

and that the only function for which M attains this lower bound is

$$\frac{AR^{2n}}{|a_1^2|\,|a_2^2| \dots |a_n^2|} Q(z).$$

If we replace $\hat{a}_i$by $R^2/\bar{a}_i$ in $Q(z)$ and multiply it by $R^n / \prod_{j=1}^n |a_j|$, then we obtain the rational function

$$R^n \frac{z-a_1}{R^2-\bar{a}_1 z} \cdot \frac{z-a_2}{R^2-\bar{a}_2 z} \cdots \frac{z-a_n}{R^2-\bar{a}_n z}$$

which has zeros at $a_1, a_2, \ldots, a_n$ and absolute value 1 on $|z| = R$. It has entered complex analysis under the name of "Blaschke product" after Wilhelm Blaschke who considered it, however, only in 1915.

3. Polynomials

Edmund Landau considered in 1906 the trinomial equation

$$a_0 + a_1 z + a_n z^n = 0$$

and proved that it has a root in the disk

$$|z| \leq 2 \left| \frac{a_0}{a_1} \right| .$$

He also considered the quadrinomial equation

$$a_0 + a_1 z + a_m z^m + a_n z^n = 0$$

and proved that it has a root in the disk

$$|z| \leq 8 \left| \frac{a_0}{a_1} \right| . \tag{1}$$

The remarkable fact about these two estimates is that the bounds depend only on a_0, a_1 and not on m, n, a_m, a_n. Landau asked the question whether a similar result holds for a general polynomial equation of the form

$$a_0 + a_1 z + a_2 z^{n_2} + a_3 z^{n_3} + \cdots + a_k z^{n_k} = 0, \tag{2}$$

where $a_1 \neq 0$, $1 < n_2 < n_3 < \cdots < n_k$. Lipót Fejér gave an answer to this question (Comptes Rendus, Paris **145** (1907), 459–461, Math. Ann., **65** (1908), 413–423, Mat. Fiz. Lapok **17** (1908), 308–324; [40], Nos. 21, 23, 24). He starts out from a then almost forgotten theorem of C. F. Gauss (also called the Gauss–Lucas theorem) according to which if $f(z)$ is a polynomial, then each root of $f'(z) = 0$ is inside the convex hull of the roots of $f(z) = 0$ unless it coincides with a root of $f(z) = 0$. In the third article quoted he gives a proof of the theorem based on the fact that if

$$f(z) = (z - z_1)^{\alpha_1} (z - z_2)^{\alpha_2} \ldots (z - z_n)^{\alpha_n},$$

where $z_1, z_2, \ldots, z_n$ are the distinct zeros of $f(z)$, then

$$\frac{f'(z)}{f(z)} = \frac{\alpha_1}{z - z_1} + \frac{\alpha_2}{z - z_2} + \cdots + \frac{\alpha_n}{z - z_n}.$$

This implies that if $f'(\zeta) = 0$, then ζ is the center of gravity of positive masses placed at the points z_j, from where Gauss' theorem follows immediately. Alternatively one can say that a unit mass placed at ζ is in equilibrium under forces of attraction inversely proportional to the distance from z_j to ζ due to masses placed at the points z_j. The theorem of Gauss follows also from this interpretation.

Fejér uses a consequence of Gauss' theorem, namely that the largest (in absolute value, of course) root of $f(z) = 0$ is not smaller than the largest root of $f'(z) = 0$. His main theorem is the following:

The polynomial equation

$$a_0 + a_1 z^{n_1} + a_2 z^{n_2} + \cdots + a_k z^{n_k} = 0, \tag{3}$$

where $a_1 \neq 0$, $1 \leq n_1 < n_2 < \cdots < n_k$, has a root in the disk

$$|z| \leq \left(\frac{n_1 n_2 \ldots n_k}{(n_2 - n_1)(n_3 - n_1) \ldots (n_k - n_1)} \right)^{\frac{1}{n_1}} \left| \frac{a_0}{a_1} \right|^{\frac{1}{n_1}}. \tag{4}$$

So the disk does not depend on the coefficients $a_2, a_3, \ldots a_k$.

A corollary of the theorem is that equation (3) has at least one solution in the disk

$$|z| \leq k \left| \frac{a_0}{a_1} \right|^{\frac{1}{n_1}}. \tag{5}$$

Here the right hand side is independent also of $n_2, n_3, \ldots, n_k$. Another corollary states that (3) has at least one root in

$$|z| \leq k \max \left(\left| \frac{a_0}{a_1} \right|, 1 \right), \tag{6}$$

where now the bound does not even depend on n_1. Fejér points out that it is very easy to obtain an estimate where in (6) the "length" k is replaced by the possibly much larger degree $n = n_k$.

Fejér's results were generalized by Paul Montel and Mieczysław Biernacki to the case when not only the first two but the first p coefficients are fixed, see Pál Turán's note in [40], p. 333 and Chapter IV of Dieudonné's expository account [29] which discusses also the results concerning polynomials of several other Hungarian mathematicians: Elemér Bálint, Jenő Egerváry, Mihály Fekete, István Lipka, Gábor Szegő, Gyula Szőkefalvi Nagy. Thus Fekete improved the estimate (4) to

$$|z| \leq \min_{1\leq r\leq k-1} \left(\frac{n_{r+1}\dots n_k}{(n_{r+1}-n_r)\dots(n_k-n_r)} \right)^{\frac{1}{n_r}} \left| \frac{a_0}{a_1} \right|^{\frac{1}{n_r}}$$

and deduced from it the following result related to the theorem of J. H. Grace: If the coefficients of the polynomial (3) satisfy a linear relation

$$\lambda_0 a_0 + \lambda_1 a_1 + \lambda_2 a_2 + \dots + \lambda_k a_k = 0,$$

where $\lambda_0 \neq 0$, then (3) has a root in the disk

$$|z| \leq 2k \max_{1\leq i\leq k} \left(\left| \frac{\lambda_i}{\lambda_0} \right|, 1 \right).$$

From (5) with $n_1 = 1$ we see that the equation (2) originally considered by Landau has a root in the disk

$$|z| \leq k \left| \frac{a_0}{a_1} \right|, \tag{7}$$

thus in (1) the factor 8 can be replaced by 3. The factor k is sharp as the equation

$$a_0 \left(1 + \frac{a_1 z}{k a_0} \right)^k = a_0 + a_1 z + \dots = 0$$

shows. Nevertheless Fejér proved (Jahresber. Deutsch. Math.-Verein., **24** (1917), 114–128; [40], No. 56) that a root of (2) can always be found in a region whose area is one-fourth of the area of the disk given by (7), namely the disk which has a diameter joining the points 0 and $-k\frac{a_0}{a_1}$ of $\mathbb{C}$.

Fejér obtains analogous results in the more general case of equation (3). Then a root can be found in each of n_1 disks contained in the disk (5). It follows that if $n_1 \geq 2$, then (3) has at least two distinct roots in (5). Using Fejér's method Fekete proved that for the equation

$$a_0 + a_1 x + \dots + a_\nu x^\nu + b_1 x^{n_1} + b_2 x^{n_2} + \dots + b_k x^{n_k} = 0, \tag{8}$$

where $a_\nu \neq 0$, $\nu < n_1 < n_2 < \cdots < n_k$, there exists a bound B, depending only on $a_0, a_1, \ldots, a_\nu$ and k, such that (8) has at least ν roots in the disk $|z| \leq B$.

A result, somewhat similar to Fejér's, where a subset can be excluded from a region in which the roots a priori lie, is due to István Lipka. It is well known that all the roots of

$$f(z) = z^n + a_1 z^{n-1} + a_2 z^{n-2} + \cdots + a_{n-1} z + a_n = 0$$

lie in the disk with center 0 and radius ξ, where ξ is the unique positive root of

$$z^n - |a_1| z^{n-1} - |a_2| z^{n-2} - \cdots - |a_{n-1}| z - |a_n| = 0$$

([129], III. 17). The unique positive root η of

$$z^{n-1} - |a_1| z^{n-2} - |a_2| z^{n-3} - \cdots - |a_{n-1}| = 0$$

satisfies $\eta < \xi$. Lipka proves that if $a_n > 0$, then all the roots of $f(z) = 0$ lie in the cogwheel-shaped region obtained by excluding from the disk $|z| \leq \xi$ the annular sectors described, using the polar representation $z = \rho e^{i\varphi}$, by the inequalities

$$\eta < \rho \leq \xi, \quad \frac{4k-1}{2n} < \varphi < \frac{4k+1}{2n}, \quad k = 0, 1, \ldots, n-1.$$

A significant number of Lipka's publications deal with the rule of signs of Descartes.

The article (Math. Ann., **85** (1922), 41–48; [40], No. 60) of Fejér, written in honor of Hilbert's 60th birthday, uses the observation that one can approach simultaneously all the trees of a forest only if one is outside the convex hull of the forest. Within the convex hull whenever one gets closer to some trees, the distance from others increases.

Let $n \geq 1$ be an integer, denote by $\mathcal{P}_n$ the set of all polynomials of the form

$$g(z) = z^n + c_1 z^{n-1} + \cdots + c_n$$

($c_j \in \mathbb{C}$), and let S be a closed set in $\mathbb{C}$ which has at least n points. Associate with each $g \in \mathcal{P}_n$ a number $A(g) \geq 0$ which is increasing in the following sense: if

$$|g(z)| < |h(z)| \quad \text{for} \quad h(z) \neq 0, \quad z \in S,$$

$$|g(z)| = |h(z)| \quad \text{for}\, h(z) = 0, \quad z \in S,$$

then $A(g) < A(h)$. Fejér's theorem states that if the polynomial $t \in \mathcal{P}_n$ is such that $A(t) \leq A(g)$ for all $g \in \mathcal{P}_n$, then the zeros of t lie in the convex hull of S.

Examples of functionals A are

$$M_\infty(g) = \max_{z \in S} |g(z)| \quad \text{and} \quad \int_S |g(x)|^p \, dz.$$

When S is the interval $[-1, 1]$, then the polynomial minimizing M_∞ is 2^{1-n} times the usual Čebishov polynomial $T_n(x) = \cos(n \arccos x)$, so Fejér's theorem yields the classical result that all the zeros are in $[-1, 1]$.

Proof. Write $t(z) = (z - z_1)(z - z_2) \ldots (z - z_n)$, and assume that z_1 is not in the convex hull of S. There exists $\zeta \in \mathbb{C}$ such that $|s - \zeta| < |s - z_1|$ for every "tree" s in the "forest" S. Set $g(z) = (z - \zeta)(z - z_2) \ldots (z - z_n)$. For every $s \in S$ such that $(s - z_2) \ldots (s - z_n) \neq 0$ (and such exist by hypothesis) one has $|g(s)| < |t(s)|$, hence $A(g) < A(t)$, which contradicts the minimality of t. ■

Fejér's article inspired a large amount of research. When Neumann János (a.k.a. John von Neumann) was still a student in the Lutheran "Gymnasium" of Budapest, his parents engaged Fekete as a tutor. Not that Neumann needed a tutor, he has read on his own Carathéodory's "Reelle Funktionen" at the age of fifteen, but Fekete needed an income. For political reasons he was not only dismissed from his job as a secondary school teacher but his title of "Privatdozent" was taken from him and he was expelled from the Mathematical Society. It is difficult to imagine Fekete, the meekest of men, as a dangerous revolutionary.

Together they wrote Neumann's first paper (Jahresber. Deutsch. Math.-Verein., **31** (1922), 125–128; [118]) which apeared when he was nineteen years old. They prove Gauss' theorem using Fejér's result by considering a polynomial $p(z) = \frac{z^n}{n} + p_1 z^{n-1} + \cdots + p_n$, taking for S the set of its roots, and for the functional A the expression

$$A(g) = \max_{z \in S} \frac{|g(z)|}{|p'(z)|}.$$

They prove that the polynomial $g(z) \in \mathcal{P}_{n-1}$, for which $A(g)$ is the smallest, is $p'(z)$ which has thus its zeros in the convex hull of S.

Another celebrated theorem concerning the location of the zeros of the derivative of a polynomial was stated by J. L. W. V. Jensen ([129], III.35), and was first proved by Gyula Szőkefalvi Nagy (Jahresber. Deutsch. Math.-Verein., **27** (1918), 44–48): Let $f(z)$ be a polynomial with real coefficients so that its zeros are either real or occur in pairs of conjugate complex numbers. Then the zeros of $f'(z)$ are either real or are contained in the union of disks with a diameter that has two conjugate zeros of $f(z)$ as endpoints.

Fekete and von Neumann base a proof of Jensen's theorem on the following analogue of Fejér's principle: Let S be a closed subset of $\mathbb{C}$ which is symmetric to the real axis. If z is a point in $\mathbb{C}$ lying outside the union of the disks with diameters having $s \in S$ and $\bar{s} \in S$ as endpoints, then there exists a point $\zeta \in \mathbb{C}$ such that $|\zeta - s| \cdot |\zeta - \bar{s}| < |z - s| \cdot |z - \bar{s}|$ To prove this, they first consider the case when S consists of just two points $s \neq \bar{s}$. Let z be outside the disk D having the segment $[s, \bar{s}]$ as diameter. If z is real, then the assertion is obvious. If z is complex, then $\bar{z}$ is also outside D, and so are some points ζ of the open segment $]z, \bar{z}[$. The required inequality follows from the area formula $A = \frac{1}{2}ab\sin\gamma$ we learned in trigonometry. The proof is similar when S consists of a finite number of pairs of points, and the general case follows by a continuity argument.

Towards the end of his life Fekete returned to the principles of Fejér and of Fekete–von Neumann alone (Proc. Nat. Acad. Sci. U.S.A., **37** (1951), 95–103, Bull. Res. Council, Israel **5A** (1955), 11–19) and with Joseph L. Walsh (J. Analyse Math., **4** (1954/55), 49–87, J. Analyse Math., **5** (1956), 47–76, Pacific J. Math., **7** (1956), 1037–1064; [194], 163, 178, 181). Then Walsh continued alone ([194], 145, 172, 187, 188, 201, 202), and in collaboration with T. S. Motzkin ([194], 151, 166, 174, 179, 189, 193), M. Zedek ([194], 170) and Oved Shisha ([194], 209, 222, 226, 227).

In the case when the functional A is a weighted maximum

$$A(g) = \max_{z \in S} w(z) \big| g(z) \big| ,$$

where $w > 0$ is continuous, Gyula Szőkefalvi Nagy brings precisions to the theorems of Fejér and Fekete–von Neumann (Acta Sci. Math. Szeged **6** (1932), 49–58). Let $g^*(z) \in \mathcal{P}_n$ be the polynomial which minimizes $A(g)$ and let $S^* \subset S$ be the set of points where $w(z)|g^*|$ achieves its maximum. Then:

a) the zeros of g^* lie in the convex hull of S^*;

b) if S is a subset of the real axis, then S^* contains $n+1$ points which separate the n real zeros of $g^*(z)$;

c) if S is symmetric with respect to the real axis and the coefficients of g^* are real, then the complex zeros of g^* lie in the union of disks with diameters that have two conjugate points of S^* as endpoints.

A theorem of Edmond Laguerre states that if $f(z)$ is a polynomial of degree n with real coefficients and real zeros, and we divide the interval between two consecutive zeros into n equal parts, then $f'(z)$ has no zero in the two extreme subintervals. Paul Montel made this result more precise by proving that if the coefficients of $f(z)$ are real, z_k and z_{k+1} are real zeros of $f(z)$, and the real part of no other zero lies between z_k and z_{k+1}, then $f'(z)$ has no zero in the intervals $(z_k, z_k + \delta)$ and $(z_{k+1} - \delta, z_{k+1})$, where $\delta = (z_{k+1} - z_k)/n$. Gyula Szőkefalvi Nagy (Math. Természettud. Értesítő **53** (1935), 781–792, Acta Sci. Math. Szeged **8** (1936), 42–52) strengthened this result by showing that the same conclusion holds if $f(z)$ has no roots in the disk with diameter $[z_k, z_{k+1}]$, i.e. he replaced an infinite strip by a bounded disk.

In a similar vein Pál Turán (Acta Sci. Math. Szeged **11** (1946/48), 106–113; [184], No. 27) considers a polynomial $f(z)$ of even degree n with real coefficients such $f(-1) = f(1) = 0$. He assumes that $f(z)$ has no zeros in the open interval $]-1, +1[$ but makes no other assumptions on the zeros of $f(z)$. If ξ is the point where $|f(z)|$ achieves its maximum in $[-1, 1]$, then he proves that $|\xi| \leq \cos\frac{\pi}{n}$ and this bound is sharp. A more complicated estimate holds if n is odd.

Gyula Szőkefalvi Nagy (Tôhoku Math. J., **35** (1932), 126–135) finds still other regions which do not contain zeros of the derivative. Let $z_1 \in \mathbb{C}$ be a zero with multiplicity p of the polynomial $f(z)$. Assume that on either side of any straight line going through z_1 there lie at most s zeros of $f(z)$. Denote by D a closed disk with radius R, having z_1 on its boundary and containing no other zero of $f(z)$. Then $f'(z)$ has no zero in the disk D_1 touching D from the inside at z_1 and having radius $\frac{p}{p+s}R$. Gyula Szőkefalvi Nagy was fundamentally a geometer and the proofs of his theorems about polynomials have a strong geometric flavor: they belong to the "Geometry of Polynomials", the title of the book by Morris Marden [113] which discusses many of Gyula Szőkefalvi Nagy's results and lists 24 of his articles in the bibliography. Marden presents the results concerning polynomials also of other Hungarian authors: Elemér Bálint, Jenő Egerváry, Pál Erdős, Tibor Faragó, Lipót Fejér, Mihály Fekete, Péter Lax, István Lipka, Endre Makai, János Neumann, György Pólya, Tibor Radó, Ottó Szász, Pál Turán and István Vincze.

The theorems of Gauss–Lucas and Jensen are distant relatives of Michel Rolle's theorem which is also a statement concerning the position of a zero of the derivative with respect to the zeros of the function. Another basic fact of real analysis is the theorem of Bernhard Bolzano or the intermediate value theorem which does not hold in the complex case: $e^{i\pi} = -1$, $e^0 = 1$ but for no z in $\mathbb{C}$ is e^z equal to 0. For polynomials the situation is more favorable. Fekete (Jahresber. Deutsch. Math.-Verein., **34** (1926), 222–233) proved that if $f(z)$ is a polynomial of degree n and $f(z_1) = w_1 \neq f(z_2) = w_2$, then every value in the set of those points from which the segment $[w_1, w_2]$ can be seen under an angle $\geq \varphi$ is assumed by $f(z)$ in the set of those points from which $[z_1, z_2]$ can be seen under an angle $\geq \varphi/n$.

Elemér Bálint (Jahresber. Deutsch. Math.-Verein., **34** (1926), 233–237) proved that if $f(z)$ is a lacunary polynomial of the form (3) having "length" k, then every value in the disk with diameter having endpoints $w_1 = f(z_1)$ and $w_2 = f(z_2)$ is assumed by $f(z)$ in a disk with center $\frac{1}{2}(z_1 + z_2)$ and having radius $\leq Ck$, where C is an absolute constant.

Assume that the polynomial $f(z)$ of degree n assumes the values $w_1, \ldots, w_n$ at the points $z_1, \ldots, z_n$. Let P be the convex hull of the points z_j and Π the convex hull of the points w_j. Finally, let D be the smallest disk such that from every point of the boundary of D the polygon P can be seen under and angle $\leq \frac{\pi}{n}$. Gyula Szőkefalvi Nagy (Jahresber. Deutsch. Math.-Verein., **32** (1923), 307–309) proved that $f(z)$ assumes in D every value lying in Π.

Let me recall that the elementary symmetric functions $S_k = S_k(x_1, \ldots, x_n)$, $0 \leq k \leq n$, are the polynomials in the n indeterminates $x_1, \ldots, x_n$ defined by

$$(x - x_1) \ldots (x - x_n) = S_0 x^n - S_1 x^{n-1} + \cdots + (-1)^k S_k x^{n-k} + \cdots + (-1)^n S_n,$$

i.e. $S_0 = 1$, $S_1 = x_1 + \cdots + x_n$, $S_2 = x_1 x_2 + x_1 x_3 + \cdots + x_{n-1} x_n$, $\ldots$, $S_n = x_1 x_2 \cdots x_n$.

Now let $x_1, \ldots, x_n$ and $y_1, \ldots, y_n$ be two n-tuples of indeterminates, set $s_k = S_k(x_1, \ldots, x_n)$, $\sigma_k = S_k(y_1, \ldots, y_n)$ and introduce the polynomial

$$G = \sum_{k=0}^{n} (-1)^k \frac{1}{\binom{n}{k}} \sigma_{n-k} s_k$$

in $2n$ indeterminates. The following result is due to J. H. Grace ([29], Th. VII, p. 11):

Let the complex numbers x_j and y_j $(1 \leq j \leq n)$ satisfy

$$G(x_1, \ldots, x_n, y_1, \ldots, y_n) = 0.$$

If all the y_j lie in the circular domain D, then at least one x_j lies in D.

This theorem and related results have been investigated in great detail by Gábor Szegő (Math. Z., **13** (1922), 28–55; [173], I, pp. 505–534). With Grace he considers polynomials of degree n having the form

$$A(x) = a_0 + \binom{n}{1} a_1 x + \cdots + \binom{n}{k} a_k x^k + \cdots + a_n x^n$$

and denotes the solutions of $A(x) = 0$ by $\alpha_1, \ldots, \alpha_n$. Substituting the roots α_j for the indeterminates y_j and using that

$$(-1)^{n-k} a_n \sigma_{n-k} = \binom{n}{k} a_k,$$

the equation $G = 0$ becomes Szegő's "Faltungsgleichung"

$$A^\sharp(x_1, \ldots, x_n) = a_0 s_0 + a_1 s_1 + \cdots + a_n s_n = 0, \tag{9}$$

where $s_k = S_k(x_1, \ldots, x_n)$. Observe that $A^\sharp(x, \ldots, x) = A(x)$. The above theorem then yields the form of Grace's theorem which Szegő calls the "Faltungssatz" and for which he gave a simple new proof ([173], I, p. 509):

Assume that the α_j all lie in the circular domain D and that $x_1, \ldots, x_n$ satisfy (9). Then at least one of the x_j lies in D.

This result has several consequences, among then the usual formulation of the theorem of Grace: Consider a second polynomial

$$B(x) = b_0 + \binom{n}{1} b_1 x \cdots + \binom{n}{k} b_k x^k + \cdots + b_n x^n$$

of the same form, and denote its zeros by $\beta_1, \ldots, \beta_n$. Replacing the variables x_j by the zeros β_j, the "Faltungsgleichung" becomes the condition

$$a_0 b_n - \binom{n}{1} a_1 b_{n-1} + \cdots + (-1)^k \binom{n}{k} a_k b_{n-k} + \cdots + (-1)^n a_n b_0 = 0$$

expressing the fact that the two polynomials are "apolar". The assertion is that if $A(x)$ and $B(x)$ are apolar, and if all the zeros of $A(x)$ lie in the circular domain D, then at least one of the zeros $B(x)$ lies in D ([129], V. 145).

Another result is the following: Assume $|\alpha_j| \leq 1$ and $|\beta_j| < 1$ for $1 \leq j \leq n$, and let

$$C(x) = \sum_{k=0}^{n} \binom{n}{k} a_k b_k x^k$$

be the polynomial obtained by 'composition". Then every solution of $C(x) = 0$ lies in the interior of the unit disk. This result was also found by Jenő Egerváry (Acta Sci. Math. Szeged, **1** (1922), 39–45).

Let me quote one more theorem. Assume that all the roots α_j of $A(x)$ lie in the circular domain D. Then every root γ of $C(x)$ has the form $\gamma = -\delta\beta_j$, where $\delta \in D$ and β_j is one of the roots of $B(x)$.

For further discussion I refer to Chapter II, Nos. 2 and 3, pp. 11–14 of [29]. The composition polynomial was studied by N. G. de Bruijn and T. A. Springer, and Szegő's method was used by Lars Hörmander to extend Grace's theorem to polynomials of several variables over an algebraically closed field, see Chapters III and IV of [113].

4. Trigonometric polynomials, Toeplitz forms and a problem of Carathéodory

The names of Lipót Fejér and Frigyes Riesz are attached jointly to a theorem, to an inequality, and to a procedure.

The Fejér–Riesz theorem is about trigonometric polynomials, i. e., expressions of the form

$$\tau(t) = \sum_{k=0}^{n} (a_k \cos kt + b_k \sin kt)$$

($b_0 = 0$), or in the nowadays more popular complex writing

$$\tau(t) = \sum_{k=-n}^{n} c_k e^{ikt},$$

where n is the "order" of $\tau(t)$. We have $a_0 = c_0$, $a_k = c_k + c_{-k}$ and $b_k = i(c_k - c_{-k})$ for $k \geq 1$ so that $2c_{-k} = a_k + ib_k$, $2c_k = a_k - ib_k$. We shall assume that the coefficients a_k and b_k are real, i.e. that $c_{-k} = \bar{c}_k$ for $k \geq 1$ and in this case we can also write

$$\tau(t) = c_0 + 2\Re e \sum_{k=1}^{n} c_k e^{ikt}.$$

The theorem asserts that if $\tau(t) \geq 0$ for all $t \in \mathbb{R}$, then there exists a rational polynomial

$$g(z) = \gamma_0 + \gamma_1 z + \cdots + \gamma_n z^n$$

of degree n such that $\tau(t)$ equals $|g(z)|^2$ if we substitute $z = e^{it}$ ([129], VI.40). Fejér says that he conjectured the result for some time, presumably on the basis of his kernel

$$(10)\quad \frac{1}{2}(n+1) + n\cos t + (n-1)\cos 2t + \cdots + \cos nt = \frac{1}{2}\left(\frac{\sin(n+1)\frac{t}{2}}{\sin\frac{t}{2}}\right)^2$$

$$= \frac{1}{2}|1 + z + z^2 + \cdots z^n|^2,$$

but the proof was communicated to him orally by F. Riesz (J. Reine Angew. Math., **146** (1916), 53–82; [40], No. 51, vol. I). Riesz in an article printed immediately after Fejér's (pp. 83–87; [156], I4), and in which he applies the theorem, only says that the theorem was proved in Fejér's article. It seems that the two could not agree to write a joint paper on the subject. Curiously, the proof employs the geometric theorem on which the Carathédory–Fejér result of Section 2 is based.

The Fejér–Riesz theorem can be considered as a parametrization of the set of positive trigonometric polynomials with real coefficients a_k, b_k: to each $\tau(t)$ there correspond $n+1$ complex parameters (or $2n+2$ real parameters if we separate the real and imaginary parts of the γ_k). It has been extended to entire functions of exponential type which are in a sense generalizations of trigonometric polynomials (see [15], 7.5.1, p. 125, where a wrong reference is given).

As a first application, consider the set of all positive trigonometric polynomials with constant term $a_0 = c_0 = 1$. Since $c_k = \frac{1}{2\pi}\int_0^{2\pi} \tau(t)e^{-ikt}\,dt$, we have $|c_k| \leq \frac{1}{2\pi}\int_0^{2\pi} \tau(t)\,dt = c_0 = 1$ and so trivially $\tau(0) \leq 2n+1$. The parametrization $\tau(t) = |g(e^{it})|^2$ permits to find exactly the maximum of

$\tau(0)$. Indeed, $c_0 = \sum_{k=0}^{n} |\gamma_k|^2$ and $\tau(0) = |\gamma_0 + \gamma_1 + \cdots + \gamma_n|^2$. If we write $\gamma_k = \alpha_k + i\beta_k$, we have to determine the maximum of

$$\left(\sum_{k=0}^{n} \alpha_k\right)^2 + \left(\sum_{k=0}^{n} \beta_k\right)^2$$

under the constraint $\sum_{k=0}^{n} \alpha_k^2 + \sum_{k=0}^{n} \beta_k^2 = 1$. The usual methods of calculus yield that the maximum is achieved when

$$\alpha_0 = \alpha_1 = \cdots = \alpha_n, \quad \beta_0 = \beta_1 = \cdots = \beta_n.$$

In this case $\alpha_0^2 + \beta_0^2 = \gamma_0^2 = \frac{1}{n+1}$. Thus $\tau(0) \leq n+1$ and $\tau(0) = n+1$ only for $\frac{2}{n+1}$ times the trigonometric polynomial (10). Considering $\tau(t + t_0)$ we obtain that $\tau(t) \leq n+1$ for $t \in \mathbb{R}$ if τ is a positive trigonometric polynomial of order n having constant term 1.

The ideas of Fejér's paper were taken up by Gábor Szegő in an article (Math. Ann., **79** (1918), 323–339; [173], I, pp. 153–170) where orthonormal systems of polynomials on the unit circle were first considered. They were to play a central role in Szegő's research as József Szabados tells us elsewhere in this volume. For $z = e^{i\theta}$, $0 < r < 1$, let

$$p(\theta, r) = \frac{1 - r^2}{|z - r|^2} = \frac{1 - r^2}{1 - 2r\cos\theta + r^2}.$$

Denote by $\tau(\theta)$ a positive trigonometric polynomial of order n which satisfies

$$\frac{1}{2\pi} \int_0^{2\pi} p(\theta, r)\tau(\theta)\, d\theta = 1.$$

Szegő proves that then $p(\theta, r)\tau(\theta) \leq n + 1 + \frac{1+r}{1-r}$, and the upper bound is achieved for

$$\tau(\theta) = \frac{1+r}{n+1-r(n-1)} \left| 1 + \frac{z-r}{1+r} \frac{z^n - 1}{z - 1} \right|^2$$

when $\theta = 0$. If $r \to 0+$ we obtain the result of Fejér proved above. The proof uses a parametrization

$$\tau(t) = \left| \gamma_0\varphi_0(z) + \cdots + \gamma_0\varphi_n(z) \right|^2, \quad z = e^{it},$$

where $\big(\varphi_k(z)\big)$ is a sequence of polynomials orthonormal on $|z| = 1$ with respect to the weight $p(\theta, r)$.

Returning to Fejér's article, let $\tau(t)$ be a trigonometric polynomial of order n whose constant term is zero. Let $-m$ be the minimum and M the maximum of $\tau(t)$. Then

$$\frac{M-\tau(t)}{M} \quad \text{and} \quad \frac{\tau(t)+m}{m}$$

are positive trigonometric polynomials with constant term 1, and so by the result proved above

$$\frac{M+m}{M} \leq n+1 \quad \text{and} \quad \frac{M+m}{m} \leq n+1,$$

i.e. $m \leq nM$ and $M \leq nm$. If now $\tau(t)$ is an arbitrary trigonometric polynomial of order n, we call the maximum of $\tau(t) - a_0$ the "height" H and the maximum of $a_0 - \tau(t)$ the "depth" h of $\tau(t)$, i.e. $-h$ is the minimum of $\tau(t) - a_0$. Observing that the constant term of $\tau(t) - a_0$ is zero, we obtain a famous result of Fejér (which can also be proved in other ways):

The height of a trigonometric polynomial of order n is at most n times its depth, and its depth is at most n times its height.

Ottó Szász continued Fejér's investigations in four articles, the last one written in collaboration with Jenő Egerváry (Sitzungsber. Bayer. Akad.-Wiss. Math.-Phys. Kl. 1917, 317–320, ibid. 1927, 185–196, Math. Z., **1** (1918), 149–162, ibid., **27** (1928), 641–652; [172], pp. 656–683, 734–757). Some of his results are discussed in the report written by Fejér (Mat. Fiz. Lapok **37** (1930), 63–90; [40], No. 74, vol. II) when Szász was awarded the Kőnig Gyula prize in 1930.

I reproduce some very elegant and elementary proofs of certain inequalities concerning the coefficients of a positive trigonometric polynomial $\tau(t)$. The equality

$$\tau(t) = c_0 + \sum_{k=1}^{n} (c_k e^{ikt} + \bar{c}_k e^{-ikt}) = \left|\gamma_0 + \gamma_1 e^{it} + \cdots + \gamma_n e^{int}\right|^2$$

yields that $c_k = \sum_{r=0}^{n-k} \gamma_{k+r}\bar{\gamma}_r$ for $k \geq 0$ (note: Szász denotes by c_k our $2c_k$). Assume now that $c_0 = a_0 = \sum_{r=0}^{n} |\gamma_r|^2 = 1$. Using the arithmetic mean-geometric mean inequality we have

$$|c_k| \leq \sum_{r=0}^{n-k} |\gamma_{k+r}\bar{\gamma}_r| \leq \frac{1}{2} \sum_{r=0}^{n-k} \left(|\gamma_{k+r}|^2 + |\gamma_r|^2\right)$$

$$= \frac{1}{2}\left(|\gamma_0|^2 + \cdots + |\gamma_{n-k}|^2 + |\gamma_k|^2 + \cdots + |\gamma_n|^2\right).$$

Since $n-k<k$ if $k>\left[\frac{n}{2}\right]$ (where $[x]$ denotes the largest integer $\leq x$), we obtain that the coefficients of a positive trigonometric polynomial with $a_0=1$ satisfy the inequalities

$$|a_k-ib_k|\leq 1 \quad \text{if}\, k>\left[\frac{n}{2}\right],$$

$$|a_k-ib_k|\leq 2 \quad \text{for} \quad 1\leq k\leq n.$$

These inequalities were proved by Fejér for cosine polynomials, i.e., when $b_k=0$ for all k.

Still under the assumption that $a_0=c_0=1$ we have, using the inequality

$$\left(\frac{x_0+x_1+\cdots+x_n}{n+1}\right)^2\leq\frac{x_0^2+x_1^2+\cdots x_n^2}{n+1},$$

that

$$\sum_{k=0}^{n}|c_k|\leq\sum_{s=0}^{n}|\gamma_s|^2+\sum_{k=1}^{n}\sum_{s=0}^{n-k}|\gamma_{k+s}\bar{\gamma}_s|$$

$$=\frac{1}{2}+\frac{1}{2}\left(\sum_{s=0}^{n}|\gamma_s|^2+2\sum_{k=1}^{n}\sum_{s=0}^{n-k}|\gamma_{k+s}\bar{\gamma}_s|\right)$$

$$=\frac{1}{2}+\frac{1}{2}\left(\sum_{s=0}^{n}|\gamma_s|\right)^2\leq\frac{1}{2}+\frac{n+1}{2}\sum_{s=0}^{n}|\gamma_s|^2=\frac{n+2}{2}$$

and so

$$\sum_{k=0}^{n}|a_k-ib_k|\leq n+1.$$

In his report Fejér lists some of the more profound results of Szász and Egerváry. In particular he mentions the inequalities

$$|a_k-ib_k|\leq 2\cos\frac{\pi}{\left[\frac{n}{k}\right]+2}\min(h,H) \quad (1\leq k\leq n)$$

valid for any trigonometric polynomial with real coefficients. If $\tau(t)\geq 0$ and $a_0=1$, then $h\leq 1$; if furthermore $k>\left[\frac{n}{2}\right]$ then $\left[\frac{n}{k}\right]=1$ and since $\cos\frac{\pi}{3}=\frac{1}{2}$ we get again $|a_k-ib_k|\leq 1$.

Before I can tell you to what use Frigyes Riesz put the Fejér–Riesz theorem, I must go back to 1911 and speak about a problem of Carathéodory (Rend. Circ. Mat. Palermo **32** (1911), 193–217; [21], Bd. III, No. LIII, pp. 78–110): Find a condition on the $2n$ real numbers

$$a_1, b_1, a_2, b_2, \dots, a_n, b_n \tag{11}$$

so that there should exist a power series

$$\frac{1}{2} + \sum_{k=1}^{\infty} (a_k + ib_k) z^k$$

which has the numbers (11) as initial coefficients, converges for $r = |z| < 1$, and whose real part

$$u(r, \theta) = \frac{1}{2} + \sum_{k=1}^{\infty} r^k (a_k \cos k\theta - b_k \sin k\theta)$$

is a positive harmonic function ($z = re^{i\theta}$). Carathéodory proved that for this to happen the following condition is necessary and sufficient:

(C) The point in $\mathbb{R}^{2n}$, whose coordinates are the numbers (11), lies in the closed convex hull K_{2n} of the curve with parametric representation

$$x_1 = \cos\theta, \quad y_1 = -\sin\theta, \quad x_2 = \cos 2\theta, \quad y_2 = -\sin 2\theta, \quad \dots,$$
$$x_n = \cos n\theta, \quad y_n = -\sin n\theta.$$

Frigyes Riesz in an article on integral equations (Ann. Sci. École Norm. Sup. Paris **28** (1911), 33–62; [156], F3, vol. II, pp. 788–827) proved that the following condition is also necessary and sufficient:

(R) For all $(\xi_1, \eta_1, \dots, \xi_n, \eta_n) \in \mathbb{R}^{2n}$ the value of $a_1\xi_1 + b_1\eta_1 + \dots + a_n\xi_n + b_n\eta_n$ is between the minimum and the maximum value of the trigonometric polynomial

$$\tau(t) = \sum_{j=1}^{n} (\xi_j \cos jt + \eta_j \sin jt).$$

To state the condition due to Otto Toeplitz let us write $\gamma_0 = 1, \gamma_k = a_k + ib_k$ and $\gamma_{-k} = \bar{\gamma}_k$ for $1 \le k \le n$. The following is necessary and sufficient for the existence in Carathéodory's problem:

(T) The "Toeplitz form"

$$H(\zeta_0, \zeta_1, \ldots, \zeta_n) = \sum_{j=0}^{n} \sum_{k=0}^{n} \gamma_{j-k} \zeta_j \bar{\zeta}_k$$

is positive semi-definite.

In his 1916 paper quoted at the beginning of this section F. Riesz observes that the three conditions are obviously equivalent since they are all necessary and sufficient for the same fact, nevertheless he proves their equivalence in a direct way. This seems to be the first time that Toeplitz forms make an appearance in Hungarian mathematics. They were to have a central importance in the research of Gábor Szegő as I will indicate briefly below.

Riesz first remarks (as he already did in his 1911 article) that the equivalence of (C) and (R) is simply a restatement of the fact that the closed convex hull of a set is the intersection of the closed half-spaces containing it.

To prove (R) $\Leftrightarrow$ (T) he starts by saying that (R) is clearly equivalent to:

(R′) Whenever the real numbers $\xi_0, \xi_1, \eta_1, \ldots, \xi_n, \eta_n$ are such that the trigonometric polynomial

$$\xi_0 + \sum_{j=1}^{n} (\xi_j \cos jt + \eta_j \sin jt) \tag{12}$$

is positive, it follows that the expression

$$\xi_0 + a_1 \xi_1 + b_1 \eta_1 + \cdots + a_n \xi_n + b_n \eta_n \tag{13}$$

is positive.

Now assume that (12) is positive. Then by the Fejér–Riesz theorem there exists a polynomial

$$g(z) = \zeta_0 + \zeta_1 z + \cdots + \zeta_n z^n$$

such that

$$\xi_0 + \sum_{j=1}^{n} (\xi_j \cos jt + \eta_j \sin jt) = \left| g(z) \right|^2 = \sum_{j=0}^{n} \sum_{k=0}^{n} \zeta_j \bar{\zeta}_k e^{i(j-k)t}. \tag{14}$$

Substituting a_j for $\cos jt$ and b_j for $\sin jt$, i.e. γ_j for e^{ijt} and $\bar{\gamma}_j = \gamma_{-j}$ for e^{-ijt} $(j \geq 0)$, the left-hand side becomes (13) and the right hand yields $H(\zeta_0, \zeta_1, \ldots, \zeta_n)$. Let us e.g., assume that (R′) holds. For any $\zeta_0, \zeta_1, \ldots, \zeta_n$ (14) yields a positive trigonometric polynomial. By assumption (13) is positive. But after substitution (13) equals $H(\zeta_0, \zeta_1, \ldots, \zeta_n)$ which is therefore ≥ 0. The proof of (T) $\Rightarrow$ (R′) is similar.

In a joint article of Fejér and Carathéodory (Rend. Circ. Mat. Palermo **32** (1911), 218–239; [40], No. 39, vol. I, pp. 693–715; [21], Bd. III, No. LIV, pp. 111–138), printed immediately following the above quoted paper of Carathéodory, the authors use the results of that paper to answer questions of the following type:

(a) Let the real numbers

$$a_0, a_1, b_1, a_2, b_2, \ldots, a_n, b_n \tag{15}$$

and a number $R > 0$ be given. Consider the set of all harmonic functions $u(r, \theta)$ given by the series

$$u(r, \theta) = a_0 + \sum_{k=1}^{\infty} r^k (a_k \cos k\theta + b_k \sin k\theta) \tag{16}$$

with the given initial coefficients which converge for $r < R$. Find the maximum of the least upper bounds of the functions belonging to the set.

(b) Let the numbers (15) be given, assume that among those with subscripts $1 \leq j \leq n$ at least one is different from zero, and let $m^* < a_0$. Find the largest value of $R > 0$ such that there exists a series (16) with the given initial coefficients which converges for $r < R$ and whose greatest lower bound is $\geq m^*$.

(c) Let $c_0, c_1, \ldots, c_n$ be given complex numbers, $c_0 \neq 0, 1$ and not all c_j with $1 \leq j \leq n$ equal to zero. Find the largest number $R > 0$ such that a power series $\sum_{j=0}^{\infty} c_j z^j$ with given initial coefficients converges for $|z| < R$ and represents a function which does not assume the values 0 and 1.

The last question is related to Landau's sharpening of the Picard theorem: Given $c_0, c_1 \neq 0$, there exists $R = R(c_0, c_1) > 0$ such that every power series $\sum_0^{\infty} c_k z^k$ which converges for $|z| < R$ assumes either the value 0 or the value 1 in $|z| < R$ ([74], III. 7, §6, p. 448).

To answer e.g., the question (a), Carathéodory and Fejér introduce the Toeplitz matrices ($\gamma_k = a_k + ib_k$, $\gamma_{-k} = \bar{\gamma}_k$)

$$\begin{pmatrix} \gamma_0 & \gamma_1 & \cdots & \gamma_k \\ \gamma_{-1} & \gamma_0 & \cdots & \gamma_{k-1} \\ \cdot & \cdot & \cdots & \cdot \\ \gamma_{-k} & \gamma_{-k+1} & \cdots & \gamma_0 \end{pmatrix}$$

and their determinants $D_k(\gamma_0, \gamma_1, \dots, \gamma_k)$. It was Toeplitz who pointed out to them that the equivalent conditions (C) and (R) are also equivalent to (T), i.e. to the inequalities $D_k(1, \gamma_1, \dots, \gamma_k) \geq 0$ for $1 \leq k \leq n$. More precisely, $(a_1, b_1, \dots, a_n, b_n) \in \mathbb{R}^{2n}$ lies in the interior of K_{2n} if $D_k(1, \gamma_1, \dots, \gamma_k) > 0$ for $1 \leq k \leq n$, and it lies on the boundary of K_{2n} if $D_k(1, \gamma_1, \dots, \gamma_k) \geq 0$ for $1 \leq k \leq n-1$ and $D_n(1, \gamma_1, \dots, \gamma_n) = 0$ With the help of this formulation a geometric argument gives the

Theorem. *Assume that the series* (16) *converges in* $r < 1$ *and has as initial coefficients the given numbers* (15). *The equation*

$$D_n\big(2(a_0 - \lambda), \gamma_1, \dots, \gamma_n\big) = 0$$

in the unknown λ *has only real solutions. Let* λ_{n*} *be the smallest among the solutions and* λ_n^* *the largest one. Then* $\inf_{r<1} u(r, \theta) \leq \lambda_{n*}$ *and* $\sup_{r<1} u(r, \theta) \geq \lambda_n^*$. *There exists a function in the set of functions* (16) *considered whose greatest lower bound equals* λ_{n*} *and one whose least upper bound equals* λ_n^*.

Gábor Szegő wrote two articles in Hungarian on Toeplitz and Hankel forms in 1917–1918 (Math. Természettud. Értesítő **35** (1917), 185–122, **36** (1918), 497–538; [173], I, pp. 69–108, 113–148). The second of these articles appeared in English in the Translations of the American Mathematical Society ((2) **108** (1977), 1–36), and it is this translation which is reproduced in the Collected Papers. Most of the results of the two articles are treated again in his great 1920–21 paper published in two parts (Math. Z., **6** (1920), 167–202, **9** (1921), 167–190; [173], I, pp. 237–272, 279–302). He returned to the subject in 1952 (Comm. Sem. Math. Univ. Lund, Tome Suppl., Festskrift Marcel Riesz (1952), 228–238; [173], III, pp. 270–280), in a joint paper with Mark Kac and W. L. Murdock of 1953 (J. Rational Mech. Anal., **2** (1953), 767–800, **3**, 802–803; [173], III, pp. 333–367), and in his 1958 book with Ulf Grenander [59].

Let f be a positive integrable function in $[0, 2\pi]$ and assume that its arithmetic mean

$$A(f) = \frac{1}{2\pi} \int_0^{2\pi} f(t)\, dt$$

satisfies $A(f) > 0$. Denote by $\mu_n(f)$ the greatest lower bound of all expressions

$$\frac{1}{2\pi} \int_0^{2\pi} f(t) \big| P_n(e^{it}) \big|^2 \, dt, \tag{17}$$

where $P_n(z)$ varies in the set of all polynomials of the form

$$P_n(z) = 1 + \zeta_1 z + \zeta_2 z^2 + \cdots + \zeta_n z^n.$$

If we introduce the Fourier coefficients

$$c_j = \frac{1}{2\pi} \int_0^{2\pi} f(t) e^{ijt} \, dt,$$

the expression (17) becomes the Toeplitz form

$$\sum_{j=0}^{n} \sum_{k=0}^{n} c_{j-k} \zeta_j \bar{\zeta}_k, \tag{18}$$

where $\zeta_0 = 1$. As we have done above, we introduce the Toeplitz matrices

$$M_n(f) = \begin{pmatrix} c_0 & c_1 & \cdots & c_n \\ c_{-1} & c_0 & \cdots & c_{n-1} \\ \cdot & \cdot & \cdots & \cdot \\ c_{-n} & c_{-n+1} & \cdots & c_0 \end{pmatrix}$$

and their determinants $D_n(f) = \det M_n(f)$. By assumption $D_0(f) = c_0 = A(f) > 0$, and $D_n(f) > 0$ for $n \geq 1$ since (18) is positive definite by its definition. A well-known theorem on Hermitian forms implies that

$$\mu_n(f) = \inf_{\zeta_0 = 1} \sum_{j=0}^{n} \sum_{k=0}^{n} c_{j-k} \zeta_j \bar{\zeta}_k = \frac{D_n(f)}{D_{n-1}(f)}.$$

Since every polynomial $P_n(z)$ is a particular polynomial $P_{n-1}(z)$, the sequence $(\mu_n(f))$ of positive numbers is decreasing and therefore

$$\mu(f) = \lim_{n \to \infty} \mu_n(f)$$

exists.

Let the real numbers $\lambda_0^{(n)}, \lambda_1^{(n)}, \ldots, \lambda_n^{(n)}$ be the eigenvalues of the matrix $M_n(f)$. Then $D_n(f) = \lambda_0^{(n)}\lambda_1^{(n)} \ldots \lambda_n^{(n)}$ and Szegő proves that

$$\mu(f) = \lim_{n\to\infty} \frac{D_n(f)}{D_{n-1}(f)} = \lim_{n\to\infty} \sqrt[n+1]{D_n(f)}$$

is equal to the geometric mean of f, i.e.

$$\lim_{n\to\infty} \sqrt[n+1]{\lambda_0^{(n)}\lambda_1^{(n)} \ldots \lambda_n^{(n)}} = e^{\frac{1}{2\pi}\int_0^{2\pi} \log f(t)\,dt}$$

if $\log f$ is integrable. From this special case he deduces a general formula which shows that the eigenvalues behave like values of f at equidistant points, namely if F is, say, continuous or has only finitely many jump discontinuities, then

$$\lim_{n\to\infty} \frac{F(\lambda_0^{(n)}) + F(\lambda_1^{(n)}) + \cdots + F(\lambda_n^{(n)})}{n+1} = \frac{1}{2\pi}\int_0^{2\pi} F\big(f(t)\big)\,dt.$$

Later Szegő generalized his result to the case when in (17) the exponent 2 is replaced by $p > 0$. Andrei N. Kolmogorov and Mark G. Krein replaced $f(t)\,dt$ by a Stieltjes measure ([2], Appendix B).

The second part of the 1920–21 article is devoted to orthogonal polynomials, so I again refer to the corresponding Chapter.

5. The Fejér–Riesz inequality

The Fejér–Riesz inequality appeared in the only article the two authors published jointly (Math. Z., **11** (1921), 305–314; [40], No. 59, vol. II, pp. 111–120; [156], D5, vol. I, pp. 625–634).

The authors first observe that if f is a regular analytic function in the closed disk $|z| \le 1$, then it follows from Cauchy's integral theorem that the integral of $|f(z)|^2$ along $|z| = 1$ is at least twice as large as the integral along a diameter, i.e.,

$$\int_{-1}^{1} |f(x)|^2\,dx \le \frac{1}{2}\int_0^{2\pi} |f(e^{i\theta})|^2\,d\theta. \tag{19}$$

Next they state that the inequality also holds for $|f|^p$ with any $p \geq 1$ instead of just $p = 2$, and prove it for $p = 1$. If f has no zeros in $|z| \leq 1$, the inequality follows from (19) applied to $\sqrt{f(z)}$. If $f(z)$ does have zeros, they asssume first that they lie all in the interior of the disk. Then there are only finitely many, say $a_1, \ldots, a_n$, and $f(z)$ can be divided by the "Blaschke product"

$$Q(z) = \frac{z - a_1}{1 - \bar{a}_1 z} \frac{z - a_2}{1 - \bar{a}_2 z} \cdots \frac{z - a_n}{1 - \bar{a}_n z}$$

mentioned in Section 3. Thus $f(z) = Q(z)g(z)$, where $\left|Q(z)\right| = 1$ for $|z| = 1$, hence $\left|Q(z)\right| \leq 1$ in the disk, and $g(z) \neq 0$ for $|z| \leq 1$. Finally they take care of the general case by considering circles $|z| = r < 1$ on which $f(z)$ has no zeros and letting $r \to 1$.

A first application is a simple proof of Hilbert's inequality

$$\sum_{j=1}^{\infty} \sum_{k=1}^{\infty} \frac{a_j a_k}{j + k} \leq \pi \sum_{j=1}^{\infty} |a_j|^2.$$

Applying the second inequality to $f'(z)$ one obtains

$$\int_{-1}^{1} \left|f'(z)\right| dz \leq \frac{1}{2} \int_0^{2\pi} \left|f'(e^{i\theta})\right| d\theta$$

which has the following geometric interpretation: let $f(z)$ map $|z| < 1$ conformally onto a domain, and $|z| = 1$ onto its boundary which we assume to be a rectifiable Jordan curve Γ. Then the image of any diameter of $|z| \leq 1$ is at most half as long as Γ. Mapping the disk onto a very elongated ellipse shows that the factor $\frac{1}{2}$ is the best possible.

R. M. Gabriel proved a similar result: let L be a closed convex curve inside $|z| \leq 1$. Then

$$\int_L \left|f(z)\right| |dz| \leq 2 \int_0^{2\pi} \left|f(e^{i\theta})\right| d\theta$$

and the constant 2 is the best possible. This does not contain the Fejér–Riesz inequality since a diameter counted twice in opposite directions can be considered a closed convex curve, and then the best factor is one. F. Carlson gave a common generalization to the two theorems: Let L be a rectifiable curve in $|z| \leq 1$ and denote by $V(z)$ the least upper bound of the sum of angles under which the line elements of L can be seen from the point z. Then

$$\int_L |f(z)|\,|dz| \le \frac{1}{\pi}\int_0^{2\pi} |f(e^{i\theta})|\,V(e^{i\theta})\,d\theta.$$

In particular, if $V(z) \le M$ for all z with $|z| = 1$, then the right-hand side is $\le \frac{M}{\pi}\int_0^{2\pi} |f(e^{i\theta})|\,d\theta$. So if L is a diameter, then $M = \frac{\pi}{2}$, and if L is a closed convex curve in $|z| \le 1$, then $M \le 2\pi$.

Marcel Riesz in a note written in honor of his brother and Fejér on the occasion of their 70th birthday (Acta Sci. Math. Szeged **12A** (1950), 53–56; [158], No. 50, pp. 794–797) proves Carlson's theorem using a double layer potential.

6. Boundary values, H^p spaces

The only article the Riesz brothers published jointly (4ème Congrès des Math. Scandinaves, Stockholm 1916 (1920), pp. 27–44; [156], D1, vol. I pp. 537–554; [158], No. 22, pp. 195–212) was inspired by the thesis of Pierre Fatou (Acta Math., **30** (1906), 335–400), the same memoir which gave Frigyes Riesz the "idea and the courage" to use the Lebesgue integral (Annales Inst. Fourier (Grenoble) **1** (1949), p. 29; [156], B16, p. 317). One of Fatou's main results was the following ([104], §5, pp. 35–42):

Let w be a function that is bounded and holomorphic in the unit disk $\mathbf{D} = \{z \in \mathbb{C} : |z| < 1\}$. Then for almost every point $e^{i\theta}$ of the circumference $\mathbf{T} = \{z \in \mathbb{C} : |z| = 1\}$ the limit $f(e^{i\theta}) = \lim_{r\to 1-} w(re^{i\theta})$ exists, i.e., those points $e^{i\theta}$ in which the limit does not exist form a set of Lebesgue measure zero on the circumference.

Fatou also proved that if w is not identically zero, then on any arc of $\mathbf{T}$ the set of those points $e^{i\theta}$ in which $f(e^{i\theta}) \ne 0$ has positive measure. He conjectured that the measure of the set on which the boundary value vanishes is zero but added that to prove this seems to be very difficult. Frigyes and Marcel Riesz say that Fatou underestimated the scope of his methods and present a proof of the conjecture "which is closely related to certain arguments of his work and is based directly on his results". Assume that $f(e^{i\theta}) = 0$ on a set $M \subset \mathbf{T}$ with measure $m > 0$. They construct a bounded holomorphic function $g(z)$ in $\mathbf{D}$ with $g(0) = 1$ such that $|g(z)|$ has boundary value $e^{A/m}$ on M and $e^{A/(m-2\pi)}$ on the complementary set

$M^c = \mathbf{T}\backslash M$, where A is a yet unspecified positive number. Then by Cauchy's theorem and by hypothesis

$$w(0) = \frac{1}{2\pi i}\int_{\mathbf{T}} \frac{f(z)g(z)}{z}\,dz = \frac{1}{2\pi i}\int_{M^c} \frac{f(z)g(z)}{z}\,dz$$

and

$$\left|\int_{M^c} \frac{f(z)g(z)}{z}\,dz\right| \le e^{\frac{A}{m-2\pi}} \int_0^{2\pi} \left|f(e^{i\theta})\right| d\theta.$$

Let $A \to \infty$. Since $m - 2\pi < 0$ we get that $w(0) = 0$, i.e. $w(z)/z$ is holomorphic and bounded in $\mathbf{D}$. Repeating the argument, we see that $z^{-n}w(z) = 0$ at $z = 0$ for all $n \in \mathbb{N}$, hence w is identically zero.

Next, the Riesz brothers consider a holomorphic function F which maps the open unit disk $\mathbf{D} \subset \mathbb{C}$ onto a bounded domain Ω of a Riemann surface whose boundary Γ is a rectifiable curve. The mapping can be extended to yield a function $\mathbf{T} \to \Gamma$ which is of bounded variation because Γ is rectifiable. Let M be a closed subset of measure zero of $\mathbf{T}$. They construct a positive integrable function $\varphi(\theta)$ on $\mathbf{T}$ which has the value $+\infty$ on M and finite values on M^c. If u is a positive harmonic function in $\mathbf{D}$ with boundary values φ, and v is the harmonic function conjugate to u, set

$$g(z) = \frac{u + iv}{1 + u + iv}.$$

Then $|g(z)| < 1$ in $\mathbf{D}$, the boundary value of $|g(z)|$ is < 1 on M^c and $= 1$ on M, so

$$\lim_{n\to\infty} \int_{|z|=1} (g(z))^n\,dF(z) = \int_M dF(z). \tag{20}$$

But

$$\int_{|z|=1} z^k\,dF(z) = -k\int_{|z|=1} z^{k-1}F(z)\,dz = 0$$

for $k > 0$ by Cauchy's theorem, and for $k = 0$ because of the factor k, so the integral on the left-hand side of (20) is zero. The integral on the right-hand side of (20) is the measure of $F(M)$, therefore the image under F of any subset of $\mathbf{T}$ of measure zero is a subset of Γ of measure zero. By a theorem credited to Stefan Banach ([71], (18.25), p. 288) a function of bounded variation is absolutely continuous if and only if it maps sets of measure zero into sets of measure zero, therefore F is absolutely continuous on $\mathbf{T}$.

The preceding proof shows the validity of the following result which is how the Theorem of F. and M. Riesz is most often quoted:

If F is a function of bounded variation on $\mathbf{T}$ and

$$\int_0^{2\pi} e^{in\theta}\, dF(e^{i\theta}) = 0$$

for $n = 1, 2, 3, \ldots$, then F is absolutely continuous.

Since functions of bounded variation correspond to signed measures, the theorem can also be stated in terms of measures. The work of the brothers Riesz was presented at the fourth Congress of Scandinavian Mathematicians in 1916 right in the middle of World War I. The Proceedings of the Congress were printed as late as 1920 and by error only in 50 copies. So the article of F. and M. Riesz, of which Paul Koosis says that every analyst should read it ([97], p. 40), was very difficult to find until a reprinted edition came out.

The Theorem of F. and M. Riesz spawned an enormous amount of research. It was generalized to several variables which was not straightforward because the obvious analogues are false. Abstract forms of the theorem became fundamental in the theory of function algebras and their generalizations. The output has not ceased and every year we see a number of articles about yet another generalization of the F. and M. Riesz Theorem.

Let me return to the article of Carathéodory and Fejér mentioned in Section 4. Let the $n+1$ complex numbers

$$c_0, c_1, \ldots, c_n \tag{21}$$

be given. As above, consider the set $\mathcal{F}(\mathbf{c})$ of all functions $f(z)$ that are holomorphic in the closed unit disk $\mathbf{D} \cup \mathbf{T}$ and whose power series expansion $\sum_{k=0}^{\infty} c_k z^k$ starts with the coefficients (21). They prove that there exists a unique function $f_*(z)$ in $\mathcal{F}(\mathbf{c})$ for which $M[f] = \max_{|z|\le 1} \left| f(z) \right|$ is a minimum. This function $f_*(z)$ is determined by the following properties: it is meromorphic with at most n poles in the complex plane, has at most n zeros in $\mathbf{D}$, and on $\mathbf{T}$ its absolute value equals the constant $M[f_*]$.

H. T. Gronwall gave an elementary proof of this theorem which uses neither the geometry of convex bodies nor Toeplitz forms (Ann. of Math. (2) **16** (1914/15), 77–81).

Very soon after the lecture in Stockholm, Frigyes Riesz wrote a paper published in 1917 in Hungarian (Math. Természettud. Értesítő **35** (1917), 605–632; [156], D2, pp. 555–582) but only in 1919 in German (Acta Math.,

42 (1920), 145–171; [156], D3, pp. 583–609). In it he also considers the above class $\mathcal{F}(\mathbf{c})$ but asks for the function that minimizes

$$I[f] = \int_0^{2\pi} \left| f(e^{i\theta}) \right| d\theta.$$

If $F(z)$ is a primitive of $f(z)$, then the expression we want to minimize is

$$T[f] = \int_0^{2\pi} \left| F'(e^{i\theta}) \right| d\theta,$$

i.e. the arc length of the image of $\mathbf{T}$ under the map $F(z)$. He proves the following

Theorem. *There exists a unique function $f^*(z)$ in the set $\mathcal{F}(\mathbf{c})$ for which $I[f]$ is minimal. This minimizing function $f^*(z)$ is characterized by the following two properties:*

1) *it is a polynomial of degree at most $2n$,*

2) *its zeros can be organized in pairs such that either the two elements of the pair are equal and they lie outside $\mathbf{D}$, or the two are reflections of each other with respect to $\mathbf{T}$.*

To prove the existence of $f^*(z)$ the obvious idea is to take a sequence $f_k(z)$ in $\mathcal{F}(\mathbf{c})$ for which $I[f_k]$ converges to the greatest lower bound I^* of $I[f]$. However, in this way one cannot prove that the limit function is regular on the *closed* unit disk. Therefore Riesz considers a sequence $F_k(z)$ of primitives such that $T[F_k]$ converges to $T^* = \inf T[F]$. He proves directly that the limit $F^*(z)$ is absolutely continuous and its derivative is the required function $f^*(z)$. He remarks that the properties of $F^*(z)$ he just proved also follow from investigations he conducted with his brother Marcel Riesz (in the Hungarian version he adds: "a Privatdozent at the University of Stockholm") but those investigations reach much more deeply into the theory of Lebesgue integration then what is necessary for the special problem considered here.

Though Frigyes Riesz says that Gronwall's treatment of the Carathéodory–Fejér result is so elementary that it could hardly be simplified, at the end of the article he shows how their theorem follows from his.

The next step concerning the Fatou–F. Riesz–M. Riesz circle of ideas is quite spectacular and its consequences are felt to the present day. It was published in Hungarian as an exchange of letters between Gábor Szegő and

Frigyes Riesz (Math. Természettud. Értesítő **38** (1921), 113–127; [156], D4, pp. 610–624; [173], 21–7, I, pp. 421–435), followed by two papers in German, one by each author (Math. Z., **18** (1923), 117–124; [156], D7, pp. 645–653; Math. Ann., **84** (1921), 232–244; [173], 21–6, I, pp. 404–416). Szegő informed Riesz that using his result on Toeplitz forms and the geometric mean (see Section 4) he proved that if w is holomorphic in $\mathbf{D}$, not identically zero, and

$$\int_0^{2\pi} \left|w(re^{i\theta})\right|^2 d\theta$$

is bounded as $r \to 1-$, then $f(e^{i\theta}) = \lim_{r\to 1-} w(e^{i\theta})$, which exists for almost every θ, is such that $\log\left|f(e^{i\theta})\right|$ is integrable in $[0, 2\pi]$. Thus in particular $f(e^{i\theta})$ cannot be zero on a set of positive measure.

In his answer Riesz first gives a proof of Szegő's result using only the Jensen formula and no Toeplitz forms. Then he quotes a result of G. H. Hardy according to which if w is holomorphic in $\mathbf{D}$ and $p > 0$, then

$$\mathcal{M}_p[r; w] = \int_0^{2\pi} \left|w(re^{i\theta})\right|^p d\theta$$

is an increasing function of r in $[0, 1)$, so it is either bounded or tends to $+\infty$ as $r \to 1-$. Hardy also proved that $\log \mathcal{M}_p[r; w]$ is a convex function of $\log r$, i.e., shares with $\mathcal{M}_\infty[r; w] = \max_\theta \left|w(re^{i\theta})\right|$ the property expressed by the Hadamard three circles theorem ([104], Kap. 6, pp. 88–97). Riesz introduces the class $H^p(\mathbf{D})$ of holomorphic functions for which $\mathcal{M}_p[r; w]$ is bounded, and proves that every $w \in H^p(\mathbf{D})$ has a product decomposition $w(z) = g(z)h(z)$, where $g(z)$ also belongs to $H^p(\mathbf{D})$ and is nowhere zero, and $h(z)$ is the already mentioned Blaschke product which satisfies $\left|h(z)\right| = 1$ for almost every $|z| = 1$ (if w has infinitely many zeros in $\mathbf{D}$, then $h(z)$ is a convergent infinite product). Since $g(z)^{p/2}$ belongs to $H^2(\mathbf{D})$, it follows that if w belongs to $H^p(\mathbf{D})$ for some $p > 0$ and is not identically zero, then its boundary values $f(e^{i\theta})$ exist for almost every θ and $\log\left|f(e^{i\theta})\right|$ is integrable. In particular, $f(e^{i\theta})$ can vanish at most on a set of measure zero.

The boundary values of functions $w \in H^p(\mathbf{D})$ form a function space $H^p(\mathbf{T})$ which for $p > 1$ is essentially $L^p(\mathbf{T})$ but for $p \leq 1$ has many important additional properties. Their generalizations to several variables have played a central role in harmonic analysis in the last fifty years (see [166], Chaps. III and IV).

Frederick Riesz was not lucky with the names of spaces. He discovered the L^p spaces and denoted them so in honor of Henri Lebesgue whose integral is essential for their definition; now everybody calls them Lebesgue spaces. In his great 1918 Acta Mathematica article he introduced complete normed spaces; the accepted terminology for them is now "Banach spaces." Then he introduced the spaces he called H^p in honor of G. H. Hardy who proved the theorem quoted above; now they are called Hardy spaces (though Ronald Coifman and Guido Weiss say: "...it could be argued fairly that the name 'Riesz' should be attached to these spaces", Bull. Amer. Math. Soc., **83** (1977), p. 570). On the other hand Bourbaki gave the name "espaces de Riesz" to lattice-ordered vector spaces but Riesz says: "...quelques auteurs français ... m'ont honoré en donnant mon nom à quelques notions (mais) je n'ai pas réussi à pénétrer suffisamment dans le langage et l'écriture créés par la société anonyme Bourbaki pour les comprendre entièrement" (Ann. Inst. Fourier (Grenoble), **1** (1949), p. 40; [156], B16, p. 338).

7. Kakeya's theorem, power series with monotone coefficients

The simplest result concerning monotone coefficients is the theorem of Eneström–Kakeya ([104], §2, p. 26; [129], II. 22):

If $a_0 \geq a_1 \geq a_2 \geq \cdots \geq a_n > 0$, then no solution of

$$P(z) = a_0 + a_1 z + a_2 z^2 + \cdots + a_n z^n = 0 \tag{22}$$

lies in the open disk $\mathbf{D} = \{ z \in \mathbb{C} : |z| < 1 \}$.

To prove it, one considers the expression

$$(1-z)P(z) = a_0 - (a_0 - a_1)z - (a_1 - a_2)z^2 - \cdots - (a_{n-1} - a_n)z^n - a_n z^{n+1}.$$

From here one concludes, as remarked by Simon Szidon (Sidon) (Acta Sci. Math. Szeged **9** (1938–1940), 244–246), as follows: introduce the positive numbers $p_k = a_k - a_{k+1}$ $(0 \leq k \leq n-1)$, $p_n = a_n$. Setting $z = 1$ we see that $a_0 = p_0 + p_1 + \cdots + p_n$, so (22) is equivalent to

$$\frac{p_0 z + p_1 z^2 + \cdots + p_n z^{n+1}}{p_0 + p_1 + \cdots + p_n} = 1.$$

The left hand side lies in the convex hull of the points z^k $(1 \leq k \leq n+1)$ and 1 does not lie in that convex hull if z, and therefore all z^k, lie in $\mathbf{D}$.

István Vincze (Mat. Lapok **10** (1959), 127–132) extended the Eneström–Kakeya theorem to complex coefficients as follows: Let the complex numbers $a_k = \alpha_k e^{i\theta} + \beta_k e^{i\varphi}$ satisfy $\alpha_0 \geq \alpha_1 \geq \cdots \geq \alpha_n \geq 0$, $\beta_0 \geq \beta_1 \geq \cdots \geq \beta_n \geq 0$ and $|\theta - \varphi| < \pi$. Then no solution of (22) lies in the disk $|z| < |a_0|/(\alpha_0 + \beta_0)$.

If $\beta_0 = 0, \theta = 0$ we have of course $|a_0|/\alpha_0 = 1$. If $\theta = 0$, $\varphi = \frac{\pi}{2}$, so that α_k and β_k are the real and imaginary part of a_k, then

$$\frac{|a_0|}{\alpha_0 + \beta_0} \geq \frac{1}{\sqrt{2}},$$

so that (22) has no solution in the disk $|z| < 1/\sqrt{2}$.

Considering the expression $z^n P(\frac{1}{z})$ one can also state the Eneström–Kakeya theorem as follows: If $0 < a_0 \leq a_1 \leq a_2 \leq \cdots \leq a_n$, then all the solutions of (22) satisfy $|z| \leq 1$. If the inequalities between the a_k are strict, then the solutions lie in $|z| < 1$. Egerváry (Acta Sci. Math. Szeged **5** (1931), 78–82) proved the following generalization of the Eneström–Kakeya theorem: Let $a_k > 0$ for $0 < k < n$, let m be an integer satisfying $0 \leq m \leq n$, and assume that there exist numbers $R \geq r > 0$ such that

$$Rra_{k+1} - (R + r)a_k + a_{k-1} > 0 \tag{23}$$

for $k = 0, \ldots, m-1, m+1, \ldots, n$ $(a_{-1} = a_{n+1} = 0)$. Then m solutions of (22) lie in the open disk $|z| < r$ and $n - m$ solutions lie in the domain $|z| > R$. If for some values of k the inequality (23) holds with ≥ 0, then the solutions can also lie on the boundaries of the two regions.

If $a_0 > 0$ and $ra_{k+1} > a_k$ for $0 \leq k \leq n-1$ and some $r > 0$ then (23) is satisfied for a sufficiently large $R > r$ and for $0 \leq k \leq n-1$, so all solutions of (22) lie in the disk $|z| < r$; for $r = 1$ this is again the Eneström–Kakeya theorem.

Denote by $\Delta^1 a_k = a_k - a_{k+1}$ the first differences and by

$$\Delta^2 a_k = a_k - 2a_{k+1} + a_{k+2} = \Delta^1 a_k - \Delta^1 a_{k+1}$$

the second differences of the sequence $a_0, a_1, \ldots, a_n$. For $R = r = 1$ the inequality (23) with $\geq$ can be written $\Delta^2 a_k \geq 0$ and also

$$a_{k+1} \leq \frac{a_k + a_{k+2}}{2}.$$

The validity of the last inequality for $0 \leq k \leq n-1$ expresses the geometric fact that the graph in the rectangular coordinate system, given by the polygon whose sides are the segments joining a point (k, a_k) with $(k+1, a_{k+1})$, is convex. We then say simply thet the sequence $a_0, a_1, \ldots, a_n, a_{n+1} = 0$ is convex.

Fejér (Jahresber. Deutsch. Math-Verein., **38** (1929), 231–238; [40], No. 70, vol. II, pp. 256–263) found a trigonometric analogue of the Eneström–Kakeya theorem. He considers a cosine polynomial

$$\tau(t) = \frac{a_0}{2} + \sum_{k=1}^{n} a_k \cos kt$$

($a_n \neq 0$) and with the help of the formula

$$\cos t \cos kt = \frac{1}{2}\{ \cos{(k-1)t} + \cos{(k+1)t}\}$$

obtains the identity

$$(1 - \cos t)\tau(t) = \frac{1}{2}\Delta^1 a_0 - \frac{1}{2}\sum_{k=1}^{n} \Delta^2 a_{k-1} \cos kt - \frac{1}{2} a_n \cos{(n+1)t}.$$

If we assume that the sequence $a_0, a_1, \ldots, a_n, a_{n+1} = 0, a_{n+2} = 0$ is convex, then it follows from this identity that $\tau(t) \geq 0$ for all t. It also follows that $a(0) > a_1 > \cdots > a_n > 0$ and that the cosine polynomal with a convex sequence of coefficients for which the first n coefficients are the smallest possible is a multiple of Fejér's signature polynomial (10), i.e.,

$$a_n \left\{ \frac{1}{2}(n+1) + n \cos t + (n-1)\cos 2t + \cdots + \cos nt \right\}.$$

Lipka (Acta Sci. Math. Szeged **9** (1938–1940), 69–77) observes first the following immediate consequence of Fejér's result: If the sequence $2a_0, a_1, \ldots, a_n > 0, a_{n+1} = 0$ is convex, then equation (22) has no solution in $|z| < 1$. The convexity condition implies that $2a_0 \geq a_1$, while in the Eneström–Kakeya theorem $a_0 \geq a_1$ is required. Lipka proves that if $2a_0 < a_1$ but $a_1, a_2, \ldots, a_n, a_{n+1}$ remains convex, then $P(z)$ has exactly one zero in $|z| < 1$. This follows from:

If each coefficient $a_1, a_3, \ldots, a_n$ and at least one of a_0, a_2 is > 0, and the sequence $2a_1, (a_0 + a_2), a_3, \ldots, a_n, a_{n+1} = 0$ is convex in such a way that no three vertices (k, a_k) are collinear, then $P(z) = 0$ has exactly one solution in $|z| < 1$.

Szegő (Trans. Amer. Math. Soc., **39** (1936), 1–17; [173], II, pp. 593–609) has some results of Eneström–Kakeya type concerning trigonometric polynomials. Consider for instance a cosine polinomial

$$\tau(t) = a_0 \cos nt + a_1 \cos (n+1)t + \cdots + a_{n-1} \cos t + a_n.$$

(i) If $a_0 > a_1 \geq \cdots \geq a_n \geq 0$, then $\tau(t)$ has only simple zeros in $-\pi < t < 0$. The positive zeros $t_1, t_2, \ldots, t_n$ satisfy the inequalities

$$\frac{2\nu - 1}{2n+1}\pi < t_\nu < \frac{2\nu + 1}{2n+1}\pi, \quad 1 \leq \nu \leq n.$$

(ii) If

$$2a_0 - a_1 > a_1 - a_2 \geq a_2 - a_3 \geq \cdots \geq a_{n-1} - a_n \geq a_n \geq 0$$

(this condition is satisfied whenever $a_0, a_1, \ldots, a_n, 0, 0$ is convex and not identically zero), then $\tau(t)$ has again n positive roots and this time they satisfy the stronger inequalities

$$\frac{2\nu - 1}{2n+1}\pi < t_\nu < \frac{\nu}{n}\pi, \quad 1 \leq \nu \leq n.$$

Szegő's results are the discrete analogues of theorems connected with a question investigated by Pólya, which we will discuss a little later.

Consider now an infinite sequence

$$c_0, c_1, \ldots, c_n, \ldots$$

generalizing the notation introduced above, set $\Delta^0 c_n = c_n$, i.e, let the sequence of differences of order zero be the original sequence itself, and for $k \geq 1$ define inductively the sequence of differences of order k by $\Delta^k c_n = \Delta^{k-1} c_n - \Delta^{k-1} c_{n+1}$. Explicitly we have

$$\Delta^k c_n = \sum_{l=0}^{n} (-1)^l \binom{k}{l} c_{n+l}.$$

The sequence (c_n) is said to be monotone of order $k \in \mathbb{N}$ if $\Delta^l c_n \geq 0$ for $0 \leq l \leq k$ and all $n \in \mathbb{N}$.

Inspired probably by the Eneström–Kakeya theorem, Fejér considered in a number of publications during the 1930's (Z. Angew. Math. Mech., **13**

(1933), 80–88; [40], No. 81, vol. II, 479–492; Trans. Amer. Math. Soc., **39** (1936), 18–59; [40], No. 89, vol. II, 581–620; Proc. Cambridge Philos. Society **31** (1935), 307–316; [40], No. 90, vol. II, 621–631; Math. Természettud. Értesítő **54** (1936), 160–176; [40], No. 92, vol. II, 640–662; Acta Sci. Math. Szeged **8** (1936) 89–115; [40], No. 94, vol. II, 679–701; Math. Természettud. Értesítő **55** (1936), 1–27; [40], No. 95, vol. II, 702–725) also in collaboration with Szegő (Prace Matematyczno Fizyczne **44** (1935), 15–25; [40], No. 91, vol. II, 631–639; [173], No. 35–3, vol. II, 579–586) power series and trigonometric series whose coefficients form a sequence which are monotone of a certain order. Let me list some of their results concerning power series

$$f(z) = c_0 + c_1 z + c_2 z^2 + \cdots + c_n z^n + \cdots$$

convergent in $|z| < 1$.

Fejér proved in the second paper quoted above that if the sequence $c_1, c_2, \ldots$ is monotone of order four then $f(z)$ is univalent ("schlicht") in $|z| < 1$. Szegő (Duke Math. J., **8** (1941), 559–564; [173], No. 41–2, vol. II, 797–802) improved this result by showing that the conclusion already holds if $c_1, c_2, \ldots$ is only assumed to be monotone of order three.

The example $1 + z + z^2 + \cdots + z^n + 0 + 0 + \cdots$ shows that the conclusion does not hold if the sequence of coefficients is monotone of order one. Both Szegő and Szidon (loc. cit.) gave examples of power series whose coefficients form a sequence monotone of order two and such that the map effectuated by the sum $f(z)$ is not univalent.

Theorem A. *If (c_n) is monotone of order two, then one has the chain of inequalities*

(24)
$$\big|f(z)\big| \geq \frac{\big|f(z) - s_0(z)\big|}{|z|} \geq \frac{\big|f(z) - s_1(z)\big|}{|z|^2} \geq \cdots \geq \frac{\big|f(z) - s_n(z)\big|}{|z|^{n+1}} \geq \cdots$$

for $|z| < 1$, where $s_n(z) = c_0 + c_1 z + \cdots + c_n z^n$ is the nth partial sum of the power series.

Observe that we have then $|s_n| \leq 2\big|f(z)\big|$ for all n. This is remarkable because Fejér has earlier given an example of a power series for which $\big|f(z)\big| < 1$ in $|z| < 1$ but $\big|s_n(z)\big|$ is unbounded (Sitzungsber. München **40** (1940), 1–17; [40], No. 32, Vol. I, 573–583; [104], §3, p. 29). Szegő (Math. Z., **25** (1926), 172–187; [173], No. 26–3, I, 758–773) proved that if $c_n > 0$, c_{n+1}/c_n increases and $\sum c_n$ diverges, then also the inequalities $|s_n| \leq 2\big|f(z)\big|$ hold.

I want to reproduce Szegö's short and beautiful proof of Theorem A. In the first place, it is sufficient to prove the first inequality in (24) since the others then follow by iteration applied to the remainders $\sum_{\nu=n+1}^{\infty} c_\nu z^\nu$. One begins by proving $\left|f(z)\right| \geq \left|f(z) - c_0\right|$, which for geometric reasons is equivalent to $\Re e f(z) \geq \frac{1}{2}c_0$ since $c_0 > 0$ by hypothesis. An Abel transformation (which is Fejér's main tool in problems about coefficients monotone of higher order) gives

$$f(z) = \sum_{n=0}^{\infty} \Delta^2 c_n . \varphi_n(z),$$

where $\varphi_n(z) = (n+1) + nz + \cdots + z^n$. We know from (10) that $\Re e \varphi_n(z) \geq \frac{n+1}{2}$. On the other hand

$$f(0) = c_0 = \sum_{n=0}^{\infty} \Delta^2 c_n . \varphi_n(0) = \sum_{n=0}^{\infty} \Delta^2 c_n (n+1),$$

hence

$$\Re e f(z) = \sum_{n=0}^{\infty} \Delta^2 c_n . \Re e \varphi_n(z) \geq \sum_{n=0}^{\infty} \Delta^2 c_n \frac{n+1}{2} = \frac{c_0}{2},$$

which proves our claim. Now

$$\left| \frac{f(z) - c_0}{f(z)} \right| \leq 1,$$

$f(0) \neq 0$ and $f(z) - c_0$ vanishes at $z = 0$, so by the Schwarz lemma

$$\left| \frac{f(z) - c_0}{f(z)} \right| \leq |z|,$$

which is the inequality we wanted to prove.

Here are some more results:

If the sequence $\{nc_n \, : \, n = 1, 2, 3, \ldots\}$ is monotone of order two, then the image of $|z| < 1$ under $f(z)$ is starlike with respect to $f(0)$.

If the sequence $\{n^2 c_n \, : \, n = 1, 2, 3, \ldots\}$ is monotone of order two, then the image of $|z| < 1$ is convex.

If the sequence (c_n) is monotone of order $k+1$, then $f^{(l)}(x) \geq 0$ for $-1 < x < 1$ and $0 \leq l \leq k$.

Fejér also proved results concerning series of the form $\sum_{n=0}^{\infty} c_n z^{2n+1}$ whose coefficients are monotone of higher order.

Turán (Proc. Cambridge Philos. Soc., **34** (1938), 134–143; [184], No. 15) continued the investigations of Fejér and Szegő. Fejér proved that is $c_n \geq 0$ and $c_{n+1} - c_n \geq 0$, then the inequalities

$$\big|f(z) - s_n(z)\big| \geq \big|f(z) - s_{n+1}(z)\big| \tag{25}$$

hold in the disk $|z| < \frac{1}{2}$. Turán proves that if $c_{k!} = 1$, $c_{k!+1} = k!$ $(k \geq 1$ and $c_n = \frac{1}{n!}$ for $n \neq k!, k!+1$, then the inequalities (25) do not hold in any disk with center 0. On the other hand, there exists $\rho > 0$ depending only on $\max \sqrt[n]{|c_n|}$ such that the means

$$S_n(z) = \frac{s_0(z) + s_1(z) + \cdots + s_n(z)}{n+1}$$

satisfy $\big|f(z) - S_n(z)\big| \geq \big|f(z) - S_{n+1}(z)\big|$ for $|z| < \rho$, $n = 1, 2, 3, \ldots$.

Consider the partial sums

$$g_n(z) = g_n^{(0)}(z) = 1 + z + z^2 + \cdots + z^n$$

of the infinite geometric progression, and the iterated partial sums defined inductively by

$$g_n^{(r)}(z) = g_0^{(r-1)}(z) + g_1^{(r-1)}(z) + \cdots + g_n^{(r-1)}(z)$$

$(r = 1, 2, \ldots; n = 0, 1, 2, \ldots)$. Fejér's proofs are based in part on the fact that the derivatives of order $\leq r$ of any $g_n^{(r)}(x)$ are strictly positive for $-1 < x < 1$, $n \geq r$. Fejér observed that to prove this it is enough to show that $D^r g_n^{(r)}(x) > 0$ for $-1 < x < 1, n \geq r$. Turán (Publ. Math. Debrecen **1** (1949), 95–97; [184], No. 42) strengthens this by showing that all the zeros of $D^r g_n^{(r)}(z)$ lie on the unit cercle $|z| = 1$; for $r = 1$ this was proved by Egerváry (Math. Z., **42** (1937), 221–230). Turán also proves that $D^r g_n^{(r)}(x) > 0$ for all $x \in \mathbb{R}$ if n is even, and that $D^r g_n^{(r)}(x)$ has a simple zero at $x = -1$ and is > 0 for all real $x \neq -1$ if n is odd.

Writing

$$\big|f(re^{i\theta})\big|^2 = f(re^{i\theta})f(re^{-i\theta}) = \sum_{n=0}^{\infty} P_n(\cos\theta) r^n$$

we have

$$P_n(\cos\theta) = \sum_{l=0}^{n} c_l c_{n-l} e^{i(2n-l)\theta}.$$

The expressions P_n are polynomials of degree n in the variable $x = \cos\theta$ which Fejér calls the generalized Legendre polynomials associated with $f(z)$ or with the sequence (c_n) (Math. Z., **24** (1925), 285–298; [40], No. 64, vol. II, 161–175). If $f(z) = (1-z)^{-\rho}$, then the $P_n(x)$ are the ultraspherical (or Gegenbauer) polynomials of index $\rho \in \mathbb{R}$. The special case $\rho = \frac{1}{2}$ yields the classical Legendre polynomials. For $\rho = 1$ we get the Čebishov polynomials

$$U_n(x) = \frac{\sin(n+1)\theta}{\sin\theta},$$

and interpreting the case $\rho = 0$ carefully one gets the Čebishov polynomials $T_n(x) = \cos n\theta$.

Fejér proves that if the terms of the sequence (c_n) are positive and decrease with n, then the the generalized Legendre polynomials satisfy the inequalities

$$|P_n(x)| \leq \frac{4}{\sin\theta} c_{[n/2]},$$

where $[x]$ is the integral part of x. In the case of the classical Legendre polynomials this becomes

$$|P_n(\cos\theta)| \leq \frac{C}{\sqrt{n}\sin\theta}. \tag{26}$$

For the classical Legendre polynomials Stieltjes gave a proof of the inequality

$$|P_n(\cos\theta) - P_{n+2}(\cos\theta)| \leq \frac{C}{\sqrt{n}}.$$

Fejér shows that Stieltjes' proof also serves to obtain an analogous inequality for his generalized Legendre polynomials.

In the case of the classical Legendre polynomials the inequality

$$|P_n(\cos\theta)| \leq \frac{C}{\sqrt{n}\sqrt{\sin\theta}},$$

stronger than (26), also holds. Fejér gives a "short and elementary" proof of this inequality. It was, however, Szegő who, with the help of his result giving

$|s_n(z)| \leq 2|f(z)|$ quoted above, proved that under the same hypotheses on (c_n) the generalized Legendre polynomials satisfy

$$|P_n(\cos\theta)| \leq 8c_{[n/2]}|f(e^{2i\theta})|$$

for $0 < \theta < \pi, n \in \mathbb{N}$. He also showed that the asymptotic formula

$$P_n(\cos n\theta) = 2c_n \Re e e^{-in\theta} f(e^{2i\theta}) + \mathrm{o}(c_n)$$

holds uniformly in $\varepsilon < \theta < 2\pi - \varepsilon$ $(\varepsilon > 0)$ whenever $c_n > 0, (c_n)$ decreases, $\lim(c_{n+1}/c_n) = 1$ and $c_n = \mathrm{O}(c_{2n})$.

Fejér ([40], vol. II, p. 699 and p. 721) proved that if the sequence of coefficients (c_n) is monotone of order two, then the arithmetic means of the partial sums of the series $\sum_{n=0}^{\infty} P_n(\cos\theta)$, whose terms are the Legendre polynomials, are all posititive.

There is a treatment of Fejér's generalized Legendre polynomials in Szegő's book ([174], p. 134).

If in $U(z) = \int_0^{\infty} f(t)\cos zt\, dt$ we substitute

$$f(t) = \sum_{n=1}^{\infty} \left(4\pi^2 n^4 e^{\frac{9}{2}t} - 6\pi n^2 e^{\frac{5}{2}t}\right) e^{-n^2\pi e^{2t}},$$

then $U(z)$ becomes the Riemann function $\xi(z)$. It is known that the Riemann hypothesis is equivalent to the fact that $\xi(z)$ has only real zeros, and this was the motivation for György Pólya to investigate which entire functions defined by trigonometric integrals have only real zeros. He consecrated a number of articles to this problem, let me just refer to J. Reine Angew. Math., **158**(1927), 6–18 ([128], vol. II), where further references can be found. Of course such entire functions arise also in other situations, e.g., in the case of Bessel functions.

In an early article (Math. Z., **2** (1918), 352–383; [128], vol. II, pp. 166–197) Pólya considers a strictly positive increasing function $f(t)$ defined in $0 \leq t < 1$ such that $\int_0^1 f(t)\, dt$ exists. By the Eneström–Kakeya theorem the polynomial

$$f(0) + f\left(\frac{1}{n}\right) z + \cdots + f\left(\frac{n-1}{n}\right) z^{n-1}$$

has all its zeros in $|z| \leq 1$, and since $\frac{1}{n}\sum_{\nu=0}^{n-1} f\left(\frac{\nu}{n}\right) e^{\frac{\nu}{n}z}$ converges to $\int_0^1 f(t)e^{zt}\, dt$, it seems plausible that all the zeros of the function $W(z)$ defined by this integral lie in $\Re e z \leq 0$. Pólya proves this not by passing to the limit but directly by imitating the proof of the Eneström–Kakeya theorem. Actually the zeros of $W(z)$ lie in the open half-plane $\Re e z < 0$ unless $f(t)$ is in the "exceptional case". i.e., it is a step-function having a finite number of jumps at points with rational abscissa.

Next assume that the coefficients of the polynomial $P(z) = a_0 + a_1 z + \cdots + a_n z^n$ are real, $a_n > 0$, and that its zeros are all in $|z| < 1$ (i.e., we are in the situation of the conclusion of the Eneström–Kakeya theorem). Then the trigonometric polynomials

$$u(t) = \Re e P(e^{it}) = a_0 + a_1 \cos t + \cdots + a_n \cos nt$$

and

$$v(t) = \Im m P(e^{it}) = a_1 \sin t + \cdots + a_n \sin nt$$

have exactly $2n$ simple roots in the interval $0 \leq t < 2\pi$, consequently all their roots are real. For the proof (cf. also [129], III. 179) let $z_1, \ldots, z_n$ be the zeros of $P(z)$, each listed as often as its multiplicity indicates. Write $e^{it} - z_\nu = \rho_\nu(t) e^{i\psi_\nu(t)}$, where $\psi_\nu(t)$ increase by 2π as t increases from 0 to 2π. Then

$$P(e^{it}) = a_n \prod_{\nu=1}^{n} \rho_\nu(t) e^{i\Psi(t)} = R(t) e^{i\Psi(t)},$$

where $\Psi(t) = \sum_{\nu=1}^{n} \psi_\nu(t)$ increases by $2n\pi$ as t goes from 0 to 2π. Since $u(t) = R(t)\cos\Psi(t)$ and $v(t) = R(t)\sin\Psi(t)$, the claim is proved because both $\cos\theta$ and $\sin\theta$ have $2n$ zeros in any half-open interval of length $2n\pi$.

Keeping the above hypotheses concerning $f(t)$, introduce with Pólya the entire functions

$$U(z) = \int_0^1 f(t) \cos zt \, dt, \quad V(z) = \int_0^1 f(t) \sin zt \, dt.$$

It follows from the Eneström–Kakeya theorem, combined with the proposition just proved, that the polynomials

$$U_n(z) = \frac{1}{n} \sum_{\nu=0}^{n-1} f\left(\frac{\nu}{n}\right) e^{\frac{\nu}{n^2}} \cos \frac{\nu z}{n} \quad \text{and} \quad V_n(z) = \frac{1}{n} \sum_{\nu=0}^{n-1} f\left(\frac{\nu}{n}\right) e^{\frac{\nu}{n^2}} \sin \frac{\nu z}{n}$$

have only real zeros. Passing to the limit, Pólya obtains that the functions $U(z), V(z)$, and more generally the functions $\alpha U(z) + \beta V(z)$, where α, β are real constants not both zero, have only real roots. If $f(t)$ is not in the "exceptional case", then the zeros are simple, furthermore $U(z)$ and

$V(z)$ have no common zero because otherwise $W(z)$ would have a purely imaginary zero, which is excluded by the first reult.

In Szegő's Transaction article discussed above, in which he presents inequalities for zeros of trigonometric polynomials, he says: "the elementary inequalities ... lead in a direct way to a theorem of Pólya, giving even a slightly more precise result". He proves the following:

Let α, β be two real constants, not both zero, and set $\alpha + i\beta = \rho e^{i\delta}$ ($\rho > 0$, $0 < \delta \leq 2\pi$). Then the entire function $\alpha U(z) + \beta V(z)$ has only real simple zeros; every interval $\big((k-\frac{1}{2})\pi\big)+\delta$, $\big((k+\frac{1}{2}\pi)+\delta\big)$, $k = 0, \pm 1, \pm 2, \ldots$, except the one containing the origin, contains exactly one zero. (If $\alpha = 0$, i.e. $\delta = \frac{\pi}{2}$, then $V(z) = 0$ at $z = 0$ and there are no zeros in $(0, \pi)$.) The only exception is when $f(t)$ is a step function with jumps at the points

$$t = 1 - \frac{2\pi h}{(k - \frac{1}{2})\pi + \delta},$$

h, k integers; then the zeros of $\alpha U(z) + \beta V(z)$ are still real and lie in the closed intervals $\big[(k - \frac{1}{2})\pi + \delta, (k + \frac{1}{2})\pi + \delta\big]$.

Both Pólya and Szegő have further results concerning the regularity with which the zeros of $U(z)$ and $V(z)$ are distributed. For instance if $f(t)$ satisfies the additional condition to be convex, then $V(0) = 0$ and $V(z)$ has exactly one simple zero in each interval $\big(k\pi, (k + \frac{1}{2})\pi\big)$, $k = 1, 2, 3, \ldots$.

Pólya points out that if $f(t) > 0$ and is decreasing, then $\int_0^\infty f(t) \sin zt\, dt$ does not vanish for any $z > 0$, and $\int_0^\infty f(t) \cos zt\, dt$ has no real zeros. A decreasing $f(t)$ figures also in the following theorem of Alfréd Rényi (C.R. Acad. Bulgare Sci., **3** (1950), 9–11; [151], No. 38, I, pp. 199–201) which generalizes results of L. Ilieff:

Let n and m be two positive integers such that $n + m$ is odd. Let $f(t)$ be n times differentiable in $0 < t \leq 1$ and satisfy the following conditions: $f(1) = 0, f^{(k)}(1) = 0$ for $1 \leq k \leq n-1, f^{(2k+1)}(0) = 0$ for $1 \leq 2k+1 \leq n-1$, the function $t^{-m} f^{(n)}(t)$ is positive, increasing and integrable in $0 \leq t \leq 1$. Then $U(z)$ and $V(z)$ have only real zeros.

8. Power series: singularities and analytic continuation

György Pólya begins his influential article "Untersuchungen über Lücken und Singularitäten von Potenzreihen" (Math. Z., **29** (1929), 549–640; [128], I, pp. 363–454) by saying that it is directed towards the "Hadamardsche Aufgabe der Funktionentheorie", i.e., its purpose is to obtain from properties of the sequence $c_0, c_1, \dots, c_n, \dots$ conclusions concerning the behavior of the function $f(z)$ represented by the power series

$$\sum_{n=0}^{\infty} c_n z^n \tag{27}$$

in the open disk where it converges.

The most classical example is of course the theorem found by Augustin Louis Cauchy, and made precise by Jacques Hadamard, according to which if

$$\limsup \sqrt[n]{|c_n|} = l,$$

then $f(z)$ has a singularity at some point z_0 with $|z_0| = 1/l$. This theorem is usually stated saying the (27) has a radius of convergence $r = 1/l$.

Equally famous is the gap theorem of Hadamard: if (λ_k) is a sequence of positive integers such that

$$\frac{\lambda_{k+1}}{\lambda_k} \geq q \tag{28}$$

for some $q > 1$ and all $k \geq 1$ then the "lacunary series"

$$\sum_{k=1}^{\infty} c_k z^{\lambda_k} \tag{29}$$

cannot be continued analytically beyond its circle of convergence, i.e., the function it represents has a singularity at every point of the said circle.

Pál Turán (*On the gap theorem of Fabry*, Hungarica Acta Math., **1** (1947), 21–29; [184], No. 30) says: "...one of the most exciting parts of the Weierstrassian theory of functions is the group of those theorems which draw conclusions from the lacunary distribution of the exponents λ_k

to the impossibility of analytical continuation over the convergence-circle, i.e.... the gap theorems". Hadamard's condition (28) was weakened to

$$\lim_{k\to\infty} (\lambda_{k+1} - \lambda_k) = \infty \tag{30}$$

and Edouard Fabry introduced the even weaker condition

$$\lim_{k\to\infty} \frac{\lambda_k}{k} = \infty \tag{31}$$

([104], §19, Satz 2, p. 83).

It seems to have been known for a long time that the deep reason why Hadamard's gap theorem holds is the inequality

$$\max_{0\leq\theta\leq 2\pi} \left| \sum_{\nu=1}^{n} c_\nu e^{i\lambda_\nu\theta} \right| \leq C(q,\delta) \max_{\alpha\leq\theta\leq\alpha+\delta} \left| \sum_{\nu=1}^{n} c_\nu e^{i\lambda_\nu\theta} \right|, \tag{32}$$

and Norbert Wiener proved in 1933 that the gap theorem under condition (30) follows from an inequality which is the L^2-norm analogue of (32). He and R. E. A. C. Paley asked: from which similar inequality does the Fabry theorem follow? Turán (loc. cit.) showed that the answer is (32) provided the factor $C(q,\delta)$ is replaced by

$$\left(\frac{48\pi}{\delta}\right)^n .$$

This more precise inequality is one of the first applications of Turán's "new method" of which more will be said below. Actually, in his book ([187], Section 20, pp. 219–220) Turán proves the generalization of Fabry's theorem to Dirichlet series, first proved by Ottó Szász (Math. Ann., **85** (1922), 99–110; [172], pp. 503–514).

Let me list some early results "of classical beauty" (Landau) due to Hungarian mathematicians about power series.

Assume that the radius of convergence of (27) is $0 < r < \infty$. Giulio Vivanti stated and Alfred Pringsheim proved that if $c_n \geq 0$ for all (large) n, then the function represented by (27) in $|z| < r$ has a singularity at the point $z = r$. Indeed, the Cauchy-Hadamard theorem and a simple calculation show that the radius of convergence of

$$\sum_{n=0}^{\infty} \frac{1}{n!} f^{(n)}\left(\frac{r}{2}\right)\left(z - \frac{r}{2}\right)^n$$

equal to $\frac{1}{2}r$ ([104], §17).

In a long paper (J. Math. Pures Appl. (6) **5** (1909), 327–413) Pál Dienes, who later wrote two books on power series ([27], [28]), showed that it is sufficient to assume that the c_n lie in an angular domain with vertex 0 and opening $< \pi$.

Ottó Szász (loc. cit.) considered

$$f(z) = \sum_{n=0}^{\infty} c_n z^n \quad \text{and} \quad g(z) = \sum_{n=0}^{\infty} \Re e c_n . z^n,$$

where he assumed that the two power series have the same radius of convergence $0 < r < \infty$. If $g(z)$ has a singularity at $z = r$, so does $f(z)$. From here the theorem of Dienes follows immediately: Assume, as we may, that $-\alpha \leq \arg c_n \leq \alpha$, where $\alpha < \frac{\pi}{2}$. Then $0 \leq \Re e c_n \leq |c_n| \leq \Re e c_n / \cos\alpha$, so the two power series have the same radius of convergence r, and $g(z)$ has a singularity at $z = r$ by the Vivanti–Pringsheim Theorem. Again Szász proved the result even for Dirichlet series.

Let $f(z)$ be the function represented by the power series (27) in the disk $|z| < r$ and assume that $\left|f(z)\right| \leq M$ there. In analogy with his theorem on trigonometric series, Fejér proved that if

$$s_n(z) = c_0 + c_1 z + \cdots + c_n z^n$$

is the partial sum of order n of (27), then

$$\left|s_0(z) + s_1(z) + \cdots + s_n(z)\right| \leq (n+1)M \tag{33}$$

for $n \geq 0$ and $|z| \leq r$ (special case of Theorem VII in Rend. Circ. Mat. Palermo **38** (1914), 79–97; [40], No. 46, I, pp. 783–802; see also [104], §1). Ottó Szász (Math. Z., **1** (1918), 163–183; [172], 446–466) and Issai Schur proved that the conditions

$$\left|s_0(z)\right| + \left|s_1(z)\right| + \cdots + |s_n| \leq (n+1)M$$

and

$$\left|s_0(z)\right|^2 + \left|s_1(z)\right|^2 + \cdots + \left|s_n(z)\right|^2 \leq (n+1)M^2$$

for $|z| \leq r$ are equivalent to (33) and to $\left|f(z)\right| \leq M$ ([104], §1).

On the other hand, modifying his example of a continuous function with a divergent Fourier series, Fejér gave an example of a function analytic in

$|z| < 1$, continuous (hence bounded) in $|z| \leq 1$, for which the partial sums of its power series expansion are unbounded (Sitzungsber. Bayrische Akad. Wiss. Math.-Phys. Kl., **40** (1910), 1–17; [40], No. 32, I, pp. 573–583; [104], §3).

Assume that the series

$$\sum_{n=0}^{\infty} c_n \tag{34}$$

whose terms are complex numbers, is Cesàro-summable of order 1, i.e., that the sequence of arithmetic means

$$\frac{s_0 + s_1 + \cdots + s_n}{n+1}$$

of the partial sums $s_k = c_0 + c_1 + \cdots + c_k$ converges as $n \to \infty$. The Tauberian theorem of Hardy and Landau states that if furthermore

$$c_n/n > -K \tag{35}$$

for some $K > 0$ and all n, then (34) is convergent. Therefore the Dirichlet–Jordan criterion for the convergence of the Fourier series of a function of bounded variation follows from Fejér's summability theorem.

Fejér replaced the Tauberian condition (35) by

$$\sum_{n=0}^{\infty} n|c_n|^2 < \infty.$$

Simple examples show that neither condition implies the other. Landau proves ([104], §13) that instead of Cesàro summability of (34) it is sufficient to assume Abel-summability.

This theorem has the following consequence: Let the function $f(z)$ represented by (27) be regular and univalent in the disk $|z| < 1$. If the area

$$\int\int_{|z|<1} \left| f'(z) \right|^2 dx\, dy = \pi \sum_{n=0}^{\infty} n|c_n|^2$$

of the image of the unit disk is finite, then $\sum_{n=0}^{\infty} c_n e^{in\varphi}$ converges for almost every value of φ (Schwarz–Festschrift, Mathematische Abhandlungen (1914), 42–53; [40], No. 49, I, pp. 813–822).

In general there is no connection between the behavior of $f(z)$ and the convergence of its power series expansion on the circle of convergence: $\sum z^n$ diverges for all $|z| = 1$, its sum $(1-z)^{-1}$ is regular at $z = -1$ and has a pole at $z = 1$; $\sum n^{-1} z^n$ converges at $z = -1$ and $-\log(1-z)$ is regular there; $\sum n^{-2} z^n$ converges for all $|z| = 1$ but by Vivanti–Pringsheim has a singularity at $z = 1$. Clearly if (27) converges at a single point of its circle of convergence $|z| = r$, then necessarily $\lim_{n\to\infty} c_n r^n = 0$. It was a remarkable result of Pierre Fatou that if $f(z)$ is regular at a point z_0 of the circle of convergence, then the necessary condition $c_n |z_0|^n \to 0$ is also sufficient, i.e., $\sum c_n z_0^n$ converges.

Marcel Riesz (J. Reine Angew. Math., **140**(1911), 89–99; Nachrichten der Königlichen Gesellschaft der Wissenschaften zu Göttingen, math.-phys. Klasse 1916, 62–65; [158], No. 9, pp. 76–86, No. 17, pp. 145–148) gave a simple proof of Fatou's theorem ([104], §13) and extended it in several ways (Ark. Mat. Astr. Fys. 11, no. 12 (1916), 1–16; [158], No. 18, 149–164). For the sake of brevity, assume without loss of generality that the radius of convergence is 1. If $\lim_{n\to\infty} c_n = 0$ and $f(z)$ is regular on the closed arc $\{e^{i\varphi} : \alpha \leq \varphi \leq \beta\}$, then the convergence of $\sum c_n e^{in\varphi}$ is uniform for $\alpha \leq \varphi \leq \beta$. If $\lim_{n\to\infty} c_n/n^k = 0$ for some $k \geq 0$, then $\sum c_n e^{in\varphi}$ is Cesàro-summable of order k, uniformly on every closed arc on which $f(z)$ is regular. If c_n/n^k is only bounded, then the Cesàro-sums of order k of $\sum c_n e^{in\varphi}$ are bounded. He also generalized these results to Dirichlet series (C.R. Acad. Sci. Paris **149** (1909, 909–912; [158], No. 7, pp. 62–64).

Answering a question raised by Gösta Mittag-Leffler at the 1908 International Mathematical Congress in Rome, Marcel Riesz proved the following theorem (Rend. Circ. Mat. Palermo **30** (1910), 339–345; [158], No. 8, 65-71; [104], §12): Let $R > 1$, $0 < \alpha < \frac{\pi}{2}$ and S the closed set defined by the inequalities $|z| \leq R$, $\alpha \leq \arg(z-1) \leq 2\pi - \alpha$. Assume that $f(z)$ is continuous on S and that it is regular on S with the exception of $z = 1$. If for $|z| < 1$ the series (27) converges to $f(z)$, then it converges uniformly to $f(z)$ on $|z| = 1$, and in particular $f(1) = \sum c_n$.

Fatou conjectured that given a power series (27) with radius of convergence 1, there exists a sequence $\varepsilon_n = \pm 1$ such that $\sum \varepsilon_n c_n z^n$ cannot be continued analytically beyond the unit circle. Pólya sent a proof of the conjecture in a letter to Adolf Hurwitz, who answered with another proof. The two letters were published in a joint article it is one of Pólya's first publications on power series (Acta Math., **40** (1916), 173–183; [128], I, pp. 17–21). Landau ([104], §20) calls the result Pólya's theorem and reproduces the proof of Hurwitz. Later Pólya (Acta Sci. Math. Szeged **12B** (1950), 199–203; [128], I, pp. 720–724) writing in honor of the 70th birthday of Fejér and F. Riesz showed that if (27) converges but is not a polynomial, then $\varepsilon_n = \pm 1$ can be chosen in such a way that $\sum \varepsilon_n c_n z^n$ satisfies no algebraic differential equation.

As is apparent from the above references, Landau's delightful little book [104] treats the work of several Hungarian mathematicians. A similar source of some material is a small book by Hadamard, whose second edition, written in collaboration with Szolem Mandelbrojt, appeared in 1926 [64]. The second edition of Landau's book was published in 1929, the same year as the fundamental paper of Pólya quoted at the beginning of this section appeared. Ludwig Bieberbach in his 1955 Ergebnisse volume [14] reports on the "very lively" progress made in the course of the preceding twenty-five years "in great part under the influence of Pólya's first gap-paper". Bieberbach discusses the work of many Hungarian mathematicians: Manó Beke, Dienes, Erdős, Fekete, Pólya, Marcel Riesz, Ottó Szász, Szegő, Turán. In particular Pólya's name appears on 36 out of the 155 pages of the book.

Pólya approaches the study of lacunary power series through the theory of entire functions of exponential type. The book [15] of Ralph P. Boas, which appeared at the same time as [14], contains a succinct presentation of the theory in Chapter 5, however, Boas himself says on p. 789 of [128], vol. I that Chapter 2 of Pólya's 1929 paper "is still a nearly complete and very readable exposition".

Let

$$f(z) = \sum_{n=0}^{\infty} c_n z^n$$

be an entire function and set

$$M(r) = \max_{|z|=r} \left| f(z) \right|.$$

The order $0 \leq \rho \leq \infty$ of $f(z)$ is defined by

$$\rho = \limsup_{r \to \infty} \frac{\log \log M(r)}{\log r}.$$

If $0 < \rho < \infty$, then the type of $f(z)$ is the number

$$\tau = \limsup_{r \to \infty} r^{-\rho} \log M(r).$$

According to the terminology introduced by Pólya, $f(z)$ is of exponential type if either $\rho = 1$ and τ is finite, or if $\rho < 1$, i.e., if

$$\left| f(z) \right| \le e^{a|z|}$$

for some $a > 0$ and large $|z|$. If

$$h(\varphi) = \limsup_{r \to \infty} r^{-1} \log \left| f(re^{i\varphi}) \right|$$

is the Phragmén–Lindelöf indicator function of $f(z)$, then the intersection of the half-planes

$$x \cos \varphi + y \sin \varphi \le h(\varphi) \quad (z = x + iy)$$

as φ varies, is the indicator diagram $\mathcal{D}$ of $f(z)$, and its reflection $\overline{\mathcal{D}}$ in the real axis is the conjugate indicator diagram. The function

$$F(z) = \frac{c_0}{z} + \frac{1!c_1}{z^2} + \frac{2!c_2}{z^3} + \cdots + \frac{n!c_n}{z^{n+1}} + \cdots \tag{36}$$

is regular outside $\overline{\mathcal{D}}$, and the series (36) converges for $|z| > \tau$. It is called the Borel–Laplace transform of $f(z)$ and is given by

$$F(z) = \int_0^\infty f(t) e^{-zt}\, dt$$

for $x > \tau$. Inversion gives gives the Pólya representation

$$f(z) = \frac{1}{2\pi i} \oint_C F(\zeta) e^{\zeta z} d\zeta,$$

where C is a contour containing $\overline{\mathcal{D}}$ in its interior.

Let (λ_k) be a sequence of real numbers such that $\lambda_0 > 0$ and $\lambda_{k+1} - \lambda_k \ge c > 0$. For $t \ge 0$ let $N(t)$ be the number of the λ_k with $\lambda_k \le t$. The *density* of the sequence is defined by

$$D = \lim_{k \to \infty} \frac{k}{\lambda_k} = \lim_{t \to \infty} \frac{N(t)}{t}$$

if it exists. Thus the condition in Fabry's gap theorem means that $D = 0$. The *upper density* of (λ_k) given by

$$\overline{D} = \limsup_{k \to \infty} \frac{k}{\lambda_k}$$

always exists, and so does the *lower density* $\underline{D}$ obtained when lim sup is replaced by lim inf. Pólya introduces further the *maximum density*

$$\overline{\Delta} = \lim_{s \to 1-} \limsup_{t \to \infty} \frac{N(t) - N(st)}{t - st}$$

and the *minimum density* $\underline{\Delta}$ in whose definition liminf replaces limsup. One always has

$$\underline{\Delta} \leq \underline{D} \leq \overline{D} \leq \overline{\Delta}$$

and all four are equal if D exists.

André Bloch pointed out that there is a strong analogy between the study of singularities of a function on the circle of convergence of its power series expansion, and the study of the Julia directions of an entire function. Let me recall that α is a Julia direction (sometimes called a Picard direction) of $f(z)$ if $f(z)$ assumes every complex value with at most one exception in the angular region $\alpha - \delta \leq \varphi \leq \alpha + \delta, r \geq 0$, where $z = re^{i\varphi}$ and $\delta > 0$ is arbitrarily small. E.g., $\pm\frac{\pi}{2}$ are Julia directions for e^z. Pólya presents a parallel treatment of the two questions. One of the main results in Chapter 3 of his 1929 paper is the following:

Theorem IV. *Let*

$$G(z) = \sum c_{\lambda_k} z^{\lambda_k} \quad (c_{\lambda_k} \neq 0)$$

be an entire function of order ρ and type τ, and let the maximum density of the exponents λ_k be $\overline{\Delta}$ (Pólya calls it the maximum density of the non-vanishing coefficients, or sometimes simply of the coefficients: "maximale Koeffizientendichte"). Writing

$$M(r; \alpha, \beta) = \max_{\alpha \leq \varphi \leq \beta} \left| G(re^{i\varphi}) \right| \tag{37}$$

for $\alpha < \beta$, define

$$\rho(\alpha, \beta) = \limsup_{r \to \infty} \frac{\log \log M(r; \alpha, \beta)}{\log r}$$

and

$$\tau(\alpha, \beta) = \limsup_{r \to \infty} r^{-\rho(\alpha,\beta)} \log M(r; \alpha, \beta).$$

If $\beta - \alpha > 2\pi\overline{\Delta}$, then $\rho(\alpha, \beta) = \rho$ and $\tau(\alpha, \beta) = \tau$.

In order to obtain a theorem concerning the singularities of a lacunary power series he uses the following result which he ascribes to Émile Borel:

Lemma a. *Let*

$$f(z) = c_0 + c_1 z + \cdots + c_n z^n + \cdots$$

be an entire function of order 1 *and type* $0 < \tau < \infty$. *We have*

$$h(\alpha) = \limsup_{r \to \infty} r^{-1} \log \left| f(re^{i\alpha}) \right| = \tau$$

if and only if the half-line $z = re^{i\alpha}$ $(0 \le r < \infty)$ *meets the circle of convergence of*

$$\frac{1}{z} F\left(\frac{1}{z}\right) = c_0 + 1!c_1 z + 2!c_2 z^2 + \cdots + n!c_n z^n + \cdots$$

in a singular point.

This leads to the famous result

Theorem IVa. *Let the maximum density of the nonvanishing coefficients of a power series with finite radius of convergence be* $\overline{\Delta}$. *Then every closed arc of the circle of convergence, whose central angle equals* $2\pi\overline{\Delta}$, *contains a singular point of the function represented by the power series.*

$\overline{\Delta} = 0$ is equivalent to $D = 0$, so Fabry's gap theorem is a special case. The function $(1 - z^k)^{-1} = \sum z^{kn}$ with $D = 1/k$ illustrates Theorem IVa nicely.

In order to obtain a theorem concerning Julia lines, Pólya uses the following result of Bieberbach:

Lemma b. *Let* $G(z)$ *be an entire function of infinite order, and* $M(r; \alpha, \beta)$ *as in* (37). *If* α *is such that*

$$\limsup_{r \to \infty} \frac{\log \log M(r; \alpha - \delta, \alpha + \delta)}{\log r} = \infty$$

for any $\delta > 0$, *then* α *is a Julia direction of* $G(z)$.

He obtains:

Theorem IVb. *If $f(z)$ is an entire function of order ∞, and the maximum density of the nonvanishing coefficients of its power series expansion is $\overline{\Delta}$, then every closed angle with opening $2\pi\overline{\Delta}$ contains a Julia direction.*

Pólya proves a theorem (Theorem II) which is of the same nature as Theorem IV, and deduces from it with the help of Lemma a the Vivanti–Pringsheim–Dienes theorem. The parallel result is:

If all the coefficients of the power series expansion of an entire function of infinite order lie in an angular domain with vertex 0 and opening $< \pi$, then the direction of the positive real axis is a Julia direction.

Related to Lemma b is the comparision of $\log M(r)$ and of $\log M(r; \alpha, \beta)$ for small $\beta - \alpha$ when the power series is lacunary. The study of this problem was pursued by Turán and by Tamás Kővári, see Chapter 21 of [187] and the references quoted there. Pál Erdős and Kővári (Acta Math. Acad. Sci. Hungar., **7** (1957), 305–317) proved that for any maximum modulus $M(r) = \max_{|z|=r} |f(z)|$ of an entire function there exists a series $N(r) = \sum \gamma_n r^n$ with $\gamma_n \geq 0$ such that $e^{-\varepsilon} < M(r)/N(r) < e^{\varepsilon}$ with $\varepsilon = 0.005$.

Let $f(z)$ be represented by the power series (27) and set $\mu_n = \inf \frac{M(r)}{r^n}$. By the Cauchy inequality $|c_n| \leq \mu_n$. Vincze (Acta. Sci. Math. Szeged **19** (1958), 129–140) proved that $\sum \frac{|c_n|}{\mu_n} = \infty$, i.e. $|c_n|$ cannot always be much smaller than μ_n.

Let $f(z) = \sum c_k z^{\lambda_k}$ have radius of convergence $R > 0$ and assume that the λ_k satisfy the Fabry gap condition. For any $z_0 \neq 0$ with $|z_0| < R$ the expansion

$$\sum a_n (z - z_0)^n \tag{38}$$

has only one singularity of $f(z)$ on its circle of convergence C, namely the point where C touches $|z| = R$. Therefore (38) cannot be a lacunary series satisfying the Fabry condition. More precisely, Kővári (J. London Math. Soc., **34** (1959), 185–194) proved with a geometric argument using Theorem IVa of Pólya that if the radius of convergence of $f(z) = \sum c_n z^n$ is 1 and the maximum density of its nonvanishing coefficients is $\overline{\Delta}$, while for $z_0 = re^{i\alpha} \neq 0$ $(r < 1)$ the maximum density of the nonvanishing coefficients of $\sum a_n (z - z_0)^n$ is $\overline{\Delta}_0$, then $\overline{\Delta} + \overline{\Delta}_0 \geq 1 - \frac{1}{\pi} \arcsin r$.

Pólya conjectured that the power series expansions of an entire function $f(z)$ at two distinct points cannot both have Fabry gaps. This was proved by Kató (Catherine) Rényi (Acta Math. Acad. Sci. Hungar., **7** (1956), 145–150). For $a \in \mathbb{C}$ denote by $Z_a(n)$ the number of terms of the sequence $f(a), f'(a), \ldots, f^{(n)}(a)$ which are equal to zero. The Fabry gap condition at a means that

$$\lim_{n\to\infty} \frac{Z_a(n)}{n} = 1.$$

Kató Rényi proves that if $a \neq b$, then

$$\liminf_{n\to\infty} \frac{Z_a(n) + Z_b(n)}{n} \leq 1.$$

In particular the power series of a periodic entire function (e.g. e^z, $\cos z$, $\sin z$) cannot have Fabry gaps at any point. In a later article (ibid., **8** (1957), 227–233) she proved that if $f(z)$ has finite order $\rho \geq 1$ then

$$\liminf_{n\to\infty} \frac{Z_a(n) + Z_b(n) - n}{n^{1-\frac{1}{\rho+\varepsilon}}} \leq 0$$

for any $\varepsilon > 0$, and if $f(z)$ has furthermore finite type $\tau \geq 0$, then

$$\liminf_{n\to\infty} \frac{Z_a(n) + Z_b(n) - n}{n^{1-\frac{1}{\rho}}} \leq \frac{|b-a|}{2} e^2 \left(\frac{\tau}{e}\right)^{\frac{1}{\rho}}.$$

Kató Rényi returned to the topic several times, and studied also lacunary power series of two variables (Colloq. Math., **11** (1964), 165–171). This is one of the first articles of a Hungarian mathematician on analytic functions of several complex variables. Turán encouraged the research in this direction, which then produced contributions by younger mathematicians, e.g., László Lempert and László Sztacho.

The proofs of Kató Rényi's theorems are related to another area of Pólya's interests: "The zeros of derivatives of a function and its analytic character". In 1942 he gave an address with this title to the American Mathematical Society (Bull. Amer. Math. Soc., **49** (1943), 178–191; [128], II, pp. 394–407). In this lecture he summarized the known results, and also stated some new results and conjectures. One of the theorems stated without proof, and used by Kató Rényi is the following: Let $f(z)$ be an entire function which is real for real z, and denote by $\mathcal{N}(n)$ the number of zeros of $f^{(n)}(z)$ in the closed interval $[0, 1]$. Then

$$\liminf_{n\to\infty} \frac{\mathcal{N}(n)}{n} = 0.$$

It was proved by Erdős and Alfréd Rényi (Acta Math. Acad. Sci. Hungar., **7** (1956), 125–141) even when $\mathcal{N}(n)$ denotes the number of zeros is $|z| \le 1$. Later (ibid., **8**(1957), 223–225) they proved that if $f(z)$ is an arbitrary entire function, and $r = H(s)$ is the inverse function of $s = \log M(r)$, then

$$\liminf_{n\to\infty} \frac{\mathcal{N}(n).H(n)}{n} \le e^2,$$

and if $f(z)$ is of finite order $\rho \ge 1$, then the right hand side can be replaced by $e^{2-1/\rho}$.

In his lecture Pólya introduced the set of all points z such that in any neighborhood of z infinitely many derivatives of $f(z)$ vanish. Kővári (Mat. Lapok **7** (1956), 106–108) gave an example of an entire function such that the zeros of all the successive derivatives are dense in $\mathbb{C}$. Erdős (ibid., **7** (1956), 214–217) proved that given an arbitrary sequence (z_k) of complex numbers, and a sequence $n_1 < n_2 < \cdots$ of integers such that the complementary sequence is infinite, there exists an entire function $f(z)$ such that $f^{(n_k)}(z_k) = 0$ for $k = 1, 2, \ldots$.

In a joint paper Alfréd and Kató Rényi (J. Analyse Math., **14** (1965), 303–310) proved that if $f(z)$ is a non-constant entire function and $P(z)$ is a polynomial of degree ≥ 3, then $f\big(P(z)\big)$ cannot be periodic. However, $f(z) = e^{\sqrt{z}} + e^{-\sqrt{z}}$ is entire and $f(z^2)$ is periodic.

The fourth chapter of Pólya's great article appeared only in 1933 and in a different journal (Ann. of Math. (2) **34** (1933), 731–777; [128], I, pp. 543–589). Let

$$f(z) = a_0 + a_1 z + \cdots + a_n z^n + \cdots \tag{a}$$

$$g(z) = b_0 + b_1 z + \cdots + b_n z^n + \cdots \tag{b}$$

$$h(z) = a_0 b_0 + a_1 b_1 z + \cdots + a_n b_n z^n + \cdots . \tag{c}$$

Hadamard's composition theorem states that if γ is a singularity of $h(z)$, then $\gamma = \alpha\beta$, where α is a singularity of $f(z)$, and β is a singularity of $g(z)$. Émile Borel proved that if the pole α is the only singularity on the circle of convergence of (a), and the pole β is the only singularity on the circle of convergence of (b), then $\alpha\beta$ is the unique singularity on the circle of convergence of (c) and it is a pole of $h(z)$. Georg Faber proved a somewhat more general statement, and Pólya announced in 1927 without proof that the product of two isolated singular points is a singular point.

To go beyond this result Pólya introduces the following definitions:

Let the power series (38) represent $f(z)$ in $|z - z_0| < R$, and let α be a singular point of $f(z)$ on the circle of convergence. The singularity α is said to be *almost isolated* for the series (38) if there exists a neighborhood of α in which there is no other singular point of $f(z)$ except possibly on the straight half-line joining z_0 with α.

A singular point α of $f(z)$ on the circle of convergence of (38) is said to be *isolable* if in any neighborhood of α there exists a simple closed curve surrounding α along which the function $f(z)$ defined by (38) can be continued analytically.

His main result is:

Theorem C. *If on the circle of convergence of the series* (a) *there is a unique singularity α of $f(z)$ which is almost isolated for* (a), *and on the circle of convergence of the series* (b) *there is a unique singularity β of $g(z)$ which is isolable, then the point $\gamma = \alpha\beta$ is singular for $h(z)$ and it is the only singularity on the circle of convergence of the power series* (c).

The proof is based on two auxiliary results which have a great interest on their own.

Theorem A. *If on the circle of convergence of a power series there is a unique singular point, and this singular point is almost isolated for the power series, then the upper density $\overline{D}$ of the nonvanishing coefficients is* 1.

Theorem B. *If the lower density $\underline{D}$ of the nonvanishing coefficients of a power series is* 0, *then the domain of existence of the function represented by the power series is a simply connected domain in* $\mathbb{C}$.

Later Pólya published proofs of the converses of Fabry's gap theorem and of Theorem B (Trans. Amer. Math. Soc., **52** (1942), 65–71; [128], I, pp. 713–719): Let (λ_k) be an increasing sequence of positive integers. If $\liminf_{k\to\infty}(\lambda_k/k) < \infty$, i.e. $\overline{D} > 0$, then there exists a power series $\sum a_k z^{\lambda_k}$ whose radius of convergence is 1 but for which the circle $|z| = 1$ is not a natural boundary; if $\limsup_{k\to\infty}(\lambda_k/k) < \infty$, i.e. $\underline{D} > 0$, then there exists a power series $\sum a_k z^{\lambda_k}$ whose radius of convergence is 1 and which defines a multivalent analytic function (hence its domain of definition is not a simply connected part of $\mathbb{C}$). Erdős gave an elementary proof of the first result (Trans. Amer. Math. Soc., **57** (1945), 102–104).

Pólya (Comment. Math. Helv., **7** (1934/35), 201–221; [128], I, pp. 593–613) also studied the following question of Hadamard type: how must the coefficients c_n be constituted in order that the function defined by (27) have the following properties: it is single-valued on the Riemann surface of $\sqrt[p]{z-1}$, it is regular at all points excluding the points of ramification $z = 1$ and $z = \infty$, and it vanishes at $z = \infty$.

Many of Pólya's results, in part with new proofs, can be found in the book of Vladimir Bernstein [13] generalized to Dirichlet series, which are the series (29) after substituting $z = e^{-s}$, where the λ_k do not have to be integers. Pólya (Nachrichten von der Gesellschaft der Wissenschaften zu Göttingen Math.-Phys. Kl. 1927, 187–194; [128], I, pp. 309–317) considers Dirichlet series with complex exponents λ_k and proves that if they satisfy the Fabry condition $k/\lambda_k \to 0$, then the domain of existence of the function defined by the series is convex. He explains that in the case when the λ_k are positive integers this yields the Fabry gap theorem.

9. Turán's "new method"

"An idea, which is used once, is a trick. If it is used a second time, it becomes a method" – say Pólya and Szegő in the Preface of [129]. Turán's idea, that too many consecutive power-sums of n complex numbers cannot simultaneously be small, occurs first as a hypothesis in "Über die Verteilung von Primzahlen (I)" (Acta Sci. Math. Szeged **10** (1941), 81–104; [184], No. 23).

In 1912 Landau stated as one of the main problems of the theory of prime numbers to prove that between x^2 and $(x+1)^2$ there is always a prime. Denoting, as usual, by $\pi(x)$ the number of primes $p \leq x$, one asks more generally for an estimate of $\pi(x + x^\theta) - \pi(x)$ as $x \to \infty$.

As every reader of these lines knows, the Dirichlet series

$$\sum_{n=1}^{\infty} \frac{1}{n^s}, \qquad s = \sigma + it \in \mathbb{C},$$

converges for $\sigma > 1$ and defines the Riemann $\zeta(s)$-function, which is analytic in the whole complex plane with the exception of $s = 1$, where it has a simple pole. It follows from the functional equation discovered by Riemann that $\zeta(s) = 0$ for $s = -2k$ $(k = 1, 2, 3, \ldots)$; these are the trivial zeros. It is known that all the other zeros lie in the strip $0 < \sigma < 1$, and that there are infinitely many zeros ρ with $\Re e\rho = \frac{1}{2}$. The million dollar question is the Riemann hypothesis: all the nontrivial zeros of $\zeta(s)$ lie on $\sigma = \frac{1}{2}$.

To approach this problem F. Carlson introduced in 1920 the function $N(\alpha, T)$ which equals the number of zeros of $\zeta(s)$ in the rectangle $\alpha \leq \sigma < 1$, $0 < t \leq T$. A. E. Ingham proved in 1937 that if

$$N(\alpha, T) = \mathrm{O}\big(T^{b(1-\alpha)} \log^B T\big) \tag{39}$$

holds uniformly for $\frac{1}{2} \leq \alpha \leq 1$, then

$$\pi(x + x^\theta) - \pi(x) \sim \frac{x^\theta}{\log x} \tag{40}$$

for $\theta > \frac{1}{b}$. Observe that according to Riemann and H. von Mangold we have $N(\frac{1}{2}, T) \sim \frac{T}{2\pi} \log \frac{T}{2\pi}$, so that b cannot be less than 2.

The Riemann hypothesis implies the Lindelöf hypothesis ([181], Chap. XIII):

$$\zeta(\frac{1}{2} + it) = \mathrm{O}\big(|T|^\varepsilon\big) \tag{41}$$

for any $\varepsilon > 0$. The converse implication does not hold (op. cit. p. 279). Ingham proved that in (39) one can take $b = 2 + 4c, B = 5$, where c is the greatest lower bound of all numbers ε for which (41) holds. Thus if the Lindelöf hypothesis is true, then $c = 0$, so in (39) one has the optimal $b = 2$: this is called the *density hypothesis* ([187], p. 359). In this case (40) is true for $\theta > \frac{1}{2}$.

Now Turán says that the behavior of $\zeta(s)$, and in particular the Lindelöf hypothesis, is inextricably connected with the distribution of primes. Therefore he proves

$$N(\alpha, T) \leq T^{2(1-\alpha)} \exp(13 \log^{0.18} T)$$

under a hypothesis that has nothing to do with prime numbers:

Let $|z_j| \leq 1$ for $1 \leq j \leq n$. Then

$$\max_{l(n) \leq \nu \leq u(n)} \big|z_1^\nu + \cdots + z_n^\nu\big| > \exp(-n^{0.09}),$$

where $l(n) = n^{3/2}(1 - n^{-0.42})$, $u(n) = n^{3/2}$. It is clearly visible on page 98 of Turán's article how this inequality is used in formula (35b).

László Kalmár called the following statement the *quasi-Riemann hypothesis:* One can find a number $\frac{1}{2} \leq \alpha < 1$ such that $\zeta(s)$ has only finitely many zeros in the half-plane $\sigma > \alpha$. Turán (Izv. Akad. Nauk SSSR, Ser. Mat., **11** (1947), 197–262; [184], No. 31) gave a necessary and sufficient condition for the quasi-Riemann hypothesis to hold. The manuscript was received on December 2, 1945, he lectured on the subject in Budapest on February 7, 1944, so he worked on the paper during the darkest days of World War II. The condition in question is the existence of numerical constants $c > 0$ and $C > 0$ such that for $t > 0$, $N \in \mathbb{N}$ the condition

$$0 < c < t^{10} \leq \frac{1}{2}N \leq N_1 < N_2 \leq N$$

implies

$$\left| \sum_{N_1 \leq p \leq N_2} e^{it \log p} \right| \leq Ct^{-\frac{1}{2}} N e^{23(\log \log N)},$$

where p is prime (cf. [187], Section 33). The statement about consecutive power-sums has been promoted from hypothesis to Lemma XII, it is a somewhat weaker form of Turán's "Second Main Theorem", see below.

So the "trick" has become a "method"! Other applications soon followed: to lacunary power series, as we saw in the preceding section, to the quasi-analyticity of functions having an expansion into trigonometric series with "small" coefficients (C.R. Acad. Sci. Paris **224** (1947), 1750–1752; [184], No. 29), to the distribution of real roots of almost periodic polynomials (Publ. Math. Debrecen **1** (1949/1950), 38–41; [184], No. 40), etc. Already in 1949 Turán lectured in Prague with the title "On a new method in the analysis with applications", and in 1953 his book with a similar title appeared simultaneously in Hungarian and in German. It lists twelve previous papers of the author in which the power-sum method is used. An expanded version in Chinese appeared in 1956. Several Hungarian mathematicians, mostly students and later collaborators of Turán, joined him in solving the fascinating problems which arose: István Dancs, Gábor Halász, János Komlós, Endre Makai, János Pintz, András Sárközy, Vera Sós, Mihály Szalay, Endre Szemerédi. But the theory also had an influence outside Hungary: F. V. Atkinson, A. A. Balkema, N. G. de Bruijn, J. D. Buchholz, J. W. S. Cassels, D. Gaier, J. M. Geysel, H. Leenman, D. J. Newman, S. Uchiyama and H. Wittich contributed to it. Furthermore Alfred J. van der Poorten and R. Tijdeman wrote their doctoral dissertations on the subject, the first at the University of New South Wales (Sydney, Australia), the second at the University of Amsterdam.

A considerably augmented English edition of *"On a New Method of Analysis and its Applications"* [187] appeared in 1984, eight years after the premature death of the author. Nine sections had their final versions written by Halász, and thirteen by Pintz. The book has two parts; Part I (16 sections) deals with minimax problems concerning power-sums, concentrating on those results which then have applications in Part II (42 sections). Each part ends with a long section on open problems.

Let me state two of the three Main Theorems. Set $z = (z_1, \ldots, z_n) \in \mathbb{C}^n$, for $\nu \in \mathbb{N}$, write $s_\nu(z) = z_1^\nu + \cdots + z_n^\nu$, and introduce the generalized power-sums

$$g_\nu(z) = b_1 z_1^\nu + \cdots + b_n z_n^\nu,$$

where the b_j $(1 \leq j \leq n)$ are complex constants.

First Main Theorem. *Assume that* $\min_{1 \leq j \leq n} |z_j| = 1$. *Then for* $m \in \mathbb{N}$ *we have*

$$\max_{m+1 \leq \nu \leq m+n} \left| g_\nu(z) \right| \geq C(m,n) \left| \sum_{j=l}^{n} b_j \right|,$$

where $C(m,n) = \left(\frac{m}{2e(m+n)}\right)^n$.

de Bruijn and Makai proved that the best posible value of $C(m,n)$ is $P(m,n)^{-1}$ with

$$P(m,n) = \sum_{j=0}^{n-1} \binom{m+j}{j} 2^j.$$

Second Main Theorem. *Assume that* $1 = |z_1| \geq |z_2| \geq \cdots \geq |z_n|$. *Then*

$$\max_{m+1 \leq \nu \leq m+n} \left| g_\nu(z) \right| \geq 2 \left(\frac{n}{8e(m+n)} \right)^n \min_{1 \leq j \leq n} |b_1 + \cdots + b_j|.$$

In special cases stronger lower bounds can be found. Thus if $b_1 = \cdots = b_n = 1$, then we have:

Theorem. *Assume that* $\min_{1 \leq j \leq n} |z_j| = 1$. *Then*

$$\min_z \max_{1 \leq \nu \leq n} \left| s_\nu(z) \right| = 1.$$

The minimum is achieved when the z_j *are* n *vertices lying on the unit circle of a regular* $(n+1)$*-gon.*

I cannot resist the temptation to prove at least the first part of the Theorem. Set $\prod_1^n(\zeta - z_j) = \zeta^n + a_1\zeta^{n-1} + \cdots + a_n$. The assumption yields $|a_n| = |z_1 \ldots z_n| \geq 1$, so $|a_l| = \max |a_j| \geq 1$. The Newton–Girard formulas (also called Newton–Waring formulas, or Newton formulas, see Heinrich Weber, Lehrbuch der Algebra, vol. I, §46)

$$s_\nu(z) + a_1 s_{\nu-1}(z) + \cdots + a_{\nu-1}s_1(z) + \nu a_\nu = 0, \quad 1 \leq \nu \leq n,$$

yield

$$l|a_l| = \bigl|s_l(z) + a_1 s_{l-1}(z) + \cdots + a_{l-1}s_1(z)\bigr|$$
$$\leq \bigl(1 + |a_1| + \cdots + |a_{l-1}|\bigr) \max_{1\leq\nu\leq n} \bigl|s_\nu(z)\bigr| \leq l|a_l| \max_{1\leq\nu\leq n} \bigl|s_\nu(z)\bigr|,$$

so indeed $\max_{1\leq\nu\leq n} \bigl|s_\nu(z)\bigr| \geq 1$. ■

After an introduction (Sections 17–19), the first applications of Part II are to Complex Function Theory (Sections 20–26). The results already mentioned on lacunary series and on quasi-analyticity can be found here. Other applications are to Borel summability, to the value distribution of entire functions satisfying a linear differential equation, to linear combinations of entire functions, etc. Following this, the topics covered are: Differential Equations (Sections 27–28), Numerical Algebra (Sections 29–31), Markov Chains (Section 32), and the largest portion of Part II (Sections 33–57) is devoted to Analytic Number Theory, Here we find the topics discussed earlier: the density hypothesis, the quasi-Riemann hypothesis, but also the remainder term in the prime number formula, the least prime in an arithmetic progression, and mainly the joint creation of Turán and Stefan Knapowski: comparative prime number theory. The primary object of the study is the analytic function $\zeta(s)$ and its cousins, the Dirichlet $L(s;\chi)$-series, however, I will not transcribe the results but refer to the accessible and eminently readable book [187].

10. Power series: behavior on the circle of convergence

Pál Turán has given a number of results and examples concerning power series which are independent of his "new method".

If the power series $\sum c_n z^n$ converges in $|z| < 1$ and represents there a function which is continuous in $|z| \leq 1$, then $\sum |c_n|^2 < \infty$. There exist, however, functions for which the series $\sum |c_n|^{2-\varepsilon}$ diverges for any $\varepsilon > 0$. This phenomenon is named after Torsten Carleman whose article appeared in 1918. But he and authors before him (E. Fabry, G. H. Hardy, S. Bernstein, J. E. Littlewood) listed in Turán's note (Bull. Amer. Math. Soc., **54** (1948), 932–936; [184], No. 37) only consider the analogous phenomenon for trigonometric series. It was Ottó Szász (Math. Z., **8** (1920), 222–236; [172], pp. 481–496) who first stated it explicitly for power series. Turán quotes two articles of Simon Szidon in which the divergence of $\sum \left| c_{n_k} \right|^{2-\varepsilon}$ is examined for an increasing sequence (n_k) of integers.

Turán gives an example for the Carleman phenomenon which requires only elementary (though not simple) calculations, and does not need van der Corput's Lemma as the example given in [203] (Chapter 5, (4.11), p. 200). Ottó Szász pointed out to Turán that completely elementary examples are also furnished by the reasoning he and S. Minakshisundaram used in their paper (Trans. Amer. Math. Soc., **61** (1947), 36–53; [172], pp. 1054–1071).

A Möbius transformation

$$z \mapsto w = \mu(z) = c\frac{z - z_0}{1 - \bar{z}_0 z},$$

where $|c| = 1$ and $|z_0| < 1$, maps the unit disk $|z| < 1$ bijectively and conformally onto $|w| < 1$, and the circle $|z| = 1$ onto $|w| = 1$. For simplicity take $c = 1$. The inverse transformation is

$$w \mapsto z = \mu^{-1}(w) = \frac{w - w_0}{1 - \bar{w}_0 w},$$

where $w_0 = -z_0 = \mu(0)$. Let

$$f_1(z) = \sum_{n=0}^{\infty} a_n z^n$$

be a function which is holomorphic in the unit disk, and set

$$f_2(w) = f_1\big(\mu^{-1}(w)\big) = \sum_{n=0}^{\infty} b_n w^n.$$

Turán (Publ. Inst. Math. (Beograd) **12** (1958), 19–26; [184], No. 103) gave an example of a series $f_1(z)$ which converges at $z = 1$ but the series $f_2(w)$

does not converge at the corresponding point $w = \mu(1)$. Turán also proved that if $f_1(z)$ is Abel-summable at $z = 1$, then $f_2(w)$ is Abel-summable at $w = \mu(1)$.

László Alpár, Turán's friend and disciple, devoted several articles to this situation. In a paper written in Hungarian (Mat. Lapok **11** (1960), 312–322) he shows that there exists an $f_1(z)$ such that $\sum |a_n| < \infty$ but $\sum |b_n|$ diverges. What makes this result interesting is the fact that $\sum |a_n| < \infty$ implies $\sum |a_n|^2 < \infty$, i.e. that $f_1(z) \in L^2$ on $|z| = 1$ and thus also $\sum |b_n|^2 < \infty$. Alpár asks whether there exists a number $0 < \varepsilon < 1$ such that $\sum |b_n|^{2-\varepsilon}$ converges. Gábor Halász (Publ. Math. Debrecen **14** (1967), 63–68) gave a negative answer by proving the following theorem:

Given z_0 and $0 < \omega(n)$ which tends to $+\infty$ as $n \to \infty$, there exists a function $f_1(z)$ such that $\sum |a_n| < \infty$ but

$$\sum_{n=0}^{\infty} |b_n|^{2-\frac{\omega(n)}{\log n}} = \infty.$$

However, as he shows in a footnote, if $k \geq 0$, then

$$\sum_{n=0}^{\infty} |b_n|^{2-\frac{k}{\log n}} < \infty.$$

In a second paper (ibid., **15** (1968), 23–31) Halász proves that if there exists a decreasing sequence (A_n) such that $|a_n| \leq A_n$ and $\sum A_n < \infty$, then also $\sum |b_n| < \infty$. If, however, $\sum A_n = \infty$, then there exists (a_n) with $|a_n| \leq A_n, \sum |a_n| < \infty$ but $\sum |b_n| = \infty$.

Alpár wrote a sequence of eight articles in French on the subject. The last one appeared in Studia Sci. Math. Hungar., **1** (1966), 379–388 and contains references to the preceding ones. Motivated by his result on Abel-summability, Turán asked whether the same holds for Cesàro-summability. Alpár gave a negative answer. More precisely, he proved that if $k \geq 0$ and $\sum a_n$ is Cesàro-summable of order k, then $\sum b_n \mu(1)^n$ is Cesàro-summable of order $k + \frac{1}{2}$ but not necessarily of smaller order. If $\alpha_n^{(k)}$ is the nth Cesàro-sum of order k of $\sum a_n$ and $\beta_n^{(k+\delta)}$ is the nth Cesàro-sum of order $k + \delta$ of $\sum b_n \mu(1)^n$, then one has a linear relation

$$\beta_n^{(k+\delta)} = \sum_{\nu=0}^{\infty} \gamma_{n\nu}^{(k,\delta)} \alpha_\nu^{(k)}.$$

The idea of Alpár was to show that the matrix $\left(\gamma_{n\nu}^{(k,\delta)}\right)$ satisfies the Toeplitz–Schur regularity conditions if $\delta \geq \frac{1}{2}$ but not if $\delta < \frac{1}{2}$.

Alpár also considered the analogous problems when instead of power series one studies expansions into a series of Faber polynomials.

As we have seen in Section 8, Fejér has given an example of a function $f(z)$ holomorphic in $|z| < 1$, continuous in $|z| \leq 1$ whose Taylor series $\sum c_n z^n$ diverges at $z = 1$. If $|c_n| \leq \frac{1}{n}$ for all n, then $\sum c_n z^n$ converges uniformly in $|z| \leq 1$. Turán (Mat. Lapok, **10** (1959), 278–282; [184], No. 113) uses the method of Fejér to show that if $\omega(n)$ is a positive sequence which tends increasingly to $+\infty$ as $n \to \infty$, then there exists a function $f(z) = \sum c_n z^n$ in $|z| < 1$, continuous in $|z| \leq 1$ such that $|c_n| \leq \frac{\omega(n)}{n}$ but $\sum c_n$ diverges.

Let $f(z)$ be an entire function of order ρ, let $b \in \mathbb{C}$, and denote by z_ν the points (counted with multiplicity), where $f(z_\nu) = b$. Denote by $\rho(b)$ the exponent of convergence of (z_ν), i.e., the number such that

$$\sum_\nu \frac{1}{|z_\nu|^\alpha}$$

converges for $\alpha > \rho(b)$ and diverges for $\alpha < \rho(b)$; a number b_0 for which $\rho(b_0) < \rho$ is called a *Borel exceptional value* of $f(z)$.

This concept can be generalized. Let φ be a positive function defined for $0 \leq x < \infty$, strictly decreasing to zero. A number $b \in \mathbb{C}$ is a φ-exceptional value of $f(z)$ if $\sum \varphi(|z_\nu|) < \infty$. In a joint article Alpár and Turán (Publ. Math. Inst. Hungar. Acad. Sci., **A6** (1960), 157–164; [184], No. 120) show that for any function φ of the above kind there exists an entire function of infinite order that has no φ-exceptional values.

Let (λ_k) be a sequence of positive integers which satisfies the gap condition

$$\lambda_{k+1} - \lambda_k \geq \gamma$$

for some fixed $\gamma \geq 1$. Assume that the power series

$$f(z) = \sum_{k=0}^{\infty} c_k z^{\lambda_k}$$

has radius of convergence 1. If writing $z = re^{i\theta}$ the limit

$$f(\theta) = \lim_{r \to 1-0} f(re^{i\theta})$$

exists almost everywhere and belongs to L^2 on an arc of length greater than $2\pi/\gamma$, then $f(\theta)$ exists everywhere and belongs to L^2 on $[0, 2\pi]$. The result can be found in [203] (Chapter V, (9.1), p. 222), where the notes contain the following remark: "Nothing seems to be known about possible extensions to classes L^p, $p \neq 2$" (p. 380).

In the book [135] published in honor of the 75th birthday of Pólya, the champion of gap theorems, there is a contribution by Erdős and Rényi (pp. 110–116) and one by Turán (pp. 404–409) addressing this problem for $q > 2$.

The first two use probability theory. They consider the exponents λ_k as random variables and prove that with probability one there exists a function $f(\theta)$ whose Fourier series $\sum c_k \cos \lambda_k \theta$ satisfies $\lambda_{k+1} - \lambda_k \to \infty$, belongs to L^2 in $|\theta| \leq \pi$, is bounded in $\delta \leq |\theta| \leq \pi$ for every $\delta > 0$, but does not belong to any L^q with $q > 2$ on $|\theta| \leq \pi$.

Turán constructs for any $q > 6$ an explicit lacunary power series such that

$$\lambda_{k+1} - \lambda_k > \frac{1}{2} \lambda_k^{1/(q+6)},$$

and for which $f(\theta) \in L^q\left(\frac{\pi}{2}, \frac{3\pi}{2}\right)$ but $f(\theta)$ is not in $L^q(0, 2\pi)$.

11. Pólya–Schur functions

One of the earliest publications of György Pólya has the title "Über ein Problem von Laguerre" (Rend. Circ. Mat. Palermo **34** (1912) 89–120; [128], II, pp. 1–32); it is in fact an exchange of letters between him and Mihály Fekete. Much later Pólya wrote a paper with almost the same title: "Über einen Satz von Laguerre" (Jahresber. Deutsch. Math.-Verein., **38** (1929), 161–168; [128], II, pp. 314–321). It is the preoccupation with problems left open by Edmond Laguerre which led to the class we now call Pólya–Schur functions. There is a nice account of their theory in the little book of Nikola Obrechkoff [125].

Let $f(z)$ be an entire function of finite order ρ, denote by (z_k) the sequence of its zeros different from 0 counted according to their multiplicities, and set $|z_k| = r_k$. The genus p *of the sequence* (r_k) is the smallest integer such that

$$\sum_{k=1}^{\infty} \frac{1}{r_k^{\nu}}$$

converges for $\nu \geq p+1$. The function has the product representation

$$f(z) = z^m e^{Q(z)} \prod_{k=1}^{\infty} \left(1 - \frac{z}{z_k}\right) e^{\frac{z}{z_k} + \frac{1}{2}\left(\frac{z}{z_k}\right)^2 + \cdots + \frac{1}{p}\left(\frac{z}{z_k}\right)^p},$$

where $Q(z)$ is a polynomial of degree $q \leq \rho$. Laguerre called the genus *of the function* the larger of the two integers p and q. Completing the proofs and weakening the hypotheses of theorems stated by Laguerre, Pólya proved the following results (Rend. Circ. Mat. Palermo **36** (1913), 279–295; Nachr. Ges. Wiss. Göttingen 1913, 325–330; [128], II, pp. 54–70, 71–75):

Let $P_l(z)$ $(l \in \mathbb{N})$ be a sequence of polynomials which converges uniformly in a disk $|z| \leq R$ to a function $F(z)$.

I. If the zeros of all the polynomials $P_l(z)$ are > 0, then $F(z)$ is an entire function of the form $e^{-\beta z}G(z)$, where $G(z)$ is of genus zero and $\beta \geq 0$. If $F(z)$ is not identically zero, then $P_l(z)$ converges to $F(z)$ in the whole plane and the convergence is uniform in every bounded domain.

More generally ([125], pp. 13–14): If the zeros of the $P_l(z)$ lie in an angular domain W with opening $< \pi$, then $F(z) = e^{-\beta z}G(z)$, where $G(z)$ has genus 0, and β lies in the domain $\overline{W}$ which is symmetric to W with respect to the real axis.

II. If the zeros of the $P_l(z)$ are all real, then $F(z)$ is a Pólya–Schur function, i.e., an entire function of genus 1 multiplied by a Gauß density function $e^{-\gamma z^2}$ $(\gamma \geq 0)$.

Actually I. follows easily from a theorem of Hurwitz ([104], p. 17) and the Hadamard factorization. The proof of II. is "weniger einfach" (less simple).

Pólya wrote the immediately following article (Rend. Circ. Mat. Palermo **37** (1914), 297–302; [128], II, pp. 76–83) in collaboration with Egon Lindwart. They prove that if $z_{l1}, z_{l2}, \ldots, z_{ll}$ are the zeros of $P_l(z)$ and if there exists $M > 0$ such that

$$\sum_{j=1}^{l} \frac{1}{|z_{lj}|^k} \leq M$$

for some $k > 0$, then $F(z)$ is an entire function of genus $\leq [k]$; if k is an integer, $F(z) = e^{\gamma z^k} G(z)$, where the genus of $G(z)$ is $\leq k-1$.

The authors list several consequences, e.g., the following suggested by Fekete: Write $z_{ls} = r_{ls}e^{i\theta_{ls}}$ and assume that the θ_{ls} belong to the union of the r closed intervals $\left[(4t-1)\frac{\pi}{2r}, (4t+1)\frac{\pi}{2r}\right]$, $t = 0, 1, \ldots, r-1$. Then $F(z)$

has genus $\leq 2r$. Furthermore if z_k denotes again the zeros $\neq 0$ of $F(z)$ and $|z_k| = r_k$, then $\sum r_k^{-2k} < \infty$.

Another corollary of the theorem of Lindwart and Pólya was given the following stronger form by Ottó Szász (Bull. Amer. Math. Soc., **49** (1943), 377–383; [172], pp. 1390–1396): Assume that the zeros of each $P_l(z)$ lie in a half-plane containing 0 on its boundary, which can vary with l. If P_l converges to $F(z)$ on a set which has a finite limit point and the coefficients of the $P_l(z)$ are bounded, then $P_l(z)$ converges to $F(z)$ uniformly on every bounded domain and

$$F(z) = e^{\alpha+\beta z+\gamma z^2} \prod \left(1 - \frac{z}{z_k}\right) e^{-\frac{z}{z_k}},$$

where $\sum 1/|z_k|^2 < \infty$.

The problem of Laguerre, investigated by Pólya, received a very general treatment in the 1949 Leiden thesis of Jacob Korevaar. He assumes that the z_{lj} belong to an arbitrary subset of $\mathbb{C}$ and characterizes $F(z) = \lim_{l\to\infty} P_l(z)$. An account of his results can be found in [31], pp. 261–272.

Then Issai Schur got into the picture. At the origin is the following theorem by E. Malo which appeared of all places in the Journal de Mathématiques Spéciales (4) **4** (1895), 7: Assume that the zeros of the polynomial

$$a_0 + a_1 z + a_2 z^2 + \cdots + a_m z^m \quad (a_m \neq 0) \tag{42}$$

are all real, and that the zeros of the polynomial

$$b_0 + b_1 z + b_2 z^2 + \cdots + b_n z^n \quad (b_n \neq 0) \tag{43}$$

are all real and of the same sign. Set $k = \min(m, n)$. Then the zeros of

$$a_0 b_0 + a_1 b_1 z + a_2 b_2 z^2 + \cdots + a_k b_k z^k \tag{44}$$

are all real. If $m \leq n$ and $a_0 b_0 \neq 0$, then the zeros of (44) are distinct.

Schur proved (J. Reine Angew. Math., **144** (1914), 75–88) a result he calls "composition theorem" and which asserts that under the same hypotheses as before the zeros of

$$0! a_0 b_0 + 1! a_1 b_1 z + 2! a_2 b_2 z^2 + \cdots + k! a_k b_k z^k$$

are all real. If $m \leq n$, $a_0 b_0 \neq 0$ the same conclusion holds as above.

From the composition theorem Malo's result follows with a neat little trick (§3 of loc. cit.).

In their joint article (J. Reine Angew. Math., **144** (1914), 89–113; [128], II, pp. 100–124; [163], II, No. 24, pp. 56–69) Pólya and Schur say that a sequence

$$\text{(A)} \qquad \alpha_0, \alpha_1, \alpha_2, \ldots, \alpha_n, \ldots$$

of real numbers is a *factor sequence of the first kind* if given any polynomial (42) whose zeros are all real, the polynomial

$$\alpha_0 a_0 + \alpha_1 a_1 z + \alpha_2 a_2 z^2 + \cdots + \alpha_m a_m z^m$$

has only real zeros. Similarly a sequence

$$\text{(B)} \qquad \beta_0, \beta_1, \beta_2, \ldots, \beta_n, \ldots$$

is a *factor sequence of the second kind* if for any polynomial (43) whose zeros are all real and have the same sign (i.e., are all positive or all negative), the polynomial

$$\beta_0 b_0 + \beta_1 b_1 z + \beta_2 b_2 z^2 + \cdots + \beta_n b_n z^n$$

has only real zeros.

Clearly a factor sequence of the first kind is also one of the second kind but not conversely. It was Laguerre who gave the first examples of factor sequences.

Pólya and Schur start with giving algebraic criteria for factor sequences. For instance (A) is a factor sequence of the first kind if and only if the polynomials

$$\alpha_0 + \binom{n}{1}\alpha_1 z + \binom{n}{2}\alpha_2 z^2 + \cdots + \alpha_n z^n$$

have only real zeros of the same sign. In one direction this follows from the fact that $(1+z)^n$ has -1 as its only zero and from Descartes' rule of signs.

Let $\gamma_0, \gamma_1, \ldots, \gamma_n, \ldots$ be a sequence of real numbers. The authors prove that

$$\frac{\gamma_0}{0!}, \frac{\gamma_1}{1!}, \frac{\gamma_2}{2!}, \ldots, \frac{\gamma_n}{n!}, \ldots$$

is a factor sequence of the first kind if and only if the following condition is satisfied: whenever (42) has only real zeros and (43) only real zeros with the same sign, the polynomial

$$\gamma_0 a_0 b_0 + \gamma_1 a_1 b_1 z + \gamma_2 a_2 b_2 z^2 + \cdots + \gamma_k a_k b_k z^k$$

has only real zeros. Since the sequence where $\alpha_n = 1$ for all n is obviously a factor sequence of the first kind, this yields Schur's composition theorem.

In order to give transcendental criteria for factor sequences, Pólya and Schur introduce two classes of entire functions with real Taylor coefficients.

A function

$$\Phi(z) = \sum_{k=0}^{\infty} \frac{\alpha_k}{k!} z^k \tag{45}$$

with real zeros having the same sign is of type (I.) if $\Phi(z)$ or $\Phi(-z)$ has the representation

$$\Phi(z) = \frac{\alpha_r}{r!} z^r e^{\beta z} \prod_{\nu=1}^{\infty} (1 + \gamma_\nu z)$$

with $\alpha_r \neq 0$, $\beta, \gamma_\nu \geq 0$ (i.e. if on some disk $|z| \leq R$ it is the uniform limit of a sequence of polynomials having only real zeros of the same sign).

A function

$$\Psi(z) = \sum_{k=0}^{\infty} \frac{\beta_k}{k!} z^k \tag{46}$$

whose zeros are all real is of type (II.) if it has the representation

$$\Psi(z) = \frac{\beta_r}{r!} z^r e^{\beta z - \gamma z^2} \prod_{\nu=1}^{\infty} (1 + \delta_\nu z) e^{-\delta_\nu z},$$

where $\beta_r \neq 0, \gamma \geq 0, \beta$ and δ_ν real (i.e. if on some disk it is the uniform limit of a sequence of polynomials having only real zeros).

Theorem. (A) *is a factor sequence of the first kind if and only if* (45) *is of type* (I.). (B) *is a factor sequence of the second kind if and only if* (46) *is of type* (II.).

Now a new motif enters in the form of the Hermite–Poulain theorem: Let the polynomials

$$P(z) = a_0 + a_1 z + a_2 z^2 + \cdots + a_n z^n \quad (a_n \neq 0)$$

and

$$Q(z) = b_0 + b_1 z + b_2 z^2 + \cdots + b_n z_n + b_{n+1} z^{n+1} + \cdots + b_{n+m} z^{n+m},$$

where $b_0, b_1, \ldots, b_n > 0$, have only real zeros. Then the polynomial

$$b_0 P(z) + b_1 P' z + b_2 P''(z) + \cdots + b_n P^{(n)}(z)$$

has only real zeros.

In a short note (Vierteljahrschr. Naturforsch. Ges. Zürich **61** (1916), 546–548; [128], II, p. 163–165) Pólya gives a geometric proof of the fact that under the same hypotheses the curve

$$F(x, y) = b_0 P(y) + b_1 x P'(y) + b_2 x^2 P''(y) + \cdots + b_n x^n P^{(n)}(y) = 0$$

has n real points of intersection with any straight line $sx - ty + u = 0$, provided that $s \geq 0$, $t \geq 0$, $s + t > 0$ and u is real.

The special case $s = 1$, $t = 0$, $u = -1$, i.e. $x = 1$, yields the Hermite–Poulain result. The case $s = 0$, $t = 1$, $u = 0$, i.e. $y = 0$, gives the Schur composition theorem. Finally, $s = t = 1$, $u = 0$, i.e. $x = y$, gives an example of Pólya–Schur according to which

$$b_0 P(z) + b_1 z P'(z) + b_2 z^2 P''(z) + \cdots + b_n z^n P^{(n)}(z)$$

has only real zeros. Conversely, the general theorem can be deduced from the three special cases by changes of variables.

In an earlier article (J. Reine Angew. Math., **145** (1915), 224–249; [128], II, pp. 128–153) Pólya made some elementary remarks related to the Hermite–Poulain theorem, and generalized it to certain pairs of entire functions. Changing slightly the hypotheses and the notation, let

$$F(z) = a_0 + a_1 z + a_2 z^2 + \cdots + a_n z^n$$

be a polynomial with only real roots, $a_0 \neq 0$ real, $a_n \neq 0$, $n \geq 1$, and let $G(z)$ be a polynomial with real coefficients having exactly r real roots. Then the following hold concerning the polynomial

$$H(z) = F(\partial) G(z) = a_0 G(z) + a_1 G'(z) + a_2 G''(z) + \cdots + a_n G^{(n)}(z):$$

(i) $H(z)$ has $r + 2k$ real zeros, $k \in \mathbb{N}$;

(ii) If $r \geq 1$, then $H(z)$ has at least one real zero with odd multiplicity, hence assumes for real z both positive and negative values;

(iii) If $r \geq 2$, then $H(z)$ has at least two distict real zeros;

(iv) If $G(z)$ has only real zeros, then the multiple zeros of $H(z)$ are also multiple zeros of $G(z)$

(v) If $r \geq 1$ and $F(z)$ has only positive zeros, then $H(z)$ has a real zero with odd multiplicity which is larger than the largest real zero of $G(z)$;

Let $\Phi(z)$ be an entire function of type (I.), where in (45) the coefficients α_k are positive, and let $\Psi(z)$ be of type (II.). Then the series

$$\sum_{\nu=0}^{\infty} \frac{\alpha_\nu}{\nu!} z^\nu \Psi^{(\nu)}(z) \quad \text{and} \quad \sum_{\nu=0}^{\infty} \frac{\beta_\nu}{\nu!} z^\nu \Phi^{(\nu)}(z)$$

converge and represent entire functions of type (II.).

The second paper referred to at the beginning of this section appeared immediately following an article of Obrechkoff who gave a simplified proof of a theorem of Hurwitz according to which the function $\sqrt{z}^\nu J_{-\nu}(2\sqrt{z})$ has exactly $[\nu]$ negative zeros if $\nu \geq 0$. In the first part of his proof Obrechkoff uses an algebraic theorem of Laguerre, in the second part also the differential equation of the Bessel function $J_{-\nu}(z)$. Pólya shows that the whole proof can be based on ideas of Laguerre if one transports them from polynomials to entire functions. He proves namely the following theorem: Let

$$g(z) = a_0 + a_1 z + a_z z^2 + a_3 z^3 + \cdots$$

be either a polynomial of degree m with only positive zeros, or an entire function of type (I.) with positive zeros. In the algebraic case set $J = [0, m]$, in the transzendental case $J = [0, \infty)$. Let $G(z)$ be an entire function of type (II.) which in J has exactly s simple zeros such that the distance between two consecutive zeros is ≥ 1. Then

$$a_0 G(0) + a_1 G(1) z + a_2 G(2) z^2 + a_3 G(3) z^3 + \cdots$$

has exactly $m - s$ strictly positive zeros in the algebraic case, and it has s negative zeros in the transzendental case.

Since

$$\sqrt{z}^\nu J_{-\nu}(2\sqrt{z}\,) = \sum_{n=0}^{\infty} \frac{(-z)^n}{n!} \frac{1}{\Gamma(n+1-\nu)},$$

setting $g(z) = e^{-z}$ and $G(z) = \left(\Gamma(z+1-\nu)\right)^{-1}$ the theorem of Hurwitz follows.

Pólya says that the content of the paper is a portion of an investigation on which he published two notes (C.R. Acad. Sci. Paris **183** (1926), 413–414, 467–468; [128], II, pp. 261–264). The full proofs appeared in the dissertation of E. Benz (Comment. Math. Helv., **7** (1934), 243–289). Let L be an operation determined by a sequence $l_0, l_1, \ldots, l_n, \ldots$, which associates with each polynomial $P(z)$ the polynomial

$$L\dot{P}(z) = l_0 P(z) + l_1 P'(z) + l_2 P''(z) + \cdots .$$

Setting $L(z) = l_0 + l_1 z + l_2 z^2 + \cdots$ one can write

$$LP(z) = L(\partial)P(z).$$

Pólya proposes to determine the class of operations L such that whenever $P(z)$ has all its roots in the convex domain K, then so does $LP(z)$. Pólya mentions that there are three equivalent necessary and sufficients conditions, the third of which involves a product decomposition of $L(z)$. When K is the lower half-plane, then $L(z)$ has to be a Pólya–Schur function with zeros in the upper half-plane. In the second note Pólya considers the case when $P(z)$ is a Dirichlet polynomial

$$P(z) = a_0 e^{\lambda_0 z} + a_1 e^{\lambda_1 z} + \cdots + a_n e^{\lambda_n z}$$

and

$$LP(z) = L(\lambda_0) a_0 e^{\lambda_0 z} + L(\lambda_1) a_1 e^{\lambda_1 z} + \cdots + L(\lambda_n) a_n e^{\lambda n z}.$$

In a joint paper of Pólya with André Bloch (Proc. London Math. Soc. (2) **33** (1932), 102–114; [128], II, pp. 336–348) the authors consider polynomials of the form

$$P(z) = 1 + \varepsilon_1 z + \varepsilon_2 z^2 + \cdots + \varepsilon_n z^n,$$

where each coefficient has one of three values -1, 0 or 1. These are now called *partition polynomials*, and for recent literature concerning them see the article by Morley Davidson (J. Math. Anal. Appl., **269** (2002), 431–443). There are 3^n such polynomials, so there are some which have a maximum number of zeros in the open interval (0,1). Denote this maximum number by π_n; clearly $0 \le \pi_n \le n$, and π_n increases with n. It is easy to see that $\pi_n = \mathrm{o}(n)$ and the authors say that the crucial question is whether π_n is of order as low as $\log n$. They find that the answer is negative. They prove that there exists a constant $A.0$ such that

$$\frac{1}{A}\frac{n^{1/4}}{(\log n)^{1/2}} < \pi_n < A\frac{n\log\log n}{\log n}$$

for $n \geq 3$.

Bloch and Pólya feel that "this question seems very particular and rather out of the way", therefore they present a number of examples and observations which motivate it. The first example is

$$(47) \qquad \left(\frac{1}{p}\right) + \left(\frac{2}{p}\right)z + \left(\frac{3}{p}\right)z^2 + \cdots + \left(\frac{p-1}{p}\right)z^{p-2},$$

where p is an odd prime number and the coefficients are the Legendre symbols. This is precisely the polynomial from which the Fekete–Pólya correspondence mentioned at the beginning of this section sets out. If p is such that (47) has no zeros in the interval $(0,1)$, then the Dirichlet series $\sum_{n=1}^{\infty}\left(\frac{n}{p}\right)\frac{1}{n^s}$ has no positive real zeros, and the problem is to decide for which primes p is this the case. Fekete conjectured that (47) has no zeros in $(0,1)$. Pólya (Jahresber. Deutsch. Math.-Verein., **28** (1919), 31–40; [128], III, pp. 76–85) disproved the conjecture by a simple calculation which showed that for $p = 67$ and for $p = 167$ the polynomial has two zeros between 0 and 1.

Another example is the partial sums S_n of the power series

$$(48) \qquad z - z^2 - z^3 + z^4 + z^5 + z^6 + z^7 - z^8 - \cdots - z^{15} + z^{16} + \cdots,$$

where on plus sign is followed by 2 minus signs, then 4 plus signs, 8 minus signs, etc. The number of zeros in (0,1) of S_n in $\log n/\log 2 + \mathrm{O}(1)$. This is related to two earlier articles of Pólya. The first (Nachr. Ges. Wiss. Göttingen 1930, 19–27; [128], I, pp. 459–467) is about the sign of the remainder term in the prime number theorem. Completing a result of Landau, he finds an upper bound for the smallest $\gamma > 0$, where $\zeta = \beta + i\gamma$ is a pole with maximal β of the function

$$\Phi(s) = \int_0^{\infty} \omega(u)u^{-s}\,du.$$

The bound is given in terms of the asymptotic behavior of the number of changes of sign of $\omega(u)$ in $0 < u \leq x$ as $x \to \infty$. The investigation is continued in a paper (Proc. London Math. Soc. (2) **33** (1932), 85–101; [128], I, pp. 521–537) which is printed preceding immediately the Bloch–Pólya article. Let (a_n) be a bounded sequence of real numbers, and set

$$D(s) = a_1 1^{-s} + a_2 2^{-s} + \cdots + a_n n^{-s} + \cdots ,$$

$$P(z) = a_1 z + a_2 z^2 + \cdots + a_n z^n + \cdots .$$

The Dirichlet series converges for $\Re e s > 1$ and the two functions are related by the formula

$$\Gamma(s)D(s) = \int_0^\infty P(e^{-x})x^{s-1}\,dx.$$

Assume that $D(s)$ is meromorphic for $\Re e s > b$, where $b < 1$ and that, if $p(n)$ is the number of zeros of

$$P_n(z) = a_1 z + a_2 z^2 + \cdots + a_n z^n$$

in the interval $(0,1)$, then there exists an increasing sequence (n_m) of integers such that

$$\lim_{m\to\infty} \frac{\log n_{m+1}}{\log n_m} = 1, \quad p(n_m) = \mathrm{O}(\log n_m) \quad \text{as} \quad m \to \infty.$$

Then either $\Gamma(s)D(s)$ is holomorphic for $\Re e s > b$ or it has a pole $\beta + i\gamma$ with maximal β such that

$$0 \le \gamma \le \pi \limsup_{m\to\infty} \frac{p(n_m)}{\log n_m}. \tag{49}$$

The example (48) is used to show that equality on the right hand side of (49) can be attained.

12. Conformal mapping, complex interpolation

The most important concept to which the names of Lipót Fejér and Frigyes Riesz are attached was not published by the two either separately or jointly. It appeared in a note of Tibor Radó (Acta Sci. Math. Szeged **1** (1922/23), 240–251), and it is the *"Fejér–Riesz procedure"* for the proof of the Riemann mapping theorem. Radó's note is reproduced in part in Fejér's collected works ([40], II, pp. 841–842) and in its entirety in the works of Riesz ([156], pp. 1483–1494).

Carathéodory, who simplified slightly the procedure, writes the following (Bull. Calcutta Math. Soc., **20** (1928), 125–134; [21], III. pp. 300–301): "About ... the main theorem of conformal mapping I must say a few words. After the insufficiency of Riemann's original proof was recognized, the miraculously beautiful but very complicated methods of proof developed by H. A. Schwarz were the only paths to this theorem. Since about twenty years in rapid succession a large series of shorter and better proofs was proposed; but it was reserved for the Hungarian mathematicians L. Fejér and F. Riesz to return to the basic idea of Riemann and to relate again the solution of the problem of conformal mapping with a solution of a variational problem. But they did not choose a variational problem which, like Dirichlet's principle, is extraordinarily difficult to treat but one for which the existence of a solution is clear. In this way a proof came about which is only a few lines long and which was immediately adopted by all newer textbooks." In Fejér's collected works Turán quotes this passage in the original German ([40], II, pp. 842–843).

Riemann's mapping theorem states that if G is a simply connected domain in $\mathbb{C}$ having at least two boundary points and a is a point in G, then there exists a univalent holomorphic function f mapping G onto the disk $\{\zeta : |\zeta| < \rho\}$ and satisfying $f(a) = 0$, $f'(a) = 1$; the radius ρ and the function f are uniquely determined. An elementary transformation shows that we may assume G to be bounded and take $a = 0$. Radó explains the Fejér–Riesz procedure as follows: Consider all bounded, holomorphic functions f which map G univalently into the ζ-plane and satisfy $f(0) = 0$, $f'(0) = 1$. Such functions exist, e.g., $f(z) = z$. Set $M(f) = \sup_{z \in G} |f(z)|$ and let ρ be the greatest lower bound of all numbers $M(f)$. There exists a sequence (f_n) such that $M(f_n) \to \rho$. By Montel's theory of normal families (or – as we say now – compactness) a subsequence of (f_n) converges uniformly on every compact subset of G to a univalent, holomorphic function f satisfying $f(0) = 0, f'(0) = 1$ and $|f(z)| < \rho$ for $z \in G$. If the image of G under f does not fill out the whole disk $|\zeta| < \rho$, then a square root transformation due to Carathéodory and Koebe yields a function $F(z)$ which has the required properties and is such that $M(F) < M(f)$, which is impossible.

Radó realized that simple connectedness is not made use of in the proof of the Fejér–Riesz procedure, and he proves with it the so-called

"Grenzkreissatz". Also this proof is simplified in the article of Carathéodory quoted above.

An often quoted result of Tibor Radó (Acta Sci. Math. Szeged **2** (1925), 101–121) states that every Riemann surface satisfies the second axiom of countability. This is interesting because Heinz Prüfer has given an example of a two-dimensional differentiable manifold which does not satisfy the axiom: the countability is a consequence of the conformal structure, Radó also pointed out that, as a consequence of countability, every Riemann surface can be triangulated.

A short note of Henri Cartan has the title "Sur une extension d'un théorème de Radó" (Math. Ann., **125** (1952), 49–50; [22], II, pp. 667–668). The theorem referred to can be found in a paper (Math. Z., **20** (1924), 1–6) whose main result asserts that there exist open Riemann surfaces F which cannot be continued, i.e., there exists no Riemann surface G such that F is conformally equivalent to a proper subdomain of G.

Radó's "theorem" in which Cartan is interested is, however, the following Lemma in Radó's article:

Let G be a simply connected domain in the unit disk D which is distinct from D. Let $f(z)$ be holomorphic in D and assume that at every boundary point of G which lies in the interior of D the function $f(z)$ has boundary value zero. Then $f(z) \equiv 0$.

Peter Thullen (Math. Ann., **111** (1935), 137–157) gave a new proof and a generalization of the theorem. Then Heinrich Behnke and Karl Stein (Math. Ann., **124** (1951), 1–16) extended it to n variables (Satz 1). They use Radó's result and even reproduce its proof. Cartan found a very simple proof of the general theorem. He uses potential theory and does not need Radó's result. His assertion, slightly different from that of Behnke-Stein, is as follows:

Let M be an n-dimensional complex analytic manifold. Let g be a continuous, complex-valued function defined on M, and assume that g is holomorphic at each point z where $g(z) \neq 0$. Then g is holomorphic on M.

If $n = 1$ we obtain Radó's result setting $g(z) = f(z)$ for $z \in G$ and $g(z) = 0$ if $z \in D \backslash G$.

Conformal mapping and interpolation is the subject of a note of Fejér (Göttinger Nachrichten 1918, 319–333; [40], II, 100–111). Let C be a continuous, simple, closed curve in $\mathbb{C}$ and $z_1^{(k)}, \ldots, z_k^{(k)}$ points on it for $k \in \mathbb{N}$. Fejér gives a sufficient condition for the following to happen:

If $f(z)$ is a function holomorphic inside C and on C itself, and if $L_k(z;f)$ is the Lagrange interpolation polynomial of F at the points $z_l^{(k)}$ $(1 \le l \le k)$, then $L_k(z;f)$ converges uniformly to $f(z)$ inside C as $k \to \infty$.

Let $\Phi(z)$ map conformally the exterior of C onto $|\zeta| > 1$ and satisfy $\Phi(\infty) = \infty$. Then Fejér's condition requires that the points $\zeta_l^{(k)} = \Phi\bigl(z_l^{(k)}\bigr)$ be the vertices of a regular k-gon inscribed in $|\zeta| = 1$. At the end of his note Fejér remarks that it would be sufficient to require that the $\zeta_l^{(k)}$ be uniformly distributed in the sense of Hermann Weyl.

The subject was taken up by László Kalmár in a prize essay he wrote as a student and which became his doctoral dissertation (Mat. Fiz. Lapok **33** (1926), 120–149). To describe his results, we change slightly our notation. Let $\Psi(z)$ be the unique holomorphic function which maps the exterior of C onto the exterior of a circle $|\zeta| = R$ and which satisfies

$$\lim_{z\to\infty} \frac{\Psi(z)}{z} = 1.$$

The uniquely determined radius $R = R_e(E)$ is the exterior mapping radius of the closure E of the inside of C. Let $\bigl(z_1^{(k)}, \dots, z_{n_k}^{(k)}\bigr)$ $(k \in \mathbb{N})$ be a sequence of n_k-tuples of points on C (the $z_j^{(k)}$, $1 \le j \le n_k$ do not have to be distinct, in that case $L_k(z;f;)$ denotes the Lagrange–Hermite interpolation polynomials). Denote by $\psi_k(z)$ that branch of the function

$$\Bigl\{\bigl(z - z_1^{(k)}\bigr)\bigl(z - z_2^{(k)}\bigr)\cdots\bigl(z - z_{n_k}^{(k)}\bigr)\Bigr\}^{1/n_k}$$

outside the curve C which satisfies

$$\lim_{z\to\infty} \frac{\psi_k(z)}{z} = 1.$$

For $0 \le a < b \le 2\pi$ denote by $\nu_k(a,b)$ the number of those points

$$\Psi\bigl(z_j^{(k)}\bigr) = Re^{i\theta_j^{(k)}}, \quad 1 \le j \le n_k,$$

whose arguments $\theta_j^{(k)}$ lie in $a \le \theta < b$. The following are equivalent:

a) $\lim_{k\to\infty} L_k(z;f) = f(z)$ uniformly for every function $f(z)$ holomorphic inside and on C, i.e., the points $z_j^{(k)}$ are "well-interpolating";

b) $\lim_{k\to\infty} \psi_k(z) = \Psi(z)$ outside C;

c)

$$\lim_{k\to\infty} \frac{\nu_k(a,b)}{n_k} = \frac{b-a}{2\pi}$$

for all $0 \leq a < b \leq 2\pi$.

For the equivalence of a) and b) the points $z_j^{(k)}$ can be chosen anywhere in E.

A different approach to finding well-interpolating points was found by Fekete using the concept of transfinite diameter introduced by him (Math. Z., **17** (1923), 246–249). Let E be a closed bounded set in $\mathbb{C}$, and $n \geq 2$ an integer. Denote by $\delta_n(E)$ the root of order

$$\binom{n}{2}$$

of the maximum of all expressions

$$\big|(z_1 - z_2)(z_1 - z_3)\cdots(z_2 - z_3)\cdots(z_{n-1} - z_n)\big| \tag{50}$$

as the $z_1, z_2, \ldots, z_n$ vary in E. The sequence of positive numbers $\delta_n(E)$ tends decreasingly to the transfinite diameter $\delta(E)$ of E.

It is a remarkable fact that $\delta(E)$ coincides with $R_e(E)$, and with the logarithmic capacity $c(E)$ (Szegő, Math. Z., **21** (1924), 203–208; [173], I, pp. 637–642). Consider furthermore the set $\mathcal{P}_n$ of all polynomials with leading coefficient 1, and denote by $M_n(E)^n$ the greatest lower bound of $\max_{z\in E} \big|P_n(z)\big|$ as $P_n(z)$ varies in $\mathcal{P}_n$. The "Čebishov constant" $M(E) = \lim_{n\to\infty} M_n(E)$ is also equal to $\delta(E)$.

For $n \geq 2$ the points (z_j^n) in E $(1 \leq j \leq n)$ for which (50) achieves its maximum are called *Fekete points.* Fekete proved that if E is the inside a continuous, simple, closed curve C together with C itself, then the Fekete points are well-interpolating (Z. Angew. Math. Mech., **6** (1926), 410–413).

This implies that the points $\Psi(z_j^n)$ are uniformly distributed. Kővári and Pommerenke obtained precise results about the distribution of Fekete points (Mathematika **15** (1968), 70–75, **18** (1971), 40–49).

13. Epilogue

And here, my friends, I cease. There is, however, much, much more. I hope I gave you a taste of some beautiful classical mathematics and the desire to read more about it. Fortunately this is easy, the works of Fejér, F. and M. Riesz, Pólya, Rényi, Szegő, Szász, Turán have appeared collected together (see Bibliography). They were not only titans of mathematics but also masters of exposition.

References

[2] Achieser, N. I., *Theory of Approximation,* Ungar (New York, 1956).

[13] Bernstein, Vladimir, *Leçons sur les progrès récents de la théorie des séries de Dirichlet,* Collection des Monographies sur la Théorie des Fonctions, Gauthiers-Villars (Paris, 1933).

[14] Bieberbach, Ludwig, *Analytische Fortsetzung,* Ergebnisse der Mathematik und Ihrer Grenzgebiete, Neue Folge, Heft 3, Springer-Verlag (Berlin–Göttingen–Heidelberg, 1955).

[15] Boas, Ralph Philip, Jr, *Entire Functions,* Pure and Applied Mathematics, Vol. V, Academic Press (New York, 1954).

[21] Carathéodory, Constantin, *Gesammelte Mathematische Schriften,* C.H. Beck'sche Verlagsbuchhandlung (München, 1954–57).

[22] Cartan, Henri, *Collected Works,* ed. R. Remmert and J-P. Serre, Springer-Verlag (Berlin–Heidelberg–New York, 1979).

[27] Dienes, Pál, *Leçons sur les singularités des fonctions analytiques,* Gauthier-Villars (Paris, 1913).

[28] Dienes, Pál, *Taylor Series,* An introduction to the theory of functions of a complex variable, Oxford University Press (1931).

[29] Dieudonné, Jean, *La théorie analytique des polynomes d'une variable (à coefficients quelconques),* Mémorial des Sciences Mathématiques, Fasc. XCIII, Gauthiers-Villars (Paris, 1938).

[31] *Entire Functions and Related Parts of Analysis,* eds. Jacob Korevaar, S. S. Chern, Leon Ehrenpreis, W. H. J. Fuchs, L. A. Rubel, Proceedings of Symposia in Pure Mathematics, Vol. XI, American Mathematical Society (Providence, RI, 1968).

[40] Fejér, Lipót, *Összegyűjtött Munkái = Gesammelte Arbeiten,* ed. Pál Turán, Akadémiai Kiadó (Budapest, 1970).

[59] Grenander, Ulf – Szegő, Gábor, *Toeplitz Forms and their Applications,* University of California Press (Berkeley and Los Angeles, 1958)/Chelsea (New York, 1984).

[64] Hadamard, Jacques – Mandelbrojt, Szolem, *La série de Taylor et son prolongement analytique,* Scientia, No. 41, Gauthiers-Villars (Paris, 1926).

[71] Hewitt, Edwin – Stromberg, Karl, *Real and Abstract Analysis,* Springer-Verlag (New York, 1965).

[74] Hurwitz, Adolf – Courant, Richard, *Funktionentheorie mit einem Anhang von H. Röhrl,* Grundlehren der Mathematischen Wissenschaften in Einzeldarstellungen, Band 3, 4th edition, Springer-Verlag (Berlin–Heidelberg–New York, 1964).

[97] Koosis, Paul, *Introduction to Hp Spaces,* London Mathematical Society Lecture Note Series, 40, Cambridge University Press (Cambridge–London–New York, 1980).

[104] Landau, Edmund, *Darstellung und Begründung einiger neuerer Ergebnisse der Funktionentheorie,* Julius Springer (Berlin, 1929).

[113] Marden, Morris, *Geometry of Polynomials,* Mathematical Surveys and Monographs, Volume 3, American Mathematical Society (Providence, RI, 1989).

[118] Neumann, János (John von Neumann), *Collected Works,* 6 volumes, ed. A. H. Taub, Pergamon Press (New York, 1961).

[125] Obrechkoff, Nikola, *Quelques classes de fonctions entières limites de polynomes et de fonctions méromorphes limites de fractions rationnelles,* Actualités Scientifiques et Industrielles, No. 891, Hermann (Paris, 1941).

[128] Pólya, György, *Collected Papers, Vol. I, Singularities of Analytic Functions,* ed. R. P. Boas. Mathematics of Our Time, Vol. 7 (1974); *Vol. II, Location of Zeros,* ed. R. P. Boas, Mathematics of Our Time, Vol. 8 (1974); *Vol. III, Analysis,* ed. J. Hersch and G.-C. Rota, Mathematics of Our Time, Vol. 21 (1984); *Vol. IV, Probability, Combinatorics, Teaching and Learning in Mathematics,* ed. G.-C. Rota, Mathematics in Our Time, Vol. 22 (1984), The MIT Press (Cambridge, Mass).

[129] Pólya, György – Szegő, Gábor, *Aufgaben und Lehrsätze aus der Analysis,* Grundlehren der Mathematischen Wissenschaften in Einzeldarstellungen, Band 19–20, Springer-Verlag (Berlin, 1925), Dover (New York, 1945), 2nd edition, Springer, 1954, 3rd edition, Springer 1964, 4th edition, Heidelberger Taschenbücher, Band 73–74, Springer, 1970. English translation: *Problems and Theorems in Analysis,* Die Grundlehren der Mathematischen Wissenschaften, Band 193, 216, Springer (Berlin–New York–Heidelberg, 1972, 1976).

[135] Pólya, *Studies in Mathematical Analysis and Related Topics - Essays in Honor of George Pólya,* ed. Gábor Szegő, Charles Loewner, Stefan Bergman, Menahem Max Schiffer, Jerzy Neyman, David Gilberg, Herbert Solomon, Stanford University Press (Stanford, California, 1962).

[140] Rademacher, Hans – Toeplitz, Otto, *Von Zahlen und Figuren,* Springer-Verlag (Berlin, 1930).

[151] Rényi, Alfréd, *Selected Papers,* ed. Pál Turán, Akadémiai Kiadó (Budapest, 1976).

[156] Riesz, Frigyes, *Összegyűjtött Munkái = Œuvres complètes = Gesammelte Arbeiten,* ed. Á. Császár, Akadémiai Kiadó (Budapest, 1960).

[158] Riesz, Marcel, *Collected Papers,* ed. L. Gårding and L. Hörmander, Springer-Verlag (Berlin–Heidelberg–New York, 1988).

[163] Schur, Issai, *Gesammelte Abhandlungen,* ed. Alfred Brauer and Hans Rohrbach, Springer-Verlag (Berlin–New York, 1973).

[166] Stein, Elias M., *Harmonic Analysis: Real Variable Methods, Orthogonality and Oscillatory Integrals,* Princeton Mathematical Series, 43, Princeton Unversity Press (Princeton, NJ, 1993).

[172] Szász, Ottó, *Collected Mathematical Papers,* ed. H. D. Lipsich, University of Cincinnati (1955).

[173] Szegő, Gábor, *Collected Papers,* ed. R. Askey, Birkhäuser (Boston–Basel–Stuttgart, 1982).

[174] Szegő, Gábor, *Orthogonal Polynomials,* American Mathematical Society Colloquium Publications, Vol. XXIII., 1939, revised 1959, 3rd edition 1967, 4th edition 1975.

[181] Titchmarsh, E. C., *The Theory of the Riemann Zeta-Function,* Clarendon Press (Oxford, 1951).

[184] Turán, Pál, *Collected Papers,* ed. Pál Erdős, Akadémiai Kiadó (Budapest, 1990).

[187] Turán, Pál, *On a New Method of Analysis and its Applications,* Published posthumously with the assistance of Gábor Halász and János Pintz, Pure and Applied Mathematics. A Wiley-Interscience Series of Texts, Monographs and Tracts, John Wiley and Sons (New York–Chichester–Brisbane, 1984).

[194] Walsh, Joseph L., *Bibliography of Joseph Leonard Walsh,* J. Approx. Theory 5, No. 1 (1972), xii–xxviii.

[203] Zygmund, Antoni, *Trigonometric Series,* Cambridge University Press (London–New York, 1959).

János Horváth
University of Maryland
Department of Mathematics
1301 Mathematics Bldg.
College Park, Maryland 20742-4015
U.S.A.

jhorvath@wam.umd.edu

BOLYAI SOCIETY
MATHEMATICAL STUDIES, 14

A Panorama of Hungarian Mathematics in the Twentieth Century, pp. 373–382.

THEODORE VON KÁRMÁN

STUART S. ANTMAN

Theodore von Kármán (szöllőskislaki Kármán Tódor) was born in Budapest in 1881 and died in Aachen in 1963. In 1902 he received his undergraduate degree in Engineering from the Royal Joseph University of Polytechnics and Economics in Budapest. In 1908, under the direction of the eminent fluid-dynamicist Ludwig Prandtl, he received his doctorate from the University of Göttingen for his work on the buckling of columns. He served there as a Privatdozent under Prandtl until 1913, when he became Professor of Aeronautics and Mechanics at the Technical University of Aachen. In 1929 he left for the California Institute of Technology in Pasadena, where he spent the rest of his life.

Von Kármán's degrees were in engineering, his academic appointments were in engineering, and virtually all of his research was devoted to engineering science and to practical questions about the design of aircraft and missiles. He was an adept experimentalist. He always identified himself as an engineer. He became a celebrity as an engineer in the United States. And yet, von Kármán had marked mathematical ability, he was intimately associated with the great mathematicians of Göttingen and respected by them (they seemed to view him as mathematics' favorite engineer (see {38}*), many of his research papers were regarded as applied mathematics par excellence, he effectively exploited his reputation as a consummate engineer to promote the mathematical training of engineers, and he greatly influenced work in applied mathematics. (The noted fluid-dynamicist W. R. Sears, a student of von Kármán, wrote, "It was clear to those of us who worked close to him that mathematics—applied mathematics—was his true love." {39, p. 176}.)

*In this article, all reference numbers, enclosed in brackets, correspond to the list at the end of this article.

In this article, I discuss a small sampling of von Kármán's scientific work that could be regarded as applied mathematics when it was published. (Discussions of his contributions to technology and of his role as administrator, government consultant, and public figure can be found in {9, 12, 13, 32}.)

The von Kármán equations for plates. At the invitation of Felix Klein {32, pp. 52–53}, von Kármán {15} prepared the 75-page article *Festigkeitsprobleme in Maschinenbau* {15} for the *Encyklopädie der mathematischen Wissenschaften* edited by Klein. (That this invitation was made when von Kármán had just received his doctorate testifies to the esteem with which he was held by the mathematical community at Göttingen.) This survey of structural mechanics, i.e., the mechanics of deformable rods and shells, derived the governing differential equations (mostly linear) and analyzed some specific problems for them.

Von Kármán began his very brief treatment of the deformation of elastic plates with a discussion of the Kirchhoff theory, which characterizes the small transverse displacement w of a thin (homogeneous, isotropic) plate (of constant thickness) under the action of a transverse force of intensity f per unit area as the solution of

$$D\Delta^2 w = f$$

where

$$\Delta^2 u := \frac{\partial^4 u}{\partial x^4} + 2\frac{\partial^4 u}{\partial x^2 \partial y^2} + \frac{\partial^4 u}{\partial y^4}$$

is the two-dimensional biharmonic operator acting on a function u, and where D is a positive constant accounting for the stiffness of the plate; D is proportional to the cube of the thickness h. Von Kármán then observed that this model is valid only if w is small relative to the thickness of the plate. To construct a theory capable of describing larger displacements, von Kármán replaced the linear relations between the in-plane strains and the displacements with the correct nonlinear relations, but retained other geometric simplifications, and took the relation between stress and strain to be linear. By this process, in the span of one page, he came up with the celebrated von Kármán equations for plates:

$$D\Delta^2 w - h[\Phi, w] = f,$$

$$\Delta^2 \Phi = -\frac{1}{2}E[w, w]$$

where

$$[u,v] := \frac{\partial^2 u}{\partial x^2}\frac{\partial^2 v}{\partial y^2} + \frac{\partial^2 u}{\partial y^2}\frac{\partial^2 v}{\partial x^2} - 2\frac{\partial^2 u}{\partial x \partial y}\frac{\partial^2 v}{\partial x \partial y}$$

is the Monge–Ampère operator acting on the functions u and v, and E is the elastic modulus. The function Φ is a stress function whose second derivatives deliver the resultant contact forces (stress resultants) in the plane of the plate.

Although the beauty of the von Kármán equations inherent in the presence of the biharmonic and Monge–Ampère operators could not fail to attract mathematicians, their semilinearity put these equations beyond the analytic resources available at the time. Indeed, in his influential expository paper {28} of 1940, von Kármán called upon mathematicians to bring their still primitive tools of nonlinear analysis to bear on these equations. (It seems to me that von Kármán presented these equations to the mathematical community in 1940 with an assurance as to their value that was lacking in 1910.) In the meantime, von Kármán {24, 26, 29} had demonstrated the crucial role of nonlinearity in the buckling of shells. (The problems discussed in these three papers continue to provide challenges for analysis.)

Friedrichs and Stoker {10} answered the call in 1941. Their lengthy work, which influenced the development of bifurcation theory in the United States, was the first rigorous mathematical analysis of the von Kármán equations. (Friedrichs, a student of Courant's at Göttingen, had been sent by Courant to work with von Kármán at Aachen {38}.) In the mid-1950's began an intensive analysis of existence, multiplicity, and bifurcation of solutions to boundary-value problems for the von Kármán and related equations (see {6, 7, 43}). The fascinating role of these equations as an inspiration for Rabinowitz's {37} global bifurcation and continuation theory is detailed in {1}.

That the von Kármán equations, obtained by an ad hoc combination of theory with insight, represent an improvement over the traditional Kirchhoff theory has inspired several directions of research in shell theory and in its mathematical analysis: (i) The derivation of the von Kármán and related equations systematically (albeit formally) as the leading term of an asymptotic expansion in a thickness parameter {5, 6, 8}. (ii) The derivation of "geometrically exact" equations for the large motion of shells {2, 11, 35} (which do not rely on any geometric approximation and which describe new phenomena. Underlying von Kármán's derivation of his equation are approximations analogous to replacing the sin function by its cubic approximation.) (iii) The still largely open problem of deriving sharp estimates for the errors between solutions of equations for shells and those for the 3-dimensional theory {3}.

Throughout his scientific career, von Kármán maintained a research interest in problems of solid mechanics. His work on the buckling of elastic structures has become a standard part of the engineering theory of elastic stability. His work on plasticity and plastic buckling have had an important influence on modern developments {14}. But von Kármán's main research and engineering efforts after 1914 were increasingly directed towards fluid dynamics.

The von Kármán vortex street. Von Kármán received his first recognition in fluid dynamics when he explained the failure of a student of Prandtl's, despite herculean efforts, to get rid of oscillations in the experimental measurements of pressure on the surface of a circular cylinder obstructing the flow of a steady stream of water {32, pp. 62 ff.}. Von Kármán first supposed that the oscillations are in fact present, and that they are caused by water rolling up into two trails of vortices (eddies) breaking off from the top and bottom of the cylinder. (Many years earlier, Helmholtz had observed the formation of vortices in the flow past a flat plate.) When the assumption that the vortices were shed simultaneously led to unacceptable instabilities, von Kármán assumed that they were shed alternately. He then determined the spacings of these alternating vortices that are stable. Specifically, he severely idealized the problem {16}: He considered the 2-dimensional irrotational flow of an invisicid incompressible fluid produced by two parallel rows of equally spaced vortices, with one row of vortices rotating in one direction and the other row in the opposite direction, and with each vortex of one row opposite a midpoint of a pair of vortices of the other row. Since all the rotation is concentrated at the singular points holding the vortices, the flow is irrotational away from them. Consequently, the conjugate of the complex velocity is the derivative of meromorphic function determined by the poles at the vortices. Von Kármán was able to ignore the source of the vortices, the cylinder, by regarding it as shifted to infinity. In other words, he was studying a steady state that could conceivably exist away from the source. He analyzed the linear stability of the flow by perturbing the locations of the vortices. Remarkably, the stable dispositions conform well to what was observed in experiment. For accessible discussions of the physical and mathematical setting of this work see {34, 36, 41}.

This work provided an explanation of a major and hitherto unknown source of drag. The collapse of the Tacoma Narrows bridge in 1940 (discussed in detail in {31}) is attributed to the resonant forcing produced by a similar vortex structure that was shed by solid fences when the bridge was subjected to a steady transverse wind.

The statistical theory of turbulence. Von Kármán, like Prandtl, had long been concerned with the puzzling phenomena of turbulence, making important contributions in {17, 19}. His most notable contribution to the subject was to endow the statistical theory of turbulence initiated by G. I. Taylor with a rich and useful mathematical structure. In the words of S. Goldstein {12, p. 349}, " ... [H]e dealt mainly with a general systematic development of [Taylor's theory in {21, 22, 23}], the last with L. Howarth. Von Kármán pointed out that the correlations between two velocity components at any two points at a distance r apart are the components of a tensor, which is a function of the vector distance between the points. In the case of isotropy, the correlation divided by the mean square velocity depends on just two scalar functions of the distance r and the time t. In an incompressible fluid, the equation of continuity yields a relation between these two scalar functions, so only one is involved. If the triple products of components of velocities at the two points are neglected, an equation can then be derived from the equations of motion for changes in this single scalar, which can be used to obtain information about the rate of decay of the turbulence. The triple correlations were first neglected in this way, but this is incorrect, as G. I. Taylor pointed out. Von Kármán in fact explictly stated that if this is incorrect the vortex filaments would have a permanent tendency to be stretched or compressed along the axis of vorticity, and thought this was not the case; Taylor pointed out that the facts showed that it was, there being a tendency for the vortex filaments to stretch on the average. Von Kármán and Howarth showed that the triple correlation tensor also involves only one scalar function for the case of isotropy for an incompressible fluid, and that the correlation between pressure and velocity is zero in this case. A partial differential equation connecting the double and triple correlation functions was then derived, and equations for the dissipation of energy and vorticity deduced." Throughout the next 15 years, von Kármán continued to contribute novel ideas to the subject of turbulence. For a technical account of some of this work see {4}.

Mathematical methods in engineering. Following in the footsteps of his father Mór (Moritz), whose role in modernizing the Hungarian educational system earned him a 'von', von Kármán did much to modernize the mathematical training of engineers in the United States and elsewhere. In the 1930's the mathematical sophistication of American engineers was far inferior to that which von Kármán picked up in Göttingen and which he found valuable in his own work. In pushing for a far richer (but not too rich an) exposure to real mathematics for engineers, von Kármán demonstrated the same political astuteness that served him so well in dealing with bureaucracies as a public figure: In two publications {25, 30} in the 1940's directed to engineers on the role of mathematics in engineering, he prominently identified himself as an engineer and put 'engineer' or 'engineering' in the titles. (These works have a flavor different from that of {18, 28} directed to mathematicians.) In these works he cited stereotypical criticisms of pure mathematicians: They are concerned with proving the existence of solutions to equations that every engineer knows to have solutions on physical grounds, and if mathematicians were ever to solve specific problems, they would employ the simplest possible geometries (just as von Kármán did for his vortex street). Having thus demonstrated that he was not a sycophant of mathematics, he was then positioned to advocate effectively for the enrichment of engineers' actual mathematical education and also for the incorporation of mathematical notions in their scientific courses. (He was thus trying to prevent American engineering students from experiencing his own unhappy exposure to engineering sciences at the Royal Joseph University, about which he said, "The conventional courses, such as hydraulics, electricity, steam engineering, or structures, were taught like baking or carpentry, with little regard for the understanding of nature's laws which underlie the sciences" {32, p. 26}.) The popular and valuable book {27}, written with M. Biot, significantly advanced this program. It contained elementary treatments of ordinary differential equations, linear algebra, Bessel functions, Fourier methods, and finite differences in the setting of classical and structural mechanics.

Aeronautics and astronautics. Whereas liquids like water are virtually incompressible, gases are not, and the effects of compressibility in gases become pronounced when they move at speeds exceeding about a fifth of the speed of sound. The type of the governing partial differential equations depends crucially upon whether the fluid is viscous, whether it is compressible, and the local speed at which it moves. The most striking effect of compressibility is the appearance of shocks (strictly speaking for an inviscid compressible fluid), which are discontinuities in the derivatives of the velocity field and in the pressure field. Von Kármán had published some early papers on gas dynamics. In the 1930's, well before high-speed flight became a reality, he advocated the creation of a comprehensive theory and began his fundamental work on it with {20}. In the 1940's, he began serious work on rockets and jet propulsion, which would be the main focus of his activities for the rest of his life. To handle the practical complexities of high-speed flight both near and away from the earth, he promoted the development of aerothermochemistry in which fluid-dynamical, thermal, and chemical effects are coupled, as for example in combustion. It was not long before many of these ideas formed the heart of graduate teaching in aeronautics. The most accessible scientific treatment of his work in this area is in his own posthumous tract {33}.

Summary. Von Kármán's work on fluid dynamics was immediately assimilated into the main stream of the general theory and forms an extensive contribution of permanent value. Accounts of much of this work can be found in standard references on fluid dynamics. Von Kármán's work on solid mechanics, on the other hand, represented pioneering attacks on nonlinear problems of great theoretical and practical importance. His analyses continue to challenge his successors, but they cannot be said to represent permanent contributions, partly because the nonlinear problems he grappled with had not yet been subsumed under a cohesive and mature theory like that of fluid dynamics.

Appreciations of von Kármán's scientific contributions are given in numerous obituaries and memorials, among which are {9, 12, 40, 42, 44} all by fluid-dynamicists. The best place to start to learn of the personal side of von Kármán is his autobiography {32}, which is valuable also for his discussion of his research.

REFERENCES

[84] Kármán, Tódor, *Collected Works of Theodore von Kármán,* Volumes 1–4, Butterworths Scientific Publications (London, 1956); Volume 5, Von Kármán Institute for Fluid Dynamics, Rhode-St. Genese (Belgium, 1975).

{1} S. S. Antman, The influence of elasticity on analysis: Modern developments, *Bull. Amer. Math. Soc.* (New Series), **9** (1983), 267–291.

{2} S. S. Antman, *Nonlinear Problems of Elasticity,* Springer, 1995.

{3} I. Babuška and L. Li, The problem of plate modeling: Theoretical and computational results, *Comp. Meths. in Appl. Mech. Engg.*, **100** (1992), 249–273.

{4} G. K. Batchelor, *The Theory of Homogeneous Turbulence,* Cambridge Univ. Pr. (1953).

{5} P. G. Ciarlet, A justification of the von Kármán equations, *Arch. Rational Mech. Anal.,* **73** (1980), 349–389.

{6} P. G. Ciarlet, *Mathematical Elasticity, Volume II: Theory of Plates,* North-Holland (1997).

{7} P. G. Ciarlet & P. Rabier, *Les Equations de von Kármán,* Springer (1980).

{8} J.-L. Davet, Justification de modèles de plaques nonlinéaires pour des lois de comportment générales, *Mod. Math. Anal. Num.*, **20** (1986), 147–192.

{9} H. Dryden, Theodore von Kármán, 1881–1963, *Biog. Mem. Nat. Acad. Sci.*, **38** (1966), 345–384.

{10} K. O. Friedrichs and J. J. Stoker, The nonlinear boundary value problem of the buckled plate, *Amer. J. Math.,* **63** (1941), 839–888.

{11} G. Friesecke, R. D. James, and S. Müller, A theorem on geometric rigidity and the derivation of nonlinear plate theory from three-dimensional elasticity, *Comm. Pure Appl. Math.,* **55** (2002), 1461–1506.

{12} S. Goldstein, Theodore von Kármán, 1881–1963, *Biog. Mem. Fellows Roy. Soc.,* **12** (1966), 335–365.

{13} M. H. Gorn, *The Universal Man, Theodore von Kármán's Life in Aeronautics,* Smithsonian Inst. (1992).

{14} J. Hutchinson, Plastic buckling, *Advances in Applied Mechanics,* Vol. 14, Academic Press (1974), 67–144.

{15} Th. von Kármán, *Festigkeitsprobleme in Maschinenbau,* in *Encyklopädie der mathematischen Wissenschaften,* Vol. IV/4, (1910), 311–385, edited by F. Klein and C. Müller, Teubner, reprinted in [84, Vol. 1, 141–207].

{16} Th. von Kármán, Über den Mechanismus des Widerstandes, den ein bewegter Körper in einer Flüssigkeit erfährt, *Göttinger Nachr.* (1911), 509–517, (1912), 547–556, reprinted in [84, Vol. 1, 324–338].

{17} Th. von Kármán, Über laminäre und turbulente Reibung, *Z. angew. Math. Mech.,* **1** (1921), 233–252, reprinted in [84, Vol. 2, 70–97].

{18} Th. von Kármán, Mathematik und technische Wissenschaften, *Naturwiss.,* **18** (1930), 12–16, reprinted in [84, Vol. 2, 314–321].

{19} Th. von Kármán, Mechanische Ähnlichkeit und Turbulenz, *Göttinger Nachr.* (1930), 58–76, reprinted in [84, Vol. 2, 322–336].

{20} Th. von Kármán and N. B. Moore, Resistance of slender bodies moving with supersonic velocities, with special reference to projectiles, *Trans. Amer. Soc. Mech. Engrs.* (1932), 303–310, reprinted in [84, Vol. 2, 376–393].

{21} Th. von Kármán, On the statistical theory of turbulence, *Proc. Nat. Acad. Sci.,* **23** (1937), 98–105, reprinted in [84, Vol. 3, 222–227].

{22} Th. von Kármán, The fundamentals of the statistical theory of turbulence, *J. Aero. Sci.*, **4** (1937), 131–138, reprinted in [84, Vol. 3, 228–244].

{23} Th. von Kármán and L. Howarth, On the statistical theory of isotropic turbulence, *Proc. Roy. Soc. A*, **164** (1938), 192–215, reprinted in [84, Vol. 3, 280–300].

{24} Th. von Kármán and H.-S. Tsien, The buckling of spherical shells by external pressure, *J. Aero. Sci.*, **7** (1939), 43–50, reprinted in [84, Vol. 3, 368–380].

{25} Th. von Kármán, Some remarks on mathematics from the engineer's viewpoint, *Mechanical Engrg.* (1940), 308–310, reprinted in [84, Vol. 4, 1–6].

{26} Th. von Kármán, L. G. Dunn, and H.-S. Tsien, The influence of curvature on the buckling characteristics of structures, *J. Aero. Sci.*, **7** (1940), 276–289, reprinted in [84, Vol. 4, 7–31].

{27} Th. von Kármán and M. Biot, *Mathematical Methods in Engineering,* McGraw-Hill (1940).

{28} Th. von Kármán, The engineer grapples with nonlinear problems, *Bull. Amer. Math. Soc.*, **46** (1940), 615–683, reprinted in [84, Vol. 4, 34–93].

{29} Th. von Kármán and H.-S. Tsien, The buckling of thin cylindrical shells under axial compression, *J. Aero. Sci.*, **8** (1941), 303–312, reprinted in [84, Vol. 4, 107–126].

{30} Th. von Kármán, Tooling up mathematics for engineering, *Quart. Appl. Math.*, **1** (1943), 2–6, reprinted in [84, Vol. 4, 189–192].

{31} Th. von Kármán, L'aérodynamique dans l'art de l'ingénieur, *Mém. Soc. Ingénieurs de France.* (1948), 155–178, reprinted in [84, Vol. 4, 372–393].

{32} Th. von Kármán, *The Wind and Beyond,* Little–Brown (1967).

{33} Th. von Kármán, *From Low-Speed Aerodynamics to Astronautics,* Pergamon (1963).

{34} L. M. Milne-Thomson, *Theoretical Hydrodynamics,* 5th edition, Macmillan (1968).

{35} P. M. Naghdi, Theory of Shells, in: *Handbuch der Physik,* Vol. VIa/2, edited by C. Truesdell, Springer (1972), pp. 425–640.

{36} L. Prandtl, *Essentials of Fluid Dynamics,* Blackie (1952).

{37} P. H. Rabinowitz, Some global results for nonlinear eigenvalue problems, *J. Funct. Anal.*, **7** (1971), 487–513.

{38} C. Reid, *Courant,* Springer (1976).

{39} W. R. Sears, Recollections of Theodore von Kármán, *J. Soc. Indust. Appl. Math.*, **13** (1965), 175–183.

{40} W. R. Sears, The scientific contributions of Theodore von Kármán, 1881–1963, *Phys. Fluids,* **7** (1964), v–viii.

{41} A. Sommerfeld, *Mechanics of Deformable Bodies,* Academic Press (1964).

{42} G. I. Taylor, Memories of von Kármán, 1881–1963, *J. Fluid Mech.,* **16** (1963), 478–480.

{43} I. I. Vorovich, *Nonlinear Theory of Shallow Shells,* Springer (1999).

{44} F. L. Wattendorf and F. J. Malina, Theodore von Kármán, 1881–1963, *Astronautica Acta,* **10** (1964), 81–92.

Stuart S. Antman

Department of Mathematics
Institute for Physical Science and Technology
and Institute for Systems Research
University of Maryland
College Park, MD 20742-4015, U.S.A.

ssa@math.umd.edu

Geometry

BOLYAI SOCIETY
MATHEMATICAL STUDIES, 14

A Panorama of Hungarian Mathematics in the Twentieth Century, pp. 385–413.

Differential Geometry

LAJOS TAMÁSSY

In the thirties of the 19th century János Bolyai and Nikolai Ivanovič Lobacevskii created the hyperbolic geometry. Thus they proved that not only the Euclidean but also other geometries may exist. Concerning its geometrical importance, this discovery can be compared to the change which replaced the Ptolemaic geocentric concept of astronomy by the heliocentric point of view of Copernicus. Hyperbolic geometry opened new horizons. Indeed, only 30 years had to pass, and in Göttingen, in the presence of the elder Gauss, Bernhard Riemann (1826–1866) announced in his habilitation lecture (Über die Hypothesen die der Geometrie zu Grunde liegen) the basic concepts of the new geometry later named after him. His main idea joins Gauss' work.

Let us consider the hypersurface

$$\phi : U^2 \to E^3, \quad (u^1, u^2) \mapsto x^i(u^1, u^2) \qquad i = 1, 2, 3 \tag{1}$$

of the Euclidean space $E^3(x)$. According to Gauss the arc length s_E of the curve $C = \phi \circ C^*$, $C^* : I \to U^2$, $t \in (a, b) = I \mapsto u^\alpha = u^\alpha(t)$, $\alpha = 1, 2$ has (in modern notation), the form

$$s = s_E = \int_a^b \sqrt{\sum g_{\alpha\beta}\big(u(t)\big)\, \dot{u}^\alpha \dot{u}^\beta}\, dt, \quad \alpha, \beta = 1, 2. \tag{2}$$

If $\phi \circ U^2 \subset E^3$ is the plane $E^2(x^1, x^2)$ (i.e. $x^3(u^1, u^2) = 0$), then s_E gives the Euclidean arc length of C expressed in the curvilinear coordinate system (u) of E^2, where the

$$g_{\alpha\beta}(u) = \sum_\sigma \frac{\partial x^\sigma}{\partial u^\alpha} \frac{\partial x^\sigma}{\partial u^\beta} \tag{3}$$

are derived from the functions (1) describing the transition to the curvilinear system (u). Riemann's idea was to give $g_{\alpha\beta}$ $\big($Det $|g_{\alpha\beta}| \neq 0\big)$ arbitrarily, and to define the arc length by the integral (2). Today this is called the Riemannian arc length s_V of the curve C. s_V produces the Euclidean arc length in the plane E^2 related to the curvilinear coordinate system (u^1, u^2), i.e. the geometry defined by s_V is Euclidean iff (3), considered as a system of partial differential equations for the given $g_{\alpha\beta}(u)$ and the unknown functions $x^\alpha(u^1, u^2)$, is solvable. However, this occurs rarely. Hence, Riemann's geometry gives the Euclidean geometry as a special case only. If we start with an n-dimensional manifold M in place of the E^2, and give on M a tensor g of type $(0,2)$ (in local coordinates by $g_{ik}(x)$), then we obtain the Riemannian manifold $V^n = (M, g)$.

This lecture of Riemann was first published only after his death, in 1868, in the volume of his collected works. However, in this lecture one can find certain signs of the Finsler geometry too. The integrand of (2) is a special positive valued function $\mathcal{L}(u, \dot{u})$ positively homogeneous of degree 1 in $\dot{u}$. If we are given such a function on M, and define the arc length in the form

$$s_F := \int_a^b \mathcal{L}\big(u(t), \dot{u}(t)\big)\, dt,$$

then we arrive at a still more general geometry. In 1918 Paul Finsler obtained such a geometry (see his Göttingen thesis "Über Kurven und Flächen in allgeneinen Räumen" written under the supervision of Constantin Carathéodory). He called this a geometry with general metric, and later it was designated by others by the shorter name of Finsler geometry. This geometry is the most general, under certain natural requirements, among those geometries for which the arc length is the integral of the infinitesimal distance. According to Shiing-shen Chern Finsler geometry is nothing other than Riemannian geometry without the quadratic restriction on the function $\mathcal{L}^2$. He sees in this the geometry of the new century. The architect of the early part of Finsler geometry was Ludwig Berwald, the excellent professor of the Charles University in Prague, who later came to a tragic end during his deportation in the Lodz (Litzmannstadt) Ghetto. He laid the foundation of this geometry between 1920 and 1940. His pupil and later private-docent of Prague University was Ottó Varga, who after the German occupation of Prague came to Kolozsvár, and later to Debrecen.

It is well known that every differential geometry, and so the Finsler and the Riemannian geometry too, has two key concepts: the notion of metric and the parallelism of vectors. The fundamental function $\mathcal{L}(x, y)$, $x \in M$, $y \in T_xM$ determines the metric of the Finsler manifold $F^n = (M, \mathcal{L})$, $\mathcal{L}(x, y) = \|y\|_F$ gives the Finsler norm of the vector $y \in T_xM$, and $\mathcal{L}(x, dx) = \|dx\|_F$ the Finsler distance between the points x and $x+dx$. Also $\mathcal{L}$ makes each tangent space T_xM into a Minkowski space (i.e. a normed vector space). The endpoints of the unit vectors of T_xM form a convex and centrally symmetric hypersurface $\mathcal{I}(x)$ called an indicatrix. In a V^n they are ellipsoids, and unit spheres in an E^n. The parallelism of the vectors y of the tangent bundle $TM = \{(x, y)\}$ is defined by a linear connection. This is a mapping $\varphi : T_xM \to T_{x+dx}M$ between the vectors of the neighbouring tangent spaces, the differential dy being linear in $y : \varphi(y) = y + dy$, $dy^i = \sum_{j,k} \Gamma^i_{jk}(x) y^j \, dx^k$. Among these mappings those which preserve the length of the vectors are most applicable (they are called metrical), and take indicatrices into indicatrices. Nevertheless while two ellipsoids (indicatrices of a Riemannian space) can always be taken into each other by a linear transformation, this is not true for indicatrices of a Finsler space. Thus in a Finsler space among the vectors of the tangent spaces there does not exist, in general, a linear and metrical connection, and this makes impossible the development of an absolute differential calculus and curvature theory similar to the Riemannian geometry. This difficulty was surmounted by Élie Cartan in 1931. He osculated the indicatrix $\mathcal{I}(x_0)$ in each direction y_0 by an ellipsoid of equation $\sum g_{ik}(x_0, y_0) y^i y^k = 1$. He considered the vectors at line-elements (x, y). These are the Finsler vectors: $\xi^i(x, y)$. Then he defined their norm by $\| \xi(x, y) \|^2 = \sum g_{ik}(x, y) \xi^i(x, y) \xi^k(x, y)$, and thus he was able to introduce a metrical and linear connection. After several additional requirements this connection is unique, and nowadays it is called Cartan connection.

At this time began Varga's scientific carrier which, with a few exceptions, focused on Finsler geometry or questions connected with it. A deep geometric thinking was a characteristic feature of his scientific activity. In one of his early works {43} he gives a completely geometric construction which leads without any further requirement to the just mentioned metrical and linear Cartan connection. Let us consider a line-element field $x^i = x^i(\tau)$, $\dot{x}^i = \dot{x}^i(\tau)$, $\tau \in [\tau_1, \tau_2] = T$ and along this a vector field $\xi^i\big(x(\tau), \dot{x}(\tau)\big)$. From each point $x(\tau_0)$, $\tau_0 \in T$ there goes out in the direction of $\dot{x}(\tau_0)$ a unique geodesic of the Finsler space F^n. These yield a 1-parameter family, γ_1, of geodesics. γ_1 can be completed by further geodesics of the F^n to a family, γ_2, such that γ_2 covers one-fold a small tube $\mathfrak{B}$ around the curve $x(\tau)$. This completion can be done in several different ways. The tangents, $r(x)$, of the geodesics at the points $x \in \mathfrak{B}$ determine a line-element field, $\big(x, r(x)\big)$, and by

$$\gamma_{ij}(x) := g_{ij}\big(x, r(x)\big) \qquad g_{ij}(x,y) := \frac{1}{2}\frac{\partial^2 \mathcal{L}^2}{\partial y^i \partial y^j}$$

a Riemannian space V^n on $\mathfrak{B}$. This V^n osculates the F^n in the sense that the geodesics of V^n resp. F^n starting out from $x \in \mathfrak{B}$ in the direction $r(x)$ osculate each other in the second order, and the $\dot{x}(\tau)$ turn out to be parallel along $x(\tau)$ in V^n. Now $\xi^i\big(x(\tau), \dot{x}(\tau)\big)$ will be called parallel in the F^n if $\xi^i(\tau) := \xi^i\big(x(\tau), \dot{x}(\tau)\big)$ is parallel along $x(\tau)$ in the V^n. Moreover, Varga shows that with an appropriate choice of the extension of γ_1 to γ_2 we can show that the Finsler connection obtained on $\mathfrak{B}$ is just the Cartan connection.

With the problem of the linear connection of Finsler vector fields, Varga has already dealt earlier in his Ph.D. thesis. Slightly before publishing his thesis, in 1933 there appeared in the C.R. Paris a short announcement by Cartan, whose contents were explained in detail in his famous booklet "Les espaces de Finsler" (Actualités Scientifiques et Industrielles No. 79, Paris, Hermann, 1934) which is considered still the foundation of the Cartan theory of Finsler spaces. Varga's thesis showed a considerable overlap with this, so only a summary was published in the Prague journal Lotos (vol. **84** (1936), 1–4). The above discussed osculating Riemannian space represents another geometric solution of the problem leading to the same result.

He was able to apply with success the osculation of an F^n by a Riemannian space to other problems also. As well known, the sectional curvature, $R(x,p)$, of a Riemannian space V^n at a point x and plane position p is the curvature of the 2-dimensional V^2 induced by V^n on the subspace ϕ^2 consisting of the geodesics of V^n tangent to p at x, and this curvature equals the Gaussian curvature of the surface representing V^2 in Euclidean three-space. This $R(x,p)$ was generalized and transferred into the Finsler space F^n partly on the basis of its formal expression, and partly on the basis of its role in certain variational problems. This generalization, the Riemann–Berwald curvature $R(x,v;X)$ (today called flag curvature) of the F^n, is defined at a line-element (x,v) and a vector X defined at this (x,v). Varga has shown in {46} that $R(x,v;X)$ too is the curvature of a 2-dimensional subspace F^2 induced by F^n on the subspace ϕ^2 consisting of geodesics of F^n tangent to the plane-position $(v,X) = p$, similarly to the case of the Riemannian geometry. He considered the geodesic C starting from x in the direction of X (belonging to both F^2 and F^n), and constructed a Riemannian space V^n osculating F^n along this C (this is a little different from the previous osculating Riemannian space). Then he proved that the curvature $\bar{R}(x)$ of the $\bar{V}^2 := V^n \restriction \phi^2$ (i.e. the restriction of V^n to ϕ^2) equals $R(x, v; X)$. On the other hand he proved that this $\bar{R}(x)$ equals the Finsler curvature $S(x, v)$ (introduced by Finsler) at x in the direction of v of the above F^2. Thus $R(x, v; X) = S(x, v)$. This shows a complete analogy to the Riemannian case.

Finsler geometry is a most natural generalization of Riemannian geometry, using notions and apparatus which may be more sophisticated than that of Riemannian geometry, but is essentially similar and closely related to it. Therefore it is of basic interest to see how and to what extent the notions and theorems of Riemannian geometry can be transferred and extended to Finsler geometry. Ottó Varga has important achievements in this direction, especially concerning Finsler spaces of scalar or constant curvature. Riemannian spaces of constant curvature which are near Euclidean spaces have an exceptional importance. Their first well known characterization was given by Beltrami according to whom V^n is of constant curvature iff it admits a geodesic mapping φ onto an affine space A^n such that φ takes every geodesic of the V^n into a straight line of A^n. This is equivalent to the vanishing of the projective curvature tensor of Weyl or the property that the difference vector $\mathcal{P}\xi - \xi$ of any vector ξ and its parallel translated $\mathcal{P}\xi$ along an infinitesimal parallelogram Π lies in the same parallelogram Π (up to quantities of the third order in the measure of the area of the parallelogram). It is clear that this last property also characterizes the projectively flat affinely connected spaces (spaces with a linear connection, but without Riemannian metric). Among Finsler spaces or affinely connected line-element spaces in the sense of O. Varga {47} there are spaces of constant- and also of scalar-curvature. Berwald called an F^n of scalar curvature $R(x, v)$ if $R(x, v; X)$ is independent of X, and of constant curvature if $R(x, v)$ is independent of v. The independence of $R(x, v)$ of v implies its independence of x too. Varga showed in {49} that Finsler spaces of scalar or constant curvature can also be characterized in a quite similar way. According to his results, F^n is of scalar curvature iff $\mathcal{P}\xi - \xi$ belongs to the vector space spanned by Π and ξ, and F^n is of constant curvature iff $\mathcal{P}\xi - \xi$ lies in Π. From his calculation it also follows that one can build the whole curvature theory on the main curvature tensor, T, introduced by him and defined by $T^i_{jk\ell} = R^i_{jk\ell} - \sum_{s,m} A^i_{s\ell} R^s_{jkm} \ell^m$, where the $R^i_{jk\ell}$ are the components of the first curvature tensor of Cartan, $A^i_{s\ell}$ those of the torsion tensor, and $\ell^m = v^m$ are the components of the unit line-element.

Varga also gave other interesting characterizations of Finsler spaces of constant curvature. An $F^n = (M, \mathcal{L})$ makes any of its k-dimensional $(k < n)$ submanifolds $N \subset M$ into a Finsler space $F^k = (N, \tilde{\mathcal{L}})$. A curve $C \subset N \subset M$ has the same arc length in F^n and in F^k, but a geodesic of F^k between two points $p, q \in N \subset M$ is not, in general, a geodesic of the F^n, for in M there may exist curves between p and q shorter than C. A submanifold in which every geodesic is at the same time a geodesic of the embedding space is called a totally geodesic submanifold. This is a generalization of the Euclidean k-dimensional planes. In a Euclidean space E^n there exists through each point and every plane position a totally geodesic submanifold (a k-plane). In a V^n or F^n this is not so. Varga showed that among the Finsler spaces this holds exactly for the spaces of constant curvature. In the case of constant negative curvature the metric induced on these totally geodesic submanifolds is Euclidean.

Of fundamental importance are his results concerning the angular-metric. For two unit vectors ξ and η at the same line-element (x_0, v_0) the angle $\varphi = \measuredangle(\xi, \eta)$ is defined by the Euclidean metric at the given line-element

$$\cos^2 \varphi = \sum_{i,k} g_{ik}(x_0, v_0)\xi^i \eta^k, \qquad g_{ik}(x, y) = \frac{1}{2} \frac{\partial^2 \mathcal{L}^2}{\partial y^i \partial y^k}.$$

These g_{ik} induce on the indicatrix $\mathcal{I}(x_0) \subset T_{x_0}M$ a Riemannian space V^{n-1}. If the direction of the unit vectors $\xi(x_0, v)$ and $\eta(x_0, \hat{v})$ coincides with the direction of their line-elements, i.e. we have $\xi(x_0, v) = av$ and $\eta(x_0, \hat{v}) = b\hat{v}$, $a, b \in R$; and moreover $\hat{v}$ is sufficiently near to v: $\hat{v} = v + dv$, then for $\measuredangle(\xi, \eta) \equiv \measuredangle(v, v + dv) = d\varphi$ we can put

$$\cos^2 d\varphi = \sum_{i,k} g_{ik}(x_0, v)\, dv^i\, dv^k = \|dv\|_V.$$

Thus the measure of the infinitesimal angle equals the Riemannian measure of the corresponding arc. So this V^{n-1} and its metric play a distinguished role in the angular metric of the F^n. Varga has shown in {51} that the curvature tensor of this V^{n-1} is the sum of the curvature tensor S (the third curvature tensor of Cartan) and the metric tensor of the bivectors of the V^{n-1}. The geometric meaning of the first and second curvature tensor R and P was already known, however, the geometric role of S was revealed by this result. He also gave a simple criterion for this V^{n-1} to be of constant curvature.

We cannot discuss in detail his numerous other results, yet we mention a few here. He had valuable results on Minkowski geometry. He gave a very clever and direct geometric derivation of the Euclidean connection in the Minkowski geometry making no use of the roundabout way of Finsler geometry, see {44}. He showed that if a hypersurface of a Minkowski space has constant normal curvature at every line-element, then the geometry induced by the Minkowski space on the hypersurface is a Finsler geometry of constant curvature with respect to the induced connection, see {54}. He achieved a number of results concerning hypersurfaces. Each of them is a little masterpiece of Finsler geometry. He studied Finsler spaces which generalize the non-Euclidean spaces, see {45}; the metrizability of affinely connected line-element spaces, i.e. the possibility of endowing a Finsler connection with a Finsler metric such that parallel vector fields have constant Finsler norms, and obtained nice results for them, see {50}. He studied Hilbert geometry (the generalization of the Cayley–Klein model of hyperbolic geometry). This not-Riemannian geometry has the important property that geodesics are straight lines. He gave an analytic characterization of all functions which represent the arclength of this geometry in {53}. He has important theorems concerning the decomposition of a Finsler space into a product of other spaces, see {52}; the coincidence of the induced and the intrinsic connection on a hypersurface of an F^n, etc.

Although the main field of Varga's activity was Finsler geometry, he also obtained notable results concerning Riemann spaces V^n of constant curvature. It was known for a long time that if through any hyperplane position of a V^n totally geodesic hypersurfaces can be laid, i.e. if the plane axiom is fulfilled, then the space is of constant curvature. But this criterion-like quality does not separate a) the spaces of constant negative and b) the spaces of constant positive curvature. Varga showed in {55} that in case of a) two hyperplanes can be laid through any plane position so that the geometry induced on them by the V^n is Euclidean (they are paraspheres), and in the case of b) a totally geodesic hyperplane can be laid through any plane position which is turned by the V^n into a V^{n-1} of constant curvature. These qualities are characteristic. He could extend the above mentioned criterion-like quality of the V^n of constant curvature to Finsler spaces, i.e. he also proved that the F^n of constant curvature are characterized by the property that through any hyperplane-position a totally geodesic hyperplane can be laid, see {56}.

He also had several works on integral geometry (Math. Z., **40** (1935), 384–405; **41** (1936), 768–784; **42** (1937), 710–736; Acta Sci. Math. Szeged,

9 (1939), 88–102; etc.). They are products of his collaboration with W. Blaschke in Hamburg in 1934–35. In these he found parameter transformation and motion invariant measures, i.e. geometric densities on different sets of geometrical configurations, and disclosed the relations existing between them, i.e. with the aid of the Crofton formulae, he established relations between integral invariants.

Varga was always intent on seizing and putting into relief the geometric meaning behind the often rather complicated formalism of Finsler geometry. This is a characteristic feature of his scientific work. His mathematical thinking was guided by simple geometric ideas. If we divide the history of Finsler geometry into three periods: I) the beginning, 1918–1940; II) development of the local theory, 1940–1970; III) use of intrinsic tools and global theory, 1970–, then we can state that Varga played a decisive role in the second period. He also created a school of differential geometry in Debrecen which continues even now, and he was one of the founders of the journal Publicationes Mathematicae Debrecen in 1949.

András Rapcsák was a colleague and collaborator of O. Varga at Debrecen University. The field of his research was line-element (or support-element) spaces and connected areas. His first investigations were related to normal coordinate systems. A fortunate choice of a coordinate system can considerably facilitate the treatment of a geometrical problem. Such good coordinate systems are in Euclidean or affine spaces the Cartesian or the polar coordinate system. In affinely connected or Riemannian spaces there exist in general no Cartesian coordinate systems, but we have an analogue of the polar coordinate system called normal coordinate system in which the equations of the geodesics (in affinely connected spaces geodesic means auto-parallel curve) starting from the origin are linear. A coordinate system is a device of investigation only. Geometrical statements must be independent of it, they must express relations between invariant geometrical objects. The importance of normal coordinates stems from the fact that by them normal affinors and with their aid a complete system of invariants can be obtained, i.e. such invariants may be gained by which any other invariant of the space is expressible. This parallels the Erlanger Program of Felix Klein. However, Jesse Douglas proved that the above normal coordinates do not exist in general in line-element spaces. Thus it seemed that insurmountable obstacles stood in the way of employing this successful method for the determination of a complete system of invariants of a line-element space, a task of fundamental importance from a theoretical point of view. It turned out that the negative result was due to the inappropriate manner of carrying over the notion of geodesics. The appropriate notion, the quasigeodesic introduced by Varga is a curve $x(s)$ whose tangents are parallel translated with respect to a field of line-elements $\big(x(s), \ell(s)\big)$, where the $\ell(s)$ are parallel along $x(s)$:

$$\frac{d^2x^i}{ds^2} = \sum_{k,j} -\overset{*}{\Gamma}{}_k{}^i{}_j(x,\ell)\frac{dx^k}{ds}\frac{dx^j}{ds}, \qquad \frac{d\ell^i}{ds} = \sum_{k,j} -\overset{*}{\Gamma}{}_k{}^i{}_j\,\ell^k\frac{dx^j}{ds}.$$

Quasigeodesics reduce in point spaces to geodesics. Quasigeodesics cover one-fold a region of the $2n$-dimensional tangent bundle TM^n, and so are suited for introducing normal coordinates in line-element spaces. With their aid Varga showed that the coefficients determining the affine connection of a line-element space, L^n, the main curvature tensor and the partial- and covariant-derivatives of these, form a complete system of invariants of the L^n, see {48}. Rapcsák proved that these normal coordinates can be characterized by the same properties by which H. S. Ruse characterized the normal coordinates in Riemannian spaces. He also succeeded in establishing an invariant Taylor series (i.e. a Taylor series, where the coefficients are invariants) in Finsler spaces, see {35}. In order to obtain a complete system of differential invariants in affinely connected point- and Riemannian-spaces normal coordinates were already applied also by Ruse, T. Y. Taylor and Oscar Veblen. Rapcsák successfully used the normal coordinate systems introduced through quasigeodesics for obtaining a complete system of invariants in Cartan spaces, see {36}. These spaces originate with and are named after Élie Cartan, and concerning their structure they are very similar to Finsler spaces. While in an F^n a vector ξ is defined at a line-element (x, v), in a Cartan space the support element of a vector ξ is a point x and a hyperplane u through x : $\xi(x, u)$. The fundamental metric function of the Cartan space has the form $\mathcal{L}(x, u)$, and it has properties similar to that of an F^n.

The geodesics (or autoparallel curves) of an n-dimensional manifold (endowed with an appropriate geometric structure) form a $2n-2$ parameter family $x^i = x^i(t; a^1, \dots, a^{2n-2})$. The members of the family are called paths. The investigation of paths in point spaces was initiated by J. Douglas in 1928. In line-element spaces the more involved theory was developed by Rapcsák {34}. Here paths form a $3n-3$ parameter family of curves. Quasigeodesics of an F^n yield an example for such a family. He developed the affine connection theory of these paths. This is a little different from that of the F^n. He established curvature and torsion tensors and gave a method to resolve the equivalence problem of these spaces. The equivalence problem poses the question of whether two differential-geometric spaces are the same, but in different coordinate systems or whether their geometries are basically different. He also posed and answered the inverse question: given in a line-element space a $3n-3$ parameter family of curves $\big(x(t), v(t)\big)$, does there exist an F^n in which these curves become quasigeodesics?

A Euclidean space E^n makes any $n-1$-dimensional submanifold H also a metric space. However a geodesic of H (with respect to the induced metric) is in general not a geodesic (straight line) of the E^n. If, however, this happens for every geodesic of H, then H is called totally geodesic, and it is a hyperplane. H is also a hyperplane if its unit normals are parallel in E^n. Both of these properties can be carried over into Finsler spaces, moreover in the first case geodesics can be replaced by quasigeodesics. In an F^n these three properties yield three different notions: hyperplanes of I, II and III kind. We must mention that here unit normal vectors of H are considered at line-elements tangent to H, i.e. H is considered as an $n-1$-dimensional line-element space. Another difference is that in F^n these hyperplanes do not exist unrestrictedly (through every point and plane-position). Rapcsák showed in {37} that A): hyperplanes of the I kind exist unrestrictedly in an F^n iff F^n is of scalar curvature and projectively flat. This last property means that in an appropriate coordinate system the geodesics coincide with the straight lines or equivalently: there exists a smooth mapping $\varphi : F^n \to E^n$ such that the image of each geodesic is a straight line. Since in a V^n the condition of the unrestricted existence of totally geodesic hyperplanes is that V^n is of constant curvature (Friedrich Schur), and this is equivalent to the property that V^n is projectively flat (Enrico Bompiani), Rapcsák's result is the Finsler geometric counterpart of this famous theorem; B): hyperplanes of the II kind exist unrestrictedly iff F^n is projectively flat and the torsion tensor A satisfies the relation $A_{\alpha\beta\gamma|0} = 0$; C): hyperplanes of the III kind exist unrestrictedly iff F^n is a V^n of constant curvature.

J. M. Wegener and E. T. Davies investigated hyperplanes H, where normals are considered at line-elements transversal to H. In this case H is a point space. If these normals are parallel along H, then H is called a hyperplane of the IV kind. Rapcsák showed in {38} that in an F^n with vanishing projective curvature hyperplanes of the IV kind exist unrestrictedly iff F^n is of constant curvature. This means a considerable sharpening of a similar result of Wegener, who considered an F^n of scalar curvature and found a sufficient condition only. Rapcsák also found conditions for the metric induced by an F^n on hyperplanes of the IV kind to be the metric of a Riemannian space of constant curvature or the metric of an E^n.

Two affinely connected (point) spaces $L_1(x)$ and $L_2(u)$ allow, in general, no path preserving mapping $\varphi : L_1 \to L_2$, where paths (i.e. geodesics) of L_1 are taken into paths of L_2. An important question is, when do they allow a path-preserving mapping? If we perform a coordinate transformation $\bar{x}^i = \bar{x}^i(u)$ on L_2 such that the $\bar{x}^i$ equal the coordinates x^i of the original point, then the question takes the form: how can we change (deform) the connection on L_1, so that the paths remain unchanged. Rapcsák's studies in {39} led in a natural way to this question on line-element spaces. He solved the problem by giving necessary and sufficient relations (not complicated, but not quite short ones) between the Weyl and Douglas tensors of the original and deformed spaces. One of his nicest results {40} gives an elegant and amazingly simple answer to the question: what relation must exist between the metric functions $\mathcal{L}(x,y)$ and $\bar{\mathcal{L}}(x,y)$ of two Finsler spaces F^n and $\bar{F}^n$ with common base manifold in order that their geodesics be the same? According to his result this is the case iff

$$\bar{\mathcal{L}}_{|i} - \sum_s \frac{\partial \bar{\mathcal{L}}_{|s}}{\partial v^i} v^s = 0. \tag{4}$$

This result seems to be independent of $\mathcal{L}$. But this is not true, for the operator $|i$ denotes the Berwald covariant derivation in F^n determined by $\mathcal{L}$. Berwald too has dealt with this problem. His answer is also very simple, but it does not contain the functions $\mathcal{L}$ and $\bar{\mathcal{L}}$, at least not in an explicit form. Continuing his considerations Rapcsák was able to answer in {41} the following important and interesting questions closely related to the above ones. A): given a $2n-2$ parameter family of curves, when does there exist an F^n (i.e. a fundamental function $\mathcal{L}$) such that the geodesics of F^n are just the curves of the given family? This means the metrizability of the path-space. B): When do the geodesics of an F^n coincide with the geodesics of a Riemannian space V^n? C): When are the geodesics of an F^n straight lines in an appropriate coordinate system? This last question is the Finslerian version of Hilbert's 4th problem. The answers are similar to (4).

Arthur Moór also belonged to the Debrecen school of Finsler geometry. He dealt not only with Finsler geometry, but also with many related fields. He was a very productive mathematician who used and applied the apparatus of differential geometry with utmost ease and efficiency. He wrote more than hundred papers all of them, with a few exceptions, in German, similarly to Ottó Varga, András Rapcsák and L. Berwald.

He started with the investigation of several special Finsler spaces, special concerning the dimension or the metric function. He gave a necessary and sufficient condition in order that the curvature scalar K of an F^2 be constant, and gave an explicit form of the metric function $\mathcal{L}(x, y; \dot{x}, \dot{y})$ along a geodesic if $K = \text{const.}$, see {15}. $\mathcal{L} = f/g$, where $f = \sum_{k=0}^{N} a_k(x, y)\dot{x}^{N-k}\dot{y}^k$ and $g = \sum_{k=0}^{N-1} b_k(x, y)\dot{x}^{N-k-1}\dot{y}^k$ (i.e. f and g are homogeneous polynomials in $\dot{x}$ and $\dot{y}$ with coefficients dependent on x and y) is a special metric (fundamental) function of an F^2 investigated first by Moór. He determined for an F^2 with such an $\mathcal{L}$ the main scalar $\mathcal{J}$ and the curvature scalar $\mathfrak{K}$ in case of $N = 2$ or 3, see {16}. If $g \equiv 1$ then $\mathcal{L}$ is not first order homogeneous in $\dot{x}$ and $\dot{y}$, but $\sqrt[N]{f}$ is. Such F^2 with $N = 3$ were investigated by J. M. Wegener. Moór investigated such F^2 in case of $N = 4$, see {17}. Of course this can be generalized to dimension n. A Finsler space with $\mathcal{L} = \sqrt[N]{f_N(y)}$, where f_N is a homogeneous polynomial in $y^1, \dots, y^n$ of order N with coefficients dependent on $x^1, \dots, x^n$ is an interesting type of Finsler space, and many special cases of it are investigated. If $N = n$ and

$$f_n(y) = y^1 y^2 \dots y^n \left(= \prod_{A=1}^{n} y^A \right), \tag{5}$$

then F^n is a Minkowski space, for $\mathcal{L} = \sqrt[n]{f_n(y)}$ is independent of x and in the case of $n = 2$ it is a pseudo-Riemannian space with an indefinite Lorentz metric and non-convex indicatrix. Let us replace each y^A, $A = 1, 2, \dots, n$ by a linear form $\sum\limits_{m=1}^{n} S^A_m(x) y^m$. An F^n with the metric function

$$\mathcal{L}(x, y) = \left(\prod_{A=1}^{n} \sum_{m=1}^{n} S^A_m(x) y^m \right)^{\frac{1}{n}} \tag{6}$$

is called a Finsler space with Berwald–Moór metric (G. S. Asanov: [11], p. 53). It has an interesting and important geometric interpretation. Suppose $\det \left| S^A_m(x) \right| \neq 0$, and denote the inverse matrix by $S^m_A(x)$: $\sum_m S^A_m(x) S^m_B(x) = \delta^A_B$. Then to each vector $y = (y^1, \dots, y^m) \in T_x M$ there exist scalars λ^A such that $y^m = \sum\limits_{A=1}^{n} S^m_A(x)\lambda^A$, and hence λ^A equals $\sum_m S^A_m(x) y^m$, which can be considered as components of y in the base $S^m_1, \dots, S^m_n$. Then $\mathcal{L}^n(x, y) = \prod_{A=1}^{n} \lambda^A = \frac{V(\mathcal{P})}{V(S_A)}$, where $V(\mathcal{P})$ means the volume of the parallelotope $\mathcal{P}$ whose edges are parallel to the base vectors and whose diagonal is the vector

y. $V(S_A)$ is the volume of the parallelotope, S_A, spanned by S_A^m. Hence the Finsler measure of the vector $y(x)$ is

$$\|y(x)\|_F = \mathcal{L}(x,y) = \left(\frac{V(\mathcal{P})}{V(S_A)}\right)^{\frac{1}{n}}.$$

Thus, in this F^n the Finsler length of a vector is measured by volumes. The length can be deduced from the area. If $f_n(y)$ of (5) is replaced by $\big(\mathcal{F}(y)\big)^n$, where $\mathcal{F}$ is an arbitrary first order homogeneous function, then (6) gets the form $\mathcal{L}(x,y) = \mathcal{F}\Big(\sum_m S_m^1(x)y^m, \ldots, \sum_m S_m^n(x)y^m\Big)$. This is the renowned 1-form metric investigated by M. Matsumoto, H. Shimada and others. It has a number of beautiful properties, e.g. it allows a linear metrical connection for the vectors of the tangent bundle (connection in a Finsler space without line-elements, i.e. point Finsler spaces). The Berwald–Moór metric is clearly a simple special case of this.

The well known theorem of A. Deicke states that an F^n with vanishing torsion vector A_i and positive fundamental function $\mathcal{L}$ is Riemannian. The Berwald–Moór metric (6) was up to now the only known example, where $A_i = 0$, yet the space is not Riemannian (for $\mathcal{L}$ is not everywhere positive). The Berwald–Moór metric has still a number of interesting properties: the signature of its metric tensor g_{ij} is $(+--\ldots)$, $\det|g_{ij}|$ is independent of y, etc.

Finsler and Cartan spaces have very similar structures, as it was mentioned a few pages earlier. In an F^n all vectors and geometric objects are defined at line-elements $(x, \dot{x})$ (i.e. at a point x and a direction or oriented line $\dot{x}$), while in a Cartan space vectors and geometric objects are given at (oriented) hyperplane-elements (x, u), where u is an $n-1$ dimensional linear subspace, i.e. a hyperplane in the tangent space T_xM through the origin having an equation $\sum \mu_i y^i = 0$, where the y^i are coordinates in T_xM and μ_i is the normal of u. Thus the coordinates of (x, u) are (x^i, μ_i). A duality between Finsler and Cartan spaces was investigated by L. Berwald. Somewhat differently from him, Moór in {18} and {21} called a Finsler and a Cartan space dual if

$$\mu_i = \frac{1}{\sqrt{g^*}} \sum_k \overset{*}{g}_{ik} \dot{x}^k \quad \text{and} \quad \dot{x}^i = \sqrt{g} \sum_k g^{ik} \mu_k \tag{7}$$

establish a $1-1$ correspondence between the line-elements $(x, \dot{x})$ of the Finsler space and the hyperplane-elements (x, u) of the Cartan space having the same underlying manifold. Here $g^{ik}(x,\mu)$ is the metric tensor of the Cartan space and $g(x,\mu)$ is its determinant. The corresponding objects of the F^n are denoted by an asterisk $*$. If between a Finsler and a Cartan space (7) establishes a duality, then $g^*_{ij}(x,\dot{x}) = g_{ij}(x,\mu)$ and $\mathcal{L}^*(x,\dot{x}) = \mathcal{L}(x,\mu)$ at the corresponding elements, and also the torsion vectors $A^*_i(x,\dot{x})$ and $A_i(x,\mu)$ vanish. With Berwald $A^*_i = A_i = 0$ was a requirement. For Moór this is a consequence of the existence of the given dual mapping. Because of $A^*_i = A_i = 0$ volume elements independent of $\dot{x}$, resp. μ, exist and the dual mapping is volume preserving. It should be mentioned that according to Deicke's theorem $A^*_i = 0$ yields that F^n is a Riemannian space, however this holds only if $\mathcal{L}^*(x,\dot{x})$ is everywhere positive, which is not the case in general.

Moór investigated Varga's osculating Riemannian space V^n (reviewed 11 pages earlier) in dual Finsler and Cartan spaces, see {19} and {20}. He proved that the Riemannian spaces osculating the dual Finsler and Cartan spaces along a 1-parameter family of line-elements

$$\big(x(t),\dot{x}(t)\big), \quad \text{resp. } \big(x(t),\mu(t)\big) \tag{8}$$

are the same. From this it follows that if $\xi^i\big(x(t),\dot{x}(t)\big)$ and $\xi^i\big(x(t),\mu(t)\big)$ are corresponding vector fields along (8) in the two dual spaces, then their invariant derivatives with respect to the dual Finsler and Cartan spaces are also the same. Moreover, under a mild further condition, the curvature tensor R^V of the osculating V^n coincides along (8) with Varga's main curvature tensor $T(x,\dot{x})$ of the F^n and also with the curvature tensor $\bar{R}(x,\mu)$ of the Cartan space.

$\dot{x}^i$ and μ_i in $(x,\dot{x})$ and (x,μ) are vector densities (of weight 0, resp. -1). Thus, by replacing of $\dot{x}$ or μ by a vector density u of certain weight p one obtains a common generalization of Finsler and Cartan spaces which can be given by a metric function $\mathcal{L}(x,u)$. Investigations concerning such so called general metric spaces $\mathfrak{R}_n$ were initiated by J. A. Schouten, J. Haantjes, E. T. Davies and R. S. Clark.

The detailed development of the geometry of the generalized spaces $\mathfrak{R}_n$ was completed by Moór in several papers. First he succeeded in expanding his above sketched duality to general vector density spaces in {21}. Two spaces $\mathfrak{R}_n$ and $\tilde{\mathfrak{R}}_n$ are called by him dual, if between the elements (x,u), resp. $(x,\tilde{u})$, of the two spaces there exists a mapping $\tilde{u}^i = \varphi^i(x,u)$ such that $g_{ik}(x,u) = \tilde{g}_{ik}(x,\tilde{u})$. Then by Varga's osculating Riemannian space method he obtained results similar to his duality theory between Finsler and Cartan spaces.

In {22} he developed the curvature theory of the $\mathfrak{R}_n$, and determined four curvature tensors. One of them, $\bar{R}_j{}^i{}_{k\ell}$, reduces in an F^n to Varga's main curvature tensor T, and another: $R_j{}^i{}_{k\ell}$ generalizes Riemann's curvature tensor of a V^n. He also found two curvature invariants: $B(x,u,\eta)$ and $\bar{B}(x,u,\eta)$ depending on u and another vector η and generalizing the Riemann–Berwald curvature tensor of the F^n. In the case of an $\mathfrak{R}_n$ of scalar curvature, i.e. when B and $\bar{B}$ are independent of η, he obtained results which are counterparts of several theorems of Berwald in Finsler spaces.

Conformal transformations

$$\tilde{g}_{rs}(x,u) = e^{2\sigma(x)} g_{rs}(x,u), \quad g_{rs} = \frac{1}{2} \frac{\partial^2 \mathcal{L}^2}{\partial u^r \partial u^s}$$

of the metric of an $\mathfrak{R}_n$ were investigated first by R. S. Clark. Moór in {26} obtained a generalization partly by starting directly from the g_{rs} which in Clark's work is deduced from the fundamental function $\mathcal{L}(x,u)$ similarly to Finsler and Riemannian geometry; partly by replacing $\sigma(x)$ by $\sigma(x,u)$. Then he developed the conformal geometry of the $\mathfrak{R}_n$ in this general setting.

A Finsler space is determined by its fundamental function $\mathcal{L}(x,y)$, and the basic geometric objects, e.g. the metric tensor

$$g_{ij}(x,y) = \frac{1}{2} \frac{\partial^2 \mathcal{L}^2}{\partial y^i \partial y^i} \tag{9}$$

are derived from $\mathcal{L}$. However one can start directly with the $g_{ij}(x,y) = g_{ji}(x,y)$. This is obviously a more general case, because for a given g_{ij} (9) is, in general, not solvable for $\mathcal{L}$. Moór was always intent on new generalizations. Between 1955 and 1960 his attention turned to these so called general metric line-element spaces which he denoted by $\mathfrak{L}_n$. Finsler spaces built directly on $g_{ij}(x,y)$ have been investigated since then by a number of authors; in particular by R. Miron, and many Roumanian and Japanese geometers. Moór investigated these spaces from several points of view, see {23}. First he constructed the connection theory of the $\mathfrak{L}_n$ developing a metrical connection with the new essential generalization that the connection coefficients are allowed to be not symmetric. Dropping the symmetry of the connection coefficients is essential. In Finsler, and thus also in Riemannian spaces, where connection coefficients are symmetric, autoparallel curves coincide with geodesics. However in case of their asymmetry these curves may be different. Moór gave several concrete examples for the $\mathfrak{L}_n$, and characterized the case when these $\mathfrak{L}_n$ reduce to an F^n. He investigated the infinitesimal transformations

$$\bar{x}^i = x^i + \xi^i(x)\delta t \quad \text{which yield} \quad \bar{y}^i = y^i + \sum_r y^r \frac{\partial \xi^i}{\partial x^r}\delta t \tag{10}$$

of the $\mathfrak{L}_n$, and determined the corresponding Lie derivation denoted by Δ. Transformations generated by the velocity vectors $\xi^i(x)$ of (10) are motions if they preserve length. This is assured by the Killing equations

$$\Delta g_{ik} = 0. \tag{11}$$

He determined the integrability condition of the Killing equations (the unknowns are the ξ^i), and determined the paths of the motions. He called a motion a translation if its paths are also autoparallels. He found a simple explicit condition for this case. In an F^n the paths of a translation intersect a geodesic under the same angle. In an $\mathfrak{L}_n$ this is not so. According to his result the condition for this is the vanishing of two tensors A_{oor} and σ_{ioo}.

The number of the Killing equations (11) is $\frac{1}{2}n(n+1)$. This is the maximum of the number r of the independent parameters a solution of (11) may have. H. C. Wang showed that if in an F^n one has $r \geq \frac{1}{2}n(n-1)+1$, then it is a V^n. In an $\mathfrak{L}_n$ this is not true. Here the relation

$$\sum_{i,j} \frac{\partial^3}{\partial y^v \partial y^s \partial y^r} g_{ij}(x,y) y^i y^j = 0$$

is a necessary, but not sufficient condition for this. Moór investigated the case $r = \frac{1}{2}n(n+1)$ and found simple additional conditions in order that a) the curvature $\bar{R} := \sum_i \bar{R}_i{}^i$, $\bar{R}_i{}^s := \sum_{k,\ell} g^{ks} \bar{R}_i{}^\ell{}_{k\ell}$ of the $\mathfrak{L}_n$ vanish; b) $\mathfrak{L}_n$ is of scalar curvature; or c) $\bar{R}_{koij}$ has an interesting special form. He published his results in a number of sometimes comprehensive papers.

F. Schur's famous result states: if the sectional curvature $R(x,p)$ of a Riemannian space is independent of the plane position p, then it is independent of the point x too, i.e. R is a constant. This is also true in an F^n of scalar curvature. However this is not so in an $\mathfrak{L}_n$. Moór found a nice characterization of the $\mathfrak{L}_n$ of scalar curvature in which Schur's theorem holds, see {24}.

Geodesic deviation plays an important role in the theory of Riemannian and also in that of Finsler spaces. Moór derived the equation of autoparallel deviation in an $\mathfrak{L}_n$ and obtained conditions for the case that autoparallel curves of an $\mathfrak{L}_2$ have an envelope, see {25}. Also he developed the anholomic geometry (where the bases of T_xM are not the tangents of the coordinate lines) of the $\mathfrak{L}_n$-s, see {28}.

Albert Einstein was able to derive gravitation from the structure of a 4-dimensional Riemannian space. To this end he used up the Riemann curvature tensor and there remained no more geometric objects which would reflect the impact of the electromagnetic field. More general spaces may offer more possibility for the incorporation of electromagnetic field. Moór and his coauthor János Horváth, the physicist from Szeged, discussed in a long paper {11} the possibility of creating a unified field theory within the frame of a Finsler space F^n.

In 1949–1953 for the purpose of studying the quantum theory of wave spaces with the aid of operator calculus H. Yukawa developed a bilocal space theory in which the objects of the space are defined at pairs of points. Moór and Horváth converted this space into a general metrical line-element space $\mathfrak{L}_n$, and gave an interpretation of Yukawa's theory in this frame (Entwicklung einer Feldtheorie begründet auf einen allgemeinen metrischen Linienelementraum I. and II. Indag. Math., **17** (1955), 421–429, 581–587).

In the theory of affine or metrical point- or line-element-spaces the absolute derivation $\frac{D}{dt} = \sum\limits_k \frac{dx^k}{dt}\nabla_k$ of covariant and contravariant vectors is performed with the same connection coefficients. In 1958–61 T. Otsuki introduced a general connection in which this does not hold. Moór called this the Otsuki connection, and devoted a number of papers to its investigation. He coupled the Otsuki connection with a recurrent metric tensor g_{ij} which satisfies $\nabla_k g_{ij} = \gamma_k(x) g_{ij}$ and thus obtained a Weyl–Otsuki space {33}. He investigated in the Otsuki and Weyl–Otsuki spaces transformation groups, geodesic deviation, and considered anholonomic coordinate systems. He also considered Finsler–Otsuki spaces, their duality, etc.

An object at a point with coordinates x^i $(i = 1, \dots, n)$ is defined by certain numbers $\Omega_1, \dots, \Omega_N$ called components of the object. If its components Ω_α, $\alpha = 1, \dots, N$ in an other coordinate system $(\bar{x})$ can be calculated from the original $\Omega_\alpha(x)$ and the coordinate transformation $\bar{x}^i = \bar{x}^i(x)$ according to a given rule, then Ω is a geometric object; e.g. vectors, tensors, connection coefficients, etc. (cf. [3]). Moór investigated many problems of the theory of geometric objects. For example: The covariant derivative $\nabla_k v^i$ of the vector v^i is a tensor $T^i{}_k$ depending on v^i, $\frac{\partial v^i}{\partial x^k}$ and the connection coefficients $\Lambda_j{}^i{}_k$. One can ask, in what form can a tensor $T^i{}_k$ be composed from

v^i, $\frac{\partial v^i}{\partial x^k}$ and $\Lambda_j{}^i{}_k$; i.e. which are the possible forms of a "covariant derivative" in the above sense. He solved a number of problems of the following type: to construct a tensor from given geometric objects of different kinds in {29}; to construct covariant derivatives from the metric tensor; connection coefficients from covariant vectors; etc. He also investigated geometric objects and covariant derivatives in line-element spaces.

A tensor T is recurrent if its covariant derivative equals the tensorial product of a covariant vector λ and the tensor itself: $\nabla T = \lambda \otimes T$. Among Riemannian manifolds V^n those which are near Euclidean space enjoy a special interest. The nearest class consists of V^n of constant curvature (i.e. non-Euclidean spaces). Next are the locally symmetric V^n characterized by $\nabla R = 0$ (where R is the curvature tensor of the V^n), and after that the V^n of recurrent curvature. They were investigated in detail by H. S. Ruse and A. G. Walker. Moór extended the investigations to affinely connected spaces L_n (spaces with parallelism, but without metric), to metrical spaces with not symmetric coefficients, generalized metric spaces $\mathfrak{R}_n$ and general line-element spaces $\mathfrak{L}_n$. He wrote 15 papers on this topic, investigated spaces with recurrent torsion, and F^n with recurrent metric tensor: $\nabla_k g_{ij} = \lambda_k g_{ij}$. Metric point spaces with this property are the Weyl spaces. They found important applications in the field theories of physics. Moór also investigated spaces with double recurrence given by the property $\nabla\nabla R = T \otimes R$, where T is a tensor of type $(0,2)$. It is easy to see that simple recurrence always yields double recurrence, but not conversely. To convey the flavour of his results in this field we mention two of them chosen arbitrarily. 1) He proved that if an affinely connected space L_n of recurrent curvature (with possibly not symmetric connection coefficients) splits into two factors: $L_n = L_r \times L_{n-r}$, then one of them is flat, i.e. its curvature tensor vanishes, see {27}. 2) An F^3 is always recurrent with respect to the total curvature tensor, provided there exists in it an absolute parallelism of the line-elements, see {30}.

Also interesting are his investigations concerning equivalent variational problems. He called two variational problems

$$\text{(12)} \quad \text{a)} \;\; \delta \int_a^b F(x, \dot{x}, \ddot{x})\, dt = 0 \qquad \text{and} \qquad \text{b)} \;\; \delta \int_a^b F^*(x, \dot{x}, \ddot{x})\, dt = 0$$

equivalent if their solutions (their geodesics) are the same, i.e. if

$$\text{(13)} \qquad \mathcal{E}_i(F) = 0 \Longleftrightarrow \mathcal{E}_i(F^*) = 0, \quad \mathcal{E}_i \equiv \frac{\partial}{\partial x^i} - \frac{d}{dt}\frac{\partial}{\partial \dot{x}^i} + \frac{d}{dt}\frac{\partial}{\partial \ddot{x}^i}.$$

A. Kawaguchi has considered and investigated spaces in which the arc length of a curve $x^i(t)$ is defined by $\int_a^b F\big(x, \dot{x}, \ldots, x^{(r)}\big)\, dt$. So (12,a,b) are variational problems of Kawaguchi spaces with $r=2$, and (13) yields their equivalence. This equivalence means that the identity mapping between the two spaces given by F and F^* is a geodesic mapping. For $r=1$ one obtains Finsler spaces. In this case $\frac{\partial F}{\partial \ddot{x}^i}=0$. Moór investigated the clearly equivalent variational problems, where

$$\text{(14)}\qquad \text{a)}\quad \mathcal{E}_i(F^*)=\lambda(x)\mathcal{E}_i(F),$$

$$\text{resp.}\quad \text{b)}\quad \mathcal{E}_i(F^*)=\sum_k \lambda_i^k \mathcal{E}_k(F),\ \det|\lambda_i^k|\neq 0$$

and obtained results concerning the form of F and F^*, see {31}, e.g. in the case of (14,a) with $\lambda\neq$ const., if F and F^* do not depend on $\ddot{x}$ (i.e. in the case of Finsler spaces) he obtained that

$$F^*(x,\dot{x})=\lambda(x)F(x,\dot{x})+\sum_k S_k(x)\dot{x}^k,\quad F(x,\dot{x})=\sum_k a_k(x)\dot{x}^k,$$

where $\lambda(x)$ and $S_k(x)$ satisfy a first order partial differential equation in which $a_k(x)$ is also involved as a coefficient function. If $\lambda=$ const., then S_k is the gradient of a scalar function S, and the difference $F^*(x,\dot{x})-\lambda F(x,\dot{x})$ is a total differential of S with respect to $\dot{x}$. He solved similar problems for (14,b) (also in the case of rank $\|\lambda_i^k\|<n$) and also for variational problems in several variables: $F\big(x^i(u),\frac{\partial x^i}{\partial u^\alpha}\big)$ $i=1,\dots,n$, $\alpha=1,\dots,m<n$, and with multiple integrals $\int\underset{U}{\cdots}\int F\big(x,\frac{\partial x}{\partial u}\big)\,du$, $u\in U$, see {32}.

Hungarian mathematicians performed successful investigations in the 20th century not only on Finsler-, but also on Riemannian-geometry. One of them was Pál Dienes.

In 1917 Tullio Levi-Cività created a suitable notion of parallelism of vectors in Riemannian spaces. He considered a surface $\phi: x^i=x^i(u^1,u^2)$, $i=1,2,3$, of a Euclidean 3-space E^3, a curve $C(t)\subset\phi: x^i(t)=x^i\big(u^\alpha(t)\big)$, and a tangent vector $\xi_0\in T_{C(t_0)}\phi$. Then he translated ξ_0 parallel in the E^3 to the nearby point $C(t_0+\Delta t)$ of the curve obtaining in E^3 a vector $\hat{\xi}(t_0+\Delta t)$ (with components $\hat{\xi}^i=\xi_0^i$, $i=1,2,3$), and he called the perpendicular projection $\bar{\xi}$ of $\hat{\xi}$ on $T_{C(t_0+\Delta t)}\phi$ parallel to ξ_0. Starting this construction again with $\bar{\xi}$, then with $\bar{\bar{\xi}}$, etc., and performing a limit process $\Delta t\to 0$, he obtained a vector field $\xi(t)$ tangent to ϕ which he called parallel along

$C \subset \phi$. This construction leads analytically to the differential equation system

$$(15)\qquad \frac{d\xi^\alpha}{dt} + \sum_{\beta,\gamma} \Gamma_\beta{}^\alpha{}_\gamma\big(x(t)\big)\,\xi^\beta \frac{dx^\gamma}{dt} = 0, \quad \alpha,\beta,\gamma = 1,2$$

where $\Gamma_\beta{}^\alpha{}_\gamma$ is the Christoffel symbol of the second kind of ϕ. This notion of parallelism opened new paths and gave a considerable impulse to the investigation of Riemannian geometry.

Pál Dienes graduated at and took his doctor degree from the Péter Pázmány University in Budapest. At the beginning of his mathematical career in Hungary he dealt with the theory of functions. In 1920 he left Hungary, and at this fortunate moment of the development of Riemannian geometry he started investigations on this geometry in Wales (U.K.).

Since any n-dimensional Riemannian manifold V^n has a local isometric embedding into E^{n+k}, the idea of parallelism just described can easily be extended to V^n. In this case (15) takes the form

$$(15')\qquad \frac{d\xi^i}{dt} + \sum_{j,k} \Gamma_j{}^i{}_k\big(x(t)\big)\,\xi^j \frac{dx^k}{dt} \equiv \frac{D\xi^i}{dt} = 0, \quad i,j,k = 1,\dots,n.$$

The equation (15′) has two important consequences. Since it contains as data the functions $\Gamma_j{}^i{}_k(x)$ only, by giving them arbitrarily in a coordinate space $X^n(x)$, we can define parallelism in it, and thus make X^n into an affinely connected space L^n (a differential geometric space with parallelism, but without metric). On the other hand, the left hand side of (15′) behaves as a vector, hence it can be considered as a derivative (the absolute derivative $D\xi^i/dt$) of the vector field $\xi^i\big(x(t)\big)$. Thus the parallelism of a vector field $\xi\big(x(t)\big)$ in an L^n can be defined by $\frac{D\xi^i}{dt} = 0$, similarly as in E^n by $\frac{d\xi^i}{dt} = 0$. Moreover, $\frac{D}{dt}$ can be extended to tensor algebra, and this leads to the absolute differential calculus (Ricci-calculus) of basic importance in differential geometry.

Parallelism, affinely connected spaces and absolute differential calculus raised a number of new questions at the beginning of the 20s of the 20th century. Dienes rendered important contributions to Riemannian geometry and its nonmetrical counterpart, the theory of affinely connected spaces. He found new possible solutions for the main problems raised by the rapid advance of the differential geometry of his time. He had original ideas and realized interesting investigations. His papers appeared in well-known leading journals. His activity on differential geometry can be divided into two well separable periods. In the first of these, mainly between 1922 and 1926, he was interested in the connection theory of tensors and vectors with or without metric (C.R. Acad. Sci. Paris, **174** (1922), 1167–1170; **175** (1922), 209–211; **176** (1923), 370–372) and their application to electromagnetics (C.R. Paris, **176** (1923), 238–241). He developed a generalization of the absolute tensor calculus of Gregorio Ricci–Curbastro and of the parallel translation of Tullio Levi-Cività. He represented a tensor A by the aid of certain elementary tensors $e^1, \ldots, e^n$ in the form

$$A_0(x) + \sum_i A_i(x)e^i + \sum_{i,j} A_{i,j}(x)e^ie^j + \ldots .$$

Then he investigated their addition, multiplication, contraction and derivation, and found conditions for the associativity, distributivity and especially commutativity of these operators. He also discussed a new metric compatible with his parallel translation, and the integration of the differential equation of that general parallel translation {4}.

He also gave another generalization of the parallel translation of a line-element δx in {5} by the relation

$$d\delta x^k = f^k(x, dx, d^2x, \ldots, d^mx; \delta x) = 0, \tag{16}$$

where $(x, dx, \ldots, d^mx)$ is an m-th order element of a curve and the f^k are first order homogeneous in δx and satisfy a certain homogeneity condition also in $dx, \ldots, d^mx$. This is a very general definition of parallel translation having some relation to Finsler and also to Kawaguchi geometry. (In Kawaguchi geometry the arc length of a curve is given by such an integral $ds = \int_a^b \mathcal{L}\big(x, \dot{x}, \ddot{x}, \ldots, x^{(m)}\big)\, dt$, where the fundamental function $\mathcal{L}$ depends not only on the first, but also on higher derivatives up to the m-th.) He gave also an intermediate value theorem in relation with this parallel translation. Let $A^k(t)$ be a vector field along a curve $x(t)$ and $A^k(t, t_0)$ the parallel translated according to the new parallel displacement of $A^k(t)$ along the curve to $x(t_0)$. Then this intermediate value theorem has the form

$$A^k(t, t_0) - A^k(t_0) = \left.\frac{DA^k}{Dt}\right|_{t=t_0} (t - t_0) + \mu(t - t_0)^2,$$

where $\frac{D}{Dt}$ is the operator of the absolute derivation related to (16). This yields a Taylor formula too if f^k is linear in δx,

His paper {6} is also connected to parallel displacement. With the aid of the parallel displacement of Levi-Cività he constructs osculating p-vectors and successive curvatures of a curve and finds explicit expressions for these curvatures. The first curvature coincides with that of Luigi Bianchi.

It is well known that in a V^n the square of the norm of a vector a^i is $|a|^2 = \sum a^i a_i$. In {7} Dienes transfers and extends this important notion on tensors in a very natural and simple way. For a tensor v^{ij} he defines

$$|v|^2 := \sum_{i,j} v^{ij} v_{ij} \equiv (vv) \quad \text{and} \quad \cos(vw) := \frac{(vw)}{|v|\,|w|}.$$

If we are given a v^{ij} and we have no metric tensor, then, in order to obtain $v_{rs} = \sum_{i,j} v^{ij} g_{ir} g_{js}$ a tensor $g_{k\ell}$ is needed, and similarly, starting with a v_{ij} a tensor $g^{k\ell}$ is necessary which may have no relation to $g_{k\ell}$. Then he describes and investigates the most general parallel translation of tensors. He requires only that the parallel translated of a tensor along a curve C from its point p to another point $q \in C$ be a tensor of the same kind depending on C alone. This yields the differential equation system $\frac{dA}{dt} + f(x, \dot{x}, \ddot{x}, \dots; A) = 0$ (suppressing the indices of the tensor A and the corresponding complicated notation at f) (cf. (16)). Then derivation of tensors is obtained in the form $\frac{\Delta A}{\Delta t} = \frac{dA}{dt} + f(x, \dot{x}, \ddot{x}, \dots; A)$. However, the usual good properties of the tensor derivation are assured only under further requirements. In an isotropic space he defines a displacement of tensors which operator can be split into a parallel translation along the curve, followed by a rotation around the endpoint.

His paper {8} represents a transition from his first period to the second one. In this paper he considers an affinely connected space L^n and an m-dimensional subspace X^n_m of it. Then the n-dimensional tangent space splits into an m-dimensional and an $n-m$-dimensional one giving rise to another subspace X^n_{n-m}. Vectors of L^n also split according to X^n_m and X^n_{n-m}. Using non-holonomic coordinate systems in them, four new connections can be derived from the connection of L^n by the aid of the split vectors of the X^n, and also the splitting of the curvature and torsion tensors of the L^n will be obtained.

In the second period of his activity in differential geometry he investigated the infinitesimal deformations of spaces and connections. He considered the infinitesimal deformation

$$(17) \qquad 'x^i = x^i + \varepsilon v^i$$

of an affinely connected space L^n related to the coordinate system (x). (17) can be considered as a point transformation and, at the same time, as a coordinate transformation giving raise to new components $\overline{T}^i$ of a tensor $T(x)$. Denoting by $'T$ the element of the tensor field $T(x)$ at $'x$ (i.e. $'T = T('x)$) and by T^* the parallel displaced of $T(x)$ to $'x$, he obtained three types of differentials:

$$\delta T \equiv {}'T - \overline{T}, \quad \Delta T = T^* - \overline{T}, \quad DT = T^* - {}'T$$

corresponding to three types of derivations. It is noteworthy that δ can also be applied to quantities which are not tensors, e.g. to connection coefficients. He applied these deformations on L^n and V^n, and also on submanifolds X^n_m of these. He described the effect of these operations on the curvature tensors and connection coefficients using, in the case of a submanifold X^n_m, a splitting into tangential and transversal components. The operator δ was used also by W. Slebodzinki and D. van Danzig, and applied to Lie derivation. It turned out to be a very good tool for the study and characterization of affine and metrical motions.

His papers reporting on these investigations (Sur la déformation des espaces à connexion linéaire générale. C.R. Acad. Sci. Paris, **197** (1933), 1084–1087; Sur la déformation des sous-espaces dans un espace à connexion linéaire générale. C.R. Acad. Sci. Paris, **197** (1933), 1167–1169; On the deformation of tensor manifolds. Proc. London Math. Soc. (2), **37** (1934), 512–519) are closely related to each other and crowned in his most comprehensive paper (On the infinitesimal deformations of tensor submanifolds. J. Math. Pures Appl. (9), **16** (1937), 111–150) written together with E. T. Davies from Southampton who can be considered as his pupil. The problems treated in these papers were problems of his time, and other mathematicians too showed interest in them e.g. J. A. Schouten and E. R. van Kampen.

István Fáry who worked mainly in Berkeley, dealt with convex geometry, cell complexes, topology, etc. However, he has a very nice, interesting result on the global differential geometry of the E^3.

One of the first results of global differential geometry states that the total curvature of a closed curve of the Euclidean two space E^2 is at least 2π. This result was extended by W. Fenchel to closed space curves of the E^3 in 1929, and to those of the E^n by K. Borsuk in 1948. Borsuk also raised the question whether the total curvature of a knot (i.e. a curve of the E^3 which is homeomorphic to the circle, but is not isotopic to it) is always

$\geqq 4\pi$. This interesting question was answered by Fáry in an affirmative way in {10}. His proof is based on the interesting observation that the total curvature of a curve of the E^3 equals the mean of the total curvatures of its orthogonal projections on the 2-dimensional linear subspaces of the E^3. He found this result in France before he settled in the USA. Another positive answer on Borsuk's problem was published by J. W. Milnor in {14} a little later, in 1950 (the two papers were accomplished independently from each other).

Another differential geometrical result is due to István Vincze who worked mainly in statistics and probability theory. Let $L \subset E^3$ be a curve related to the arc length s; $Q_1(s_1)$, $Q_2(s_2)$ points of L, and $S(Q_1, Q_2)$ the center of mass of the arc $\widehat{Q_1Q_2}$ in case of a homogeneous load-distribution. If $Q_1 \to P_0 \in L$, $Q_2 \to P_0$ independently of each other, then S runs over a two parameter family of points, i.e. a surface $F \subset E^3$. Let us denote by $K(Q_1, Q_2)$ the Gauss curvature and by $H(Q_1, Q_2)$ the mean curvature of F at $S(Q_1, Q_2)$. Then, according to his result {58},

$$\lim_{Q_1,Q_2 \to P_0} K(Q_1, Q_2) = -\tau^2, \quad \text{and} \quad \lim_{Q_1,Q_2 \to P_0} H(Q_1, Q_2) = -\frac{3}{5}\frac{1}{\rho^2}\frac{d}{ds}(\varrho^2\tau),$$

where ϱ is the curvature and τ the torsion of L. He also investigated the case of a closed curve L loaded with a density $\mu(s)$ and he determined, among other things, the surface area of F. His investigation is related to a problem of Alfréd Rényi connected to the cosmological theory of O. Yu. Schmidt (cf. I. Vincze, {57}).

Jenő Egerváry was interested mainly in matrix theory and differential equations, but as professor of mathematics at Budapest Technical University he also dealt with different problems of algebra, geometry of E^3, analysis, and certain physical and technical problems. However, Egerváry has interesting results in the differential geometry of the Euclidean n-space E^n too.

Let $L(s)$ be a smooth curve of the E^n given by $x^i = x^i(s)$ or, in vectorial form, by $\boldsymbol{x}(s) = x^i(s)\boldsymbol{e}_i$ with s as arc length, and $\boldsymbol{e}_i$ as an orthonormal base of E^n. It is an elementary fact that the (first) curvature $\frac{1}{\varrho_1(s)} = \kappa_1(s)$ of L is the limit of the ratio of the angle of the tangents at s_0 and s and of $|s - s_0|$ in case of $s \to s_0$; i.e. $\kappa_1(s)$ is the angular velocity of the tangent of the curves. Similarly, the second curvature κ_2 (the torsion τ) is the angular velocity of the second normal. Further curvatures $\kappa_3, \ldots, \kappa_{n-1}$ of the L in E^n or even in a Riemannian space V^n (see W. Blaschke, {2}) are defined as the coefficients appearing in the Frenet formulas expressing the derivatives of the vectors of the moving frame linearly by the vectors of the frame itself.

Egerváry considered the Gram determinants

$$G_k = \det |h^{ab}|_k, \quad h^{ab} = \langle \boldsymbol{x}^{(a)}, \boldsymbol{x}^{(b)} \rangle, \quad \begin{matrix} a, b = 1, \dots k \\ k = 1, 2, \dots, n, \end{matrix} \quad G_0 := 1$$

of order k, where $\boldsymbol{x}^{(a)} = \frac{d^a \boldsymbol{x}}{ds^a}$, and $\langle \, , \, \rangle$ denotes the Euclidean scalar product, and proved that

$$\frac{d\eta_k}{ds} = \lim_{s \to s_0} \frac{\eta_k}{|s - s_0|} = \frac{\sqrt{G_{k+1} G_{k-1}}}{G_k}(s_0), \quad k = 1, \dots, n-1,$$

where η_k is the angle of the osculating k-planes taken at s and s_0. It turns out that these $\frac{d\eta_k}{ds}$ coincide with the curvatures $\kappa_1, \dots, \kappa_{n-1}$ of the curve L. This can easily be seen in the simple case of $k = 1$. Indeed

$$G_1 = \langle \boldsymbol{x}', \boldsymbol{x}' \rangle = |\boldsymbol{x}'|^2 = 1, \quad G_2 = \begin{vmatrix} \langle \boldsymbol{x}', \boldsymbol{x}' \rangle & \langle \boldsymbol{x}', \boldsymbol{x}'' \rangle \\ \langle \boldsymbol{x}'', \boldsymbol{x}' \rangle & \langle \boldsymbol{x}'', \boldsymbol{x}'' \rangle \end{vmatrix} = \begin{vmatrix} 1 & 0 \\ 0 & |\boldsymbol{x}''|^2 \end{vmatrix} = |\boldsymbol{x}''|^2,$$

for the parameter is the arc length, and because of $\boldsymbol{x}' \perp \boldsymbol{x}''$. Thus, taking into account $G_0 = 1$, we obtain $\frac{d\eta_1}{ds} = |\boldsymbol{x}''| = \kappa_1$. These $\frac{d\eta_k}{ds}$ also satisfy the Frenet formulas. So Egerváry's result revealed the geometric meaning of the formally defined curvatures κ_k both in E^n and V^n. Moreover, he could express $\frac{d\eta_k}{ds}$ also by volumes of simplexes with vertices on L and by distances of the vertices. In this form curvatures do not need the differentiability of the curve. Egerváry's results and ideas found applications in the book of L. M. Blumenthal {3}), and are also related to some works of G. Alexits.

One can consider a curve L as the set of points represented by $x^i = x^i(t) \in C^0$, $t \in I = (a, b)$. After Peano's investigation it became clear that this set may be a cube. This can not happen if one requires the mapping $I \to L$ to be $1:1$. Nevertheless this excludes so simple and important cases as multiple points. A new set theoretical and topological theory of curves in metrical (distance) space without any differentiability was initiated by K. Menger {12}, {13} and P. Urysohn {42}. These investigations were presented to the Hungarian mathematical community by György Alexits {1}.

György Alexits, known for his works on approximation theory and orthogonal series, joined these investigations in the 30s. He investigated the curvatures of a curve in the general distance and semi-distance spaces. He called them linear curvatures in order to distinguish them from the curvature of a surface or of a space used in Riemannian geometry. A distance space is a metric space with its well known three axioms. The space is a semi-distance space if the triangle axiom is not required (e.g. in a Minkowski space with non convex indicatrix). It was K. Menger who first started investigations on the (first) curvature of a curve in distance spaces. Alexits introduced the notion of the k-th linear curvature, and devoted several papers, first alone, and then together with Egerváry, to their investigation (La torsion des espaces distanciés. Comp. Math., **6** (1939), 471–477; Der Torsionsbegriff in metrischen Räumen. Mat. Fiz. Lapok, **46** (1939), 13–28 in Hungarian with German summary). It turned out that in the case of a Euclidean space his linear curvatures reduce to Egerváry's $\frac{d\eta_k}{ds}$ and in a Riemannian space to Blaschke's curvatures, see {9}.

References

[3] Aczél, János – Gołąb, Stanislaw, *Funktionalgleichungen der Theorie der geometrischen Objekte,* PWN (Warszawa, 1960).

[11] Asanov, G. S., *Finsler geometry, relativity and gauge theories,* Reidel (Dordrecht, 1985).

{1} Gy. Alexits, La nouvelle théorie des courbes, *Mat. Fiz. Lapok,* **44** (1937), 1–37 (in Hungarian, with French summary).

{2} W. Blaschke, Frenets Formeln für den Raum von Riemann, *Math. Z.,* **6** (1920), 94–99.

{3} L. M. Blumenthal, *Theory and Applications of Distance Geometry,* Clarendon Press (Oxford, 1953).

{4} P. Dienes, Sur la structure mathématique du calcul tensoriel, *J. Math. Pures Appl. (9),* **3** (1924), 79–106.

{5} P. Dienes, Sur les différentielles secondes et les dérivations des tenseurs, *Proc. Rome Academy* (1924), 265–269.

{6} P. Dienes, Déterminants tensoriels et la géométrie des tenseurs, *C.R. Acad. Sci. Paris,* **178** (1924), 682–685.

{7} P. Dienes, On tensor geometry, *Ann. Math. Pura Appl. (IV),* **3** (1926), 247–295.

{8} P. Dienes, On the fundamental formulae of the geometry of tensor submanifolds, *J. Pures Appl. (9),* **11** (1932), 255–282.

{9} E. Egerváry and G. Alexits, Fondements d'une théorie générale de la courbe linéaire, *Comment. Math. Helv.,* **13** (194), 257–276.

{10} I. Fáry, Sur la courbure totale d'une courbe gauche fraisant un nœud, *Bull. Soc. Math. France,* **77** (1949), 128–138.

{11} J. Horváth, Entwicklung einer einheitlichen Feldtheorie begründet auf die Finslersche Geometrie, *Z. Phys.,* **131** (1952), 544-570.

{12} K. Menger, Grundzüge einer Theorie der Kurven, *Math. Ann.,* **95** (1926), 277-306.

{13} K. Menger, Untersuchungen über allgemeine Metrik, *Math. Ann.,* **103** (1930), 466–501.

{14} J. W. Milnor, On the total curvature of knots, *Ann. Math.,* **52** (1950), 248–257.

{15} A. Moór, Espaces métriques dont le scalaire de courbure est constant, *Bull. Sci. Math.,* **74** (1950), 13-32.

{16} A. Moór, Finslersche Räume mit der Grundfunktion $\mathcal{L} = \frac{f}{g}$, *Comment. Math. Helv.,* **24** (1950), 188–194.

{17} A. Moór, Finslersche Räume mit algebraischen Grundfunktionen, *Publ. Math. Debrecen,* **2** (1952), 178–190.

{18} A. Moór, Über die Dualität von Finslerschen und Cartanschen Räumen, *Acta Math.,* **88** (1952), 347–370.

{19} A. Moór, Über oskulierende Punkträume von affinzusammenhängenden Linienelementmannigfaltigkeiten, *Ann. Math.,* **56** (1952), 397–403.

{20} A. Moór, Die oskulierenden Riemannschen Räume regulärer Cartanscher Räume, *Acta Math. Acad. Sci. Hungar.,* **5** (1954), 59–72.

{21} A. Moór, Metrische Dualität der allgemeinen Räume, *Acta Sci. Math. Szeged,* **16** (1955), 171–196.

{22} A. Moór, Allgemeine metrische Räume von skalarer Krümmung, *Publ. Math. Debrecen,* **4** (1956), 207–228.

{23} A. Moór, Entwicklung einer Geometrie der allgemeinen metrischen Linienelementräume, *Acta Sci. Math. Szeged,* **17** (1956), 85–120.

{24} A. Moór, Über den Schurschen Satz in allgemeinen metrischen Linienelementräumen, *Indag. Math.,* **19** (1957), 290–301.

{25} A. Moór, Über die autoparallele Abweichung in allgemeinen metrischen Linienelementräumen, *Publ. Math. Debrecen,* **5** (1957), 102–118.

{26} A. Moór, Konformgeometrie der verallgemeinerten Schouten–Haantjesschen Räume I. and II, *Indag. Math.,* **20** (1958), 94–113.

{27} A. Moór, Untersuchungen in Räumen mit rekurrenter Krümmung, *J. Reine Angew. Math.,* **199** (1958), 91–99.

{28} A. Moór, Über nicht-holonome allgemeine metrische Linienelementräume, *Acta Math.,* **101** (1959), 201–233.

{29} A. Moór, Über Tensoren, die aus angegebenen geometrischen Objekten gebildet sind, *Publ. Math. Debrecen,* **6** (1959), 15–25.

{30} A. Moór, Untersuchungen über Finslerräume von rekurrenter Krümmung, *Tensor, N.S.,* **13** (1963), 1–18.

{31} A. Moór, Über äquivalente Variationsprobleme erster und zweiter Ordung, *J. Reine Angew. Math.,* **223** (1966), 131–137.

{32} A. Moór, Untersuchungen über äquivalente Variationsprobleme von mehreren Veränderlichen, *Acta Sci. Math. Szeged,* **37** (1975), 323–330.

{33} A. Moór, Otsukische Übertragung mit rekurrentem Masstensor, *Acta Sci. Math. Szeged,* **40** (1989), 129–142.

{34} A. Rapcsák, Theorie der Bahnen in Linienelementmannigfaltigkeien und eine Verallgemeinerung ihrer affinen Theorie, *Acta Sci. Math. Szeged,* **16** (1955), 251–265.

{35} A. Rapcsák, (Invariante Taylorsche Reihe in einem Finslerschen Raum, *Publ. Math. Debrecen,* **4** (1955–1956), 49–60.

{36} A. Rapcsák, Über das vollständige System von Differentialinvarianten im regulären Cartanschen Raum, *Publ. Math. Debrecen,* **4** (1955–1956), 276–293.

{37} A. Rapcsák, Eine neue Charakterisierung Finslerscher Räume skalarer und konstanter Krümmung und projektiv-ebene Räume, *Acta Math. Acad. Sci. Hung.,* **8** (1957), 1–18.

{38} A. Rapcsák, Metrische Charakterisierung der Finslerschen Räume mit verschwindender projektiver Krümmung, *Acta Sci. Math. Szeged,* **18** (1957), 192–204.

{39} A. Rapcsák, Über die bahntreuen Abbildungen affinzusammenhängender Räume, *Publ. Math. Debrecen,* **8** (1961), 225–230.

{40} A. Rapcsák, Über die bahntreuen Abbildungen metrischer Räume, *Publ. Math. Debrecen,* **8** (1961), 285–290.

{41} A. Rapcsák, Über die Metrisierbarkeit affinzusammenhängender Bahnräume, *Ann. Mat. Pura Appl. (4),* **57** (1962), 233–238.

{42} P. Urysohn, Sur la ramification des lignes cantoriennes, *C.R. Acad. Sci. Paris,* **175** (1922), 481–483.

{43} O. Varga, Zur Herleitung des invarianten Differentials in Finslerschen Räumen, *Monatsh. Math. Phys.,* **50** (1941), 165–175.

{44} O. Varga, Zur Begründung der Minkowskischen Geometrie, *Acta Sci. Math. Szeged,* **10** (1943), 149–163.

{45} O. Varga, Über eine Klasse von Finslerschen Räumen, die die nichteuklidischen verallgemeinern, *Comment. Math. Helv.,* **19** (1946), 367–380.

{46} O. Varga, Über das Krümmungsmass in Finslerschen Räumen, *Publ. Math. Debrecen,* **1** (1949), 116–122.

{47} O. Varga, Über affinzusammenhängende Mannigfaltigkeiten von Linienelementen, insbesondere deren Äquivalenz, *Publ. Math. Debrecen,* **1** (1949), 1–17.

{48} O. Varga, Normalkoordinaten in allgemeinen Räumen und ihre Verwendung zur Bestimmung sämtlicher Differentialinvarianten, *Comptes Rendus de I. Congrès des Math. Hongrois, Budapest* (1950), 147–162.

{49} O. Varga, Eine geometrische Charakterisierung der Finslerschen Räume skalarer und konstanter Krümmung, *Acta Math. Acad. Sci. Hungar.,* **2** (1951), 143–156.

{50} O. Varga, Bedingungen für die Metrisierbarkeit von affinzusammenhängenden Linienelementmannigfaltigkeiten, *Acta Math. Acad. Sci. Hungar.*, **5** (1954), 7–16.

{51} O. Varga, Die Krümmung der Eichfläche des Minkowskischen Raumes und die geometrische Deutung des einen Krümmungstensors des Finslerschen Raumes, *Abh. Math. Sem. Univ. Hamburg,* **20** (1955), 41–51.

{52} O. Varga, Über die Zerlegbarkeit von Finslerschen Räumen, *Acta Math. Hungar.*, **11** (1960), 197–203.

{53} O. Varga, Zur Begründung der Hilbertschen Verallgemeinerung der nichteuklidischen Geometrie, *Monatsh. Math. Phys.*, **66** (1962), 265–275.

{54} O. Varga, Über Hyperflächen konstanter Normalkrümmung in Minkowskischen Räumen, *Tensor N.S.*, **13** (1963), 246–250.

{55} O. Varga, Über eine Kennzeichnung der Riemannschen Räume konstanter negativer und positiver Krümmung, *Ann. Mat. Pura Appl.*, **53** (1961), 105–117.

{56} O. Varga, Über eine Characterisierung der Finslerschen Räume konstanter Krümmung, *Monatsh. Math. Phys.*, **65** (1961), 277–286.

{57} I. Vincze, Determination of distributions with the aid of mean values, *Magyar Tud. Akad. Mat. Fiz. Oszt. Közl.*, **4** (1954), 513–523 (in Hungarian).

{58} I. Vincze, Bemerkung zur Differentialgeometrie der Raumkurven, *Publ. Math. Debrecen,* **4** (1955), 61–69.

Lajos Tamássy

University of Debrecen
Department of Mathematics
Debrecen 4010
Pf. 12
Hungary

tamassy@math.klte.hu

BOLYAI SOCIETY
MATHEMATICAL STUDIES, 14

A Panorama of Hungarian Mathematics in the Twentieth Century, pp. 415–425.

The works of Kornél Lánczos on the Theory of Relativity

ZOLTÁN PERJÉS

1. Introduction

Lánczos was a true acolyte of Einstein, and he kept returning to the study of relativity theory throughout his life. About thirty scientific publications, one-third of his full list, were written on this very subject. He had an open mind both to the geometrical and physical issues abundantly encountered in the theory.

He realized, for example {8}, that the redshift of light signals traveling in gravitational fields is, in fact, a Doppler-shift phenomenon as opposed to being a genuine gravitational effect. This contradicts {18} a statement, often made in general relativity. However, the Riemann tensor of the gravitational field does not appear anywhere in the redshift when expressed in terms of the four-velocities of the emitter and the observer.

He enlivened his radiant personal communication by characteristic gestures with his long fingers. The overall impression that he made arguing with the perpetual whirl of his hands was that he may have been born to live in an extraordinary neurotic state, a kind of like the *Angelman syndrome.*

Lánczos investigated a large number of research problems many of which are still of interest today. These include the formulation of junction conditions on the boundary surface of adjacent space-time regions. Known as the Lánczos–Israel boundary conditions {12, 7}, they are often used in constructing relativistic models of astrophysical bodies. He has written several papers on higher-derivative gravitational theories and their variational principles. His further contribution to the subject covers the asymptotic and exact symmetries in the formulation of conservation laws, gravitational radiation, tetrad methods and the Cauchy problem. Of special interest today is his work on the problem of motion, the potential he introduced for the Weyl tensor and his thoughts on a unified description of nature.

In a brief review of these latter works below, familiarity with the basic mathematical conventions of relativists will be an advantage for the reader. But for an added convenience, here is a summary. Space-time in general relativity is four-dimensional. Coordinates and geometrical quantities are labeled with indices assuming the values 0, 1, 2 or 3. The *Einstein convention* is used for the contraction of indices. For example, the contraction of the symmetric tensor density $\mathcal{T}_{ik}$ and the vector n^k is written, suppressing the symbol of summation over the repeated pair of indices,

$$\sum_{k=0}^{3} \mathcal{T}_{ik} n^k = \mathcal{T}_{ik} n^k. \tag{1.1}$$

Symmetrization of indexed quantities is denoted by enclosing the indices in parentheses. Skew symmetrization is similarly indicated by square brackets. Indices barred from the symmetrization are enclosed in bars.

Taking a covariant derivative (or a partial derivative) is indicated by a semicolon (or a comma, respectively) in the subscript. An example is the covariant derivative of the Lánczos potential L_{abc} when skewed in a pair of indices:

$$L_{ab[c;d]} = \frac{1}{2!}(L_{abc;d} - L_{abd;c}).$$

A convenient table to help comparison of the fluctuations in the notation is found on the red pages of {15}.

2. The problem of motion

In his 1941 paper {10}, Lánczos gave a detailed description of the motion of an isolated body in general relativity. The gist of this work is as follows.

A material body is said to be isolated in the sense that it is moving in vacuo. Other bodies may be present, but his treatment is not concerned with them. The gravitational equations inside the world tube of the particle are

$$G_{ik} = -8\pi T_{ik}.$$

Let Σ be the boundary of the world tube, with outward pointing unit normal n^i. Outside the boundary, the vacuum equations hold,

$$G_{ik} = 0$$

and G_{ik} is discontinuous across the hypersurface Σ. This discontinuity, however, is subject to the boundary condition on Σ, [*Cf.* (1.1)],

$$T_{ik}n^k = 0.$$

Inside the tube the conservation law holds for the matter

(2.1) $$T^{ik}_{;k} = 0.$$

To write this in a detailed form, we introduce the tensor density

$$\mathcal{T}^{ik} = (-g)^{-1/2}T^{ik}.$$

We then have

(2.2) $$\mathcal{T}^{ik}_{,k} = \Gamma^i$$

(where $\Gamma^i = -\Gamma^i_{rs}\mathcal{T}^{rs}$).

We now choose two space-like hypersurfaces $x^0 = a$ and $x^0 = b$, which enclose a section V_4 of the world tube. By Green's theorem,

$$\int_{x^0=b} \mathcal{T}^{i0} d^3x - \int_{x^0=a} \mathcal{T}^{i0}\, d^3x = \int_{V_4} \Gamma^i\, d^4x.$$

Passing to the limit $b \to a$ we get

(2.3) $$\frac{d}{dx^0}\int \mathcal{T}^{i0}\, d^3x = \int \Gamma^i\, d^3x,$$

where the integrals are taken on any section $x^0 = a$.

By using the identity

$$(x^j\mathcal{T}^{ik})_{,k} = \mathcal{T}^{ij} + x^j\Gamma^i$$

we can derive other relations,

(2.4) $$\frac{d}{dx^0}\int x^\alpha \mathcal{T}^{i0}\, d^3x = \int (\mathcal{T}^{i\alpha} + x^\alpha\Gamma^i)\, d^3x$$

(2.5) $$\frac{d}{dx^0}\int (x^\alpha \mathcal{T}^{\beta 0} - x^\beta \mathcal{T}^{\alpha 0})\, d^3x = \int (x^\alpha\Gamma^\beta - x^\beta\Gamma^\alpha)\, d^3x$$

where the Greek indices refer to the three-space and range through the values 1, 2 and 3.

We now introduce a more compact new notation. The *four-momentum* of the body is

$$p^i = \int T^{i0}\, d^3x.$$

The *angular momentum* of the body will be denoted

$$M^{\alpha\beta} = \int (x^\alpha T^{\beta 0} - x^\beta T^{\alpha 0})\, d^3x$$

and the *center of mass* of the body is

$$\bar{x}^\alpha = \frac{1}{p^0}\int x^\alpha T^{00}\, d^3x.$$

We may then write Eqs. (2.3) and (2.5) in the form

$$\frac{dp^i}{dt} = \int \Gamma^i\, d^3x \tag{2.6}$$

$$\frac{dM^{\alpha\beta}}{dt} = \int (x^\alpha \Gamma^\beta - x^\beta \Gamma^\alpha)\, d^3x$$

where $t = x^0$. From (2.4),

$$\frac{d}{dt}(p^0\bar{x}^\alpha) = p^\alpha + \int x^\alpha \Gamma^0\, d^3x.$$

Finally we find the physically very suggestive equation for the three-velocity of the center of mass

$$\frac{d\bar{x}^\alpha}{dt} = \frac{p^\alpha}{p^0} + \frac{1}{p^0}\int (x^\alpha - \bar{x}^\alpha)\Gamma^0\, d^3x. \tag{2.7}$$

In Eq. (2.6), the rate of change in the four-momentum and angular momentum is expressed in terms of quantities which may be regarded as the gravitational force and torque acting on the body. The last term in Eq. (2.7) expresses the amount by which the direction of the four-momentum diverges from the direction of the four-velocity. In case of a very small body, it is reasonable to neglect this term.

Synge {18} carried out an in-depth analysis of the above approach. He commented that the principal weakness of the treatment was due to its non-invariant character.

3. The Lánczos potential

In 1962, Lánczos {11} proposed that the Weyl tensor C_{abcd} (the traceless part of the Riemann tensor) in four dimensions can be given locally in terms of a potential L_{abc}. Later, however, Bampi and Caviglia {2} showed that his argument was flawed. These authors provided a new proof for the local existence of the Lánczos potential L_{abc}, which holds independently of the value of the metric signature. Yet another, spinorial derivation was presented by {6}.

The Lánczos potential has the symmetries

$$L_{abc} = -L_{bac}, \qquad L_{[abc]} = 0. \tag{3.1}$$

The Weyl tensor is given in terms of it in the form

$$\begin{aligned} C_{abcd} = {} & 2L_{ab[c;d]} + 2L_{cd[a;b]} - g_{a[c}\big(L_{|b|d]} + L_{|d|b]}\big) \\ & + g_{b[c}\big(L_{|a|d]} + L_{|d|a]}\big) + \frac{2}{3} g_{a[c}g_{d]b}L^r_r \end{aligned} \tag{3.2}$$

where

$$L_{ab} = 2L^c_{a[c;b]}.$$

This construction is analogous to that of the Maxwell tensor in terms of the vector potential. The Weyl tensor is invariant under the gauge transformations

$$L'_{abc} = L_{abc} + \chi_a g_{bc} - \chi_b g_{ac}$$

with χ_a an arbitrary four-vector. As with electromagnetism, various gauges can be introduced; one may set $L^{ab}_{\ \ b} = 0$ (this is known as the *algebraic gauge*) or $L_{abc}{}^{;c} = 0$ (the *differential gauge*).

The Lánczos potential can be utilized in the linearized theory of gravitation. Writing the metric in the form

$$g_{ab} = \eta_{ab} + h_{ab} \tag{3.3}$$

where η_{ab} is the flat metric and h_{ab} the perturbation and introducing the de Donder gauge

$$h^{ab}{}_{,b} = \frac{1}{2} h^{,a} \tag{3.4}$$

with $h = h^{ab}\eta_{ab}$, the *linearized* Lánczos potential reads

$$L_{abc} = \frac{1}{2}\left(h_{c[a,b]} - \frac{1}{6}\eta_{c[a}h_{,b]}\right). \tag{3.5}$$

Recently, there has been a revival of interest in the Lánczos potential. Novello and Neto {16} employed the linearized Lánczos potential as a model of a spin-2 field. {3} formulated the linearized gravitation theory in terms of the Lánczos potential. In a series of papers {1}, Edgar and his collaborators investigated the existence conditions of a Lánczos potential, using the Newmann–Penrose formalism.

4. The Lorentzian signature

Lánczos was deeply worried about the feature of general relativity that the signature of space-time was Lorentzian. He often argued with Einstein who was not as much concerned about this character of space-time geometry. According to Pythagoras' theorem, the distance of two neighboring points on the plane has the form

$$ds^2 = dx^2 + dy^2 \tag{4.1}$$

in Cartesian coordinates. In two more dimensions, one has

$$ds^2 = dx^2 + dy^2 + dz^2 + du^2. \tag{4.2}$$

However, in the physical world, one finds instead

$$ds^2 = dx^2 + dy^2 + dz^2 - du^2. \tag{4.3}$$

Although Einstein himself was not overtly worried about this state of affairs, he surmised that some deep mystery lurked behind the Lorentzian signature in (4.3). In Einstein's view, the essential feature of the theory was that the Riemann tensor describing the curvature of space-time is a covariant quantity and as such, it could be used in any coordinate system. From his point of view, the signature of the metric played a secondary role. However, in Lánczos's mind, the indefinite metric could not be a genuine ingredient of differential geometry since the latter is built upon the notion of small neighborhoods. With a Lorentzian metric, he argues, one cannot speak of two points being close together. In Riemannian geometry proper, zero separation means that the points coincide. However, in space-time, a pair of points can be at zero distance yet separated from each other in space by million light years. A photon that now hits one's eye from the Andromeda nebula left 3 million years before. And yet, the distance between the photon and our eyes has been zero during the whole travel. Why was there no interaction between the photon and us during the course of these 3 million years, save the moment of arrival – he asks {13}. Lánczos believed that if the notion of neighborhood lost its meaning, then differential geometry did not make sense.

In retrospect, the Lorentzian nature of the metric signature kept many researchers worried for a long time. Let us not forget that it had been a common practice right from the outset to use an imaginary time coordinate and in this way preserve, at least formally, the familiar definite form of the line element. But by now, a different viewpoint has, gradually, been adopted. This is best appreciated when recognizing that the Lorentzian geometry of space-time is a source of an abundantly rich structure and beauty. While today the topology of the manifold still occupies a central position in differential geometry, the edifice of an independent and elaborate theory of relativistic causality has grown to coexist with it. Causality theory yields deep insights into the global properties of space-time {5}. General relativity would be a good deal less appealing without the variegated world of causal phenomena.

In many other respects, Lánczos's scientific aspirations bore the influence of contemporary schools of thinking. He followed Einstein in spending much effort on attempting to geometrize all other properties of matter, including electromagnetism and quantum physics. He asked: *"With the enormous perspective allowing to interpret all material properties as special properties of space, how can it be that we can only derive* gravitation *from these very complicated relations, but both electromagnetism and quantum phenomena remain outside the scope?"*

He argued that there was a juncture in Einstein's argument leading to relativity theory where we must really take a different route to get the desired universal description of matter. This is *not* the description of the geometrical space by fundamental differential equations – a feature that he insisted on keeping. But in Lánczos's view, it is at the choice of the Lagrangian whence one must depart. He criticizes the Einstein-Hilbert Lagrangian $L = R$ on the grounds that it gives rise to a dimensioned action. In fact, this dimension is cm^2. Weyl pointed out in 1918 that it is nonsensical to seek the minimum of a dimensioned quantity since this can take any values with the appropriate choice of units. One can make the action dimensionless by choosing the Lagrangian to be quadratic in the curvature.

In 1938 Lánczos showed that the most general allowable Lagrangian can be brought to the form

$$L = R_{ik}R^{ik} - \sigma R^2 \tag{4.4}$$

where R_{ik} is the Ricci tensor, R the Einstein scalar and σ a free constant. He then considered the metric g_{ik} and the Ricci tensor as independent quantities. In this way he obtained 20 second-order differential equations for 20 unknowns. Originally, Lánczos hoped to unify gravitation and electromagnetism in this theory. To his disappointment, however, in the weak-field approximation the field equations reduce to the vacuum conditions $R_{ik} = 0$ of general relativity. Thus no room is left here for the electromagnetic field.

Later, in the sixties, Lánczos came to the idea that the class of solutions of his theory that is relevant to physics is *not* the one containing the weak-field limit. On the contrary, as he then held it, the required fields are strong periodic wave-like solutions. He conjectured that the period is the Planck length, $L_P = 10^{-32}$ cm. How come then that the world as we know is isotropic? To answer this, he resorted to the physics of the *isometric crystals* whose three principal axes are mutually orthogonal and they are equal in length. In an isotropic universe, the Ricci tensor and the metric are related by

$$R_{ik} = \lambda g_{ik}. \tag{4.5}$$

This is essentially Einstein's cosmological equation. In general relativity, the constant λ has the dimension $(\text{length})^{-2}$. There λ is extremely small because the mean *'length'* of curvature of the universe is large. For Lánczos, on the other hand, this characteristic length is extremely small, whence λ must be large.

In this strong-field approximation, the computation of the Ricci tensor is quite unlike the procedure for weak fields. For the latter, the connection quantities are small, thus only the linear terms in the connection are kept. In the strong-field case, however, it is the *linear* terms in the curvature that can be dropped and the quadratic terms dominate. The metric does not determine the effect of the background geometry. Instead, it is the mean square of the first derivatives of the metric that does this.

Lánczos abandons the Lorentz signature of the metric and explores a genuine Riemannian geometry. He notes that the field equations derived from the Lagrangian (4.4) imply a constant scalar curvature R. He next observes that with the special choice $\sigma = 1/2$ in the Lagrangian, the Ricci-tensor can be substituted for by

$$P_{ik} = R_{ik} - \mu g_{ik} \tag{4.6}$$

where μ is a constant. If one chooses $\mu = \lambda$ then one would have $P_{ik} = 0$. The meaning of this is that the submetric does not give any contribution to the slow perturbations. In such circumstances, nothing would correspond to the Minkowski-like constants $(1, 1, 1, -1)$. Hence he concludes that macroscopically there must be a small deviation from perfect isotropy. He describes this deviation by adopting the diagonal elements of the macroscopic metric to have the mean values

$$(1 + \varepsilon, 1 + \varepsilon, 1 + \varepsilon, 1 - 3\varepsilon). \tag{4.7}$$

This form satisfies that the trace of the deviations is zero. He then asserts that the four 1's in the diagonal are unobservable and the effective metric becomes

$$[\eta_{ik}] = \mathrm{diag}\,(1, 1, 1, -3).$$

We see that this effective metric is indefinite.

The weak perturbations of the effective metric are to describe electromagnetism in this theory. Denoting these metric perturbations by h_{ik}, one can use the traceless and divergence-free property of the tensor P_{ik} to derive the following relations:

$$h_{ik}\eta^{ik} = 0 \tag{4.8}$$

$$h_{ik,m}\eta^{km} = 0 \tag{4.9}$$

where a comma in the subscript denotes partial derivative.

Lánczos proposes that the perturbations can be described by a vector φ_i as follows,

$$h_{ik} = \varphi_{i,k} + \varphi_{k,i}. \tag{4.10}$$

In his view, the vector φ_i should be interpreted as the four-potential of the Maxwell field,

$$F_{ik} = \varphi_{i,k} - \varphi_{k,i}.$$

According to this interpretation, the scalar constraint (4.8) is the Lorentz gauge condition and the vector constraint (4.9) becomes the wave equation.

Thirty years later, theoreticians picture the microstructure of space-time much the same way as Lánczos did. It is being acknowledged that the smooth nature of the manifold according the description of differential geometry is inadequate on a microscopic scale. Quantum fluctuations ripple the structure of the geometry more and more forcefully as we go down to the Planck scale.

Despite the general acceptance of this picture of quantum space-time, Lánczos's theory has submerged in oblivion, much the same way as Einstein's late unified theories did. In conclusion, let me try to outline the weaknesses of this bold attempt which are discernible from a perspective of the past three decades. At several points in this theory, apparently arbitrary assumptions have been made. The first of these is the assumption that the strong wave solutions of the field equations are *periodic.* One would not expect such periodicity in a stochastic description of the quantum fluctuations. Of course, this assumption of Lánczos enormously eases the task of describing the geometry in mathematical terms. A description, which is likely to provide a more realistic picture of physics, would require at least the full machinery of relativistic quantum field theory, and possibly more from beyond that theory. Another objection I would like to raise is to the way electromagnetism is geometrized in the model. Much of the standard model of fundamental interactions was unknown three decades ago. Today a minimal task would be seen to provide a coherent description of gravitational and electroweak phenomena. It is hard to assess if the past thirty years brought the dreams of theoreticians about a glorious completion any closer to coming true.

References

{1} F. Andersson and S. B. Edgar, Curvature-free asymmetric metric connections and Lánczos potentials in Kerr–Schild space-times, *J. Math. Phys.,* **39** (1998), 2859, F. Andersson and S. B. Edgar, Spin coefficients as Lánczos scalars: Underlying spinor relations, *J. Math. Phys. (in press),* S. B. Edgar and A. Höglund, The Lánczos potential for the Weyl curvature tensor, *Proc. Roy. Soc. Lond. A,* **453** (1997), 835.

{2} F. Bampi and G. Caviglia, Third-order tensor potentials for the Riemann and Weyl tensors, *Gen. Rel. Gravitation,* **15** (1983), 375.

{3} D. Cartin, Linearized general relativity and the Lánczos potential, *preprint,* gr-qc/9910082 (1999),

{4} G. Györgyi, C. Lánczos, *Fizikai Szemle* **24,** no. 6 (1974), 166.

{5} S. W. Hawking and G. F. R. Ellis, *The large-scale structure of space-time,* Cambridge Univ. Press (1973).

{6} R. Illge, On potentials for several classes of spinor and tensor fields in curved space-times, *Gen. Rel. Gravitation,* **20** (1988), 551.

{7} W. Israel, Thin shells in general relativity, *Nuovo Cimento,* **66** (1966), 1.

{8} C. Lánczos, Über die Rotverschiebung in der de Sitterschen Welt, *Z. Physik* **17** (1923), 168.

{9} C. Lánczos, A remarkable property of the Riemann–Christoffel tensor in four dimensions, *Ann. of Math.,* **39** (1938), 842.

{10} C. Lánczos, The dynamics of a particle in general relativity, *Phys. Rev.,* **59** (1941), 813.

{11} C. Lánczos, The splitting of the Riemann tensor, *Rev. Mod. Phys.,* **34** (1962), 379.

{12} C. Lánczos, Flächenhafte Verteilung der Materie in der Einsteinschen Gravitationstheorie, *Ann. Physik,* **74** (1974), 518.

{13} C. Lánczos, Einstein and the future, *Fizikai Szemle,* **24,** no. 6 (1974), 161.

{14} G. Marx, C. Lánczos, *Fizikai Szemle,* **24,** no. 9 (1974), 277.

{15} C. W. Misner, K. Thorne and J. A. Wheeler, *Gravitation,* Freeman (1973).

{16} M. Novello and N. P. Neto, Einstein theory of gravity in Fierz variables, *preprint,* CBPF-NF-012/88 (1988).

{17} J. Stachel, "Lánczos's early contributions to relativity and his relationship with Einstein", in: *Proceedings of the Cornelius Lánczos International Centenary Conference,* North-Carolina State University (1993).

{18} J. L. Synge, *Relativity – The general theory,* North-Holland (1960).

Zoltán Perjés
KFKI Research Institute for Particle
and Nuclear Physics
Budapest 114
P.O.Box 49
H–1525 Hungary

BOLYAI SOCIETY
MATHEMATICAL STUDIES, 14

A Panorama of Hungarian Mathematics in the Twentieth Century, pp. 427–454.

Discrete and Convex Geometry

IMRE BÁRÁNY*

1. Introduction

Mathematical research in Hungary started with geometry: with the work of the two Bolyais early in the 19th century. The father, Farkas Bolyai, showed that equal area polygons are equidecomposable. The son, János Bolyai, laid down the foundations of non-Euclidean geometry. The study of geometric objects has been continuing ever since. The present chapter of this book is devoted to describing what investigations took place in Hungary in the 20th century in the field of convex and combinatorial geometry. This includes incidence geometries, finite geometries, and stochastic geometry as well. The selection of the material is, of course, a personal one, and some omissions are inevitable (though most likely unjustified). The choice is made difficult by the wide variety of topics that were to be included.

Besides mathematics, or discrete and convex geometry (to be more precise), this survey is about people, is about mathematicians. Whenever appropriate, I have tried to say a few words not only about the mathematics but the person as well. There are quite a lot of them, but the heroes of the story are two giants who stand out head and shoulder above the rest. They are László Fejes Tóth and Pál Erdős. Both of them helped to create the school of Hungarian discrete geometry, and both of them were extremely successful problem solvers and exceptionally prolific problem raisers. Yet they were of different taste, style, and character. Their questions, their results, and, in general, their mathematics have, to a large extent, determined what discrete and convex geometry in Hungary means, and how and

*Partially supported by Hungarian Science Foundation Grant T 032452 and T 037846.

in what directions the Hungarian school of geometry has developed. I have tried to group geometric research in Hungary around topics, and not so much around people. Sometimes this has been difficult and occasionally I have had to take the liberty of proceeding differently.

2. The beginnings

Geometric research in Hungary, after the work of the two Bolyais, besides containing many scattered results, is concentrated around two topics. One is geometric constructions, and the other one is equidecomposability of equal area polygons.

Geometric constructions has always been a popular topic in high school mathematics in Hungary, and the theory of geometric constructions had been popular subject in geometric research as well. The starting point is a theorem of Hilbert's from Grundlagen der Geometrie (Teubner, 1899) saying that usual geometric constructions (with ruler and compass) can be accomplished by a ruler alone if a circle, together with its centre, is drawn in the plane. (He also showed that the centre is necessary.) Hilbert proved, further, that the compass is needed only to copy segments on a line, that is, the ruler and the ability of copying segments suffice for the usual geometric constructions. Continuing Hilbert's work, József Kürschák proved (Math. Ann., **55** (1902), 597–598) that whatever can be constructed using a ruler and the ability of copying segments, can be constructed by a ruler and the ability of copying a fixed segment, say the unit segment. The algebraist Mihály Bauer proved (Ungar. Ber., **20** (1905), 43–47), that in Kürschák's "unit segment" theorem the unit segment cannot be replaced by any fixed, unmovable segment, or even by any fixed, unmovable polygon. Richárd Obláth continues these investigations. He simplifies Kürschák's proof (Math. Phys. Lapok, **18** (1909), 174–176). In Obláth (Monatsh. Math., **26** (1915), 295–298) it is proved that in Hilbert's theorem the circle can be replaced by an arbitrary arc of the circle (its centre is, of course, necessary). This was also shown by Gyula Szőkefalvi-Nagy (Tohoku J. Math., **40** (1934), 76–78). Many decades later, Obláth proved (Matematikai Lapok, **2** (1951), 219–221) that, the circle and its centre can be replaced by an arbitrary arc together with the two points that split the arc into three arcs of equal length. Gyula Szőkefalvi-Nagy wrote a book on geometric constructions: A geometriai szerkesztések elmélete (Kolozsvár, 1943). Pál Szász and Gyula Strommer had also worked in this area.

Farkas Bolyai's equidecomposablity theorem has been a popular subject as well. Mór Réthy (Ann. Math., **38** (1891), 405–428) extends the result from polygons to some other planar regions, and Zsigmond Spiegl (Math. Phys. Lapok, **2** (1893), 17–30) gives a new proof of the result. 60 years later Tamás Varga found a short and transparent proof of Farkas Bolyai's theorem (Mathematikai Lapok, **5** (1954, 101–114).

In (Ungar. Ber., **15** (1898), 196–197) Kürschák gave a purely geometric proof of the fact that the area of the regular twelvegon P is exactly three times the area of the square Q whose side length is the radius of circle, circumscribed about P. The proof is accomplished by decomposing P and three copies of Q into finitely many congruent pieces. W. Csillag (Ungar. Ber., **19** (1903), 70–73) gives an alternative proof, based on a remark of Kürschák.

Concerning best approximation problems, József Kürschák gave the first elementary proof (see [41], page 6) of the following inequalities. Let R resp. r the circumradius and inradius of a convex n-gon K_n with area A and perimeter L. Then

$$nr^2 \tan\frac{\pi}{n} \le A \le \frac{1}{2}nR^2 \sin\frac{2\pi}{n},$$

and

$$2nr \tan\frac{\pi}{n} \le L \le 2nR \sin\frac{\pi}{n},$$

and equality holds in all inequalities if and only if K_n is the regular n-gon.

There are further geometric results from the turn of the century. For instance, Lipót Klug (Monatsh. Math., **10** (1889), 84–87) proves the following interesting generalization of Pythagoras's theorem. Denote by $[x_1, \dots, x_k]$ the $(k-1)$-dimensional volume of the simplex with vertices $x_1, \dots, x_k \in \mathbb{R}^n$, where $k \le n$. If $v_1, \dots, v_n$ is a system of n pairwise orthogonal vectors in $\mathbb{R}^n$, then

$$\sum \left[v_{i_1}, \dots, v_{i_k}\right]^2 = (n-k+1) \sum \left[0, v_{j_1}, \dots, v_{j_{k-1}}\right]^2,$$

where the summation is taken over all k, resp. $(k-1)$ membered subsets of $\{1, \dots, n\}$. It is readily seen that the case $k = 2$ is a simple consequence of the Pythagoras theorem.

Gusztáv Rados (Ungar. Ber., **22** (1907), 1–12) considers regular star-n-gons inscribed in the unit circle. (A star-n-gon is obtained by connecting two vertices of a reguler n-gon when there are exactly $k-1$ vertices between them; k must be relative prime to n.) Their number is clearly $\frac{1}{2}\varphi(n)$ (where $\varphi(n)$ is Euler's totient function). The star-n-gon is subdivided by its edges into cells, the cell containing the centre is called the "kernel". Denote the sum of the areas of the $\frac{1}{2}\varphi(n)$ kernels by I_n, and the area of the regular convex n-gon, circumscribed about the unit circle by C_n. Then the ratio I_n/C_n is a rational number which is equal to $\frac{1}{4}\big(\varphi(n)+\mu(n)\big)$ where $\mu(n)$ is the Möbius function. As expected, the proof uses geometry and number theory. It is shown further that if n is a prime, then $C_{2n}I_n = n^2/2$.

Gyula Vályi (1855–1913) was a respected professor in Kolozsvár, who lectured on various subjects. He was almost blind. He did some work in geometry. For instance, (Math. Phys. Lapok, **10** (1901), 309–321), he considers the foot-triangle, $A_1B_1C_1$, of the triangle ABC where A_1 is the foot of the altitude starting at A and B_1 and C_1 are defined similarly. The foot triangle has its own foot triangle A_2, B_2, C_2, and so on. Can the nth foot triangle be similar to the original triangle? Vályi shows that there are exactly $2^n(2^n-1)$ (non-similar) triangles that are similar to their nth foot-triangle. Vályi together with Gyula Kőnig was also interested in perspective triangles and tetrahedra.

Dénes Kőnig's (1884–1944) main interest was graph theory, but had a few nice results in geometry as well. For instance, his joint paper with Adolf Szűcs (Mat. Természettud. Ért., **31** (1913), 545–558) investigates the orbit of a point in the 3-dimensional cube when it starts moving in direction v, and is reflected like light whenever it meets the boundary of the cube. They show that the orbit is periodic if and only if the ratio of any two components of v is rational (v is a rational vector, for short), the orbit is everywhere dense if and only if v is not orthogonal to any rational vector, and if v is orthogonal to exactly one rational vector, then the orbit lies on the boundary of a polyhedron and is everywhere dense there. He proves Helly's theorem (Math. Zeitschrift, **14** (1922), 208–210); the proof is identical with Helly's original proof that only appeared in 1923. (Helly found his famous theorem in 1913 but could not publish it because of the First World War. The first proof, by Radon, appeared in 1921.)

In this book, there are two long chapters about Lipót Fejér and his work in analysis. In his student years, Fejér had been attracted to geometry where he surprised his colleagues by beautiful elementary proofs. One survives, see [140] or [40], Vol. II, pp. 844–847: Assume ABC is an acute triangle.

Then its foot-triangle has the the smallest perimeter among all triangles XYZ where the point X comes from the line through B, C, the point Y from the line through C, A, and Z from the one through A, B.

Of course, János Bolyai's ground-breaking result on non-Euclidean geometry, whose real importance was understood quite late, was a central topic in the mathematical life of the time. In 1897, János Bolyai's Appendix appeared in Hungarian translation for the first time. Even more significantly, there was to be a volume on the achievements of mathematics, edited by Poincaré, which was to contain a chapter on "Géométrie de Lobatschewsky". The title of this chapter finally became "Géométrie de Bolyai et Lobatschewsky", thanks to the work of a committee consisting of G. Rados, B. Tőttössy, J. Kürschák, and L. Kopp. In another development, under the auspices of Gyula Kőnig and Mór Réthy, the Tentamen of Farkas Bolyai appeared a second time in two volumes, the first in 1897, the second in 1904.

3. Packings and coverings by circles

László Fejes Tóth has been working in geometry since 1939. His interests are very broad: packing and covering, approximation, isoperimetric inequalities for polytopes, and much more. We start by describing his ground-breaking research in the theory of packings and coverings.

One of László Fejes Tóth's early results is a new proof (the first correct proof, according to C. A. Rogers) of a theorem of Thue from 1882:

Thue's theorem ([41], page 58). *The density of any packing of congruent circles in the plane is at most* $\pi/\sqrt{12}$.

Here and in what follows packing means a collection of pairwise (internally) disjoint sets, while a covering is a collection of sets whose union contains the set which is to be covered. Density has just the usual definition: on a bounded set D it is the total area of the circles divided by Area D, and one takes the limit if the set in question is not bounded. Dual to Thue's theorem is that of Kershner.

Kershner's theorem ([41], page 58). *The density of any covering of the plane with congruent circles is at least* $2\pi/\sqrt{27}$.

László Fejes Tóth has proved many extensions and generalizations of these results. The theory of "packings and coverings" he developed in 2 and 3-dimensions is the content of his book Lagerungen in der Ebene, auf dem Kugel und im Raum [41]. The extension of the theory to higher dimension is carried out in the book by C. A. Rogers, Packing and covering, Cambridge, 1964. This extension is rather restricted since the higher dimensional packing and covering problems are much more difficult, and as a consequence, there are only few results about them.

We describe now some generalizations of Thue's and Kershner's theorem that are due to László Fejes Tóth, see [41], page 67.

Theorem. *Every packing of (at least two) congruent circles in a convex domain has density at most* $\pi/\sqrt{12}$.

Theorem. *Every covering of a convex domain by (at least two) congruent circles has density at least* $2\pi/\sqrt{27}$.

When one is only interested in the asymptotic behaviour of an extremal system of circles, the shape of the domain does not matter much. So it is quite natural to consider hexagons instead of general convex domains. The dual problems of packing and covering can be unified in the following way. How to place congruent circles in the plane, when the density is given a priori, and we want to maximize the area covered by the circles. The answer is in the following theorem, see [41], page 80.

Theorem. *Given a hexagon of area H and a system of congruent circles of total area T, let A denote the area of that part of the hexagon that is covered by the circles. Then $A \leq A^*$ where A^* is the area of the intersection of the circle of area T and a concentric regular hexagon of area H.*

This result is a special case of the so-called Moment Theorem (see [41], page 81 and Section 5 below for this particular application) which was invented in connection with isoperimetric problems for polyhedra (see later). The Moment Theorem has found several further extensions and applications in the works of P. M. Gruber and Gábor Fejes Tóth.

4. Packings and coverings by incongruent circles

The problems become more involved when incongruent circles are used. In a joint work of L. Fejes Tóth and J. Molnár (Math. Nachr., **18** (1958),

235–243), any kind of circle of radius r from the interval $[a, b]$ can be used, and the question is how to choose and arrange such circles to obtain a densest possible packing, or a thinnest possible covering. Upper and lower bounds are given for the densities in question. József Molnár constructed examples of packings and coverings, using only circles of radius a and b, that are almost optimal. An interesting remark from [41], page 79 says that if the ratio b/a is close to one, then the density is maximal if the system is the densest lattice packing of congruent circles. This line of research has been continued by Károly Böröczky, Aladár Heppes, András Bezdek, Károly Bezdek, József Molnár, Gábor Fejes Tóth, and others.

Another result concerning packings with incongruent circles (see [41], page 75) says that if a hexagon H contains n non-overlapping circles with radii $r_1, \dots, r_n$, then $(r_1 + \dots r_n)^2 \leq n \,\mathrm{Area}\, H \sqrt{12}$. This means, roughly speaking, that the total area of circles, packed in a hexagon, is maximal if they are congruent and each is touched by six others.

It is perhaps worth mentioning here that in this field there is still much to be discovered.

5. Packings and coverings by convex sets in the plane

In connection with packings and coverings it is quite natural to consider not only circles but other convex bodies as well. The following far-reaching generalization of Thue's theorem is due to László Fejes Tóth ([41], page 85).

Theorem. *Let $K \subset \mathbb{R}^2$ be a convex body, and let P_6 be a hexagon, of minimum area, circumscribed about K. If n congruent (non-overlapping) copies of K are packed in a convex hexagon H, then $n \,\mathrm{Area}\, P_6 \leq \mathrm{Area}\, H$.*

In the proof one grows the congruent copies, K_i, of K into (non-overlapping) convex polygons R_i with $K_i \subset R_i$. Next, Euler's theorem on planar graphs implies that the total number of sides of the R_is is at most $6n$. Then Dowker's theorem (see a little later) and an application of the Jensen inequality finishes the proof. For details see [41]. When K is a circle, then $\mathrm{Area}\, P_6 = \frac{\pi}{\sqrt{12}} \mathrm{Area}\, K$, which gives Thue's theorem. A very general corollary of the previous theorem says the following:

Corollary. *No packing of congruent, centrally symmetric convex bodies in the plane can have density larger then that of the densest lattice packing of the same convex body.*

Rogers proved that the above theorem remains valid for non-symmetric convex bodies K as well provided only translated copies of K are allowed to form the packing. A beautiful and short proof of this fact was given by L. Fejes Tóth (Mathematika, **30** (1983), 1–3). In the covering version of the previous theorem an extra (and most likely only technical) condition is needed, namely, that the covering is "non-crossing". Two copies K_i and K_j are said to be crossing if removing their intersection from their union, both sets split into disjoint parts.

Theorem ([41], page 86). *Let $K \subset \mathbb{R}^2$ be a convex body, and let p_6 be a hexagon, of maximum area, inscribed in K. If n congruent and pairwise non-crossing copies of K cover a convex hexagon H, then $n\,\mathrm{Area}\,p_6 \geq \mathrm{Area}\,H$.*

An interesting generalization of the last three theorems is due to Gábor Fejes Tóth (Acta Math. Acad. Sci. Hungar., **23** (1972), 263–270). It goes as follows.

For a convex body $K \subset R^2$ of unit area let $f_K(x)$ denote the maximum area of the intersection of K and a hexagon of area x. Further, let $\bar{f}_K(x)$ be the least concave function greater than or equal to $f(x)$. Given a convex hexagon of area H and a system of congruent non-crossing copies of K with total area T, let A denote the area of that part of the hexagon which is covered by the copies of K. Then

$$A \leq T \bar{f}_K(H/T).$$

This bound is sharp if K is centrally symmetric. In this case, for given density T/H an arrangement of a large number of copies of K for which A is arbitrarily close to the upper bound is generally not lattice-like, but is given by an appropriate combination of two lattice arrangements. This phenomenon shows a remarkable analogy with the phase transition of crystals, as Gábor Fejes Tóth and László Fejes Tóth remark (Computers Math. Applic., **17** (1989), 251–254).

This is the point where a result of István Fáry (1922–1984) should be mentioned. Besicovitch proved that every convex body K in the plane contains a centrally symmetric hexagon whose area is at least $2/3\,\mathrm{Area}\,K$. Fáry gave an alternative proof of this and characterized the case of equality

(Bull. Soc. Math. France, **78** (1950), 152–161). Moreover, he used it in the following result on packings and coverings in the plane (see [40], page 100).

Theorem. *Assume $K \subset \mathbb{R}^2$ is a convex body. Let $\delta_L(K)$ and $\theta_L(K)$ denote the density of the densest lattice packing and the density of the thinnest lattice covering of K (by translates of K) in the plane. Then*

$$\delta_L(K) \geq \frac{2}{3} \quad \text{and} \quad \theta_L(K) \leq \frac{3}{2},$$

with equality if and only if K is a triangle.

We mention, however, that the question whether $\theta(K) \leq 3/2$ for non-lattice coverings by translates of a triangle K is still wide open. There are plenty of such questions in this area.

6. Packings and Coverings on the Sphere

Packing and covering of the 2-dimensional sphere by spherical caps is a problem, analogous to the previous. For instance, Thammes's question asks for the densest packing of n circles (caps) on the sphere. The dual problem is that of the thinnest covering by n circles (caps) of the sphere. L. Fejes Tóth gives upper resp. lower bounds for the densities in question. Set $\omega_n = \frac{n}{n-2}\frac{\pi}{6}$.

Theorem ([41], page 114). *When $n \geq 3$ congruent caps are packed on the sphere, then their density is at most*

$$\frac{n}{2}\left(1 - \frac{1}{2\sin\omega_n}\right).$$

When the sphere is covered by $n \geq 3$ congruent caps, then their density is at least

$$\frac{n}{2}\left(1 - \frac{1}{\sqrt{3}\tan\omega_n}\right).$$

These inequalities solve the question of densest packing, resp. thinnest covering of the sphere for $n = 3, 4, 6, 12$ (see [41]). The proof shows at the same time, that the extremal systems are formed by the circles, inscribed in, resp. circumscribed about, the faces of the regular mosaics with symbols $\{k, 3\}$ with $k = 2, 3, 4, 5$. (These mosaics, for $k > 2$, correspond to a regular polytope in 3-space whose facets are regular k-gons.) Also, the above result gives an alternative proof of the theorems of Thue and Kershner (when n goes to infinity), and shows that the extremal configuration corresponds to the regular mosaic $\{6,3\}$ which is the usual tiling of $\mathbb{R}^2$ by regular hexagons.

These questions are closely related to isoperimetric problems about polytopes. For instance, assume that V is the volume, F is the surface area of a convex polytope $P \subset \mathbb{R}^3$ with f faces. What's the smallest value of the so-called isoperimetric quotient F^3/V^2? Steiner conjectured that if the polytope is combinatorially equivalent to a regular polytope, then the isoperimetric quotient is minimal for the corresponding regular polytope. (Steinitz had doubts about this conjecture.) László Fejes Tóth proved the validity of the conjecture for the case of the tetrahedron, cube, and dodecahedron ($f = 4, 6, 12$) with the following theorem (see [41], page 135, and [42], page 283).

Theorem. *Under the above conditions*

$$\frac{F^3}{V^2} \geq 54(f-2)\tan\omega_f(4\sin^2\omega_f - 1).$$

The proof of Steiner's conjecture goes via Lindelöf's theorem (stating that any polytope, extremal to the isoperimetric quotient is circumscribed about the sphere), and the so-called Moment Theorem (see [42], page 219). The Moment Theorem has various forms, here we give the one used in the plane ([41], page 81). We assume that $g : R^+ \to R^+$ is an increasing function, $H \subset \mathbb{R}^2$ is a convex hexagon and $a_1, \dots, a_n$ are points in H. Set finally $d(x) = \min\{|x-a_1|, \dots, |x-a_n|\}$.

Moment Theorem. *Under these conditions*

$$\int_H g\big(d(x)\big)\,dx \leq n\int_h g\big(|x|\big)\,dx,$$

where h is a regular hexagon, of area Area H/n, *centered at the origin.*

There are several other isoperimetric problems concerning packings and coverings. For instance, in a given a hexagon $H \subset \mathbb{R}^2$ n non-overlapping convex bodies are placed. What is the minimum of the total perimeter of these n convex bodies, if each has area at least a? This is the *perimeter problem.* Or what is the maximum of the total area of the n convex bodies, if each has perimeter at most p? This is the so called *area problem,* which was solved by László Fejes Tóth, [42], page 175. He and Aladár Heppes solved the perimeter problem, see [42], page 174. The extremal configuration consists, in both cases, either of arbitrarily arranged circles, or of certain "smooth polygons" inscribed in the faces of (a bounded piece) of the regular mosaic $\{6,3\}$. Heppes proved further (Publ. Math. Inst. Hungar. Acad. Sci., **8** (1964), 365–371) that in the area problem the same conclusion holds without assuming the convexity of the pieces.

7. Packings and coverings in the hyperbolic plane

Packing and covering problems arise naturally in hyperbolic spaces as well. However, it is impossible to define the density of a packing or covering that would satisfy the most natural and simple requirements. This was pointed out by an ingenious example due to Károly Böröczky. László Fejes Tóth gives an alternative notion of densest packing and thinnest covering, that of "solidity'. A packing of convex bodies is called *solid* if altering the positions of finitely many of the bodies, and leaving the remaining ones unmoved, the new packing obtained this way is always congruent with the original one. The definition is analogous for coverings. A solid packing (and covering) on the sphere and on the Euclidean plane is automatically the densest (thinnest, resp.). So the following theorems, due to Margit Imre (Acta Math. Hung., **15** (1964), 115–121), generalize the corresponding spherical and Euclidean results. We mention that the regular mosaic $\{n,3\}$ forms a tessellation of the sphere for $n = 2,3,4,5$, of the Euclidean plane for $n = 6$, and of the hyperbolic plane for $n > 6$.

Theorem. *For every $n \geq 2$, the circles inscribed in the faces of the regular mosaic with symbol $\{n,3\}$ form a solid packing.*

For every $n \geq 3$, the circles circumscribed about the faces of the regular mosaic with symbol $\{n,3\}$ form a solid covering.

We mention in passing that András Bezdek extended the first statement from the above theorem by showing that for $n > 7$ the packing is "super-solid", which means that removing n of the circles and putting back only $n-1$ so as to form a packing, one gets back the original system minus one circle. The question is still open for $n = 6, 7$.

The Hungarian school of discrete geometry was founded by László Fejes Tóth and was strongly influenced by his results, insight and inspiring questions. László Fejes Tóth had the exceptional ability of addressing the right question to the right people. Several young mathematicians-to-be started working in discrete geometry because of an intriguing problem of László Fejes Tóth, and became discrete geometers under his guidance and influence. The Hungarian school of discrete geometry has been incessantly working on the "Fejes Tóth" theory of packings and coverings. The following names must be mentioned here: Imre Bárány, András Bezdek, Károly Bezdek, Károly Böröczky, Károly Böröczky Jr., Gábor Fejes Tóth, Zoltán Füredi, Aladár Heppes, Jenő Horváth, Gábor Kertész, Endre Makai Jr., Emil Molnár, József Molnár, and János Pach.

8. Packings and coverings in higher dimensions

In higher dimensions the problem of densest packing and thinnest covering (of congruent balls, say) is much harder. The densest sphere packing problem in 3-dimensional space goes back to Kepler and is part of problem 18 of Hilbert's famous set of unsolved problems. In the early 50's László Fejes Tóth made a significant step toward the solution of this problem. He proposed a strategy which, if carried out succesfully, solves the problem by reducing it to a finite optimization problem. A decade later, he even foresaw the possibility of using computers in the solution. In [42], page 300, László Fejes Tóth writes that "...this problem can be reduced to the determination of the minimum of a function of a finite number of variables, providing a programme realizable in principle. In view of the intricacy of this function we are far from attempting to determine the exact minimum. But, mindful of the rapid development of our computers, it is imaginable that the minimum may be approximated with great exactitude."

We close this section by recalling two general covering theorem of Erdős and Rogers. One is from (J. London Math. Soc., **28** (1953), 287–293) and gives the first nontrivial lower bound on the density of any covering, by congruent balls, of $\mathbb{R}^d$. The second result (Acta Arithmetica, **7** (1962), 281–285) is very general and has been used often. Its proof is a powerful combination of random methods and maximal lattice packing.

Theorem. *For every convex body $K \subset \mathbb{R}^d$ there exists a covering of $\mathbb{R}^d$ by translates of K whose density is less than $d(\log d + \log\ \log d + 4)$ and so that no point is covered more than $ed(\log d + \log\ \log d + 4)$ times.*

9. Approximation

Approximation problems of convex bodies by special classes of convex bodies, usually by polytopes with few vertices or faces belong to the theory of convex sets. László Fejes Tóth has made pioneering research in this direction and used his results in the theory of packings and coverings.

Assume $K \subset \mathbb{R}^2$ is a convex body. Denote by T_n (resp. t_n) the area of the maximal (minimal) area convex n-gon inscribed in (circumscribed about) K. Answering a question of Kershner, Dowker (Bull. AMS., **50** (1944), 120–122) proved that the sequence $T_3, T_4, \dots$ is concave, and the sequence $t_3.t_4, \dots$ is convex. In other words,

$$T_{n-1} + T_{n+1} \leq 2T_n \quad \text{and} \quad t_{n-1} + t_{n+1} \geq 2t_n.$$

László Fejes Tóth and József Molnár extended this result to inscribed (circumscribed) polygons with maximal (minimal) perimeter: (Molnár, Matematikai Lapok, **26** (1955), 210–218; Fejes Tóth, Math. Phys. Semesterber., **6** (1958/59), 253–261).

It had been generally known or assumed that, among all d-dimensional convex bodies, the Euclidean ball can be approximated worst. When we measure approximation by an affinely invariant quantity (like missed area) the extreme case should be the set of ellipses.

An early example of this phenomenon is given by a theorem of Ernő Sas. Answering a question of L. Fejes Tóth, he proved the following, (see E. Sas, Compositio Math., **6** (1939), 468–470), and [41], page 36, as well. For a convex body $K \subset \mathbb{R}^2$, T_n still denotes the maximal area of an inscribed convex n-gon. Then

$$T_n \geq \operatorname{Area} K \frac{n}{2\pi} \sin \frac{2\pi}{n},$$

with equality if and only if K is an ellipse.

Dezső Lázár, a promising young mathematician (who became a victim of holocaust in 1942) proved the following result, (Lázár, Acta Univ. Szeged, **11** (1947), 129–132) and [41], page 40, as well. We keep the previous notation t_n and T_n. Then

$$\frac{T_n}{t_n} \geq \cos^2 \frac{\pi}{n},$$

with equality if and only if K is an ellipse. The analogous statement for the best approximation in perimeter was proved by László Fejes Tóth ([41], page 30).

There is a strong relation between best approximation of a convex body $K \subset \mathbb{R}^2$ and its affine perimeter, when approximation is measured by an affinely equivariant quantity. This was pointed out by László Fejes Tóth. The affine perimeter was defined by Blaschke as $\int \kappa^{1/3} ds$ where the integration goes on the boundary of K according to arc-length. Fejes Tóth proves, among other similar results, the following one. Again, K and t_n have the same meaning as above.

Theorem. *Assume K has twice differentiable boundary, and its affine perimeter is λ. Then $\lambda^3 \leq 8 t_n n^2 \sin^2(\pi/n)$.*

This result implies Blaschke's affine isoperimetric inequality, which says, that $\lambda^3 \leq 8\pi^2 \operatorname{Area} K$. Fejes Tóth proves a further remarkable property of the affine perimeter: in a given triangle ABC, among all curves connecting A to B within the triangle, and bounding, together with the side AB, a fixed area, the largest affine arc-length goes with the cone-section that touches the side AC at A, and BC at B.

L. Fejes Tóth proves [41], page 89, the following result concerning affine perimeter and packings by convex bodies:

Theorem. *Assume n convex bodies are packed in a hexagon H, and let Λ denote the sum of the affine perimeters of the convex bodies. Then*

$$\Lambda^3 \leq 72 n^2 \operatorname{Area} H.$$

10. The Erdős–Szekeres theorem

Discrete geometry in Hungary was born when Erdős and his friends (Tibor Gallai, György Szekeres, Pál Turán, Eszter Klein, and many others) were very young and became interested in all kinds of combinatorial questions. A good example of this is the so-called Erdős–Szekeres theorem, which grew out of the following observation of Eszter Klein from 1934. From every set of five points in general position in the plane one can choose four that are in convex position, where k points are said to be in convex position if none of them is contained in the convex hull of the other $k-1$. Erdős immediately generalized the question and, together with Szekeres, proved the following result.

Erdős–Szekeres theorem. *For every $k \geq 4$ there exists a finite number $N(k)$ such that every set $X \subset \mathbb{R}^2$ in general position with $|X| \geq N(k)$ contains k points that are in convex position.*

This appeared in Erdős, Szekeres (Compositio Math., **2** (1935), 463–470), in the proof they rediscover Ramsey's theorem. They published a second paper on this problem 26 years later (Ann. Univ. Budapest, Sectio Math., **3-4** (1961), 53–62). They prove, denoting by $N(k)$ the smallest integer the theorem is valid for, the following bounds

$$2^{k-2} + 1 \leq N(k) \leq \binom{2k-4}{k-2}.$$

Earlier Endre Makai Sr. and Pál Turán showed that $N(5) = 9$, and Eszter Klein's observation gives $N(4) = 5$. This is in accordance with the conjecture that $N(k) = 2^{k-2} + 1$ which has become known as the Happy End Problem because Eszter Klein and Szekeres got married, escaped from Hungary to Australia via Shanghai (because of the holocaust) and have been living happily ever since. (Actually, there is no other evidence than $N(4) = 5$ and $N(5) = 9$ for the conjecture.)

Later, Erdős asked whether among sufficiently many points (in general position) in the plane one can always find the vertices of an *empty k-gon*, that is, k points in convex position such that their convex hull contains no more points from the original set. This turned out to be true for $k = 3, 4, 5$, false for $k > 6$. Very annoyingly, the problem is still open for $k = 6$.

There are several further extensions, generalizations, and applications of the Erdős–Szekeres theorem that are beyond the scope of this survey. For instance, the theory of order types (started by Goodman and Pollack) grew out of an attempt to prove the Happy End Conjecture. The recent overview of these developments by W. Morris and V. Soltan (Bull. AMS., **37** (2000), 437–458) lists more than 200 references. The Hungarian school of discrete geometers, namely Imre Bárány, Tibor Bisztriczky, Gábor Fejes Tóth, Zoltán Füredi, Gyula Károlyi, János Pach, József Solymosi, Géza Tóth, have been actively pursuing Erdős–Szekeres type phenomena.

11. Repeated distances, distinct distances in the plane

Erdős was interested in all kinds mathematics, he knew very well that mathematics develops by asking questions, as they constitute the raw material mathematicians can work on. He himself was a prolific problem raiser, often more proud of a good question he asked than a theorem he proved. He once said that he had never been jealous of a result of someone else, but he had often been jealous of a good problem someone else asked. He raised several questions a day, some based on new insight or new theorems, some in the hope of getting closer to the solution of the some old problem, sometimes the question came just out of curiosity. With the following two questions (Erdős, Amer. Math. Monthly, **53** (1946), 248–250), he struck gold:

At most how many times can a given distance occur among a set of n points in the plane?

What is the minimum number of distinct distances determined by a set of n points in the plane?

To be more formal, let X be a set of n points in the plane, and let $f(X)$ denote the number of pairs $x, y \in X$ such that their distance $|x-y|$ is equal to one, and let $g(X)$ denote the number of distinct distances $|x-y|$, $x, y \in X$. Define

$$f(n) = \max f(X), \quad \text{and} \quad g(n) = \min g(X).$$

With this notation, Erdős's question is to find, or at least estimate, $f(n)$ and $g(n)$. These two questions have turned out both extremely hard and extremely influential.

Erdős proves, in the same paper, that $f(n) \leq cn^{3/2}$. In the proof Erdős uses a simple geometric argument to show that the graph of unit distances (with vertex set X) does not contain the complete bipartite graph $K_{2,3}$. Since such a graph cannot have more than $cn^{3/2}$ edges, the upper bound on $f(n)$ follows immediately. This is the first application of extremal graph theory in combinatorial geometry, that has been followed by many others. The effect is mutual and mutually beneficial: a question in combinatorial geometry often leads to a problem in extremal graph or hypergraph theory. Erdős did pioneering work in this direction. The best upper bound to date is $f(n) \leq cn^{4/3}$ (due to Spencer, Szemerédi, Trotter). Here is another formulation of the "unit distances" question: given n points in the plane and the n unit circles centred at these points, how many point-circle incidences can occur among them? In this form, the question immediately leads to incidence problems to be discussed in Section 12.

Again in the same paper, Erdős gives the lower bound, (which is conjectured to be the proper order of magnitude of $f(n)$):

$$f(n) > n^{1+c/\log\log n}.$$

The construction is just the $\sqrt{n} \times \sqrt{n}$ grid; the proof uses a little number theory. The same construction gives, for the number of distinct distances, that

$$g(n) \le \frac{cn}{\sqrt{\log n}}.$$

This is again the conjectured value of $g(n)$. Moser gave the lower bound

$$g(n) > cn^{2/3}$$

which has been improved several times by methods combining geometry and combinatorics. The current best lower bound (due to Katz and G. Tardos, based on earlier work of Solymosi and Cs. Tóth) is $cn^{.864\ldots}$. (A recent result of Imre Ruzsa shows that the current techniques cannot give anything of the form $n^{8/9}$.)

The problem changes if one strengthens the non-collinearity condition on X by assuming, say, that the points are in convex position, or that X is in general position. The convexity condition gave rise to the theory of *forbidden submatrices.* For the general position case, Erdős, Füredi, Pach, Ruzsa (Discrete Math., **111** (1993), 189–196) show that

$$g^{\text{gen}}(n) \le ne^{\sqrt{c\log n}},$$

while the lower bound $(n-1)/3$ is due to Szemerédi. In the same paper Erdős et al. show that, if X contains no three points on a line and no four on a circle, then the inequality $g(X) \le C|X|$ does not hold for any constant C. The proof uses a celebrated result of Freiman from additive number theory.

Erdős also asked, in his 1946 Monthly paper, how often the maximal, minimal distance can occur among pairs of points of a set $X \subset \mathbb{R}^2$. The minimal distance problem has been completely solved in $\mathbb{R}^2$, but not in higher dimensions. The maximal distance can occur n times in $\mathbb{R}^2$, and $2n-2$ times in $\mathbb{R}^3$ (the latter result is due to Heppes and Grünbaum). For higher dimensions, the Lenz construction (see in the next chapter) gives asymptotically optimal point sets. A more general question concerns the distribution of distances. Erdős, Lovász, Vesztergombi (Discrete Comp. Geom., **4** (1989), 341–349) investigate the graph determined by the k largest distances.

Concerning the possible distribution of distances, Ilona Palásti constructed examples of point sets $X \subset \mathbb{R}^2$ with $|X| = k$ for $k = 4, 5, 6, 7, 8$ where the $k(k-1)/2$ distances occur with very special distribution: one distance occurs once, another twice, a third three times, etc. See for instance Palásti (Discrete Math., **76** (1989), 155-156). In general, she was working on geometric problems proposed by Erdős, we will encounter another result of hers in Section 14.

12. Repeated and distinct distances elsewhere

Of course the same questions can be asked in any dimension. Denoting the corresponding functions by $f_d(n)$ and $g_d(n)$, Erdős proved (Publ. Math. Inst. Hung., **5** (1960), 165–169) that

$$cn^{4/3} \leq f_3(n) \leq cn^{5/3}.$$

By now there are better estimates for $f_3(n)$. The behaviour of $f_d(n)$ for $d > 3$ is simple, because of the so-called Lenz construction, (see the same paper of Erdős): half of the points are on the circle $(x, y, 0, 0)$ with $x^2 + y^2 = 1/2$, the other half on the circle $(0, 0, u, v)$ with $u^2 + v^2 = 1/2$. This gives that $f_4(n)$ is asymptotically $n^2/4$. Even more precise information on $f_d(n)$ is available. The question of distinct distances does not, however, become simpler. Here Erdős proved, still in the 1946 Monthly paper, that

$$cn^{3/(3d-2)} \leq g_d(n) \leq cn^{2/d}.$$

Many of these results have been improved since, and many by the Hungarian school of combinatorial geometry: József Beck, Zoltán Füredi, Endre Makai Jr., János Pach, Imre Ruzsa, László Székely, Endre Szemerédi, Csaba Tóth, Gábor Tardos.

Erdős, together with Hickerson and Pach (Amer. Math. Monthly, **96** (1989), 569–577) consider the same problem on the 2-dimensional unit sphere S^2 and show that every distance $d \in (0, 2)$ can occur $cn \log^* n$ times, and the special distance $\sqrt{2}$, surprisingly, occurs $cn^{4/3}$ times; this bound is optimal. (Here $\log^* n$ is the number one has to take logarithm from n to get below 2.)

A minor modification of the Lenz construction shows, further, that the maximal distance in $\mathbb{R}^d$, $d \geq 4$ can occur asymptotically

$$\frac{n^2}{2}\left(1 - \frac{1}{\lfloor d/2 \rfloor}\right)$$

times. The maximal distance question is related to the famous Borsuk conjecture stating that every set $S \subset \mathbb{R}^d$ can be partitioned into $d+1$ sets of smaller diameter. So the modified Lenz construction was an indication that the Borsuk conjecture might be false. This turned out to be the case later, from dimension 1000 onwards (but with a different example).

It is natural to ask the same questions about angles, directions instead of distances, and Erdős, of course, was asking, popularizing, and answering such questions. For details, see the survey by Erdős, Purdy: Extremal problems in combinatorial geometry (Handbook of Combinatorics, North Holland, (1995)). The following intriguing problem of Erdős is again of a similar kind: How many similar copies of a regular pentagon can an n element planar point set contain? The answer, by Erdős and Elekes (Intuitive Geometry, Colloq. Math. Soc. János Bolyai **63**, 85–104, North-Holland, 1994) is surprising: the construction of a *pentagonal lattice* in $\mathbb{R}^2$ contains cn^2 regular pentagons. Far reaching generalizations of this construction were given by Miklós Laczkovich and Imre Ruzsa.

The two questions asked by Erdős in 1946 started a novel and exciting research field in discrete geometry that has given rise to many beautiful results and hundreds of new problems. Erdős himself writes in his 80th birthday volume: "My most striking contribution to geometry is, no doubt, my problem on distinct distances".

13. Incidences

In the Educational Times in 1893, J. J. Sylvester raised the following question. Assume n points are given in the plane, not all of them on a line. Is it true then that they determine an ordinary line, that is, a line containing exactly two of the given n points. It seems that the problem lay dormant until Erdős revived it some 40 years later. Soon after that Tibor Gallai (1912–1992) found a beautiful proof which appeared (Amer. Math. Monthly, **51** (1944), 169–171) as a solution to a question posed by Erdős.

The following Euclidean Ramsey theorem, probably the first of its kind, is also due to Gallai: Given a finite set $P \subset \mathbb{R}^d$, and a colouring of $\mathbb{R}^d$ by r colours, there always exists a monochromatic and homothetic copy of P. Gallai never published this result which appeared first in R. Rado (Sitzungsber. Preuss. Akad. Wiss., Phys.-Math., **16/17** (1933), 589–596).

Now back to the Sylvester–Gallai theorem, which clearly implies that n points (not all of them on a line) determine at least n lines. A far reaching combinatorial generalization of this fact (including the case of finite projective planes) was proved by Erdős, de Bruijn (Indag. Math., **10** (1948), 421–423): Suppose $\{A_1, \dots, A_m\}$ are proper subsets of a ground set $\{a_1, \dots, a_n\}$ Suppose also that each pair a_i, a_j occurs in one and only one A. Then $m \geq n$.

Motivated, among others, by the Sylvester–Gallai theorem, Erdős conjectured that given n points in the plane, the number of lines containing at least $\sqrt{n}$ of the points is at most $c\sqrt{n}$ (where c is some positive constant). This was proved by Szemerédi and Trotter, and independently and about the same time by József Beck. In fact, Szemerédi and Trotter proved a much stronger conjecture of Erdős which says that the number of incidences between n points and m lines in the plane cannot exceed $O\big(m^{2/3}n^{2/3}+m+n\big)$. A minor modification of Erdős's construction for the upper bound for $f(n)$ shows that this bound is best possible (apart from the implied constant). This conjecture of Erdős, which is now called Szemerédi–Trotter theorem, has turned out to be a central result in the theory of complexity of line arrangements. It is not only point-line incidences that are important, but point-curve incidences as well. The curves here should by defined by fixed degree polynomials. This type of problems have been considered by Szemerédi, Beck, Pach, Székely, Tóth. We have seen above that the "unit distance" problem of Erdős can be formulated as a question on incidences between points and unit circles.

Incidence problems are closely related to the complexity of geometric objects. For instance, a set of n lines dissects the plane into cells. The complexity of a cell is the number of lines incident to the cell. In computational geometry, interest is frequently focused on the complexity of a cell, or the total complexity of some cells, or the sum of the complexities of all cells. The smaller this complexity is, the simpler the description of the system. Miraculously, or maybe not so miraculously, the complexity bounds are often close to the corresponding incidence bounds. Here is a sample theorem (due to Clarkson et al. (1990)):

Theorem. *Given a system of n lines in the plane, and some m distinct cells they determine, the total number of edges bounding one of these m cells is at most $c\big(m^{2/3}n^{2/3}+n\big)$.*

This estimate is best possible. This is shown, again, by a small modification of Erdős's $\sqrt{m}\times\sqrt{m}$ grid construction.

This is perhaps the point where the problem of halving lines should be mentioned. Given a set $X\subset\mathbb{R}^2$ of n points in general position (with n even), how many pairs $x,y\in X$ determine a halving line? That is, a line that has $(n-2)/2$ points of X on both sides. Denote this number by $h(X)$ and define $h(n)=\min h(X)$. What's the value of $h(n)$?

This innocent looking question is still unsolved. László Lovász proved in 1972, that the number of halving lines is at most $(2n)^{3/2}$, the lower bound $cn\log n$ is due to Erdős et al. (Proc. Internat. Symp., Fort Collins, Colo. (1973), 139–149, North-Holland). The best bounds, currently known are $O\big(n^{4/3}\big)$ (upper bound, by Tamal Dey) and $\Omega\big(ne^{\sqrt{\log n}}\big)$ (lower bound, by Géza Tóth). The dual to the halving lines problem is that of the complexity of the mid-level of an arrangement of n lines. This turned out to be important in computational geometry. Higher dimensional variants and analogous questions have been intensively investigated by the Hungarian school of discrete geometry, namely by Bárány, Füredi, Lovász, Pach, Szemerédi, Tardos, Tóth.

The following theorem, due to Erdős and Péter Komjáth (Discrete Comp. Geom., **5** (199)), 325–331), is just a sample of similar results from an interesting mixture of discrete geometry, combinatorics, and set theory.

Theorem. *The continuum hypothesis is equivalent to the existence of a colouring of the plane, with countably many colours, with no monochromatic right angled triangles.*

14. Miscellaneous results in combinatorial geometry

We have mentioned Tibor Gallai's result on ordinary lines. Gallai mainly worked in combinatorics, graph theory and was extremely modest, and had not published much. (But, according to Erdős, he should have published a theorem that he had proved which later became known as Dilworth's theorem.) However, a question of Gallai which appeared first in Fejes Tóth's book [41], page 97, motivated by combinatorial analogues, has proved to be very important and has become the starting point of a whole theory. This question is related to Helly's theorem: Assume that a system of unit circles in the plane has the property that any two of them have a point in common. Does this condition imply the existence of a set $F \subset \mathbb{R}^2$ of at most k points such that F intersects every circle in the family. (The answer is yes: Danzer proved that $k = 4$ always works, and cannot be improved, earlier Ungár and Szekeres showed $k \leq 7$, and L. Sztachó proved $k \leq 5$.)

József Molnár was mainly working in the theory of packings and coverings. He has an interesting Helly-type result as well. The question is the incidence structure of a finite family of convex sets in $\mathbb{R}^n$, which is only solved for $n = 1$. Molnár proves (Matematikai Lapok, **8** (1957), 108–117) the following generalization of Helly's topological theorem.

Theorem. *Let $\mathcal{C}$ be a finite family of connected compact sets in $\mathbb{R}^2$, $|\mathcal{C}| \geq 3$. Assume any two of the sets have connected intersection, and any three have nonempty intersection. Then there is a point common to all sets in $\mathcal{C}$.*

Danzer and Grünbaum proved that if every angle spanned by three points of a set $X \subset \mathbb{R}^d$ is at most $\pi/2$, then X has at most 2^d elements. (The cube shows that this bound is sharp.) They conjectured that, for $n \geq 3$, the size of X is at most $2n - 1$ if all angles spanned by three points of X are strictly smaller than $\pi/2$. This conjecture turned out to be absolutely wrong: Erdős and Füredi (Combinatorial Mathematics, North Holland Math. Studies **75** (1983), 275–283) constructed a set, X, of $n = 1.15^d$ points in $\mathbb{R}^d$ such that all angles spanned are acute. The construction is a random subset of the vertices of the unit cube, with a few unsuitable vertices deleted. A similar construction (in the same paper) gives a set X of size $(1 + \delta)^d$ with all distances within X are almost all equal: any two of them are at distance $\big(1 + O(\sqrt{\delta}\,)\big)$.

Ákos Császár has been working mainly in measure theory and topology. In 1949 he constructed a "polyhedron without diagonals", that is, a 3-dimensional polyhedron P with triangular faces and straight edges such that each pair of vertices is connected by an edge. P has seven vertices and is homeomorphic to the torus (see Császár, Acta Sci. Math. Szeged, **13** (1949), 140–142). This beautiful construction has become known as Császár's torus in the literature.

In (Acta Sci. Math. Szeged, **11** (1948), 229–233) István Fáry proves that every every planar graph can be drawn in the plane so that its edges are non-crossing straight line segments. (Actually, this follows from a remarkable theorem of Koebe from 1936, but the connection was not known at the time.) Erdős considered the problem of straight line planar representation of graphs with few crossing edges. For instance, Alon and Erdős show (Discrete Comp. Geom., **4**, (1989), 287–290) that any straight line planar drawing of a graph with n vertices and $6n-5$ edges contains three pairwise disjoint edges. This type of problems about *geometric graphs* was initiated by Erdős and Perles. By now, due to the work of János Pach and his students, the theory of geometric graphs is an exciting new field on the boundary of geometry and graph theory, rich with beautiful results and intriguing questions.

In connection with Sylvester's Orchard problem (Educational Times, **59** 1893) Ilona Palásti, together with Füredi (Proc. AMS., **92** (1984), 561–566) constructs a set of n lines, A_n, such that the number of triangles determined by the cell decomposition defined by A_n is $\frac{1}{3}n(n-3)$. A_n is a simple arrangement (no three lines concur), and it is known that the number of triangles determined by a simple arrangement of n lines is at most $\frac{1}{3}n^2 + O(n)$. So A_n is an asymptotically optimal arrangement.

15. Finite geometries

The outstanding Hungarian number theorist and algebraist, László Rédei, had made several interesting excursions to geometry. The first is closer to algebra than to geometry and is, in fact, about polynomials and finite geometries. Let p be a prime and U a subset of p elements of the affine plane over $GF(p)$. What Rédei (together with Megyesi) proves in [149] is that U determines at least $(p+3)/2$ directions unless it is a line. Further research in this direction is due to Blokhuis, Szőnyi, Lovász, and Schrijver.

We mention in passing that the analogous question (due to Erdős) for the Euclidean plane was solved by Péter Ungár (J. Comb. Theory Ser. A., **20** (1967)). His result says that $2n$ non-collinear points in the plane determine at least $2n$ distinct directions. The proof uses allowable sequences, or order types, if you like.

Rédei gave a new proof (J. London Math. Soc., **34** (1959), 205–207) of a result of Delone stating that, given a 2-dimensional lattice L, there always exists a lattice parallelogram P, such that $L \cap P$ consists of the vertices of P and these four vertices lie in four different quadrants of the plane. (The origin need not belong to L.) The "book-proof" of this theorem was found by János Surányi (Acta Sci. Math. Szeged, **22** (1961), 85–90), together with several applications.

János Surányi has been working mainly in number theory, and in geometry of numbers, in particular. He gave beautiful combinatorial geometric proofs of Wilson's theorem and Fermat's little theorem (Matematikai Lapok, **23** (1972), 25–29; joint work with K. Härtig).

We have encountered the name of Endre Makai Sr., in connection with the Erdős–Szekeres theorem. In (Mat. Fiz. Lapok, **50** (1943), 47–50) he gave an elementary proof of the fact that an empty lattice triangle has area 1/2.

Ferenc Kárteszi's field of interest was projective and later finite geometries. He ran a popular seminar on this subject. He and his disciples (G. Korchmáros, E. Boros, G. Kiss, M. H. Nguyen, T. Szőnyi and others) extended the notion of affine regular n-gon to finite geometries, see for instance, G. Kiss (Pure Math. Appl. Ser. A, **2** (1991), 59–66). An interesting result of Kárteszi (Publ. Math. Debrecen, **4** (1955), 16–27) says that, given n points in the plane, no three on a line, no point can be contained in more than $n^3/24$ of the triangles, spanned by the points.

16. Stochastic geometry

Crofton defined the mass of a set of lines in $\mathbb{R}^2$ as $\int dp d\phi$ where p and ϕ are the polar coordinates of the projection of the origin onto the line. Pólya was an analyst whose interests were very broad. For instance, he shows in (J. Leipz. Ber., **69** (1917), 457–458) that, if a mass distribution on lines is positive, additive, and independent of the position, then it is, apart from a constant factor, necessarily the one defined by Crofton. This fact has obvious implication on how to define a natural probability distribution on a (compact) subset of lines in the plane.

Alfréd Rényi (1920–1971) was a probabilist with broad interests in mathematics. He was a very influential mathematician and an able organizer. He is the founding father, and first director, of the Mathematical Institute of the Hungarian Academy of Sciences which now carries his name. He is the author of severals short popular books on mathematics, including Dialogues on Mathematics that has been translated into seven languages.

He wrote two papers on stochastic geometry: the motivation came from the so-called four-point problem of J. J. Sylvester (1863) who asked the probability that four points randomly chosen on the plane form the vertices of a convex quadrilateral. Rényi, together with Sulanke (Z. Wahrscheinlichkeitstheorie **2** (1963), 75–84, and **3** (1964), 138–147) modifies the question: drop n uniform, random, independent points $x_1, \dots, x_n$ in a convex body $K \subset \mathbb{R}^2$, let K_n be their convex hull. What's the expectation of the number of vertices, area, and perimeter, of K_n? They determine these expectations for smooth enough convex bodies and for polygons. For instance, when K is a polygon with k vertices, then the expected number of vertices of K_n is equal to

$$\frac{2}{3} k \log n(1 + o(1)).$$

When K is smooth with curvature κ, then the expected number of vertices is

$$\left(\frac{2}{3}\right)^{2/3} \Gamma\left(\frac{5}{3}\right) \int_{bdK} \kappa^{1/3} n^{1/3} (1 + o(1)) .$$

These two papers initiated a new direction that have resulted in hundreds of papers on the study of the so-called *random polytopes.*

The second paper contains the following interesting, and purely geometric, result: Let P be a convex polygon with vertices $v_1, \dots, v_k$. Write $\triangle_i$ for the triangle with vertices v_{i-1}, v_i, v_{i+1}. Then the product

$$\prod_1^k \frac{\operatorname{Area} \triangle_i}{\operatorname{Area} P}$$

is the largest when P is an affinely regular k-gon. Actually, László Fejes Tóth theorem from Section 8 (or rather its proof, see L. Fejes Tóth (Matematikai Lapok, **29** (1977/81), 33–38)) gives the stronger inequality that

$$\sum_1^k \left(\frac{\operatorname{Area} \triangle_i}{\operatorname{Area} P}\right)^{1/3}$$

is the largest when P is an affinely regular k-gon.

17. Miscellaneous results in convex geometry

Gyula Pál was working mainly in convex geometry. He was born in Hungary and later moved to Denmark. In an often cited paper J. Pál (Kgl. Danske

Videnskab. Selskab Med. **3** (1920), 1–35) he proves two interesting results. The first is that for every compact set $S \subset \mathbb{R}^2$ there is a convex set, $K \subset \mathbb{R}^2$, of constant width with $S \subset K$ and having the same diameter as S. The other result is about universal covers: every set $S \subset \mathbb{R}^2$ of diameter at most one is contained in a regular hexagon of width 1. This shows that the regular hexagon of width 1 is a universal cover for sets of diameter one. (This universal cover theorem can be used to show the validity of the Borsuk conjecture in the plane.) In the same paper, Pál constructs another universal cover with slightly smaller area than the hexagon.

The following nice result on universal covers is due to Károly Bezdek (Amer. Math. Monthly, **96** (1989), 789–806, joint work with R. Connelly). Let $\mathcal{C}$ be the class of closed planar curves of length one; a set $K \subset \mathbb{R}^2$ is *universal translation cover* for $\mathcal{C}$, if every curve in $\mathcal{C}$ is contained in a translated copy of K. Now the cited result says that every convex body of constant width $\frac{1}{2}$ is a universal translation cover for $\mathcal{C}$. Moreover, every universal translation cover for $\mathcal{C}$ which is convex and has minimal perimeter is of constant width $\frac{1}{2}$.

We mention here that Jenő Egerváry (1891–1958), who mainly worked in algebra and matrix theory, proved an isoperimetric result on curves in $\mathbb{R}^3$ (Publ. Math. Debrecen, **1** (1949), 65–70): he finds, among such curves of length one that have at most three coplanar points, the one whose convex hull has minimal volume.

In connection with geometric constructions, we encountered the name of Gyula Szőkefalvi-Nagy (1982–1959). He worked in various fields of mathematics. He considered the minimal ring containing a convex curve in the plane in (Acta Sci. Math. Szeged, **10** (1943), 174–184). In another paper (Acta Math. Hung., **5** (1954), 165–167) he proves that, given finitely many planes (not all parallel with a line) in 3-space, the set of points with sum of distances to the planes equal to $d > d_0$ form the boundary of a convex polytope. Here $d_0 > 0$ is a constant that depends only on the set of given planes.

Béla Szőkefalvi-Nagy (1914–1998) was an analyst whose research field was Hilbert spaces and operators on Hilbert spaces. He liked geometry and had written about 6 papers in geometry. (One of them is mentioned below, together with his coauthor Rédei.) In a paper (Bull. Soc. Math. France, **69** (1941), 3–4) he constructs, in dimension 4 and higher, convex polytopes, different from the simplex, that have no diagonals. This is an early example of the so-called neighbourly polytopes. Szőkefalvi-Nagy's most famous result in convex geometry states that the Helly number of axis parallel boxes (in $\mathbb{R}^d$) is 2. That is, if in a family of axis parallel boxes in $\mathbb{R}^d$, every two boxes have a point in common, then there is a point common to every box in the family. See Szőkefalvi-Nagy (Acta Sci. Math. Szeged, **14** (1954), 169–177). This paper turned out to be very influential, and the Helly number of various families of convex sets has been thoroughly investigated, for instance in the work of V. Boltjanski and János Kincses.

László Rédei and Béla Szőkefalvi-Nagy proved an interesting result in convex geometry. It is a Heron-type formula which expresses the product of the areas of two convex polygons as a polynomial of the distances between the vertices of the two polygons. For details see Rédei, Szőkefalvi-Nagy (Publ. Math. Debrecen, **1** (1949), 42–50).

Another result, again from convex geometry, of Rédei is joint with István Fáry and is about the maximal volume of a centrally symmetric convex set contained in a fixed convex body $K \subset \mathbb{R}^d$ (see Fáry, Rédei, Math. Ann., **122** (1950), 205–220). If the centrepoint is $x \in K$, then this maximal body is exactly $K \cap (2x - K)$. Fáry and Rédei show that the level sets of the function $x \to \mathrm{Vol}\big(K \cap (2x - K)\big)$ are convex, the function has a unique maximum, and compute it when K is the d-dimensional simplex.

György Hajós (1912–1972) was a very influential person in Hungarian mathematical life. He is the author of the textbook "Introduction to Geometry" that was used at Eötvös University for teaching geometry to several generations of mathematicians and high-school teachers of mathematics. On his famous Monday evening seminar one could learn clarity of ideas, precision in proofs, and rigour in presentation. He published surprisingly few papers, but there is one among them that made Hajós world-famous. It contains the solution of a long-standing conjecture of Minkowski (Hajós, Math. Z., **47** (1941), 427–467). The conjecture which is now Hajós's theorem states that in every lattice tiling of $\mathbb{R}^d$ by congruent d-dimensional cubes, there always exists a "stack" of cubes in which each two adjacent cubes meet along a full facet. The theorem has several equivalent forms and Hajós's proof is algebraic.

Hajós and Heppes construct a three-dimensional (non-convex) polyhedron P whose supporting planes intersect exactly at the vertices of the polyhedron, (see Hajós, Heppes, Acta Math. Hung., **21** (1970), 101–103). Here a supporting plane is a plane that contains at least one point of P and P is contained in the one of the halfspaces bounded by the plane.

István Vincze was a statistician who was interested in convex geometry. In 1939, motivated by a sharpening of the planar isoperimetric inequality due to Bonnesen and Fenchel, he considered the following question. Given a convex body $K \subset \mathbb{R}^2$, and a point $x \in K$, let $R(x)$ denote the radius of the smallest disk centered at x which contains K. Similarly, let $r(x)$ denote the radius of the largest disk, centered at x, which is contained in K. The function $x \to R(x) - r(x)$ attains its minimal value at a unique point $x_0 \in K$, and the circular ring about x_0 with radii $R(x_0)$ and $r(x_0)$ is called the minimal ring containing the boundary of K. Vincze (Acta Sci. Math. Szeged, **11** (1947), 133–138) proved that

$$\min\{R(x) : x \in K\} \geq \frac{\sqrt{3}}{2} R(x_0), \quad \text{and} \quad \max\{r(x) : x \in K\} < 2r(x_0).$$

Both inequalities are best possible.

References

[40] Fejér, Lipót, *Összegyűjtött Munkái = Gesammelte Arbeiten,* ed. Pál Turán, Akadémiai Kiadó (Budapest, 1970).

[41] Fejes Tóth, László, *Lagerungen in der Ebene, auf der Kugel und im Raum,* Grundlehren der Mathematischen Wissenschaften in Einzeldarstellungen, Band LXV, Springer-Verlag (Berlin–Göttingen–Heidelberg, 1953).

[42] Fejes Tóth, László, *Regular Figures,* Pergamon (London), The MacMillan Co. (New York, 1964). *Reguläre Figuren,* Akadémiai Kiadó (Budapest, 1965).

[140] Rademacher, Hans – Toeplitz, Otto, *Von Zahlen und Figuren,* Springer-Verlag (Berlin, 1930).

[149] Rédei, László, *Lückenhafte Polynome über endlichen Körpern,* Akadémiai Kiadó (Budapest) – Deutscher Verlag der Wissenschaften (Berlin) – Birkhäuser (Basel, 1970). *Lacunary polynomials over finite fields,* Akadémiai Kiadó (Budapest) – North Holland (Amsterdam, 1973).

Imre Bárány

Rényi Institute of the
Hungarian Academy of Sciences
P.O.B. 127
1364 Budapest
Hungary

and

Department of Mathematics
University College London
Gower Street
WC1E 6BT London
U.K.

barany@renyi.hu

Stochastics

BOLYAI SOCIETY
MATHEMATICAL STUDIES, 14

A Panorama of Hungarian Mathematics in the Twentieth Century, pp. 457–489.

Probability Theory

PÁL RÉVÉSZ

In the early sixties György Pólya gave a talk in Budapest where he told the following story.

He studied in his teens at the ETH (Federal Polytechnical School) in Zürich, where he had a roommate. It happened once that the roommate was visited by his fiancée. From politeness Pólya left the room and went for a walk on a nearby mountain. After some time he met the couple. Both the couple and Pólya continued their walks in different directions. However, they met again. When it happened the third time, Pólya had a bad feeling. The couple might think that he is spying on them. Hence he asked himself what is the probability of such meetings if both parties are walking randomly and independently. If this probability is big then Pólya might claim that he is innocent.

To give an answer to the above question one has to build a mathematical model. Pólya's model is the following.

Consider a random walk on the lattice $\mathbb{Z}^d$. This means that if a moving particle is in $x \in \mathbb{Z}^d$ in the moment n then at the moment $n+1$ the particle can move with equal probability to any of the $2d$ neighbours of x independently of how the particle achieved x. (The neighbours of an $x \in \mathbb{Z}^d$ are those elements of $\mathbb{Z}^d$ which have $d-1$ coordinates equal to those of x and the remaining coordinate differs by $+1$ or -1.) Let S_n be the location of the particle after n steps (i.e., at the moment n) and assume that $S_0 = 0$.

Then Pólya {55} proved

Theorem 1.

$$\mathbf{P}\{S_n = 0 \ \ i.o.\} = \begin{cases} 1 & if \ d \le 2, \\ 0 & if \ d \ge 3. \end{cases}$$

(i.o. = infinitely often).

This theorem clearly means that a random walker returns to his or her starting point infinitely often with probability 1 in the plane. It easily implies that two independent random walkers will meet infinitely often with probability 1 in the plane. Hence Pólya was innocent.

Another nice problem studied by Pólya is the following:

Let an urn contain M red and $N-M$ white balls. Draw a ball at random, replace the drawn ball and at the same time place into the urn R extra balls with the same colour as the one drawn. ($R=\pm 1, \pm 2, \dots$ in case of negative R we remove from the urn R balls of the same colour.) Then we draw again a ball and so on. What is the probability of the event that in n drawings we obtain a red ball exactly k times? Let this event be denoted by A_k. Of course we assume that at every drawing each ball of the urn is selected with the same probability. Pólya evaluated the distribution $\{\mathbf{P}(A_k)\}$. It is called Pólya distribution (see F. Eggenberger–G. Pólya, {17}).

Felix Hausdorff in 1913 asked how far can a particle go from its starting point in n steps. In the case $d=1$ A. I. Khinchine (1923) proved that the distance can be $(2n \log\log n)^{1/2}$ but not more. In fact

Theorem 2.

$$\limsup_{n\to\infty} \frac{S_n}{(2n\log\log n)^{1/2}} = -\liminf_{n\to\infty} \frac{S_n}{(2n\log\log n)^{1/2}} = 1 \quad \text{a.s.}$$

(a.s. = almost surely).

Consequently the distance of the particle from its starting point will be infinitely often more than $(1-\varepsilon)(2n\log\log n)^{1/2}$ but it will be only finitely many times more than $(1+\varepsilon)(2n\log\log n)^{1/2}$ for any $\varepsilon > 0$. Clearly this theorem implies Pólya's theorem in case $d=1$.

Paul Lévy asked how can we obtain an even sharper version of Khinchine's theorem. Pál Erdős' answer in {18} is the following.

Theorem 3. *Let $a(n)$ be an increasing function and $d=1$. Then*

$$\mathbf{P}\{\, S_n \ge n^{1/2}a(n) \ \ i.o.\} = \begin{cases} 1 & \text{if } A=\infty, \\ 0 & \text{if } A<\infty, \end{cases}$$

where

$$A = \sum_{n=1}^{\infty} \frac{a(n)}{n} \exp\left(-\frac{a^2(n)}{2}\right).$$

For example Theorem 3 implies that $S_n \geq (2n \log\log n)^{1/2}$ a.s. i.o.

The theory of random walks became one of the most popular topics of probability theory and undoubtedly Erdős was one of the most important contributors to this topic, especially in the multidimensional case. Now we formulate some of the results of Erdős on random walks.

Consider the last return $R(n)$ of a random walk to its starting point before its n-th step, i.e., let

$$R(n) = \max\{k \,:\, 0 \leq k \leq n,\ S_k = 0\}.$$

Let $d = 1$. Then Theorem 1 clearly implies that

$$\lim_{n\to\infty} R(n) = \infty \quad \text{a.s.}$$

and

$$\mathbf{P}\{\, R(n) = n \ \text{ i.o.}\} = 1.$$

Kai-Lai Chung and Erdős in {6} asked how small $R(n)$ can be. They proved

Theorem 4. *Let $d = 1$ and let $f(x)$ be an increasing function for which $\lim_{x\to\infty} f(x) = \infty$, $x/f(x)$ is increasing and $\lim_{x\to\infty} x/f(x) = \infty$. Then*

$$\mathbf{P}\left\{R(t) \leq \frac{t}{f(t)} \ \text{ i.o.}\right\} = \begin{cases} 1 & \text{if } I = \infty, \\ 0 & \text{if } I < \infty, \end{cases}$$

where

$$I = \int_1^\infty \frac{dx}{x\big(f(x)\big)^{1/2}}.$$

This theorem clearly implies that $R(n)$ can be smaller than $n(\log n)^{-2}$ infinitely often but $R(n)$ can be smaller than $n(\log n)^{-2-\varepsilon}$ $(\varepsilon > 0)$ only finitely many times.

Arieh Dvoretzky and Erdős in {15} asked: how many points will be visited by a random walk in $\mathbb{Z}^d$ $(d \geq 2)$ during its first n steps. Let $V(n)$ be the number of different vectors among $S_1, S_2, \ldots, S_n$, i.e., $V(n)$ is the number of visited points. They proved

Theorem 5.

$$\lim_{n\to\infty} \frac{V(n)}{\mathbf{E}V(n)} = 1 \quad \text{a.s.},$$

where

$$\mathbf{E}V(n) \sim \begin{cases} \dfrac{\pi n}{\log n} & \text{if } d = 2, \\ n\gamma_d & \text{if } d \geq 3, \end{cases}$$

γ_d *is a sequence of strictly positive constants, and* $\mathbf{E}$ *denotes the expected value.*

The following question can be considered as the converse of the above question of Dvoretzky and Erdős: How many times is a "typical" point of $\mathbb{Z}^d$ visited up to time n? By Theorem 1 in case $d \geq 3$ the answer is a finite random variable (r.v.). In case $d = 2$ the answer is a random sequence converging to ∞ as $n \to \infty$. In fact Erdős and S. J. Taylor {34} proved:

Theorem 6. *Let* $\xi(n)$ *be the number of visits of* $0 \in \mathbb{Z}^2$ *before* n, *i.e.,*

$$\xi(n) = \#\{k : 0 \leq k \leq n,\ S_k = 0\}.$$

Then

$$\lim_{n\to\infty} \mathbf{P}\{\,\xi(n) < x \log n\} = 1 - e^{-\pi x}.$$

As we have said, in case $d \geq 3$, by Theorem 1 any fixed point is visited only finitely many times. However, some randomly chosen point will be visited many times. Let

$$\xi(x,n) = \#\{k : 0 \leq k \leq n,\ S_k = x\}$$

and

$$\zeta(n) = \max_{x\in\mathbb{Z}^d} \xi(x,n).$$

Then Erdős and Taylor, in their above mentioned paper, proved:

Theorem 7. *Let* $d \geq 3$. *Then*

$$\lim_{n\to\infty} \frac{\zeta(n)}{\log n} = \gamma_d \quad \text{a.s.}$$

where γ_d *is the same constant as in Theorem* 5.

It is easy to see that the path of a random walk crosses itself infinitely many times with probability 1 for any $d \geq 1$. We mean that there exists an infinite sequence $\{U_n, V_n\}$ of pairs of positive integer valued r.v.'s such that $S(U_n) = S(U_n + V_n)$, and $0 \leq U_1 < U_2 \ldots$, $(n = 1, 2, \ldots)$, where $S(U_n) = S_{U_n}$. However, we ask the following question: will selfcrossings occur after a long time? For example, we ask whether the crossing $S(U_n) = S(U_n + V_n)$ will occur for every $n = 1, 2, \ldots$ if we assume that V_n converges to infinity with great speed and U_n converges to infinity much slower. In fact Erdős and Taylor {35} proposed the following two problems:

Problem A. Let $f(n) \uparrow \infty$ be a positive integer-valued function. What are the conditions on the rate of increase of $f(n)$ which are necessary and sufficient to ensure that the paths $\{S_0, S_1, \ldots, S_n\}$ and $\left\{S_{n+f(n)}, S_{n+f(n)+1} \cdots\right\}$ have points in common for infinitely many values of n with probability 1?

Problem B. A point S_n of a path is said to be "good" if there are no points common to $\{S_0, S_1, \ldots, S_n\}$ and $\{S_{n+1}, S_{n+2}, \ldots\}$. For $d = 1$ or 2 there are no good points with probability 1. For $d \geq 3$ there might be some good points: how many are there?

As far as Problem A is concerned, they (Erdős–Taylor) proved

Theorem 8. *Let $f(n) \uparrow \infty$ be a positive integer-valued function and let E_n be the event that the paths*

$$\{S_0, S_1, \ldots, S_n\} \quad \text{and} \quad \left\{S_{n+f(n)+1}, S_{n+f(n)+2}, \cdots\right\}$$

have points in common. Then

(i) *for $d = 3$, if $f(n) = n\big(\varphi(n)\big)^2$ and $\varphi(n)$ is strictly increasing, one has*

$$\mathbf{P}\{E_n \ \text{i.o.}\} = 0 \quad \text{or} \quad 1 \tag{1}$$

depending on whether $\sum_{k=1}^{\infty} \big(\varphi(2^k)\big)^{-1}$ converges or diverges,

(ii) *for $d = 4$, if $f(n) = n\chi(n)$ and $\chi(n)$ is strictly increasing, we have (1) depending on whether $\sum_{k=1}^{\infty} \big(k\chi(2^k)\big)^{-1}$ converges or diverges,*

(iii) *for $d \geq 5$, if*

$$\sup_{m \geq n} \frac{f(m)}{m} \geq C \frac{f(n)}{n}$$

(for some $C > 0$), we have (1) depending on whether

$$\sum_{n=1}^{\infty} \big(f(n)\big)^{(2-d)/2}$$

converges or diverges.

For Problem B they (Erdős–Taylor) have as an answer:

Theorem 9. *For $d \geq 3$ let $G^{(d)}(n)$ be the number of integers r $(1 \leq r \leq n)$ for which $(S_0, S_1, \dots, S_r)$ and $(S_{r+1}, S_{r+2}, \dots)$ have no points in common. Then*

(i) $d = 3$. *For any $\varepsilon > 0$*

$$\mathbf{P}\big\{G^{(3)}(n) > n^{1/2+\varepsilon} \ \ i.o.\big\} = 0.$$

(ii) $d = 4$.

$$\mathbf{P}\left\{0 = \liminf_{n\to\infty} \frac{G^{(4)}(n)\log n}{n} \leq \limsup_{n\to\infty} \frac{G^{(4)}(n)\log n}{n} \leq C\right\} = 1$$

with some positive constant C.

(iii) $d \geq 5$.

$$\lim_{n\to\infty} \frac{G^{(d)}(n)}{n} = \tau_d \quad \text{a.s.}$$

where τ_d is an increasing sequence of positive numbers with $\tau_d \uparrow 1$ as $d \to \infty$.

Random walk is the simplest mathematical model for Brownian motion that is well known in physics. However, it is not a very realistic model. In fact the assumption, that the particle goes at least one unit in a direction before turning, is hardly satisfied by the real Brownian motion observed in nature.

In a more realistic model the particle makes instantaneous steps, that is a continuous time scale is used instead of a discrete one. Such a model is the Wiener process or frequently called Brownian motion.

Now we mention a few results, on the Wiener process.

The first one, which characterizes the modulus of continuity, was proved by P. Lévy {48}.

Theorem 10. *Let $W(t) \in \mathbb{R}^1$ $(t \geq 0)$ be a Wiener process. Then*

$$\lim_{h \to 0} \frac{\sup_{0 \leq t \leq 1-h} |W(t+h) - W(t)|}{(2h \log 1/h)^{1/2}} = 1 \quad \text{a.s.}$$

A stronger version of this Theorem was given by Chung, Erdős and T. Sirao {7}.

Theorem 11. *Let $f(x)$ be a continuous increasing function and let*

$$\mathcal{J}(f) = \int_1^\infty f^2(t) \exp\left(-\frac{1}{2} f^2(t)\right) dt.$$

Then

$$\sup_{0 \leq t \leq 1-h} |W(t+h) - W(t)| \begin{cases} \leq h^{1/2} f(h^{-1}) & \text{if } \mathcal{J}(f) < \infty, \\ \geq h^{1/2} f(h^{-1}) & \text{if } \mathcal{J}(f) = \infty \end{cases}$$

a.s. *if h is small enough and $W(t) \in \mathbb{R}^1$.*

For example

$$\sup_{0 \leq t \leq 1-h} |W(t+h) - W(t)| \leq h^{1/2} \left(2 \log h^{-1} + (5+\varepsilon) \log \log h^{-1}\right)^{1/2}$$

a.s. if $\varepsilon > 0$ and h is small enough and the above inequality a.s. does not hold if $\varepsilon \leq 0$.

Beside numerous physical applications of the Wiener process it has important applications in mathematical analysis. The most important ones are in the theory of partial differential equations.

A very classical result claims that the sample paths of a Wiener process are nowhere differentiable with probability one. It has been well known for a long time that there exist continuous, nowhere differentiable functions. However, one thinks that it is a very rare phenomenon. The fact that a Wiener process is nowhere differentiable can be interpreted by saying that almost all continuous functions are nowhere differentiable. A similar fact is that almost all continuous functions are nowhere monotone.

A function $f(x)$ $(0 < x < 1)$ is said to be monotone at x_0 if there exists an $0 < \varepsilon = \varepsilon(x_0) < \min(x_0, 1 - x_0)$ such that

$$f(u) < f(x_0) \quad \text{for any} \quad x_0 - \varepsilon < u < x_0$$

and

$$f(u) > f(x_0) \quad \text{for any} \quad x_0 < u < x_0 + \varepsilon$$

or

$$f(u) > f(x_0) \quad \text{for any} \quad x_0 - \varepsilon < u < x_0$$

and

$$f(u) < f(x_0) \quad \text{for any} \quad x_0 < u < x_0 + \varepsilon.$$

It is not easy at all to construct a continuous nowhere monotone function. However, Dvoretzky, Erdős and Shizuo Kakutani in {16} proved

Theorem 12. *With probability one the sample paths of a Wiener process are nowhere monotone.*

Let $X_1, X_2, \ldots$ be independent, identically distributed r.v.'s (i.i.d.r.v.'s) with $\mathbf{E}X_1 = 0$, $\mathbf{E}X_1^2 = 1$ and let F be their distribution function. Let $Y_1, Y_2, \ldots$ be i.i.d. normal $(0,1)$ r.v.'s and put $S_n = \sum_{i=1}^n X_i$, $T_n = \sum_{i=1}^n Y_i$. Then one of the most classical result of probability theory, the central limit theorem, states that

$$\lim_{n\to\infty} \mathbf{P}\{n^{-1/2} S_n < y\} = \frac{1}{(2\pi)^{1/2}} \int_{-\infty}^{y} e^{-u^2/2}\, du$$

or equivalently

$$\lim_{n\to\infty} \left(\mathbf{P}\{n^{-1/2} S_n < y\} - \mathbf{P}\{n^{-1/2} T_n < y\} \right) = 0$$

i.e., the limiting behaviours of S_n and T_n are the same. In other words, as times goes on, S_n forgets about the distribution function F where it comes from.

A similar phenomenon was observed by Erdős and Mark Kac in {20} and {21}. They investigated the limits $(n \to \infty)$ of the distribution functions

$$\mathbf{P}\Big\{ n^{-1/2} \max_{1\le k\le n} S_k < y \Big\},$$

$$\mathbf{P}\Big\{ n^{-1/2} \max_{1\le k\le n} |S_k| < y \Big\},$$

$$\mathbf{P}\Big\{ n^{-2} \sum_{k=1}^{n} S_k^2 < y \Big\},$$

$$\mathbf{P}\Big\{ n^{-3/2} \sum_{k=1}^{n} |S_k| < y \Big\}$$

and they observed that these limit distributions also do not depend on F. Hence a program for finding the above limits may be carried out in two steps. First, they should be evaluated for a specific distribution F, and then one should show that the above considered functionals of $\{S_k\}$ do not remember the initially taken distribution. They called this method of proof the invariance principle and their papers initiated a new methodology for proving limit laws in probability theory. A very essential new step in this methodology is the invention of the strong invariance principle due to V. Strassen {75} which is capped by János Komlós–Péter Major–Gábor Tusnády {45}, see also Major {49}.

Their most important result is

Theorem 13. *Let $X_1, X_2, \ldots$ be a sequence of i.i.d.r.v.'s defined on a rich enough probability space $\{\Omega, \mathcal{S}, \mathbf{P}\}$. Assume that $R(t) = \mathbf{E}\exp(tX_1)$ exists in a neighbourhood of $t = 0$. Then there exists a Wiener process $\{W(t), t \geq 0\}$ defined on Ω such that*

$$\left| S_n - W(n) \right| = O(\log n) \quad \text{a.s.}$$

The analogues of Problems A and B can also be formulated for a Wiener process. Very important results in this area were obtained by Dvoretzky, Erdős, Kakutani and Taylor in the fifties.

Up to now we mentioned the names of two Hungarian probabilists, Erdős and Pólya. They were certainly Hungarian but Pólya lived mostly abroad and Erdős between '38 and '56 also was abroad. The first Hungarian probability school in Hungary was founded by Károly Jordán. His work is closely related to statistics. Hence his work is treated in the Mathematical Statistics Section.

The *first revolution* in probability in Hungary took place after the second world war when Alfréd Rényi founded his probability school.

Rényi originally was interested in number theory rather than in probability. He wanted to study the method of Yu. V. Linnik (Linnik's large sieve) and he went to Leningrad in '46 to study with him. There he gained insight into connections between number theory and probability, and when he returned to Hungary in '47 he also was already a real probabilist.

His first papers which belong to pure probability deal with the Poisson process. They start with an article written jointly with János Aczél and the physicist Lajos Jánossy {44} and an article with the same title written by Rényi alone (*ibid.*, **2** (1951), 83–98). Let $X(t)$ be a process defined for

$t \geq 0$ with $X(0) = 0$ and such that $X(t) - X(s)$ takes positive integer values for $t > s$. Write $W_k(t) = \mathbf{P}\{X(t) = k\}$, and make the following assumptions:

A) $X(t)$ is homogeneous, i.e., $\mathbf{P}\{X(t_2) - X(t_1) = k\} = W_k(t)$ whenever $t_2 - t_1 = t$.

B) $X(t)$ has independent increments, i.e., $X(t_n) - X(t_{n-1}), X(t_{n-2}) - X(t_{n-3}), \ldots, X(t_2) - X(t_1)$ $(n = 3, 4, \ldots)$ are independent random variables $(t_1 < t_2 \leq t_3 < t_4 \leq \ldots \leq t_{n-1} < t_n)$.

C) The "events" are rare, i.e.

$$\lim_{t \to 0} \frac{W_1(t)}{1 - W_0(t)} = 1.$$

From these hypotheses the authors deduce that

$$W_k(t) = \frac{(\lambda t)^k}{k!} e^{-\lambda t}$$

for some $\lambda > 0$, i.e., $X(t)$ is a Poisson process. The three assumptions are natural and minimal, in particular the differentiability of $W_k(t)$ is not required. The proof uses functional equations – a speciality of Aczél – instead of differential equations.

The three authors prove that if condition C is omitted, then $X(t)$ is a "composed Poisson process", i.e., of the form

$$X(t) = X_1(t) + 2X_2(t) + \ldots + nX_n(t) + \ldots,$$

where the $X_n(t)$ are independent Poisson processes. Again using functional equations, they determine $W_k(t)$.

The fact that a Poisson process is Markovian is proved and applied in a number of papers of Rényi. One of these is: Rényi, Lajos Takács {73}.

A further result in this direction investigates the discontinuities of a process of independent increments. It turns out that the number of discontinuities of "any size" are independent. More precisely András Prékopa and Rényi {72} proved:

Theorem 14. *If the process ξ_t of independent increments is weakly continuous, i.e., for every $\varepsilon > 0$*

$$\lim_{\Delta \to 0} \mathbf{P}\{|\xi_{t+\Delta} - \xi_t| \geq \varepsilon\} = 0$$

uniformly in $t \in (0,1)$ and $I_1, I_2, \ldots, I_r$ are pairwise disjoint subintervals of $[0,1]$ with positive distances from the origin, then the random variables

$$\nu(I_1), \nu(I_2), \ldots, \nu(I_r)$$

are independent, where $\nu(I)$ denotes the random variable giving the number of discontinuities of magnitudes $h \in I$.

In his paper {71} Rényi investigated the converse of these results. Consider a point process $\xi(E)$ on the real line which is Poisson in the sense that the distribution of $\xi(E)$ is Poisson for certain subsets E of the real line. Then Rényi proved that $\xi(E)$ is a Poisson process indeed, i.e., $\xi(E_1)$ and $\xi(E_2)$ are independent if E_1 and E_2 are disjoint.

The Poisson process was also investigated by Pólya {56}. He considered a stone at the bottom of a river which lies at rest for such long periods that its successive displacements are practically instantaneous. Then he observed that the total displacement within a time interval $(0,t)$ might be treated as a compound Poisson process.

Another favorite topic of Rényi was the study of the properties of empirical processes (see Rényi {59} and György Hajós, Rényi {43}). The main results of these papers show that an ordered sample can be described via exponential random variables. Let $0 < E_1 < E_2 < \ldots < E_n$ be an ordered sample obtained from a sequence of independent, exponential random variables with mean 1. Further let

$$\delta_{k+1} = (n-k)(E_{k+1} - E_k).$$

It is easy to see that $\delta_1, \delta_2 \ldots, \delta_n$ are independent, exponentially distributed with mean 1, and

$$E_k = \frac{\delta_1}{n} + \frac{\delta_2}{n-1} + \cdots + \frac{\delta_k}{n-k+1}.$$

Hence $\{E_k\}$ is a Markov chain. This simple observation was enough for Rényi to get some important results on order statistics and empirical distributions. For example he proved

Theorem 15.

$$\lim_{n\to\infty} \mathbf{P}\{nE_k < x\} = \int_0^x \frac{t^{k-1}e^{-t}}{(k-1)!}\, dt.$$

Replacing the sequence $E_1 < E_2 < \ldots < E_n$ by an ordered sample of uniform $(0,1)$ random variables, similar results can be obtained. For further details on Rényi's work in this regard we refer to the Section on Mathematical Statistics.

Rényi as a probabilist never forgot his number theorist origin. He worked occasionally in pure number theory but was mostly interested in applications of probability to number theory. He realized that a great difficulty in the application of probability to number theory comes from the fact that it is meaningless to say: choose an integer from the set of all positive integers with equal probability. As an example he says in {60}: let $U(n)$ denote the number of different prime divisors of n; let $\pi_k(N)$ denote the number of those natural numbers $n \leq N$ for which $U(n) = k$. By a theorem of Erdős {19}

$$\frac{\pi_k(N)}{N} \sim \frac{(\log\log N)^k}{k! \log N} := p_k.$$

Hence it looks natural to say that the probability that a randomly chosen integer has k divisors is p_k. However, this sentence is meaningless in Kolmogorov's theory of probability. Rényi decided to build a new theory where this sentence was going to have a meaning i.e., where unbounded measures may occur. Rényi says: "Unbounded measures occur in statistical mechanics, quantum mechanics, in some problems of mathematical statistics, in integral geometry, in number theory etc. At first glance it seems that unbounded measures can play no role in probability theory. But if we observe more attentively how unbounded measures are really used in all the cases mentioned above, we see that unbounded measures are used only to calculate conditional probabilities as the quotient of the values of the unbounded measure of two sets. Clearly in a theory in which unbounded measures are allowed, conditional probability must be taken as the fundamental concept."

Hence Rényi built up such an axiomatic theory. Further number theoretical application of this theory can be found in Rényi {63, 68}.

The *second revolution* in probability in Hungary took place in '56 when Erdős started to work intensively together with Rényi and his students.

Once in 1958 Rényi arrived in his Department in the University and asked his assistants: "Do you know why trees do not grow up to the heavens?" With this question he announced that he and Erdős started to deal with random graphs. In fact their problem is the following (Erdős–Rényi {22}):

Let us suppose that n labelled vertices $P_1, P_2, \ldots, P_n$ are given. Let us choose at random an edge among the $\binom{n}{2}$ possible edges, so that all these edges are equiprobable. After this let us choose an other edge among the remaining $\binom{n}{2} - 1$ edges, and continue this process so that if already k edges are fixed, any of the remaining $\binom{n}{2} - k$ edges have equal probabilities to be chosen as the next one. We shall study the "evolution" of such a random graph $\Gamma_{n,N}$ if the number of chosen edges, N, is increased. In this investigation we endeavour to find what is the "typical" structure at a given stage of evolution (i.e., if N is equal, or asymptotically equal, to a given function $N(n)$ of n). By a "typical" structure we mean such a structure the probability of which tends to 1 if $n \to \infty$ when $N = N(n)$.

If N is very small compared with n, namely if $N = o\left(\sqrt{n}\right)$ then it is very probable that $\Gamma_{n,N}$ is a collection of isolated points and isolated edges, i.e., that no two edges of $\Gamma_{n,N}$ have a point in common. As a matter of fact the probability that at least two edges of $\Gamma_{n,N}$ shall have a point in common is

$$1 - \frac{\binom{n}{2N}(2N)!}{2^N N!\binom{\binom{n}{2}}{N}} = O\left(\frac{N^2}{n}\right).$$

If however $N \sim c\sqrt{n}$ where $c > 0$ is a constant not depending on n, then the appearance of trees of order 3 will have a probability which tends to a positive limit as $n \to +\infty$, but the appearance of a connected component consisting of more than 3 points will be still very improbable. If N is increased while n is fixed, the situation will change only if N reaches the order of magnitude of $n^{2/3}$. Then trees of order 4 (but not of higher order) will appear with a probability not tending to 0.

Clearly if N becomes larger and larger then we obtain more and more connected graphs. For example if N is about n then the graph contains a cycle of order k for any $k \geq 3$.

In this paper on their aim the authors write:

In the present paper we consider the evolution of a random graph in a more systematic manner and try to describe the gradual development and step-by-step unravelling of the complex structure of the graph $\Gamma_{N,N}$ when N increases while n is a given large number.

We succeeded in revealing the emergence of certain structural properties of $\Gamma_{n,N}$. However a *great deal* remains to be done in this field.

A typical result of this paper characterizes the size of the greatest tree of $\Gamma_{n,N}$. In fact they prove

Theorem 16. *Let $\Delta_{n,N}$ denote the number of points of the greatest tree which is a component of $\Gamma_{n,N}$. Suppose $N = N(n) \sim cn$ with $c \neq 1/2$. Let ω_n be a sequence tending arbitrarily slowly to $+\infty$. Then we have*

$$\lim_{n\to+\infty} \mathbf{P}\left(\Delta_{n,N(n)} \geq \frac{1}{\alpha}\left(\log n - \frac{5}{2}\log\log n\right) + \omega_n\right) = 0$$

and

$$\lim_{n\to+\infty} \mathbf{P}\left(\Delta_{n,N(n)} \geq \frac{1}{\alpha}\left(\log n - \frac{5}{2}\log\log n\right) - \omega_n\right) = 1$$

where

$$e^{-\alpha} = 2ce^{1-2c}$$

i.e., $\alpha = 2c - 1 - \log 2c$ *and thus* $\alpha > 0$.

Hence the answer of the question of Rényi is: The trees cannot grow up to the heavens because when the size of a tree will be larger than $\log n$ then a triangle will appear in it.

Erdős and Rényi returned to this problem occasionally {25} and many of their students attacked the "great deal" remained to be done in this field (e.g. Komlós, Endre Szemerédy {46}, Béla Bollobás, T. I. Fenner, A. M. Frieze {5}, Lajos Pósa {57}, Bollobás {4}).

Another problem of Rényi which initiated intensive research is the so-called stochastic geyser problem.

Let $X_1, X_2, \ldots$ be i.i.d. positive and bounded r.v.'s let $\{S_n\}$ be their partial sum sequence; can one then determine the distribution function of X_i with probability one, observing only the sequence $\{[S_n]\}$? This problem was motivated by the following story: Robinson Crusoe had a geyser on his island, which kept on erupting at random time points. After he had observed the number of eruptions per day for a long time, it occurred to him that he should now be able to predict the geyser's behaviour, i.e., he should be able to estimate the distribution function of the time length between two eruptions.

It is trivial that the sequence $\{[S_n]\}$ determines with probability one the expectation $\mathbf{E}X_1$ and the variance $\operatorname{Var} X_1$. However, the determination of higher moments even, is not trivial at all.

We now formulate a more general form of the geyser problem. Let $X_1, X_2, \ldots$ be i.i.d.r.v. and let $F(\cdot)$ be their distribution function. Put

$$V_n = S_n + R_n,$$

where $\{R_n\}$ is also a random variable sequence, not necessarily independent of S_n. Then we can ask whether it is possible to determine the distribution function $F(\cdot)$ with probability one via some Borel function of $\{V_n;\ n = 1, 2, \dots\}$. In statistical terminology $\{R_n;\ n = 1, 2, \dots\}$ can be viewed as a random error sequence when trying to observe S_n in order to estimate $F(\cdot)$. Answering this question Pál Bártfai {2} proved

Theorem 17. *Assume that the moment generating function* $R(t) = \int e^{tx} dF(x)$ *of* X_1 *exists in a neighbourhood of* $t = 0$ *and* $R_n \overset{\text{a.s.}}{=} o(\log n)$. *Then, given the values of* $\{V_n;\ n = 1, 2, \dots\}$, *the distribution function* $F(\cdot)$ *is determined with probability one, i.e., there exists a r.v.* $L(x) = L(V_1, V_2, \dots; x)$ *measurable with respect to the* σ*-algebra generated by* $\{V_n\}$ *such that for any given real* x, $L(x) \overset{\text{a.s.}}{=} F(x)$.

Bártfai also conjectured that his result is the best possible in the sense that if $R_n = O(\log n)$ and $R(t)$ exists in a neighbourhood of $t = 0$ then the distribution function $F(\cdot)$ is not determined by $\{V_n\}$.

This conjecture was proved later on by Komlós–Major–Tusnády (1975). In fact Theorem 13 easily implies Bártfai's conjecture.

Theorem 17 claims the existence of a functional L which determines F but it does not give any hint how to find it. Later on Erdős and Rényi gave a surprising solution of this problem. In fact they posed a problem in {27} which has nothing to do with the above mentioned question.

In connection with a teaching experiment in mathematics, Tamás Varga posed a problem which has also nothing to do with the Stochastic Geyser Problem. A solution of a more general form of it, however, turned out to be also an answer to the latter. The experiment goes like this: his class of secondary school children is divided into two sections. In one of the sections each child is given a coin which they then throw two hundred times, recording the resulting head and tail sequence on a piece of paper. In the other section the children do not receive coins but are told instead that they should try to write down a "random" head and tail sequence of length two hundred. Collecting these slips of paper, he then tries to subdivide them into their original groups. Most of the time he succeeds quite well. His secret is that he had observed that in a randomly produced sequence of length two hundred, there are, say, head-runs of length seven. On the other hand, he had also observed that most of those children who were to write down an imaginary random sequence are usually afraid of putting down runs of longer than four. Hence, in order to find the slips coming from the coin tossing group, he simply selects the ones which contain runs longer than five.

This experiment led Varga to ask: What is the length of the longest run of pure heads in n Bernoulli trials?

An answer to this question was given by Erdős and Rényi in {27}; they proved

Theorem 18. *Let $X_1, X_2, \ldots$ be i.i.d.r.v., each taking the values ± 1 with probability $1/2$. Put $S_0 = 0$, $S_n = X_1 + \ldots + X_n$. Then for any $c \in (0,1)$ and for almost all $\omega \in \Omega$ there exists an $n_0 = n_0(c, \omega)$ such that*

$$\max_{0 \le k \le n - [c \lg n]} \left(S_{k+[c \lg n]} - S_k \right) = [c \lg n],$$

if $n > n_0$, where $\lg$ is the logarithm of base 2.

That is, this theorem guarantees the existence of a run of length $[c \lg n]$ for every $c \in (0,1)$ with probability one if n is large enough.

On the other hand, in the same paper, they also showed for $c > 1$ that the above equality can only hold for a finite number of values of n with probability one. They proved

Theorem 19. *With the above notation one has*

$$\max_{0 \le k \le n - [c \lg n]} \frac{S_{k+[c \lg n]} - S_k}{[c \lg n]} \overset{\text{a.s.}}{\to} \alpha(c),$$

where $\alpha(c) = 1$ for $c \le 1$, and, if $c > 1$, then $\alpha(c)$ is the only solution of

$$\frac{1}{c} = 1 - h\left(\frac{1+\alpha}{2}\right),$$

with $h(x) = -x \lg x - (1-x) \lg (1-x)$, $0 < x < 1$; the herewith defined $\alpha(\cdot)$ is a strictly decreasing continuous function for $c > 0$ with $\lim_{c \searrow 1} \alpha(c) = 1$ and $\lim_{c \to \infty} \alpha(c) = 0$.

They also gave the following generalization of Theorem 19:

Theorem 20. *Let $X_1, X_2, \ldots$ be i.i.d.r.v.'s with mean zero and a moment generating function $R(t) = \mathbf{E} e^{tX_1}$, finite in a neighbourhood of $t = 0$. Let*

$$\rho(x) = \inf_t e^{-tx} R(t),$$

the so-called Chernoff function of X_1. Then for any $c > 0$ we have

$$\max_{0\le k\le n-[c\log n]} \frac{S_{k+[c\log n]} - S_k}{[c\log n]} \overset{\text{a.s.}}{\to} \alpha(c),$$

where

$$\alpha(c) = \sup\{x : \rho(x) \ge e^{-1/c}\}.$$

Clearly this theorem gives a concrete functional L to determine F (c.f. Theorem 17).

Now, we turn to some number theoretical applications of probability, an area investigated frequently by Erdős and Rényi.

Let $q \ge 2$ be an integer. Then, as it is well-known, every real number x $(0 \le x \le 1)$ can be represented in the form

$$x = \sum_{n=1}^{\infty} \frac{\varepsilon_n(x)}{q^n},$$

where the n-th "digit" $\varepsilon_n(x)$ may take values $0, 1, \ldots, q-1$. The classical Borel strong law of large numbers claims that for almost all real numbers $0 \le x \le 1$ the relative frequency of the numbers $0, 1, 2, \ldots, q-1$ among the first n digits of the q-adic expansion of x tends to $1/q$ as $n \to \infty$.

A possible generalization of the q-adic expansion is the so-called Cantor's series. Let $q_1, q_2, \ldots$ be an arbitrary sequence of positive integers, restricted only by the condition $q_n \ge 2$. Then the Cantor's series of x is

$$\sum_{n=1}^{\infty} \frac{\varepsilon_n(x)}{q_1 q_2 \cdots q_n}$$

where the n-th digit $\varepsilon_n(x)$ may take the values of $1, 2, \ldots, q_n - 1$.

Let $f_n(k, x)$ denote the number of those digits among $\varepsilon_1(x), \ldots, \varepsilon_n(x)$ which are equal to k $(k = 0, 1, 2, \ldots)$, i.e., put

$$f_n(k, x) = \sum_{\varepsilon_j(x)=k,\ 1\le j\le n} 1.$$

Let us put further

$$Q_n = \sum_{j=1}^{n} \frac{1}{q_j}$$

and

$$Q_{nk} = \sum_{j=1,\ q_j>k}^{n} \frac{1}{q_j}.$$

Then a possible generalization of Borel's theorem is proved by Rényi {61} claims that for almost all $0 < x < 1$ we have

$$\lim_{n\to\infty} \frac{f_n(k,x)}{Q_{nk}} = 1$$

for those values of k for which

$$\lim_{n\to\infty} Q_{nk} = \infty.$$

For those values of k for which Q_{nk} is bounded, $f_n(k,x)$ is bounded for almost all x.

Erdős and Rényi {23} studied the behaviour of

$$\max_k f_n(k,x)$$

i.e., that of the frequency of the most frequent number among the first n digits in the case when $\lim_{n\to\infty} Q_n = \infty$. It turns out that the behaviour of $\max_k f_n(k,x)$ is very sensitive to the properties of the sequence $\{q_n\}$.

In the case when

$$\sum_{n=1}^{\infty} \frac{1}{q_n} < \infty,$$

the Cantor series has a strikingly different behaviour. It is studied by Erdős, Rényi {24}.

Other investigated expansions of the real numbers are the so-called Engel's and Sylvester's series (Erdős, Rényi, Péter Szüsz {28}, Rényi {70}) and the Cantor's products (Rényi {64}).

A very general expansion is treated by Rényi {62}.

As we have said earlier, Rényi first met probability through Linnik and he was especially interested in Linnik's large sieve. He returned later to this problem occasionally.

In the original formulation of Linnik the large sieve asserts that if we take any sequence S_N consisting of $Z \leq N$ positive integers and if Y denotes the number of those primes $p \leq \sqrt{N}$ for which all the elements of the sequence

S_N are contained in $p(1-\varepsilon)$ residue class mod p, where $0 < \varepsilon < 1$, then one has

$$Y \leq \frac{20\pi N}{\varepsilon^2 Z}.$$

Rényi {58} also proved that if Z is not too small compared with N then the elements of the sequence S_N not only occupy "almost all" residue classes mod p with respect to most primes $p \leq \sqrt{N}$ but are almost uniformly distributed in the p residue classes mod p for most primes $p \leq \sqrt{N}$.

Even more precise results are given by Rényi ({65} and {66}). The last paper in this subject (Erdős, Rényi {26}) gives a number of applications of the large sieve method.

Among the numerous very different subjects which Rényi was interested in, finally we mention two further ones. One of these is the measure of dependence (Rényi {67}). In this paper he discusses and compares certain quantities which are used to measure the strength of dependence between two random variables. He formulates seven rather natural postulates which should be fulfilled by a suitable measure of dependence. It is shown that most of the known measures of dependence fulfill these conditions. Among them the so-called maximal correlation is studied in detail.

Finally we mention a very elementary problem. In classical probability theory many identities and inequalities are known for probabilities of arbitrary events. The best known theorem of this type is the Poincaré theorem: Let $A_1, A_2, \ldots, A_n$ be arbitrary events and

$$S_0 = 1, \quad S_k = \sum_{1 \leq i_1 < i_2 < \ldots < i_k \leq n} \mathbf{P}\{A_{i_1} A_{i_2} \ldots A_{i_k}\} \quad (k = 1, 2, \ldots, n).$$

Then

$$\mathbf{P}\{\bar{A}_1 \bar{A}_2 \ldots \bar{A}_n\} = \sum_{k=0}^{n} (-1)^k S_k.$$

Rényi {69} was interested to find a general method of proof of such formulas.

Let $F_1 = f_1(A_1, \ldots, A_n), F_2 = f_2(A_1, \ldots, A_n), \ldots, F_N = f_N(A_1, \ldots, A_n)$ where f_i's are Boole functions. Consider the linear inequality

$$\sum_{i=1}^{N} \alpha_i \mathbf{P}\{F_i\} \geq 0$$

where $\alpha_1, \alpha_2, \ldots, \alpha_N$ are given real numbers. The result of Rényi tells us that if we want to prove the above inequality for any sequence of events

$A_1, A_2, \dots, A_n$ then it is enough to verify it in the 2^n special cases when some of them are the empty set and the others are the complete probability space. Later on János Galambos and Rényi {37} proved a similar theorem for quadratic inequalities.

Till his death (1970) Rényi was the spiritual and in some sense the administrative head of the Hungarian probability school. After his death Erdős took over as the spiritual leader. He suggested the most important problems to the Hungarian probabilists. One of those was a continuation of the Erdős–Rényi paper: On a new law of large numbers.

In fact Theorems 18, 19, 20 do not give an exact answer of the question of Varga. In a joint paper Erdős and Pál Révész {29} proved in a very precise sense that the length of the longest head run in n Bernoulli trials is $\lg n$.

Theorem 21. *Let Z_N be the length of the longest head-run till N. Then*

$$Z_N \geq [\lg N - \lg\lg\lg N + \lg\lg e - 2 - \varepsilon]$$

for all but finitely many N if $\varepsilon > 0$ but

$$Z_N \leq [\lg N - \lg\lg\lg N + \lg\lg e - 1 + \varepsilon] \quad \text{i.o. a.s.}$$

However, for some N the r.v. Z_N can be larger than the above given bounds.

Theorem 22. *Let $\{a_n\}$ be a sequence of positive numbers and let*

$$A(\{a_n\}) = \sum_{n=1}^{\infty} 2^{-a_n}.$$

Then

$$Z_N \leq a_N \quad \text{a.s.}$$

for all but finitely many N if $A(\{a_n\}) < \infty$ but

$$Z_N > a_N \quad \text{i.o. a.s.}$$

if $A(\{a_n\}) = \infty$.

It is also interesting to ask what the length is of the longest run containing at most T ($T = 1, 2, \ldots$) tails. Denote by $Z_N(T)$ this length. Then the four results of Theorems 21 and 22 can be generalized for this case. Here we mention only one of them:

$$Z_N(T) \geq [\lg N + T \lg\lg N - \lg\lg\lg N - \lg T! + \lg\lg e - 1 + \varepsilon] \quad \text{i.o. a.s.}$$

for any $\varepsilon > 0$.

Among the further Hungarian results going in this direction we mention only a few.

Komlós, Tusnády {47}.

Sándor Csörgő {12}.

Miklós Csörgő, J. Steinebach {11}.

Tamás F. Móri {53}.

Paul Deheuvels, Erdős, Karl Grill, Révész {13}.

Móri {54}.

Endre Csáki, Antónia Földes, Komlós {9}.

Erdős and Taylor {36} beside their many interesting new results proposed a number of unsolved problems. In the eighties new efforts were taken to solve these problems. One of them is the so-called covering problem.

We say that the disc

$$Q(r) = \{x \in \mathbb{Z}^2,\ \|x\| \leq r\}$$

is covered by the random walk $\{S_k\}$ in time n if for each $x \in Q(r)$ there exists an integer $k \leq n$ such that $S_k = x$. Let $R(n)$ be the largest integer for which $Q\big(R(n)\big)$ is covered in time n. Erdős and Taylor presented the conjecture that $R(n)$ is about $\exp\big((\log n)^{1/2}\big)$. This fact was proved by Erdős, Révész {32} and by Peter Auer, Révész {1}. The fundamental result is the following:

$$\exp\big((\log n)^{1/2}(\log\log n)^{-1/2-\varepsilon}\big) \leq R(n) \leq \exp\big(2(\log n)^{1/2} \log\log\log n\big) \quad \text{a.s.}$$

for all but finitely many n.

Having the above inequality we can say that the Erdős–Taylor conjecture is correct, i.e., the radius of the largest circle around the origin, covered in time n is about $\exp\left((\log n)^{1/2}\right)$. It is natural to ask: how big is the radius $r(n)$ of the largest circle in $\mathbb{Z}^2$ not surely around the origin, which is covered in time n. One expects that $r(n)$ cannot be much larger than $R(n)$. However, by Erdős–Révész {33} we have

Theorem 23. *Let*

$$\psi_0 = \frac{1}{50} \quad \textit{and} \quad \chi_0 = 0,42.$$

Then for any $0 < \psi < \psi_0 < \chi_0 < \chi$ *we have*

$$n^\psi \le r(n) \le n^\chi \quad \text{a.s.}$$

for all but finitely many n.

A survey on covering problems is: Révész {74}.

Let $\{S_n\}$ be a random walk on $\mathbb{Z}^d$ and let

$$\xi(x,n) = \#\{k : 0 \le k \le n,\ S_k = x\}, \quad (x \in \mathbb{Z}^d)$$

$$\zeta(n) = \max_{x\in\mathbb{Z}^d} \xi(x,n).$$

A point $z_n \in \mathbb{Z}^d$ is called a favourite value at moment n if the particle visits z_n most often during the first n steps, i.e.,

$$\xi(z_n, n) = \zeta(n).$$

Erdős liked to write papers on different subjects of mathematics with the title: "Problems and results on ...". He has only one such paper in probability (Erdős, Révész {31}).

In this paper, among others, there are a few problems mentioned on the favourite values. Here we recall one of those.

One can easily observe that for infinitely many n there are two favourite values and also for infinitely many n there is only one favourite value with probability one. More formally speaking let F_n be the set of favourite values, i.e.,

$$F_n = \left\{ z : \xi(z,n) = \zeta(n) \right\}$$

and let $|F_n|$ be the cardinality of F_n. Then the question is: whether 3 or more favourite values can occur i.o., i.e.,

$$\mathbf{P}\left\{ |F_n| = r \ \text{ i.o.} \right\} = 1?$$

It turned out that this innocent looking question is very hard. In fact it is still open. The strongest result is due to Bálint Tóth {76} who proved

$$\mathbf{P}\{|F_n| \geq 4 \text{ i.o.}\} = 0.$$

Let

$$f_n = \max\{|x| : x \in F_n\}$$

be the largest favourite value. The question how big f_n can be was studied by Erdős and Révész {30} who proved that

$$\limsup_{n\to\infty} \frac{f_n}{(2n \log\log n)^{1/2}} = 1 \quad \text{a.s.}$$

Bass and Griffin {3} studied the much harder question: how small f_n can be.

The mentioned Erdős–Taylor paper also contains very nice results and problems on the properties of

$$\zeta(n) = \max_{x\in\mathbb{Z}^d} \xi(x, n).$$

The one dimensional results are well known. In fact we have

$$\limsup_{n\to\infty} \frac{\xi(0,n)}{(2n \log\log n)^{1/2}} = \limsup_{n\to\infty} \frac{\zeta(n)}{(2n \log\log n)^{1/2}} = 1 \text{ a.s.}$$

if $d = 1$. In case $d = 2$ Erdős and Taylor proved

Theorem 24. *Let $f(x)$ resp. $g(x)$ be a decreasing resp. increasing function for which*

$$f(x)\log x \nearrow \infty, \quad g(x)(\log x)^{-1} \searrow 0.$$

Then

$$\xi(0,n) \leq \pi^{-1} g(n) \log n \quad \text{a.s.}$$

for all n large enough if and only if

$$\int_2^\infty \frac{g(x)}{x \log x} e^{-g(x)} dx < \infty$$

and

$$\xi(0,n) \geq f(n) \log n \quad \text{a.s.}$$

for all n large enough if and only if

$$\int_2^\infty \frac{f(x)}{x\log x}dx < \infty.$$

Also in case $d = 2$ they presented the following conjecture

$$\lim_{n\to\infty} \frac{\zeta(n)}{(\log n)^2} = \frac{1}{\pi} \quad \text{a.s.}$$

This conjecture was proved by Amir Dembo, Yuval Peres, Jay Rosen, Ofer Zeitouni {14}.

Theorem 4, in case $d = 1$, gives a lower estimate of the last return $R(n)$ of a random walk to its starting point before its n-th step. Let

$$\begin{aligned} R^*(n) = \max\big\{ k \,:\, 1 < k < n &\text{ for which there exists a} \\ 0 < j < n - k &\text{ such that } \xi(j+k) - \xi(j) = 0 \big\} \end{aligned}$$

be the length of the longest zero-free interval. Remember that

$$\xi(n) = \#\{k \,:\, 0 \le k \le n,\ S_k = 0\}.$$

It is easy to see that replacing $R(n)$ by $R^*(n)$, Theorem 4 remains true in its original form. For example we have

$$R^*(n) \ge n - \frac{n}{(\log n)^2} \quad \text{i.o. a.s.}$$

However for some n, $R^*(n)$ can be much smaller than the above lower estimate. Endre Csáki, Erdős and Révész in {8} asked how small $R^*(n)$ can be. As an answer of this question we proved:

Theorem 25. *Let $f(n)$ be an increasing function for which*

$$f(n) \nearrow \infty, \quad \frac{n}{f(n)} \nearrow \infty \quad (n \to \infty).$$

Then

$$\mathbf{P}\left\{R^*(n) \le \beta \frac{n}{f(n)} \quad \text{i.o.}\right\} = \begin{cases} 1 & \text{if } J = \infty, \\ 0 & \text{if } J < \infty \end{cases}$$

where

$$J = \sum_{n=1}^{\infty} \frac{f(n)}{n} \exp\big(-f(n)\big)$$

and $\beta = 0.85403\ldots$ *is the root of the equation*

$$\sum_{k=1}^{\infty} \frac{\beta^k}{k!(2k-1)} = 1.$$

As a consequence of this theorem we mention that

$$\liminf_{n\to\infty} \frac{\log\log n}{n} R^*(n) = \beta \quad \text{a.s.}$$

The path of a random walk between two zeros is called an excursion. Then Theorems 4 and 23 tell us that for any $\varepsilon > 0$ the length of the longest excursion not surely completed before n is

$$R^* \begin{cases} \leq n - \dfrac{n}{(\log n)^{2+\varepsilon}} & \text{a.s. if } n \text{ is big enough,} \\ \geq n - \dfrac{n}{(\log n)^2} & \text{i.o. a.s.,} \\ \geq (1-\varepsilon)\beta \dfrac{n}{\log\log n} & \text{i.o. a.s.,} \\ \leq (1+\varepsilon)\beta \dfrac{n}{\log\log n} & \text{a.s. if } n \text{ is big enough.} \end{cases}$$

Besides studying the length of the longest excursion $R^*(n)$, it looks interesting to say something about the second, third ... etc. longest excursions. Let $R_1^*(n) \geq R_2^*(n) \geq \ldots \geq R_{\xi(n)+1}^*(n)$ be the length of the second, third etc. longest excursions. Then we have

Theorem 26. *For any fixed* $k = 1, 2, \ldots$ *we have*

$$\liminf_{n\to\infty} \frac{\log\log n}{n} \sum_{j=1}^{k} R_j^*(n) = k\beta \quad \text{a.s.}$$

Theorem 4 tells us that for some n nearly the whole random walk $\{S_k\}_{k=0}^n$ is one excursion. Theorem 24 tells us that for some n the random walk consists of at least $\beta^{-1}\log\log n$ excursions. These results suggest the question: For which values of $k = k(n)$ will the sum $\sum_{j=1}^k R_j^*(n)$ be nearly equal to n? In fact we formulated two questions:

Question 1. For any $0 < \varepsilon < 1$ let $\mathcal{F}(\varepsilon)$ be the set of those functions $f(n)$ $(n = 1, 2, \ldots)$ for which

$$\sum_{j=1}^{f(n)} R_j^*(n) \geq n(1 - \varepsilon)$$

with probability 1 except finitely many n. How can we characterize $\mathcal{F}(\varepsilon)$?

Question 2. Let $\mathcal{F}(o)$ be the set of those functions $f(n)$ $(n = 1, 2, \ldots)$ for which

$$\lim_{n \to \infty} n^{-1} \sum_{j=1}^{f(n)} R_j^*(n) = 1 \quad \text{a.s.}$$

How can we characterize $\mathcal{F}(o)$?

Studying the first question we have

Theorem 27. *For any $0 < \varepsilon < 1$ there exists a $C = C(\varepsilon) > 0$ such that*

$$C \log \log n \in \mathcal{F}(\varepsilon).$$

Concerning Question 2, we have the following result:

Theorem 28. *For any $C > 0$*

$$f(n) = C \log \log n \notin \mathcal{F}(o)$$

and for any $h(n) \nearrow \infty$ $(n \to \infty)$

$$h(n) \log \log n \in \mathcal{F}(o).$$

Up to now we mostly concentrated on the results of Erdős and Rényi and their students. For a recent review of Erdős's work in probability and statistics we refer to Miklós Csörgő {10}.

We now consider the works of a few Hungarian probabilists whose results are not so strongly connected to the Erdős–Rényi school. In fact we give a short survey of the works of Béla Gyires, Pál Medgyessy and József Mogyoródi.

Gyires' most important contribution to probability theory is the foundation of the theory of stationary matrix valued processes and the solution of some extrapolation problems {38, 40, 41}. In this theory block Toeplitz matrices generated by matrix valued functions play an important role. He generalized results due to Szegő and to Helson and Lowdenslager.

In {39} he proves an interesting generalization of a central limit theorem for a sequence $\zeta_n = \xi_1 + \cdots + \xi_n$, where $\xi_k = \theta^{(k)}_{\eta_{k-1},\eta_k}$ with mutually independent random variables $\theta^{(k)}_{i,j}$ $(i,j = 1,\ldots,p;\ k = 1,2,\ldots)$ which are independent of the Markov chain $\{\eta_n\}$ with states $1,\ldots,p$. He shows that if the chain is ergodic and the conditional distribution functions $\mathbf{P}\{\xi_k < x \mid \eta_{k-1} = i\}$ have zero mean and finite second moments and satisfy a condition of the Lindeberg type then it satisfies the central limit theorem.

He also developed a systematic investigation of the decomposability problems of distribution functions. The main results can be found in his book {42}. The problem is to give conditions for a distribution function F to be a mixture of a given stochastic kernel G with a weight function H from a certain given set of distribution functions.

Medgyessy was mostly interested in the decomposition of superpositions of distribution functions. The superposition of distributions is a frequently used operation in probability. Let $F_1, F_2, \ldots, F_N$ resp. $p_1, p_2, \ldots, p_N$ be sequences of distributions resp. real numbers. The function

$$G(x) = \sum_{k=1}^{N} p_k F_k(x)$$

will be called a superposition of F_k. Assume also that F_i's $(i = 1,2,\ldots,N)$ are elements of a class of distributions containing a finite number of parameters. Then the problem is the following: Given the superposition $G(x)$ determine the parameters of the components F_k when their analytic form is known (only their parameters should be determined by the aid of $G(x)$).

Working through 8 years on this topic Medgyessy wrote a book {50}.

Mogyoródi was also a student of Rényi. However, his research area moved away from Rényi's school. He was mostly interested in martingales and Orlicz and Hardy spaces.

First we recall the definition of the Orlicz space. A random variable X defined on a probability space $\{\Omega, \mathcal{S}, \mathbf{P}\}$ belongs to the Orlicz space $L^\Phi(\Omega, \mathcal{S}, \mathbf{P})$ if there exists a constant $a > 0$ such that $\mathbf{E}\Phi\big(a^{-1}|X|\big) \le 1$ where Φ is a Young-function. The L^Φ-norm of X is:

$$\|X\|_\Phi = \inf\big\{a : a > 0,\ \mathbf{E}\Phi\big(a^{-1}|X|\big) \le 1\big\}.$$

Now we give the definition of the Hardy space $\mathcal{H}_\Phi$. Let $\mathcal{F}_0 \subset \mathcal{F}_1 \subset \ldots$ be a sequence of σ-fields with $\lim_{n\to\infty} \mathcal{F}_n = \mathcal{S}$. Consider the martingale $X_n = \mathbf{E}(X \mid \mathcal{F}_n)$ and the martingale-differences $d_i = X_{i+1} - X_i$. We say that $X \in L^1$ belongs to $\mathcal{H}_\Phi$ if

$$S = \left(\sum_{i=1}^{\infty} d_i^2\right)^{1/2} \in L^\Phi.$$

In general a sequence $\Theta = (\Theta_1, \Theta_2, \ldots)$ belongs to the Banach space $\delta\mathcal{H}_\Phi$ if

$$\left(\sum_{i=1}^{\infty} \Theta_i^2\right)^{1/2} \in L^\Phi.$$

Mogyoródi in {51} gave a characterization of the linear functionals on Hardy spaces, similar to Riesz' characterization of the linear functionals on Hilbert spaces. It is known that if (Φ, Ψ) is a pair of conjugate Young-functions and A is a bounded linear functional on L^Ψ then there exists a random variable $X \in L^\Phi$ such that $AY = \mathbf{E}XY$ for any $Y \in L^\Psi$. The main result of this paper concludes that under mild conditions we have:

$$AY = \lim_{n\to\infty} \mathbf{E}X_n Y_n,$$

where $X_n = \mathbf{E}(X \mid \mathcal{F}_n)$ and $Y_n = \mathbf{E}(Y \mid \mathcal{F}_n)$.

In an other paper {52} Mogyoródi investigates the properties of $X_n^* = \max_{0\le k\le n} X_k$ and those of $X^* = \sup_{k\ge 0} X_k$.

Let $\Phi(\cdot)$ be a Young-function with derivative $\varphi(\cdot)$ and let $\xi(x) = x\varphi(x) - \varphi(x)$.

One of the main results of the paper concludes that for any $\rho > 1$ we have

$$\mathbf{E}\xi\left(\frac{X_n^*}{\rho\|M_n\|_\Phi}\right) \le \frac{1}{\rho - 1},$$

where $\{M_n\}$ is the martingale in the Doob decomposition of X_n, i.e., $M_n = X_n + A_n$, where $\{A_n\}$ is a unique and predictable sequence of random variables.

Acknowledgement. The author is indebted to Gyula Pap who wrote the part on Gyires's contributions. For further comments on the work of Gyires, we refer to the Section on Mathematical Statistics.

REFERENCES

{1} P. Auer and P. Révész, On the relative frequency of points visited by random walk on $\mathbb{Z}^2$, *Coll. Math. Soc. J. Bolyai,* **57**, *Limit theorems in probability and Statistics* (1990), 27–33.

{2} P. Bártfai, Die Bestimmung der zu einem wiederkehrenden Prozess gehörenden Verteilungsfunktion aus den mit Fehlern behafteten Daten einer Einzigen Realisation, *Studia Sci. Math. Hung.*, **1** (1966), 161–168.

{3} Bass and Griffin, The most visited site of Brownian motion and simple random walk, *Z. Wahrscheinlichkeitstheorie verw. Gebiete,* **70** (1985), 417–436.

{4} B. Bollobás, *Random Graphs,* Academic Press, London, 1985.

{5} B. Bollobás, T. I. Fenner and A. M. Frieze, An algorithm for finding Hamiltonian paths and cycles in random graphs, *Combinatorica,* **7** (1987), 327–342.

{6} K.-L. Chung and P. Erdős, On the application of the Borel–Cantelli lemma, *Trans. Amer. Math. Soc.*, **72** (1952), 179–186.

{7} K.-L. Chung, P. Erdős and T. Sirao, On the Lipschitz condition for Brownian motion, *J. Math. Soc. Japan,* **11** (1959), 263–274.

{8} E. Csáki, P. Erdős and P. Révész, On the length of the longest excursion, *Z. Wahrscheinlichkeitstheorie verw. Gebiete,* **68** (1985), 365–382.

{9} E. Csáki, A. Földes and J. Komlós, Limit theorems for Erdős–Rényi type problems, *Studia Sci. Math. Hung.*, **22** (1987), 231–232.

{10} M. Csörgő, A glimpse of the impact of Pál Erdős on probability and statistics, *The Canadian Journal of Statistics,* **30** (2002), 493–556.

{11} M. Csörgő and J. Steinebach, Improved Erdős–Rényi and strong approximation laws for increments of partial sums, *Ann. Probab.*, **9** (1981), 988–996.

{12} S. Csörgő, Erdős–Rényi laws, *Ann. Statist.*, **7** (1979), 772–787.

{13} P. Deheuvels, P. Erdős, K. Grill and P. Révész, Many heads in a short block, *Math. Stat. and Probab. Th. Proc. of the 6th Pannonian Symp.* Vol. A (1986), 53–67.

{14} A. Dembo, Y. Peres, J. Rosen and O. Zeitouni, Thick points for planar Brownian motion and the Erdős–Taylor conjecture on random walk, *Acta Math.* (to appear).

{15} A. Dvoretzky and P. Erdős, Some problems on random walk in space, *Proc. Second Berkeley Symposium* (1950), 353–368.

{16} A. Dvoretzky, P. Erdős and S. Kakutani, Nonincrease everywhere of the Brownian motion process, *Proc. 4th Berkeley Sympos. Math. Statist. and Probab.* Vol II. (1961), 103–116. Univ. California Press, Berkeley.

{17} F. Eggenberger and G. Pólya, Über die Statistik verketteter Vorgänge, *Z. Angew. Math. Mech.*, **3** (1923), 279–289.

{18} P. Erdős, On the law of the iterated logarithm, *Ann. of Math.*, **43** (1942), 419–436.

{19} P. Erdős, On the integers having exactly k prime factors, *Ann. of Math.*, **49** (1948), 53–66.

{20} P. Erdős and M. Kac, On certain limit theorems of the theory of probability, *Bull. Amer. Math. Soc.*, **52** (1946), 292–302.

{21} P. Erdős and M. Kac, On the number of positive sums of independent random variables, *Bull. Amer. Math. Soc.*, **53** (1947), 1011–1020.

{22} P. Erdős and A. Rényi, On random graphs I. *Publ. Math. Debrecen,* **6** (1959), 290–297.

{23} P. Erdős and A. Rényi, Some further statistical properties of the digits in Cantor's series, *Acta Math. Acad. Sci. Hung.*, **10** (1959), 21–29.

{24} P. Erdős and A. Rényi, On Cantor's series with convergent $\sum 1/q_n$, *Ann. Univ. Sci. Budapest. Eötvös Sect. Math.*, **2** (1959), 93–109.

{25} P. Erdős and A. Rényi, On the strength of connectedness of a random graphs, *Acta Math. Acad. Sci. Hung.*, **12** (1961), 262–267.

{26} P. Erdős and A. Rényi, Some remarks on the large sieve of Yu. V. Linnik, *Ann. Univ. Sci. Budapest. Eötvös Sect. Math.*, **11**, (1968), 3–13.

{27} P. Erdős and A. Rényi, On a new law of large numbers, *J. Analyse Math.*, **23** (1970), 103–111.

{28} P. Erdős, A. Rényi and P. Szüsz, On Engel's and Sylvester's series, *Ann. Univ. Sci. Budapest. Eötvös Sect. Math.*, **1** (1958), 7–32.

{29} P. Erdős and P. Révész, On the length of the longest head-run, in: *Topics in Information Theory. Coll. Math. Soc. J. Bolyai,* **16** (1976), 219–228 (ed. Imre Csiszár, P. Elias).

{30} P. Erdős and P. Révész, On the favourite points of a random walk, *Math. Structures Comput. Math. Modelling,* **2** Sofia (1984), 152–157.

{31} P. Erdős and P. Révész, Problems and results on random walks, *Math. Stat. and Probab. Th., Proc. of the 6th Pannonian Symp.* Vol. B (1986), 59–65.

{32} P. Erdős and P. Révész, On the area of the circles covered by a random walk, *J. of Multivariate Anal.*, **27** (1988), 169–180.

{33} P. Erdős and P. Révész, Three problems on the random walk in $\mathbb{Z}^d$, *Studia Sci. Math. Hung.*, **26** (1991), 309–320.

{34} P. Erdős and S. J. Taylor, Some problems concerning the structure of random walk paths, *Acta Math. Acad. Sci. Hung.*, **11** (1960), 137–162.

{35} P. Erdős and S. J. Taylor, Some intersection properties of random walk paths, *Acta Math. Acad. Sci. Hung.*, **11** (1960), 231–248.

{36} P. Erdős and S. J. Taylor, Some problems concerning the structure of random walk paths, *Acta Math. Acad. Sci. Hung.*, **11** (1960), 137–162.

{37} J. Galambos and A. Rényi, On quadratic inequalities in probability theory, *Studia Sci. Math. Hung.*, **3** (1968), 351–358.

{38} B. Gyires, Eigenwerte verallgemeinerter Toeplitzscher Matrizen, *Publ. Math. Debrecen,* **4** (1956), 171–179.

{39} B. Gyires, Eine Verallgemeinerung des zentralen Grenzwertsatzes, *Acta Math. Acad. Sci. Hung.*, **13** (1962), 69–80.

{40} B. Gyires, On the uncertainty of matrix-valued predictions, *Proc. Colloq. Inform. Theory, Debrecen* 1967, pp. 253–268 (1968).

{41} B. Gyires, The extreme linear predictions of the matrix-valued stationary stochastic processes, *Mathematical statistics and probability theory* Vol. B (Bad Tatzmannsdorf, 1986), pp. 113–124, Reidel, Dordrecht, 1987.

{42} B. Gyires, *Linear approximations in convex metric spaces and the application in the mixture theory of probability theory,* World Scientific Publishing Co. Inc., River Edge, NJ, 1993.

{43} Gy. Hajós and A. Rényi, Elementary proofs of some basic facts concerning order statistics, *Acta Math. Acad. Sci. Hung.,* **5** (1954), 1–6.

{44} L. Jánossy, A. Rényi and J. Aczél, On composed Poisson distributions, I. *Acta Math. Acad. Sci. Hung.,* **1** (1950), 209–224.

{45} J. Komlós, P. Major and G. Tusnády, An approximation of partial sums of independent R.V.'s and the sample D.F. I and II., *Z. Wahrscheinlichkeitstheorie verw. Gebiete,* **32** (1975), 111–131. and **34** (1976), 33–58.

{46} J. Komlós and E. Szemerédy, Limit distributions for the existence of Hamilton cycles in a random graph, *Discrete Math.,* **43** (1983), 55–63.

{47} J. Komlós and G. Tusnády, On sequences of "pure heads", *Ann. Probab.,* **3** (1975), 608–617.

{48} P. Lévy, *Théorie de l'addition des variable aléatoires,* Gauthier–Villars, Paris (1937).

{49} P. Major, An improvement of Strassen's invariance principle, *Ann. Probab.,* **7** (1979), 55–61.

{50} P. Medgyessy, *Decomposition of superpositions of distribution functions,* Akadémiai Kiadó, Budapest (1961).

{51} J. Mogyoródi, Linear functionals on Hardy spaces, *Ann. Univ. Sci. Budapest. Eötvös Sect. Math.,* **26** (1983), 161–174.

{52} J. Mogyoródi, Maximal inequalities and Doob's decomposition for non-negative supermartingales, *Ann. Univ. Sci. Budapest Eötvös. Sect. Math.,* **26** (1983), 175–183.

{53} T. F. Móri, Large deviation results for waiting times in repeated experiments, *Acta Math. Acad. Sci. Hung.,* **45** (1985), 213–221.

{54} T. F. Móri, Maximum waiting time when the size of the alphabet increases, *Math. Statist. and Probab. Th. Proc. of the 6th Pannonian Symp.* Vol. B (1986), 169–178.

{55} G. Pólya, Über eine Aufgabe der Wahrscheinlichkeitsrechnung betreffend die Irrfahrt im Strassennetz, *Math. Ann.,* **84** (1921), 149–160.

{56} G. Pólya, *Zur Kinematik der Geschiebebewegung,* Mitt. Versuchsinst. f. Wasserbau an der ETH, Zürich (1937), 1–21.

{57} L. Pósa, Hamiltonian circuits in random graphs, *Discrete Math.,* **14** (1976), 359–364.

{58} A. Rényi, On the representation of an even number as the sum of a prime and of an almost prime, *Izv. Akad. Nauk SSSR Ser. Mat.*, **12** (1948), 57–78.

{59} A. Rényi, On the theory of order statistics, *Acta Math. Acad. Sci. Hung.*, **4** (1953), 191–231.

{60} A. Rényi, On a new axiomatic theory of probability, *Acta Math. Acad. Sci. Hung.*, **6** (1955), 285–335.

{61} A. Rényi, On the distribution of the digits in Cantor's series, *Mat. Lapok,* **7** (1956), 77–100.

{62} A. Rényi, Representations for real numbers and their ergodic properties, *Acta Math. Acad. Sci. Hung.*, **8** (1957), 477–493.

{63} A. Rényi, Probabilistic methods in number theory, *Shuxue Jinzhan,* **4** (1958), 465–510.

{64} A. Rényi, On Cantor's products, *Colloq. Math.*, **6** (1958), 135–139.

{65} A. Rényi, On the probabilistic generalization of the large sieve of Linnik, *MTA Mat. Kut. Int. Közl.*, **3** (1958), 199–206.

{66} A. Rényi, New version of the probabilistic generalization of the large sieve, *Acta Math. Acad. Sci. Hung.*, **10** (1959), 217–226.

{67} A. Rényi, On measure of dependence, *Acta Math. Acad. Sci. Hung.*, **10** (1959), 441–451.

{68} A. Rényi, On the evaluation of random graphs, *MTA Mat. Kut. Int. Közl.*, **5** (1960), 17–61.

{69} A. Rényi, A general method for proving theorems in probability theory and some applications, *MTA III. Oszt. Közl.*, **11** (1961), 79–105.

{70} A. Rényi, A new approach to the theory of Engel's series, *Ann. Univ. Sci. Budapest. Eötvös Sect. Math.*, **5** (1962), 25–32.

{71} A. Rényi, Remarks on the Poisson process, *Studia Sci. Math. Hung.*, **2** (1967), 119–123.

{72} A. Rényi and A. Prékopa, On the independence in the limit of sums depending on the same sequence of independent random variables, *Acta Math. Acad. Sci. Hung.*, **7** (1956), 319–326.

{73} A. Rényi and L. Takács, On processes generated by Poisson process and on their technical and physical applications, *MTA Mat. Kut. Int. Közl.*, **1** (1952), 139–146.

{74} P. Révész, Covering problems, *Theory Probab. Appl.*, **38** (1993), 367–379.

{75} V. Strassen, An invariance principle for the law of the iterated logarithm, *Z. Wahrscheinlichkeitstheorie verw. Gebiete,* **3** (1964), 211–226.

{76} B. Tóth, No more than three favourite sites for simple random walk, *Ann. Probab.*, **29** (2001), 484–503.

Pál Révész
Alfréd Rényi Institute of Mathematics
Hungarian Academy of Sciences
P.O.B. 127
1364 Budapest
Hungary
reveszp@renyi.hu

BOLYAI SOCIETY
MATHEMATICAL STUDIES, 14

A Panorama of Hungarian Mathematics in the Twentieth Century, pp. 491–521.

Mathematical Statistics

ENDRE CSÁKI*

1. Introduction

The word "statistics" originated from the Latin word "status" and according to Kendall and Stuart (*The Advanced Theory of Statistics. Vol. 1. Distribution Theory,* Hafner Publishing Co., New York (1958)) "Statistics is the branch of scientific method which deals with the data obtained by counting or measuring the properties of populations". So the main task of Statistics is to collect data and make conclusions, usually called Statistical Inference. Thus statistical methods have been used for a long time also in Hungary, e.g., in Hungarian Central Statistical Office founded in 1867 and also in other organizations. The data are usually subject to random fluctuations and so the theory of statistical inference should be based on rigorous mathematical concepts treating random phenomena, i.e., on the Theory of Probability. Mathematical Statistics is the theory of statistical methods based on rigorous mathematical concepts of Probability. In this way we can consider Károly (Charles) Jordan as the founder of the probability and statistics school in Hungary, who wrote the first book on Mathematical Statistics in Hungary.

K. Jordan was born in 1871 in Budapest. He started his activity in mathematics, probability and statistics in particular, around 1910. He wrote 5 books and 83 scientific papers. His book on Mathematical Statistics appeared in 1927 in Hungarian and also in extended form in French (*Statistique Mathématique,* Gauthier-Villars (1927)). His books "Calculus of

*Supported by the Hungarian National Foundation for Scientific Research, Grant No. T 029621 and T 037886.

Finite Differences", appeared in 1939, and "Chapters on Classical Probability Theory", appeared in 1956 in Hungarian, are very important frequently cited works in probability and in the theory of difference equations. He had a number of students and a great influence in developing the theory of probability and statistics in Hungary in the first half of twentieth century.

Another prominent Hungarian probabilist and statistician was Ábrahám Wald who made significant contributions in these subjects. He was born in Kolozsvár, Hungary in 1902. In 1938 he went to the United States where he turned his main interest toward Statistics. Perhaps he is most well-known as a founder of sequential analysis and also the theory of statistical decision functions, but his basic results in other areas such as hypothesis testing, goodness of fit tests, tolerance limits, analysis of variance, nonparametric statistics, sampling, etc. are also very important.

One of the most distinguished mathematicians of 20th the century, János (John von) Neumann has also contributed to statistics. We refer to Section 5 for his results which appeared in The Annals of Mathematical Statistics.

Mathematical Statistics in Hungary became a vigourous subject in the fifties when the Mathematical Institute of the Hungarian Academy of Sciences was founded, featuring also a Department of Mathematical Statistics. First of all, the works and school in probability and statistics of Alfréd Rényi should be mentioned. His main works are in Probability Theory and Applications (see the Probability Theory Section), but he has also important contributions in Statistics. István Vincze, Károly Sarkadi, Lajos Takács and their collaborators made also significant contributions. The works of Béla Gyires in statistics at the Kossuth Lajos University, Debrecen, should also be mentioned.

2. Early statistics in Hungary

In the first half of the 20th century the outstanding works of K. Jordan in both theoretical and applied statistics are to be mentioned. His contributions to applied statistics concern a number of subjects such as meteorology, chemistry, population statistics, industry, etc. Even in his applied works he was very careful to base his investigations on rigorous theoretical disciplines. Since his works started well before Kolmogorov's fundamental works to establish rigorous mathematical probability theory, Jordan himself had to work on theoretical foundations of statistics, i.e., he had to develop a rigorous probability theory needed for the application in statistics. In a series of papers {23}, {24} etc. he gave a rigorous definition of probability and proved some fundamental theorems. This can be considered as a forerunner of Kolmogorov's theory. His book {31} is based on the author's experience of fifty years of research and thirty years of teaching. It is written in his lucid style and reflects his profound knowledge of the history of probability and his significant contributions to probability theory. He contributed also to other subjects in mathematics such as geometry and, first of all, to the theory of difference equations which he also needed in his research in probability and statistics. His book [78] contains 109 sections and gives a detailed account of the theory and application of statistics, equipped also with a number of illuminating examples. In order to give a flavour of the content, here is a selection of section titles: Definition of mathematical probability — Theorem of total probability — Mathematical expectation — Theorem of Bernoulli — Poisson limit — Theorem of Tchebychef — Theory of least squares — Elements of calculus of differences — Statistical classifications — Mean values — Standard deviation — Construction of statistical functions — Normal distribution — Asymmetric distributions — Approximation of functions — Method of moments — Method of least squares — Interpolations — Correlations — Independence — Correlation for nonnormal distribution — Correlation ratio — Theory of sampling — Contingency tables — Rank correlations.

In his far-reaching statistical investigations K. Jordan had to develop certain numerical methods such as interpolation, least square methods, etc. An outline of his contributions in this area is based on nice accounts of Jordan's life and works by L. Takács {48} and by B. Gyires {18}.

In {25}, {28} and {29} he formulated and proved the following result: Let $A_1, \ldots, A_n$ arbitrary events and put

$$B_j = \sum_{1 \le i_1 < \cdots < i_j \le n} \mathbf{P}(A_{i_1} \cap \cdots \cap A_{i_j}).$$

Then the probability P_k that exactly k events occur among them is given by

$$P_k = \sum_{j=k}^{n} (-1)^{j-k} \binom{j}{k} B_j, \qquad k = 0, 1, \ldots, n.$$

In {26} Jordan gave an interpolation formula

$$f(a+xh) = \sum_{m=0}^{n-1} C_m(x) \sum_{k=1}^{m+1} B_{mk} I_k + R_{2n},$$

where

$$C_m(x) = (-1)^m \binom{x+m-1}{2m}$$

and

$$B_{mk} = (-1)^{k+1} \binom{2m+1}{k+m} \frac{2k-1}{2m+1}.$$

I_k is obtained by linear interpolation:

$$I_k = \frac{x+k-1}{2k-1} f(a+kh) + \frac{k-x}{2k-1} f(a-kh+h).$$

Moreover, R_{2n} is a remainder for which

$$|R_{2n}| \le h^{2n} \left| \binom{n-\frac{1}{2}}{2n} D^{2n} f(a+\xi h) \right|$$

with some $-n+1 < \xi < n$.

A further contribution of K. Jordan concerns the following least square problem. Let $Y_0, Y_1, \ldots, Y_{N-1}$ be observations corresponding to $x = 0, 1, \ldots, N-1$. Find polynomials $f_n(x)$ of degree n such that

$$S_n = \sum_{x=0}^{N-1} \left(Y_x - f_n(x) \right)^2$$

is minimum. The solution of this problem given by C. Jordan in {27} is as follows: Consider the expansion

$$f_n(x) = \sum_{m=0}^{n} a_m U_m(x),$$

where the polynomials $U_m(x)$ are orthogonal with respect to $x = 0, 1, \ldots, N-1$, i.e.,

$$\sum_{x=0}^{N-1} U_i(x) U_j(x) = 0, \qquad i \neq j.$$

The Newton expansion of $U_m(x)$ has the form

$$U_m(x) = C_m \sum_{i=1}^{m} (-1)^{m+i} \binom{m+i}{m} \binom{N-i-1}{m-i} \binom{x}{i},$$

where the coefficients C_m can be chosen as

$$C_m = \left[(m+1) \binom{N}{m+1} \right]^{-1}.$$

The values of a_m which minimize S_n are independent of n. The Newton expansion of $f_n(x)$ is given by

$$f_n(x) = \sum_{m=0}^{n} \sum_{i=0}^{m} C_{mi} \Theta_m \binom{x}{i},$$

where

$$C_{mi} = (-1)^{m+i} (2m+1) \binom{m+i}{m} \frac{\binom{N-i-1}{m-i}}{\binom{N+m}{m}}$$

and

$$\Theta_m = \sum_{x=0}^{N-1} U_m(x) Y_x.$$

The mean square deviation is

$$\sigma^2 = \frac{1}{N} \sum_{x=0}^{N-1} \left(Y_x - f_n(x) \right)^2 = \frac{1}{N} \sum_{x=0}^{N-1} Y_x^2 - \Theta_0^2 - |C_{10}|\Theta_1^2 - \cdots - |C_{n0}|\Theta_n^2.$$

In {30} he introduced the notion of surprisingness. If the events $A_1, A_2, \ldots, A_i$ occur respectively $k_1, k_2, \ldots, k_i$ times in n trials ($k_1 + k_2 + \cdots + k_i = n$), its probability being $P_{k_1,k_2,\ldots,k_i}$, then define the surprise index by

$$S = 1 - \frac{P_{k_1,k_2,\ldots,k_i}}{P_{m_1,m_2,\ldots,m_i}},$$

where $P_{m_1,m_2,\ldots,m_i}$ is the probability of the most probable system $m_1, m_2, \ldots, m_i$. This can be used in hypothesis testing to control the type 1 error by constructing a critical region which contains the points of the sample space with small probabilities, i.e., high surprise index. This approach is used to introduce Pearson's chi-square and other tests.

In the mid fifties of the last century one of the main tasks of the Statistics Department of the Mathematical Institute was to introduce statistical applications in practice, industrial quality control in particular. This is well reflected in producing the book {53} edited by I. Vincze, the head of the department. This book is written for engineers and technicians who wish to acquire familiarity with the theoretical foundations and with the applications of statistical quality control. An introduction into the elements of probability theory and mathematical statistics is given; the viewpoint of the quality control engineer is stressed and this also determines the choice of examples. Seven authors participated in writing the book; Part I, Theoretical foundations was written by K. Sarkadi and I. Vincze. Part II, Chapter 1 was written by Ágnes Fontányi and Mrs. Éva Vas and deals with statistical methods for the control of a manufacturing process. An interesting method developed by Á. Fontányi, K. Sarkadi and Mrs. É. Vas which uses order statistics, is discussed in detail. Part II, Chapter 2 (written by Károly Kollár) treats the statistical methods of acceptance control. The theory is adequately discussed, and sampling plans, with due references to the American sources, are given. Part III has the title "Applications of statistical quality control". Chapters 1 and 2 (written by Tibor Tallián) deal with problems of organizing quality control in a plant. Chapter 3 (written by M. Borbély) discusses specific applications in the textile industry. A mathematical appendix to part I, a collection of statistical tables and a bibliography conclude the book.

3. Sequential analysis

A. Wald in the mid forties of the 20th century developed a statistical procedure called sequential method. Here the number of observed elements are not specified in advance, in certain situations the experimentation should be continued until there is enough evidence as to which decision should be made. This method can be described as follows. (cf. {67} and [192]).

Let X be a random variable having a density function $f(x)$. Consider two hypotheses, $H_0 : f(x) = f_0(x)$ and $H_1 : f(x) = f_1(x)$, where $f_0(x)$ and $f_1(x)$ are two different density functions. The sequential probability ratio test for testing H_0 against H_1 is given as follows: Put

$$Z_i = \log \frac{f_1(X_i)}{f_0(X_i)},$$

where X_i denotes the i-th observation on X. Let $\log A > 0$ and $\log B < 0$ be two constants depending on error probabilities. At each stage of the experiment, at the m-th trial consider the partial sum

$$S_m = Z_1 + \cdots + Z_m$$

and continue experimentation as long as $\log B < S_m < \log A$. The first time $S_m \notin (\log B, \log A)$, the experimentation is terminated. Accept H_1, resp. H_0 if $S_m \geq \log A$, resp. $S_m \leq \log B$. It is proved that with probability one, this sequential probability ratio test terminates in a finite (random) number of steps. Let ν be the number of observations required by this test. Then ν is a random variable, for which Wald showed that

$$\mathbf{E}\big(e^{S_\nu t}\varphi(t)^{-\nu}\big) = 1$$

for all points t on the complex plane for which the moment generating function $\varphi(t) = \mathbf{E}(e^{Zt})$ exists and its absolute value is not less than 1. Here $\mathbf{E}$ stands for expectation and $Z = \log\big(f_1(X)/f_0(X)\big)$. This is a celebrated identity, called Wald's identity in the literature today.

In order to investigate the number of observations required by this test, Wald shows also that

$$\mathbf{E}(S_\nu) = \mathbf{E}(\nu)\mathbf{E}(Z) = \mathbf{E}(\nu)\mathbf{E}\left(\log\frac{f_1(X)}{f_0(X)}\right).$$

Based on this identity, it is shown that if both the absolute value of the expectation and the variance of Z are small, then the expectation of ν can be approximated by

$$\mathbf{E}(\nu) \sim \frac{(1-\gamma)\log B + \gamma \log A}{\mathbf{E}(Z)},$$

where γ is the probability that H_1 is accepted, i.e. $S_\nu \geq \log A$. Let $\mathbf{E}_i$, $i = 1, 2$ denote the expectation under H_i, and let α be the probability of an error of the first kind (H_1 is accepted when H_0 is true), β be the probability of an error of the second kind (H_0 is accepted when H_1 is true). Then Wald gives the following inequalities for the expectation of ν:

$$\mathbf{E}_0(\nu) \geq \frac{1}{\mathbf{E}_0(Z)}\left((1-\alpha)\log\frac{\beta}{1-\alpha} + \alpha\log\frac{1-\beta}{\alpha}\right),$$

$$\mathbf{E}_1(\nu) \geq \frac{1}{\mathbf{E}_1(Z)}\left(\beta\log\frac{\beta}{1-\alpha} + (1-\beta)\log\frac{1-\beta}{\alpha}\right).$$

The denominators

$$\mathbf{E}_0(Z) = \int f_0(x) \log \frac{f_1(x)}{f_0(x)}\, dx \qquad \text{and} \qquad \mathbf{E}_1(Z) = \int f_1(x) \log \frac{f_1(x)}{f_0(x)}\, dx$$

are the same quantities as introduced and called I-divergence a few years later by S. Kullback (*Information Theory and Statistics,* Wiley, New York (1959)). So this can be considered as the first statistical application of information theory (see the Information Theory section). Further investigations on the expectation $\mathbf{E}(\nu)$ was given by A. Wald {68} and {69}. In subsequent papers {47} and {72} used the sequential method also for estimation problems.

4. Statistical decision functions

In one of his first papers in Statistics {63} he presented some of the main concepts of decision theory developed later in his book (*Statistical Decision Functions,* John Wiley and Sons, 1950). The basic idea can be described as follows.

Assume that experimentations are carried out on a random phenomenon, i.e., we have random observations $X = (X_1, X_2, \dots)$ on a random variable having a distribution function F. Usually F is unknown, but it is assumed to be known that F is a member of a given class Ω of distribution functions. Moreover, there is a space D, called decision space, whose elements d represent the possible decisions that can be made by the statistician in the problem under consideration. Let $W(F, d, x)$ be the loss when F is the true distribution function, the decision d is made and x is the observed value of X. A distance on the space D can be defined by

$$\rho(d_1, d_2) = \sup_{F,x} \left| W(F, d_1, x) - W(F, d_2, x) \right| .$$

A decision function $\delta(x)$ is a function which associates with each x a probability measure on D. Usually, this is a randomized decision function. In the particular case, when $\delta(x)$ for each x assigns the probability one to a single point d in D, the decision function is called nonrandomized. The aim of the statistician is to choose d so that W is in some sense minimized. Practically all statistical problems, including estimation, testing hypotheses, and the design of experiments, can be formulated in this way.

Given the sample point x and given that $\delta(x)$ is the decision function adopted, the expected value of the loss is given by

$$W^*(F, \delta, x) = \int_D W(F, d, x)\, d\delta(x).$$

The function

$$r(F, \delta) = \int_R W^*(F, \delta, x)\, dF(x)$$

is called the risk when F is true and δ is adopted. Wald's fundamental idea consists in considering this risk as the outcome of a zero-sum two-person game played by the statistician against nature. The main theorems refer to conditions under which the decision problem is strictly determined or has a Bayes and/or minimax solution.

In {11} it is shown that when Ω and D are finite and each element of Ω is atomless, then for any decision function $\delta(x)$ there exists an equivalent nonrandomized decision function $\delta^*(x)$, i.e. $r(F, \delta^*) = r(F, \delta)$ for all $F \in \Omega$.

In a series of papers Wald and his collaborators investigated the properties of statistical decision functions.

Wald and Wolfowitz (Characterization of the minimal complete class of decision functions when the number of distributions and decisions is finite, *Proceedings of the Second Berkeley Symposium on Mathematical Statistics and Probability,* University of California Press, Berkeley and Los Angeles (1951), 149–157) consider a statistical decision problem with the space of distributions consisting of a finite number m of distinct probability distributions on a Euclidean space, and the space of decisions being also finite. Then every admissible decision function is a Bayes solution with respect to some a priori probability distribution, but the converse is not true. The concept of a Bayes solution with respect to a sequence $(\xi_i)_{i=1}^h = (\xi_{i1}, \ldots, \xi_{im})_{i=1}^h$ of a priori probability distributions is introduced. This is defined as follows: When $h = 1$ it is a Bayes solution with respect to ξ_1. When $h > 1$ it is a Bayes solution with respect to ξ_h if one restricts oneself only to those decision functions which are Bayes solutions with respect to the sequence $\xi_1, \ldots, \xi_{h-1}$. The main result of the paper is the following: A decision function is admissible if and only if it is a Bayes solution with respect to a sequence of $h \leq m$ a priori probability distributions $\xi_1, \ldots, \xi_h$ such that $\sum_{i=1}^h \xi_{ij} > 0$ for $j = 1, \ldots, m$. The proof involves a rather elaborate study of the intersections of a convex body with its supporting planes.

5. Asymptotic theory of testing and estimation

In a series of papers Wald worked out a theory on the asymptotic properties of tests and estimations. In {64} Wald showed that, under certain regularity conditions, the test based on maximum likelihood estimation is asymptotically most powerful and gives some examples for the most powerful tests. The connection between most powerful tests and shortest confidence intervals is treated in {65}.

Wald's asymptotic theory was developed in {66} generalizing and extending his previous works on the subject. He considered random vectors and multidimensional parameter space. The main feature of this paper is to reduce the general problem to the normal case and show optimum properties for the normal distribution. The general model is as follows: Let

$$f(x_1, x_2, \ldots, x_r; \theta_1, \ldots, \theta_k)$$

be a density function involving k unknown parameters $\theta_1, \ldots, \theta_k$ lying in a parameter space Ω. For a subset ω of Ω denote by H_ω the hypothesis that the parameter point lies in ω. Consider independent observations $X_1, X_2, \ldots, X_n$, where each X_i is a vector in the r-dimensional space and has density f. The maximum likelihood estimator $(\hat{\theta}_{1n}, \ldots, \hat{\theta}_{kn})$ of the parameters is the values of $\theta_1, \ldots, \theta_k$ for which $\prod_{i=1}^n f(X_i; \theta_1, \ldots, \theta_k)$ becomes a maximum. Wald considers asymptotic properties of tests constructed by the help of maximum likelihood estimators. He introduced the idea of most stringent test, which can be described briefly as follows (cf. {73}): define the envelope power function of a family of tests as the supremum at each parameter point of the powers of the tests and define the "shortcoming" of a test at a parameter point as the amount by which the power of the test falls short of the envelope power there. We may then define the maximum shortcoming of the test as the supremum over the parameter values of its shortcoming. A sequence of tests is asymptotically most stringent if the maximum amount by which its maximum stringency can be reduced tends to zero as n increases. Note that an asymptotically most powerful test is asymptotically most stringent.

Concerning estimation problems in {70}, Wald studies the asymptotic properties of maximum likelihood estimators in the case of stochastically dependent observations. Let (X_i), $i = 1, 2, \ldots$, be a sequence of random variables. It is assumed that for any n the first n variables admit a joint probability density function $f(x_1, \ldots, x_n; \theta)$ involving an unknown parameter θ. It is shown that under certain restrictions on the joint probability distribution the maximum likelihood equation has at least one root which is a consistent estimate of θ, and any root of the maximum likelihood equation which is a consistent estimate of θ is shown to be asymptotically efficient. Therefore the consistency of the maximum likelihood estimate implies its asymptotic efficiency, since this estimate is always a root of the maximum likelihood equation.

6. Randomness

It is an important problem in Statistics to test whether a sequence $(X_1, X_2, \ldots, X_N)$ of variables is a random one, i.e. they are independent and identically distributed (abbreviated i.i.d.). Tests for randomness are important in the analysis of time series, its investigations usually based on serial correlation coefficient

$$R_h = \frac{\sum_{i=1}^N X_i X_{h+i} - \left(\sum_{i=1}^N X_i\right)^2/N}{\sum_{i=1}^N X_i^2 - \left(\sum_{i=1}^N X_i\right)^2/N}.$$

Here h is a given positive integer and for $h+i > N$ the term X_{h+i} is to be replaced by X_{h+i-N}.

Wald and Wolfowitz in {75} proposed the following procedure: Let a_i be the observed value of X_i, $i = 1, \ldots, N$. Consider the subpopulation where the set $(X_1, \ldots, X_N)$ is restricted to permutations of $a_1, \ldots, a_N$ and for any particular permutation assign the probability $1/N!$ This determines the probability distribution of R_h in the subpopulation. They propose a randomness test based on this distribution. It is shown moreover that under some mild restrictions, the limiting distribution is normal. Further investigations for tests based on permutations of the observations is given in {76}. They show that under some conditions the weighted sum

$$L_N = \sum_{i=1}^N d_i X_i$$

has normal limiting distribution. As consequences of this result, they conclude that a number of statistics such as rank correlation coefficient, Pitman's two-sample statistics, Hotelling's generalized T statistics, etc. are asymptotically normal.

Related statistics were investigated by J. von Neumann with R. H. Kent, H. R. Bellinson and B. I. Hart (The mean square successive difference, *Ann. Math. Statist.*, **12** (1941), 153–162). Minimizing the effect of the trend on dispersion they considered

$$\delta^2 = \frac{\sum_{i=1}^N (X_i - X_{i+1})^2}{N-1}$$

as a competitor of the sample variance

$$s^2 = \frac{\sum_{i=1}^N (X_i - \overline{X})^2}{N},$$

where

$$\overline{X} = \frac{\sum_{i=1}^N X_i}{N}.$$

It was shown that if the X_i are independent normal with mean μ and variance σ^2, then the density of δ^2 is given by

$$p(\delta^2) = \frac{1}{\sigma^2\sqrt{3}} e^{-2\delta^2/3\sigma^2} J_0\left(\frac{i\delta^2}{3\sigma^2}\right),$$

where J_0 is the Bessel function of order zero.

It was also shown that the efficiency of δ^2 compared to s^2, the best estimation of σ^2, is $2(n-1)/(3n-4)$. The advantage of using δ^2 instead of s^2, is that it is robust in the sense that it has small effect when the mean of the observation is not constant, but can be changed in time. The ratio $\eta = \delta^2/s^2$ can also be used in testing randomness and this was investigated in further papers of J. von Neumann, see {61} and {62}.

7. Nonparametric tests, order statistics

Consider a random sample

$$(X_1, X_2, \dots, X_n)$$

of size n, coming from a population with (theoretical) distribution function $F(x) = \mathbf{P}(X_1 < x)$. In this context, the above random variables are independent and identically distributed. The order statistics are the rearrangement of sample elements according to their magnitude:

$$X_1^* \le X_2^* \le \cdots \le X_n^*.$$

The empirical or sample distribution function is defined by

$$F_n(x) = \frac{1}{n}\sum_{i=1}^{n} \mathbf{I}\{X_i < x\} = \begin{cases} 0, & \text{if } x \le X_1^*, \\ \frac{k}{n}, & \text{if } X_k^* < x \le X_{k+1}^*, \\ 1, & \text{if } X_n^* < x. \end{cases}$$

Here $\mathbf{I}\{A\}$ stands for the indicator of the event A. Order statistics and empirical distribution functions are widely used in statistics, nonparametric statistics, in particular. Basic results are due to V. Glivenko {13} and F. P. Cantelli {5}: with probability one

$$\lim_{n\to\infty} D_n = 0,$$

where

$$D_n = \sup_{x\in R} \left| F_n(x) - F(x) \right|.$$

This theorem expresses the important fact that with probability one, the empirical distribution tends uniformly to the theoretical distribution. Hence, it can be effectively used for goodness of fit problems, i.e., to test whether a sample comes from a population with given distribution. This test is applicable thanks to a result of A. N. Kolmogorov {32} who determined the limiting distribution of D_n:

$$\lim_{n\to\infty} \mathbf{P}\big(\sqrt{n} D_n < y\big) = \sum_{k=-\infty}^{\infty} (-1)^k e^{-2k^2y^2}, \qquad y > 0,$$

provided $F(x)$ is continuous. Later N. V. Smirnov {46} proved a one-sided version:

$$\lim_{n\to\infty} \mathbf{P}\big(\sqrt{n} D_n^+ < y\big) = \lim_{n\to\infty} \mathbf{P}\big(\sqrt{n} D_n^- < y\big) = 1 - e^{-2y^2}, \qquad y > 0,$$

where

$$D_n^+ = \sup_{x\in R} \big(F_n(x) - F(x)\big), \quad D_n^- = \sup_{x\in R} \big(F(x) - F_n(x)\big).$$

Note that in the above theorems the distributions of D_n, $D_n^{\pm}$ do not depend on the underlying distribution function $F(x)$.

In his fundamental paper, dedicated to Kolmogorov's fiftieth birthday, A. Rényi {38} proposed to modify the above statistics by considering the "relative error" of the empirical distribution. He defined the following statistics:

$$R_n^+(a) = \sup_{a \le F(x)} \frac{F_n(x) - F(x)}{F(x)}$$

and

$$R_n(a) = \sup_{a \le F(x)} \frac{\bigl|F_n(x) - F(x)\bigr|}{F(x)}.$$

Now these are called Rényi statistics in the literature. In the said paper, Rényi determined the limiting distributions:

$$\text{(1)} \qquad \lim_{n\to\infty} \mathbf{P}\bigl(\sqrt{n}R_n^+(a) < y\bigr) = \sqrt{\frac{2}{\pi}} \int_0^{y\sqrt{\frac{a}{1-a}}} e^{-t^2/2}\, dt, \qquad y > 0,$$

$$\text{(2)} \qquad \lim_{n\to\infty} \mathbf{P}\bigl(\sqrt{n}R_n(a) < y\bigr) = \frac{4}{\pi} \sum_{k=0}^{\infty} \frac{(-1)^k}{2k+1} e^{-\frac{(1-a)\pi^2(2k+1)^2}{8ay^2}}, \qquad y > 0.$$

Rényi considered also the more general statistics

$$R_n^+(a,b) = \sup_{a \le F(x) \le b} \frac{F_n(x) - F(x)}{F(x)}$$

and

$$R_n(a,b) = \sup_{a \le F(x) \le b} \frac{\bigl|F_n(x) - F(x)\bigr|}{F(x)}.$$

The proofs of (1) and (2) given by Rényi are based on his method presented in the same paper. Since, as remarked above, the statistics are distribution free, it is no loss of generality assuming that the sample comes from exponential distribution, i.e.

$$\text{(3)} \qquad P(X_i < x) = 1 - e^{-x}, \qquad x > 0.$$

Rényi proves that in this case the variables X_k^* can be expressed in the form

$$\text{(4)} \qquad X_k^* = \frac{\delta_1}{n} + \frac{\delta_2}{n-1} + \cdots + \frac{\delta_k}{n-k+1}, \qquad k = 1, 2, \ldots, n,$$

where the variables $\delta_1, \delta_2, \ldots, \delta_n$ are independent having exponential distribution as in (3). Based on this simple fact, with some clever manipulations, Rényi derives the above limiting distributions and also some other limit distributions such as the asymptotic normality of the sample quantile.

{22} based on (4), present elementary and simple method to derive certain distributions and conditional distributions concerning order statistics.

Another ingenious method concerning empirical distributions was developed by Lajos Takács, based on his celebrated ballot theorem. The classical ballot theorem due to M. J. Bertrand {3}, V. André {1} and É. Barbier {2} says that if in a ballot one candidate scores a votes, the other candidate scores b votes and $a \geq \mu b$ with a non-negative integer μ, then the probability that throughout the counting the number of votes registered for the first candidate is always greater than μ times the number of votes registered for the second candidate is given by

$$\mathbf{P} = \frac{a - \mu b}{a + b}.$$

L. Takács in a series of papers {49} and {50} etc. and in his book {51} presented a generalization of the ballot theorem and applied it in various problems, such as empirical distribution functions, queueing, dams, etc.

Let $\chi_1, \chi_2, \ldots, \chi_n$ be non-negative, cyclically interchangeable random variables and let $\tau_1 < \tau_2 < \cdots < \tau_n$ be the order statistics of a random sample, uniformly distributed on the interval $(0, t)$, and assume also that $\{\chi_r\}$ and $\{\tau_r\}$ are independent. Define

$$\chi(u) = \sum_{\tau_r \leq u} \chi_r, \qquad 0 \leq u \leq t.$$

Then

$$\mathbf{P}\big(\chi(u) \leq u,\ 0 \leq u \leq t \mid \chi(t) = y\big) = 1 - \frac{y}{t}, \qquad 0 \leq y \leq t.$$

Based on this extension of the ballot theorem, Takács (An application of a ballot theorem in order statistics, *Ann. Math. Statist.*, **35** (1964), 1356–1358) derives the exact distributions of the statistics

$$T_n^+(a, b, c) = \sup_{a \leq F(u) \leq b} \big(F_n(u) - cF(u)\big)$$

and

$$R_n^+(a,b,c) = \sup_{a \le F(u) \le b} \left(\frac{F_n(u) - cF(u)}{F(u)} \right).$$

Takács (On the comparison of a theoretical and an empirical distribution function, *J. Appl. Probab.*, **8** (1971), 321–330) proved a ballot-type theorem equivalent to the following: Let $X_1, \ldots, X_n$ be i.i.d. random variables with distribution $P(X_1 = i) = q$, $i = 1, \ldots, n$, $P(X_1 = n+1) = p$, where $p + nq = 1$, $0 < p < 1$. Let $X_1^* \le X_2^* \le \cdots \le X_n^*$ be its order statistics. For $1 \le l \le n$, let A denote the event: There exist at least l distinct positive integers $k_1, k_2, \ldots, k_l$ for which $X_{k_i}^* = k_i$ $(i = 1, 2, \ldots, l)$. Then

$$\mathbf{P}(A) = q^l n(n-1)\ldots(n-l+1).$$

Based on this result, the distribution of the number of intersections of $cF(x)$ with $F_n(x) + a/m$ was determined. K. Sarkadi {45} presented a simple elegant combinatorial proof of this theorem.

In the two-sample case B. V. Gnedenko and V. S. Korolyuk {14} developed a method based on random walk models. Let

$$(X_1, X_2, \ldots, X_m) \quad \text{and} \quad (Y_1, Y_2, \ldots, Y_n)$$

be two samples coming from continuous distributions. Let $F(x)$ and $G(x)$, resp. be their theoretical distribution functions and let $F_m(x)$ and $G_n(x)$, resp. be their empirical distribution functions. Testing the null hypothesis $H_0 : F(x) = G(x)$, a number of statistics has been investigated and their distributions, limiting distributions and other characteristics have been determined in the statistical literature. The idea of Gnedenko and Korolyuk was as follows: let

$$Z_1^* < Z_2^* < \cdots < Z_{m+n}^*$$

denote the order statistics of the union of the two samples and define

$$\theta_i = \begin{cases} +1 & \text{if } Z_i^* = X_j \text{ for some } j, \\ -1 & \text{if } Z_i^* = Y_j \text{ for some } j \end{cases} \tag{5}$$

$i = 1, 2, \ldots, m+n$. Put

$$S_0 = 0, \quad S_i = \theta_1 + \ldots \theta_i, \quad i = 1, 2, \ldots, 2n.$$

Then $(S_0, S_1, \ldots, S_{m+n})$ is a random walk path with $S_{m+n} = m - n$ and under H_0 each of them has the same probability. This idea of Gnedenko and

Korolyuk enables one to determine the distributions of certain statistics by reducing the problems to combinatorial enumeration.

In a series of papers Vincze and his collaborators presented a number of results in this subject. His first result concerns the joint distribution of the maximum and its location in the case $m = n$: Let $x_0^{(n)}$ be the first point where $F_n(x) - G_n(x)$ takes its (one-sided) maximum for the first time. Then, under H_0, I. Vincze {54} showed that

$$\mathbf{P}\left(\max_{-\infty<x<\infty}(F_n(x) - G_n(x)) = \frac{k}{n},\ \frac{1}{2}\left(F_n\left(x_0^{(n)}\right) + G_n\left(x_0^{(n)}\right) = \frac{r}{2n}\right)\right.$$

$$= \mathbf{P}\left(\max_{1\le i\le r-1} S_i < k,\ S_r = k,\ \max_{r+1\le i\le 2n} S_i \le k\right)$$

$$= \frac{k(k+1)}{r(2n-r+1)} \frac{\binom{r}{\frac{r+k}{2}}\binom{2n-r+1}{n-\frac{r+k}{2}}}{\binom{2n}{n}},$$

$k = 1, 2, \dots, n$; $r = k, k+2, \dots, 2n-k$.

$$\lim_{n\to\infty} \mathbf{P}\left(\sqrt{\frac{n}{2}} \max_{-\infty<x<\infty}(F_n(x) - G_n(x)) < y,\ \frac{1}{2}\left(F_n\left(x_0^{(n)}\right) + G_n\left(x_0^{(n)}\right) < z\right)\right.$$

$$= \sqrt{\frac{2}{\pi}} \int_0^y \int_0^z \frac{u^2}{\big(v(1-v)\big)^{3/2}} \exp\left(-\frac{u^2}{v(1-v)}\right) du\, dv.$$

Similar results were given for the absolute maximum and its location.

Using his extension of the ballot theorem, L. Takács {52} gives joint distributions of the maximum and its location for different sample sizes. In the case when m divides n, Takács gives the joint exact distribution of the statistics

$$T^-(m,n) = \sup_{x\in R}\left(G_n(x) - F_m(x)\right) = \max_{1\le r\le n}\left(G_n(Y_r^*) - F_m(Y_r^*)\right),$$

and of $\rho^-(m,n)$, the smallest $1 \le r \le n$ for which the maximum is attained.

In a subsequent paper {55} I. Vincze proposed to use generating functions to determine distributions and joint distributions. He determined, e.g., the generating function of the above joint distribution.

Further results in this topic (in the case $m = n$):

Let

$$(X_1^* < X_2^* < \cdots < X_n^*), \qquad (Y_1^* < Y_2^* < \cdots < Y_n^*)$$

denote the ordered samples. Then

$$\gamma_n = \sum_{i=1}^{n} \mathbf{I}\{X_i^* > Y_i^*\}$$

is the so-called Galton statistics. For random walk paths defined above, γ_n is the number of i's such that $S_{2i-1} > 0$, $i = 1, \ldots, n$. K. L. Chung and W. Feller {6} showed that γ_n is uniformly distributed, i.e.,

$$\mathbf{P}(\gamma_n = g) = \frac{1}{n+1}, \qquad g = 0, 1, 2, \ldots, n.$$

The proof of Chung and Feller was based on generating function, while {40} gave a combinatorial proof by showing that there exists a bijection between random walk paths with $\gamma_n = 0$ and $\gamma_n = g$.

E. Csáki and I. Vincze in {8} considered the number of times the random walk crosses zero (number of intersections):

$$\lambda_n = \sum_{i=1}^{n-1} \mathbf{I}\{S_i = 0, \; S_{i-1}S_{i+1} < 0\}$$

and showed

$$\mathbf{P}(\lambda_n = \ell - 1) = \frac{2\ell}{n} \frac{\binom{2n}{n-\ell}}{\binom{2n}{n}}, \qquad \ell = 1, 2, \ldots, n.$$

The joint exact and limiting distribution of (γ_n, λ_n) was also given:

$$\mathbf{P}(\gamma_n = g, \; \lambda_n = \ell - 1)$$

$$= \frac{1}{\binom{2n}{n}} \frac{\ell^2}{2g(n-g)} \binom{2g}{g - \ell/2} \binom{2n-2g}{n - g - \ell/2}$$

for ℓ even. A similar result was given for ℓ odd. For the limiting distribution it was shown that

$$\lim_{n\to\infty} \mathbf{P}\big(\gamma_n \le zn, \lambda_n \le y\sqrt{2n}\,\big)$$

$$= \sqrt{\frac{2}{\pi}} \int_0^y \int_0^z \frac{u^2}{\big(v(1-v)\big)^{3/2}} \exp\left(-\frac{u^2}{2v(1-v)}\right) du\, dv.$$

Another use of the generating function method is found in {9} where the joint distribution of the maximum and the number of intersections was given in the form

$$\sum_{n=1}^{\infty}\binom{2n}{n}\mathbf{P}\left(\max_x \bigl|F_n(x)-G_n(x)\bigr| = \frac{k}{n},\ \lambda_n = \ell-1\right)z^n$$

$$= 2\left(\frac{w-w^k}{1-w^{k+1}}\right)^{\ell}, \qquad \ell, k = 1, 2, \ldots,$$

where

$$w = \frac{1-\sqrt{1-4z}}{1+\sqrt{1-4z}}, \qquad |z| < \frac{1}{4}.$$

Vincze's idea in determining joint distributions was to construct tests based on a pair of statistics (instead of one single statistic) in order to improve the power of the tests. For details see {56}. This idea however deserves further investigations even today.

In {37} the two-sample problem is treated for different sample sizes by investigating

$$\max_x \bigl(nF_n(x) - mG_m(x)\bigr)$$

and related quantities.

The power of the Kolmogorov–Smirnov two-sample test is treated in {57}. In {58} the analogues of Gnedenko–Korolyuk distribution is given both for discontinuous random variables and for the two-dimensional case.

K. Sarkadi in {43}, by using the well-known inclusion-exclusion principle in combinatorics, gives an alternative method of deriving the exact distribution of the Kolmogorov–Smirnov statistics for both the one-sample and the two-sample cases.

In the two-sample case an important contribution was made by A. Wald and J. Wolfowitz {74}, who constructed a test based on the number of runs. Consider two samples $(X_1, X_2, \ldots, X_m)$ and $(Y_1, Y_2, \ldots, Y_n)$ as before, and the variables θ_i defined by (5). A subsequence $\theta_{s+1}, \theta_{s+2}, \ldots, \theta_{s+r}$ is called a run, if $\theta_{s+1} = \theta_{s+2} = \ldots \theta_{s+r}$ but $\theta_s \neq \theta_{s+1}$ when $s > 0$ and $\theta_{s+r} \neq \theta_{s+r+1}$ when $s + r < m + n$. Let U be the number of runs in the sequence $(\theta_1, \theta_2, \ldots, \theta_{m+n})$. The exact distribution, mean and variance of

U under the null hypothesis $F(x) = G(x)$ was given for continuous F and it was shown that U is asymptotically normal with mean and variance

$$E(U) = \frac{2mn}{m+n} + 1,$$

$$\mathrm{Var}\,(U) = \frac{2mn(2mn - m - n)}{(m+n)^2(m+n-1)}.$$

Hence using either the exact (for small sample sizes) or the asymptotic (for large sample sizes) distribution, a test can be constructed with critical region $U < u_0$, so that $\mathbf{P}(U < u_0) = \beta$, where β is a predetermined level of significance. In other words the null hypothesis $H_0 : F(x) = G(x)$ is rejected if the number of runs in the combined sample is too small. Wald and Wolfowitz have also shown that the test is consistent against any alternatives $F(x) \neq G(x)$.

Z. W. Birnbaum and I. Vincze {4} proposed a test based on order statistics, which can replace Student's t test. Let $X_1, \ldots, X_n$ be a random sample from a population with continuous distribution function $F(x)$. Let $X_1^* < X_2^* < \ldots X_n^*$ be their order statistics. For a given $0 < q < 1$ the q-quantile is defined by

$$\mu_q = \inf\left\{x : F(x) = q\right\}$$

and the corresponding sample quantile is defined as the order statistic X_k^* such that

$$\left|\frac{k}{n} - q\right| \leq \frac{1}{2n}.$$

Consider the statistic

$$S_{n,k,r,s} = \frac{X_k^* - \mu_q}{X_{k+s}^* - X_{k-r}^*}$$

that can be used for testing the location parameter when the scale parameter is unknown for a general distribution. Exact and limiting distributions are derived for this statistic under some mild conditions on the distribution function F.

B. Gyires in {20} investigated asymptotic results for linear rank statistic defined as

$$S = \sum_{j=1}^{m} f_j\left(x_{R(X_j)}^{(j)}\right),$$

where $X_1, \ldots, X_n$ are i.i.d. continuous random variables, $R(X_j)$ denotes the rank of X_j, $N = m + n$, $x_i^{(j)}$ $(i = 1, \ldots, N,\ j = 1, \ldots, m)$ are real numbers in $(0, 1)$, and f_j are continuous functions on $[0, 1]$ with bounded variation. Let

$$V = \sum_{j=1}^{m} f_j(\eta_j),$$

where the η_j are i.i.d. uniform $(0, 1)$ random variables. An upper bound of

$$\left|\varphi(t) - N^m \prod_{j=1}^{m} (n + j)^{-1} \varphi_V(t)\right|$$

is given, where $\varphi(t)$ and $\varphi_V(t)$ are the characteristic functions of S and V, respectively. The bound is then exploited to prove that, as $n \to \infty$ with m remaining fixed, S converges weakly to V if and only if the discrepancy of the sequence $\left(x_i^{(j)}\right)_{i=1}^{n}$ from what is called a uniform sequence tends to zero. Application of this result to certain two-sample rank tests are also given.

Further results on asymptotic properties for linear rank and order statistics can be found in {16}, {17}, {19} and {21}. In these papers Gyires gives a necessary and sufficient condition for linear order statistics to have a limit distribution and he studies the case when the limit distribution is normal in particular. Limit distributions are also given for linear order statistics in the case when the observations are not necessarily independent. A doubly ordered linear rank statistic is also investigated. The methods employed by Gyires uses matrix theory, in particular Gábor Szegő's result concerning the eigenvalues of Toeplitz and Hankel matrices. For further comments in this regard we refer to the Section on Probability Theory.

8. Goodness of fit tests

An important problem in Mathematical Statistics is to test whether a random sample comes from a well-defined family of distributions. E.g., tests for normality or other goodness of fit tests are aimed to decide whether a sample comes from normal, or other distributions usually involving nuisance parameters, i.e., we are faced with a composite hypothesis. The most commonly used goodness of fit tests are Pearson's χ^2-tests. In the case of simple hypothesis $H_0 : F(x) = F_0(x)$ with given F_0 this is based on the statistics

$$\chi^2 = \sum_{i=1}^{k} \frac{(\nu_i - Np_i)^2}{Np_i},$$

where the range of the variable is divided into a number k of class intervals, N is the sample size, ν_i stands for the number of sample elements in ith class and p_i is the probability that a sample element falls into the ith class. H. B. Mann and A. Wald {34} investigated the problem of optimal choice of class intervals. They show that

$$k = k_N = 4 \left(\frac{2(N-1)^2}{c^2} \right)^{1/5}$$

and $p_i = 1/k$, $i = 1, \dots, k$ is in certain sense optimal, where c is a constant depending on the probability of the critical region.

E. Csáki and I. Vincze in {10} proposed a modification of the Pearson's χ^2-statistic:

$$\overline{\chi}^2 = \sum_{i=1}^{k} \left(\frac{\overline{X}_{(i)} - E_i}{\sigma_i} \right)^2 \nu_i,$$

where E_i and σ_i^2, resp. are the expectation and variance, resp. of the observations in the ith class and $\overline{X}_{(i)}$ are the mean value of the observations in the ith class. It was shown that (for fixed k) the limiting distribution of $\overline{\chi}^2$ statistic is *chi − square* with k degrees of freedom (instead of $k-1$ degrees of freedom of Pearson's χ^2).

For simple hypotheses the one-sample Kolmogorov–Smirnov type tests discussed in Section 6 are also applicable for goodness of fit problems. In case of composite hypotheses, i.e., when parameters are unknown, a usual procedure is to estimate the parameters and apply a modified χ^2-test. But in some cases this has disadvantages. K. Sarkadi {39}, {41} in the case of normality test, presented a method which reduces the problem of composite hypothesis to a simple one. Assume first that we want to test normality based on the sample $(X_1, \dots, X_n, X_{n+1})$ in the case when the expectation is unknown and the variance is known. Define

$$\overline{Y} = \frac{\overline{X} - X_{n+1}}{\sqrt{n+1}},$$

where

$$\overline{X} = \frac{X_1 + \cdots + X_n}{n}.$$

Put

$$Y_i = \overline{Y} + X_i - \overline{X}, \qquad i = 1, \ldots, n.$$

If $X_1, \ldots, X_{n+1}$ are independent random variables having normal distribution with expectation μ and variance σ^2, then $Y_1, \ldots, Y_n$ are independent random variables, each normally distributed with expectation zero and variance σ^2. This way the normality test with unknown expectation reduces to the normality test with expectation 0.

Similarly, if the variance is unknown and the expectation is known (assuming to be equal to zero without loss of generality), so that $(X_1, \ldots, X_{n+1})$ are i.i.d. mean zero normal random variables with unknown variance, Sarkadi gives the following transformation:

$$Y_i = X_i \frac{s'}{s}, \qquad i = 1, \ldots, n,$$

where

$$s = \sqrt{\frac{X_1^2 + \cdots + X_n^2}{n}},$$

$$s' = \psi_n \left(\frac{|X_{n+1}|}{s} \right),$$

and the function $\psi_n(t)$ is defined by the following relation:

$$\int_{\psi_n^2}^{\infty} u^{(n-1)/2} \exp\left(-u/2\right) du = \frac{2^{n/2+1} \Gamma\left(\frac{n+1}{2}\right)}{\sqrt{n\pi}} \int_{-\infty}^{t} \left(1 + \frac{u^2}{n}\right)^{-(n+1)/2} du.$$

It is shown that $(Y_1, \ldots, Y_n)$ are independent standard normal variables. Hence testing normality in the case of composite hypothesis is reduced to that of simple hypothesis.

Similarly, if $(X_1, X_2, \ldots, X_{n+2})$ are independent random variables each having normal distribution with expectation μ and variance σ^2, Sarkadi gives a transformation based on this sample, resulting in $(Y_1, Y_2, \ldots, Y_n)$, independent standard normal variables. The advantage of Sarkadi's transformation is that random numbers are avoided and he also shows that the transformation is optimal in some sense.

Sarkadi (The asymptotic distribution of certain goodness of fit test statistics, *Lecture Notes in Statistics* **8**, Springer, New York (1981), 245–253) investigated goodness of fit statistics of the form

$$W_n = \frac{\left(\sum_{i=1}^n a_{in} X_i^*\right)^2}{\sum_{i=1}^n \left(X_i - \overline{X}\right)^2},$$

where $X_1^* < \cdots < X_n^*$ are order statistics, $a_{1n}, \ldots, a_{nn}$ are appropriately chosen constants, and $\overline{X}$ is the sample mean. Sufficient conditions are given for W_n to have asymptotically normal distribution. It is shown that many statistics proposed for testing goodness of fit are of the above type with different values of a_{in}. Asymptotic properties of these tests are discussed and some of the tests are shown to be inconsistent for specific alternatives. In the case when $a_{in} = m_{in}/\left(\sum_{j=1}^n m_{jn}^2\right)^{1/2}$, with $m_{in} = E(X_i^*)$, this is the Shapiro–Francia test for which K. Sarkadi {44} proved consistency.

In {33} a goodness of fit test is proposed for testing uniformity. The test statistic is

$$J = n^2 \sum_{i=1}^n d_i^2 - n\left(\sum_{i=1}^n d_i\right)^2,$$

where $d_i = \left(X_i^* - i/(n+1)\right)/i(n-i+1)$ and the X_i^* are order statistics from a sample of size n. The Monte Carlo method is used to compare the test with some competitors.

9. Cramér–Fréchet–Rao inequality

Let $X = (X_1, X_2, \ldots, X_n)$ be a sample from a distribution having (joint) density $p(x;\theta) = p(x_1, x_2, \ldots, x_n; \theta)$ with respect to a measure μ, where θ is a parameter and we want to estimate its function $g(\theta)$. Let $t(X)$ be an unbiased estimator of $g(\theta)$, i.e. $\mathbf{E}_\theta\left(t(X)\right) = g(\theta)$. M. Fréchet {12}, C. R. Rao {36} and H. Cramér {7} gave the following inequality:

$$\operatorname{Var}_\theta\left(t(X)\right) \geq \frac{\left(g'(\theta)\right)^2}{I(\theta)},$$

with

$$I(\theta) = \int \left(\frac{\partial p}{\partial \theta}\right)^2 p(x;\theta)\, dx.$$

I. Vincze {59} and {60} for fixed θ, θ' considered the mixture

$$p_\alpha = p_\alpha(x;\theta,\theta') = (1-\alpha)p(x;\theta) + \alpha p(x;\theta'), \quad 0<\alpha<1$$

with α being a new parameter. Then

$$\hat{\alpha} = \frac{t(X)-g(\theta)}{g(\theta')-g(\theta)}$$

is an unbiased estimator of α.

It follows that

$$\operatorname{Var}_\alpha(\hat{\alpha}) \geq \frac{1}{J_\alpha(\theta,\theta')},$$

where

$$J_\alpha(\theta,\theta') = \int \frac{\big(p(x;\theta')-p(x;\theta)\big)^2}{p_\alpha(x;\theta,\theta')}\, d\mu.$$

Then

$$(1-\alpha)\operatorname{Var}_\theta\big(t(X)\big) + \alpha \operatorname{Var}_{\theta'}\big(t(X)\big) \geq \frac{1}{J_\alpha(\theta,\theta')} - \alpha(1-\alpha)$$

and in the case when $\operatorname{Var}_\theta\big(t(x)\big)$ does not depend on θ, Vincze concluded the following lower bound:

$$\operatorname{Var}\big(t(X)\big) \geq \sup_\alpha \sup_{\theta'} \alpha(1-\alpha)\big(g(\theta')-g(\theta)\big)^2 \left(\frac{1}{\alpha(1-\alpha)J_\alpha} - 1\right).$$

In certain cases this gives a reasonably good bound. This problem was further investigated by M. L. Puri and I. Vincze {35} and Z. Govindarajulu and I. Vincze {15}. It was shown among others that for the translation parameter of the uniform distribution this lower bound is of order n^{-2}, which is attainable.

10. Estimation problems

An interesting estimation problem is treated in {71}. Let $X_1, X_2, \ldots$ be an infinite sequence of random variables, such that for each n the variables $X_1, \ldots, X_n$ admit a continuous joint probability density $f_n(x_1, \ldots, x_n \mid \theta, \xi_1, \ldots, \xi_n)$, where $\theta, \xi_1, \ldots, \xi_n$ are unknown parameters, all of which are restricted to finite intervals. Then $t_n(X_1, \ldots, X_n)$ is said to be a uniformly consistent estimate of θ if $P\left(|t_n - \theta| < \delta\right) \to 1$ as $n \to \infty$, for any $\delta > 0$, uniformly in θ and the ξ's. Necessary and sufficient conditions for the existence of a uniformly consistent estimate are given. An information function is defined for the present case, and a sufficient condition for the nonexistence of a uniformly consistent estimate is given in terms of the information function. In the particular case when the X_i are independent, the total information contained in the first n observations is equal to the sum of the amounts of information contained in each observation separately.

Another interesting estimation problem is treated by K. Sarkadi. In statistics the following selection procedure often occurs. Let $\mu_1, \ldots, \mu_n$ be parameters characterizing different populations. The parameters are unknown but we know their unbiased estimators $X_1, X_2, \ldots, X_n$, i.e., $E(X_i) = \mu_i$. Given these estimators, one population is selected according to some predetermined decision rule. Suppose, e.g., the population of the lowest value X_i is selected, because this proves to be the highest quality among the possible choices. In this case $\min_i X_i$ as the estimator of the parameter of the corresponding population is obviously biased. This problem was treated by K. Sarkadi {42} who proved that though no unbiased estimation with finite variance exists in general, he suggests randomized estimations with arbitrarily small bias. The variance of the estimator however tends to infinity when the bias tends to zero.

References

[78] Jordán, Károly, *Matematikai Statisztika (Mathematical Statistics),* Természet és Technika, Volume 4, Athenaeum (Budapest, 1927, in Hungarian), and *Statistique mathématique,* Gauthier-Villars (Paris, 1927, in French).

[192] Wald, Ábrahám, *Sequential Analysis,* John Wiley and Sons (New York) – Chapman and Hall (London, 1947).

{1} V. André, Solution directe du problème résolu par M. J. Bertrand, *C.R. Acad. Sci. Paris,* **105** (1887), 436–437.

{2} É. Barbier, Généralisation du problème résolu par M. J. Bertrand, *C.R. Acad. Sci. Paris,* **105** (1887), 407.

{3} M. J. Bertrand, Solution d'un problème, *C.R. Acad. Sci. Paris,* **105** (1887), 369.

{4} Z. W. Birnbaum and I. Vincze, Limiting distributions of statistics similar to Student's t, *Ann. Statist.,* **1** (1973), 958–963.

{5} F. P. Cantelli, Sulla determinazione empirica delle leggi di probabilità, *Giorn. Ist. Ital. Attuari,* **4** (1933), 421–424.

{6} K. L. Chung and W. Feller, Fluctuations in coin tossing, *Proc. Nat. Acad. Sci. USA,* **35** (1949), 605–608.

{7} H. Cramér, *Mathematical Methods of Statistics,* Princeton University Press (Princeton, 1946).

{8} E. Csáki and I. Vincze, On some problems connected with the Galton test, *Publications of the Math. Inst. Hung. Acad. Sci.,* **6** (1961), 97–109.

{9} E. Csáki and I. Vincze, Two joint distribution laws in the theory of order statistics, *Mathematica, Cluj* **5** (1963), 27–37.

{10} E. Csáki and I. Vincze, On limiting distribution laws of statistics analogous to Pearson's chi-square, *Mathematische Operationsforschung und Statistik, Ser. Statistics,* **9** (1978), 531–548.

{11} A. Dvoretzky, A. Wald and J. Wolfowitz, Elimination of randomization in certain statistical decision procedures and zero-sum two-person games, *Ann. Math. Statist.,* **22** (1951), 1–21.

{12} M. Fréchet, Sur l'extension de certaines évaluations statistiques au cas de petits échantillons, *Rev. Inst. Internat. Statist.,* **11** (1943), 182–205.

{13} V. Glivenko, Sulla determinazione empirica delle leggi di probabilità, *Giorn. Ist. Ital. Attuari,* **4** (1933), 92–99.

{14} B. V. Gnedenko and V. S. Korolyuk, On the maximum discrepancy between two empirical distribution functions, *Dokl. Akad. Nauk SSSR,* **80** (1951), 525–528; English translation: *Selected Transl. Math. Statist. Probab., Amer. Math. Soc.,* **1** (1951), 13–16.

{15} Z. Govindarajulu and I. Vincze, The Cramér–Fréchet–Rao inequality for sequential estimation in non-regular case, *Statistical Data Analysis and Inference* (Ed. Y. Dodge), North Holland (Amsterdam, 1989), 257–268.

{16} B. Gyires, On limit distribution theorems of linear order statistics, *Publ. Math. Debrecen,* **21** (1974), 95–112.

{17} B. Gyires, Linear order statistics in the case of samples with non-independent elements, *Publ. Math. Debrecen,* **22** (1975), 47–63.

{18} B. Gyires, Jordan Károly élete és munkássága, *Alkalmazott Matematikai Lapok,* **1** (1975), 274–298.

{19} B. Gyires, Normal limit-distributed linear order statistics, *Sankhyā*, Ser. A **39** (1977), 11–20.

{20} B. Gyires, Linear rank statistics generated by uniformly distributed sequences, *Colloq. Math. Soc. János Bolyai* **32**, North-Holland (Amsterdam, 1982), 391–400.

{21} B. Gyires, Doubly ordered linear rank statistics, *Acta Math. Acad. Sci. Hungar.*, **40** (1982), 55–63.

{22} G. Hajós and A. Rényi, Elementary proofs of some basic facts in the theory of order statistics, *Acta Math. Acad. Sci. Hung.*, **5** (1954), 1–6.

{23} K. Jordan, A valószínűség a tudományban és az életben (Probability in science and life, in Hungarian), *Természettudományi Közlöny,* **53** (1921), 337–349.

{24} K. Jordan, On probability, *Proceedings of the Physico-Mathematical Society of Japan,* **7** (1925), 96–109.

{25} K. Jordan, A valószínűségszámítás alapfogalmai (Fundamental concepts of the theory of probability) *Mathematikai és Physikai Lapok,* **34** (1927), 109–136.

{26} K. Jordan, Sur une formule d'interpolation *Atti del Congresso Internazionale dei Matematici,* Bologna, Vol. **6** (1928), 157–177.

{27} C. Jordan, Approximation and graduation according to the principle of least squares by orthogonal polynomials, *Ann. Math. Statist.*, **3** (1932), 257–357.

{28} Ch. Jordan, Le théorème de probabilité de Poincaré, généralisé au cas de plusieurs variables indépendantes, *Acta Scientiarum Mathematicarum (Szeged),* **7** (1934–35), 103–111.

{29} Ch. Jordan, Problèmes de la probabilité des épreuves répétées dans le cas général, *Bull. de la Société Mathématique de France,* **67** (1939), 223–242.

{30} K. Jordan, Következtetések statisztikai észlelésekből (Statistical inference, in Hungarian), *Magyar Tud. Akad. Mat. Fiz. Oszt. Közleményei,* **1** (1951), 218–227.

{31} K. Jordan, *Fejezetek a klasszikus valószínűségszámításból,* Akadémiai Kiadó (Budapest, 1956, in Hungarian), *Chapters on the classical calculus of probability* (1972, in English).

{32} A. N. Kolmogorov, Sulla determinazione empirica di una legge di distribuzione, *Giorn. Ist. Ital. Attuari,* **4** (1933), 83–91.

{33} P. Kosik and K. Sarkadi, A new goodness-of-fit test, *Probability Theory and Mathematical Statistics with Applications* (Visegrád, 1985), Reidel (Dordrecht, 1988), 267–272.

{34} H. B. Mann and A. Wald, On the choice of the number of class intervals in the application of the chi square test, *Ann. Math. Statist.*, **13** (1942), 306–317.

{35} M. L. Puri and I. Vincze, On the Cramér–Fréchet–Rao inequality for translation parameter in the case of finite support, *Statistics,* **16** (1985), 495–506.

{36} C. R. Rao, Information and accuracy attainable in the estimation of statistical parameter, *Bull. Calcutta Math. Soc.*, **37** (1945), 81–91.

{37} J. Reimann and I. Vincze, On the comparison of two samples with slightly different sizes, *Publications of Math. Inst. Hung. Acad. Sci.*, **5** (1960), 293–309.

{38} A. Rényi, On the theory of order statistics, *Acta Math. Acad. Sci. Hung.*, **4** (1953), 191–231.

{39} K. Sarkadi, On testing for normality, *Magyar Tud. Akad. Mat. Kutató Int. Közl.*, **5** (1960), 269–275.

{40} K. Sarkadi, On Galton's rank order test, *Magyar Tud. Akad. Mat. Kutató Int. Közl.*, **6** (1961), 127–131.

{41} K. Sarkadi, *Proc. Fifth Berkeley Sympos. Math. Statist. and Probability* **I**, Univ. California Press (Berkeley, Calif., 1967), 373–387.

{42} K. Sarkadi, Estimation after selection, *Studia Sci. Math. Hung.*, **2** (1967), 341–350.

{43} K. Sarkadi, On the exact distributions of statistics of Kolmogorov–Smirnov type, *Collection of articles dedicated to the memory of Alfréd Rényi, II. Period. Math. Hungar.*, **3** (1973), 9–12.

{44} K. Sarkadi, The consistency of the Shapiro–Francia test, *Biometrika,* **62** (1975), 445–450.

{45} K. Sarkadi, A direct proof for a ballot type theorem, *Colloq. Math. Soc. János Bolyai,* **32**, North-Holland (Amsterdam, 1982), 785–794.

{46} N. V. Smirnov, On the empirical distribution function (in Russian), *Mat. Sbornik,* **6 (48)** (1939), 3–26.

{47} C. Stein and A. Wald, Sequential confidence intervals for the mean of a normal distribution with known variance, *Ann. Math. Statist.*, **18** (1947), 427–433.

{48} L. Takács, Charles Jordan, 1871–1959, *Ann. Math. Statist.*, **32** (1961), 1–11.

{49} L. Takács, Ballot problems, *Z. Wahrsch. verw. Geb.*, **1** (1962), 154–158.

{50} L. Takács, Fluctuations in the ratio of scores in counting a ballot, *J. Appl. Probab.*, **1** (1964), 393–396.

{51} L. Takács, *Combinatorial Methods in the Theory of Stochastic Processes,* Wiley (New York, 1967).

{52} L. Takács, On maximal deviation between two empirical distribution functions, *Studia Sci. Math. Hung.*, **10** (1975), 117–121.

{53} I. Vincze, *Statisztikai minőségellenőrzés: az ipari minőségellenőrzés matematikai statisztikai módszerei* (*Statistical quality control: The mathematical-statistical methods of industrial quality control,* in Hungarian), Közgazdasági és Jogi Könyvkiadó, Budapest (1958).

{54} I. Vincze, Einige zweidimensionale Verteilungs- und Grenzverteilungssätze in der Theorie der geordneten Stichproben, *Publications of the Math. Inst. Hung. Acad. Sci.*, **2** (1958), 183–209.

{55} I. Vincze, On some joint distribution and joint limiting distribution in the theory of order statistics, II, *Publications of the Math. Inst. Hung. Acad. Sci.*, **4** (1959), 29–47.

{56} I. Vincze, Some questions connected with two sample tests of Smirnov type, *Proc. of the Fifth Berkeley Symp. on Math. Stat. and Prob.*, Univ. of Calif. Press Vol. 1 (1967), 654–666.

{57} I. Vincze, On the power of the Kolmogorov–Smirnov two-sample test and related nonparametric tests, *Studies in Mathematical Statistics,* Publishing House of the Hung. Acad. Sci. (Budapest, 1968), 201–210.

{58} I. Vincze, On some results and problems in connection with statistics of the Kolmogorov–Smirnov type, *Proceedings of the Sixth Berkeley Symposium on Mathematical Statistics and Probability,* **I**, Univ. California Press (1972), 459–470.

{59} I. Vincze, On the Cramér–Fréchet–Rao inequality in the non–regular case, *Contributions to Statistics. Hájek Memorial Volume,* Academia (Prague, 1979), 253–262.

{60} I. Vincze, On nonparametric Cramér–Rao inequalities, *Order Statistics and Nonparametrics* (Ed. P. K. Sen and I. A. Salama) North Holland (Amsterdam, 1981), 439–454.

{61} J. von Neumann, Distribution of the ratio of the mean square successive difference to the variance, *Ann. Math. Statist.,* **12** (1941), 367–395.

{62} J. von Neumann, A further remark concerning the distribution of the ratio of the mean square successive difference to the variance, *Ann. Math. Statist.,* **13** (1942), 86–88.

{63} A. Wald, Contributions to the theory of statistical estimation and testing hypotheses, *Ann. Math. Statist.,* **10** (1939), 299–326.

{64} A. Wald, Asymptotically most powerful tests of statistical hypotheses, *Ann. Math. Statist.,* **12** (1941), 1–19.

{65} A. Wald, Asymptotically shortest confidence intervals, *Ann. Math. Statist.,* **13** (1942), 127–137.

{66} A. Wald, Tests of statistical hypotheses concerning several parameters when the number of observations is large, *Trans. Amer. Math. Soc.,* **54** (1943), 426–482.

{67} A. Wald, Sequential tests of statistical hypothesis, *Ann. Math. Statist.,* **16** (1945), 117–186.

{68} A. Wald, Some improvements in setting limits for the expected number of observations required by a sequential probability ratio test, *Ann. Math. Statist.,* **17** (1946), 466–474.

{69} A. Wald, Differentiation under the expectation sign in the fundamental identity of sequential analysis, *Ann. Math. Statist.,* **17** (1946), 493–497.

{70} A. Wald, Asymptotic properties of the maximum likelihood estimate of an unknown parameter of a discrete stochastic process, *Ann. Math. Statist.,* **19** (1948), 40–46.

{71} A. Wald, Estimation of a parameter when the number of unknown parameters increases indefinitely with the number of observations, *Ann. Math. Statist.,* **19** (1948), 220–227.

{72} A. Wald, Asymptotic minimax solutions of sequential point estimation problems, *Proceedings of the Second Berkeley Symposium on Mathematical Statistics and Probability,* University of California Press (1951), 1–11.

{73} A. Wald, *Selected Papers in Statistics and Probability by Abraham Wald,* McGraw-Hill Bokk Company, Inc. (New York–Toronto–London, 1955).

{74} A. Wald and J. Wolfowitz, On a test whether two samples are from the same population, *Ann. Math. Statist.*, **11** (1940), 147–162.

{75} A. Wald and J. Wolfowitz, An exact test for randomness in the nonparametric case based on serial correlation, *Ann. Math. Statist.*, **14** (1943), 378–388.

{76} A. Wald and J. Wolfowitz, Statistical tests based on permutations of the observations, *Ann. Math. Statist.*, **15** (1944), 358–372.

Endre Csáki

Alfréd Rényi Institute of Mathematics
Hungarian Academy of Sciences
P.O.B. 127
1364 Budapest
Hungary

csaki@renyi.hu

BOLYAI SOCIETY
MATHEMATICAL STUDIES, 14

A Panorama of Hungarian Mathematics in the Twentieth Century, pp. 523–535.

Stochastics: Information Theory

IMRE CSISZÁR

Information Theory has been created by Claude Shannon as a mathematical theory of communication. His fundamental paper {19} appeared in 1948. This was one of the major discoveries of the 20th century, establishing theoretical foundations for communication engineering and information technology. The key ingredients of Shannon's work were (i) a stochastic model of communication, (ii) the view of information as a commodity whose amount can be measured without regard to meaning, and (iii) the emphasis of coding as a means to enhance information storage and transmission, in particular, to achieve reliable transmission over unreliable channels.

Today, the mathematical discipline built on these ideas is often called Shannon theory, while the term information theory is frequently used in a much broader sense. In the terminology we use here, information theory – abbreviated as IT – is a branch of stochastic mathematics whose characteristic tools are mathematical expressions interpreted as measures of information. Problems relevant to information transmission and storage that involve coding represent a central but by no means the only subject of this theory. In fact, IT ideas turned out to be very useful in various fields of pure and applied mathematics, such as combinatorial analysis, ergodic theory, mathematical statistics, probability theory, etc. (But statistical investigations using the measure of statistical information introduced by Ronald A. Fisher in 1925 are not considered pertaining to IT.)

Though IT was created effectively by Shannon alone, ideas of other scientists did influence its birth and early development. Among these scientists there were several Hungarians.

Shannon's entropy, the basic measure of the amount of information, has a close relationship to the concept of entropy in physics. The first to relate information and (physical) entropy was the Hungarian physicist *Leo*

Szilárd (Z. Physik, Vol. 53, 1929, p. 840). A key inequality of IT known as Kraft's inequality is sometimes attributed also to Szilárd (e.g. in [154]), though I could not confirm the correctness of this attribution. Of course, Shannon did rely on the theory of communication available at the time, e.g., he gave special credit to Norbert Wiener for the "formulation of communication theory as a statistical problem". A famous Hungarian contributor to communication theory, explicitly mentioned by Shannon in {20} (on page 11) was *Dénes Gábor* (Nobel laureate, inventor of holography).

As a forerunner of IT, the work of *Ábrahám Wald* on sequential analysis [192] deserves special emphasis, although the IT aspect of this outstanding contribution to statistics has been recognized only later. The IT approach to statistics is generally associated with the name of Solomon Kullback whose book [98] systematically develops statistical applications of the information-theoretic measure of distance of probability distributions, now called information divergence (I-divergence), or Kullback–Leibler distance (also known as relative entropy or information gain). It was, however, Wald who first made essential use of I-divergence, without giving it a name. Kullback (loc. cit., p. 2) writes: "Although Wald did not explicitly mention information in his treatment of sequential analysis, it should be noted that his work must be considered a major contribution to the statistical applications of information theory." For further details on Wald's work in this regard we refer to the Section on Mathematical Statistics.

Information theory in Hungary

Within Hungary, research in IT was initiated by *Alfréd Rényi,* in the fifties (but the first to write about IT in Hungary was *Albert Korodi,* electrical engineer, former coworker of Szilárd). Rényi wrote cca. 25 research papers on IT and, not less importantly, started teaching IT at the Loránd Eötvös University, Budapest; his Probability Theory textbook [152] includes an Appendix on IT. The Eötvös University is still rather exceptional in having IT in the curriculum of mathematics students; elsewhere, IT is mostly pursued in electrical engineering departments. Among the mathematicians covered in this volume, in addition to Rényi it was *István Vincze* who devoted several papers to IT; these mainly address statistical applications of IT. The extraordinarily rich life-work of *Paul Erdős* also contains some papers related to IT, though this certainly was a side-issue for him. Today, the IT research group in Budapest established by Rényi enjoys international reputation. Its contributions are outside the scope of this volume, but some of them directly continuing and complementing Rényi's work will be briefly mentioned below.

Rényi always preferred brand new problems to already much investigated ones, and also in IT he was looking for new directions as opposed to the mainstream subject of coding theorems. His works on IT were mostly concentrated around the following subjects: (i) amount of information for non-discrete distributions ("dimensional entropy") (ii) information measures different from Shannon's ("Rényi informations") (iii) axiomatic characterization of information measures (iv) asymptotic evaluation of the amount of information provided by a statistical experiment. Rényi also initiated a systematic development of search theory that he regarded as a part of IT.

Dimensional entropy

Rényi's first paper on IT, joint with *János Balatoni,* appeared in 1956, see the Selected Papers [151], paper [151, article 121]; also in the sequel, Rényi's papers will be referred to by their number in [151]. The main contribution of the paper [151, article 121] was to clarify the relationship of Shannon's entropy formulas for discrete and continuous distributions, via the concept of "dimensional entropy." Rényi returned to this subject in several subsequent papers, in particular, the results of the first publication were substantially strengthened three years later [151, article 160]; they are reviewed below as appearing there.

The entropy of a discrete random variable ξ whose distribution is $P = (p_1, p_2, \ldots)$ is defined as

$$H(\xi) = H(P) = -\sum p_k \log_2 p_k,$$

and the entropy of a real-valued or vector-valued random variable with density $f(x)$ is defined as $-\int f(x) \log_2 f(x)\, dx$. While Shannon's results convincingly support the interpretation of discrete entropy as a measure of information content, this interpretation does not directly carry over to continuous entropy. Rényi gave a precise meaning to the interpretation of continuous entropy as a "measure of information content up to an infinitely large additive constant."

Rényi defined the (information theoretic) dimension and the d-dimensional entropy of a real-valued random variable ξ via the discrete approximations $\xi^{(n)} = [n\xi]/n$, $n = 1, 2, \ldots$ of ξ, by

$$d(\xi) = \lim_{n\to\infty} \frac{H(\xi^{(n)})}{\log_2 n}$$

$$H_d(\xi) = \lim_{n\to\infty} \Big(H(\xi^{(n)}) - d \log_2 n \Big)$$

(the dimension resp. d-dimensional entropy of ξ is undefined if the corresponding limit does not exist). He proved the following: Suppose that $\xi^{(1)} = [\xi]$ has finite entropy. Then, if ξ is discrete, $d(\xi) = 0$ and $H_0(\xi) = H(\xi)$, while if ξ has a density then $d(\xi) = 1$ and $H_1(\xi)$ is given by Shannon's integral entropy formula. Moreover, if the distribution of ξ is a mixture of a discrete and a continuous component, the latter having a density, then ξ has dimension equal to the weight of the continuous component; the corresponding d-dimensional entropy was also determined. Extensions of these results to vector-valued random variables were also treated; the information theoretic dimension of an $\mathbb{R}^d$-valued ξ having a density equals the geometric dimension d, and the corresponding d-dimensional entropy is given by Shannon's integral formula. On the other hand, even for ξ taking values in the unit interval, the dimension need not exist if ξ has a singular distribution.

Rényi [151, article 160] also considered discrete approximations other than $\xi^{(n)}$ above. To any partition π of the range X of ξ into disjoint subsets X_k there corresponds an approximation ξ_π defined by $\xi_\pi = k$ if $\xi \in X_k$. Extensions of "dimensional entropy" results to approximations of this kind were given for the case when X was the unit interval and the X_k's were intervals of length $\le \varepsilon$ with $\varepsilon \to 0$. At the same time *Vincze* (Matematikai Lapok, vol. 10, 1959, pp. 255–266) considered approximations of a real-valued ξ corresponding to partitions into intervals "of equal interest," namely to partitions $\pi_{n,\Phi}$ of the real line into intervals of Φ-measure $\frac{1}{n}$, where Φ is a given probability measure interpreted as the distribution of our interest. He showed (assuming regularity conditions) that

$$\lim_{n\to\infty} \Big(H(\xi_{\pi_{n,\Phi}}) - \log_2 n \Big) = - \int \log_2 \frac{dF}{d\Phi}\, dF,$$

where F is the distribution of ξ. The integral here is the I-divergence of F from Φ, thus Vincze's result provided an interesting new interpretation of I-divergence, as a measure of information relative to the distribution of our interest.

The concept of dimensional entropy was extended to a class of stochastic processes by M. Rudemo {18}. Using this extension, Rényi immediately established a maximum entropy property of Poisson processes, see [151, article 228]: The maximal dimension of a homogeneous point process in $(0, T)$ with density λ is equal to λT, and the Poisson process has the largest λT dimensional entropy.

Let us mention some later developments, complementing Rényi's work on dimensional entropy. *Imre Csiszár* in {4} proved the following: For a measure space (X, μ) where μ is a non-atomic σ-finite measure on X, consider partitions π_ε of X into subsets of equal μ-measure ε, and let ξ be an X-valued random variable. Then, subject to mild regularity conditions, $H(\xi_{\pi_\varepsilon}) - \log_2 \frac{1}{\varepsilon}$ converges as $\varepsilon \to 0$ to the generalized entropy of ξ with respect to μ, that equals standard (continuous) entropy when $X = \mathbb{R}^d$ and μ is the Lebesque measure, and equals negative I-divergence when μ is a probability measure. *József Fritz* {11} proved the extension of Rényi's Poisson process result to point processes on arbitrary (nonatomic, separable) measure spaces. Rényi did point out a relationship of his dimension (of random variables) to Hausdorff dimension (of sets), see [151, article 175]; *Péter Gács* in {12} suggested a modified definition of information theoretic dimension that leads to an even closer relationship of this kind.

Rényi informations

Rényi's most widely known contribution to IT was to show that certain quantities different from Shannon's information measures come also into account as alternatives to the latter. These "informations of order α" are now called Rényi informations. Rényi's first publication about them appeared in 1960, where he noted that "information quantities of order α were already investigated in the literature from other viewpoints," the new results consisted in "showing that some reasonable postulates can be satisfied only by them and by Shannon's entroy, and ... how the known results on Shannon's entropy generalize to information measures of order α." Rényi's best known work on this subject is his Berkeley Symposium contribution [151, article 180], one of Rényi's most often cited papers.

Shannon's entropy $H(P)$ of a probability distribution $P = (p_1, \dots p_n)$, measuring the average amount of information provided by a random experiment whose possible outcomes have probabilities $p_1, \dots, p_n$, equals the (weighted) arithmetic mean of the individual informations $I_k = \log_2 \frac{1}{p_k}$ associated with these outcomes.

The arithmetic mean is a special case (φ linear) of means of form $\varphi^{-1}\big(\sum p_k\varphi(I_k)\big)$, where φ is some strictly monotonic function. Rényi argued that means with non-linear φ might also be used, provided they satisfy the intuitive requirement of additivity for independent experiments. The exponential functions $\varphi(x) = \exp\big\{(1-\alpha)x\big\}\,(\alpha \neq 1)$ meet that requirement, and accordingly, Rényi defined his entropy of order $\alpha \neq 1$ as $H_\alpha(P) = \frac{1}{1-\alpha}\log_2 \sum p_k^\alpha$.

Since $H_\alpha(P)$ converges to Shannon's entropy $H(P)$ as $\alpha \to 1$, the latter is regarded as entropy of order $\alpha = 1$. Via similar considerations, Rényi also defined I-divergence of order α (he used the term "information gain"), of which standard I-divergence is the limit as $\alpha \to 1$. He also extended his previous "dimensional entropy" results to entropy of order α.

The theory of generalized information measures initiated by Rényi's work has now an extensive literature. Since poorly motivated generalizations have also been published, it is important to note the Rényi did not endorse those. He emphasized, see his 1965 survey paper [151, article 242], that only such quantities deserve to be called information measures that can be effectively used in solving concrete problems. Rényi was able to find interesting problems whose solution involved entropy of order $\alpha \neq 1$, namely in the theory of random search (see the subsection on that topic). Later, coding problems were also found that led to Rényi entropy, see Lorain Campbell {2} and Csiszár {5}. The latter paper gives an operational characterization of Rényi's information measures (including an "order α" analogue of Shannon's mutual information, somewhat different from that suggested by Rényi) within the standard Shannon theory framework.

In 1959, Yuri Linnik presented an information-theoretic proof of the central limit theorem that intrigued Rényi. He observed that the essence of Linnik's idea was that convergence of a sequence of probability distributions P_n to a limiting distribution P may be proved by showing that the I-divergence of P_n from P converges to 0. Rényi hoped to simplify Linnik's very difficult proof by using an I-divergence of order α, say with $\alpha = 2$; this was a major cause of his interest in generalized information measures. Rényi's hope to simplify Linnik's proof did not come true, and it is still an open question under what conditions does the I-divergence of the distribution of the normalized sum of n independent random variables from the standard normal distribution converge to 0 (only the case of identically distributed summands comes close to be satisfactorily settled, see Andrew Barron's paper {1}. On the other hand, Rényi showed in his Berkeley Symposium paper [151, article 180] that the "information theoretic method" leads to a simple proof of the convergence of n-step conditional distributions of a stationary (finite state) Markov chain to the stationary distribution, and I-divergence of order α and some more general expressions are equally suitable for that purpose. The last observation motivated Csiszár {3} to introduce a general class of information-type measures of distance of probability distributions, corresponding to arbitrary convex functions f; these f-divergences turned out to have many applications in statistics, see Igor Vajda's book [188]. Rényi's followers to prove limit theorems for Markov chains via the information theoretic method include David Kendall {14} and Fritz {10}; for more recent applications of this idea, in the theory of interacting particles systems, see Liggett's book [110].

Axiomatic characterizations

Shannon's main justification for his information measures was their usefulness in communication problems, but he also showed that his entropy was uniquely characterized by certain postulates that a measure of amount of information was intuitively expected to satisfy. Later, starting with Aleksandr Jakovlevich Khinchin (Usp. Mat. Nauk, Vol. 8, 1953, pp. 3–51) and D. K. Fadeev (Usp. Mat. Nauk, Vol. 11, 1956, pp. 227–231), several mathematicians put forward axiomatic characterizations using weaker or "more natural" postulates. Rényi also contributed to this direction of research that he considered conceptually important for IT. An instructive exposition of his view appears in his survey paper [151, article 242]. He says that what he calls the axiomatic and pragmatic approaches to the problem of measuring information "are compatible and even complement each other and therefore both deserve attention. Both of the mentioned approaches may and should be used as a control of the other."

Axiomatic considerations appeared already in Rényi's first IT paper [151, article 121]. In [151, article 159], he pointed out that the key step in Fadeev's characterization (loc. cit.) was a number theoretic result that had been previously proven by Erdős {9}, and he gave a new simple proof of that result. It says that an additive number theoretic function must be equal to constant times $\log n$ if it satisfies $\lim_{n\to\infty}\big(F(n+1)-F(n)\big)=0$. Actually, the latter hypothesis may be weakened to $\liminf_{n\to\infty}\big(F(n+1)-F(n)\geq 0$, a fact later also used in a characterization of Shannon entropy, see *Zoltán Daróczy* and *Imre Kátai,* {8}.

Axiomatic characterizations play a substantial role in Rényi's Berkeley Symposium paper [151, article 180] on information measures of order α. The postulates there involve (generalized) means, and information measures are assumed to be assigned also to "incomplete distributions" (where the sum of probabilities is less than 1). When characterizing I-divergences of order α, the additivity postulate is shown to admit only means corresponding to a linear or exponential function φ, while when characterizing entropies, the same remains an unproven assumption. As reported in [151, article 242], that deficiency could be removed: Daróczy in {7} showed that entropy of order α could be satisfactorily characterized by Rényi's postulates, even without recourse to incomplete distributions.

The majority of the (substantial) contributions of Hungarian mathematicians to axiomatic characterizations of information measures (see the book of *János Aczél* and Daróczy [5]) is out of the scope of this volume. It should be noted that today this subject is not considered of primary importance for IT but, on the other hand, research in this direction has strongly contributed to the development of the theory of functional equations.

Random search

Search theory is a subject whose systematic development was initiated by Rényi. He regarded it as part of IT, which is not the prevailing view today. Still, Rényi's work on random search certainly has an IT flavor, in particular, it provides operational justification for Rényi entropies of order α.

Rényi had been fascinated by the guessing game called "twenty questions" in the US. In this game, consecutive questions (in the US, at most 20) answerable by yes or no are asked about an unknown object, in order to identify it from the answers. In Rényi's lectures, this game was a standard example to visualize the basic ideas of IT. His research about random search was motivated by his interest in what happens if the questions are selected not by a well designed strategy but at random. Rényi showed that consecutive random selections from the set of all possible questions admit identification with only slightly more questions than an optimal strategy. This visualizes a key idea of IT, the efficiency of random coding, although this fact was not emphasized by Rényi.

In 1961/62, Rényi published four papers on random search, of main interest is [151, article 193]. There, identification of elements of a set H of size n via k questions, each with r possible answers, is considered. Each question corresponds to a labeling of the elements of H by numbers randomly selected from $\{1, \ldots, r\}$ with probabilities $p_1, \ldots, p_r$, independently of each other; the answer to this question, for any fixed $x \in H$, gives the label assigned to x. The k questions identify x if for no other $y \in H$ are all k answers the same as for x. Problem: when r is fixed and $n \to \infty$, how large a k is needed in order that with a prescribed probability, either (i) a particular $x \in H$ be identified or (ii) all elements of H be identified by k random questions as above. When $P = (p_1, \ldots, p_r)$ is the uniform distribution, Rényi's result was that $k \sim \frac{\log_2 n}{\log_2 r}$ questions are needed in case (i), and $k \sim \frac{2 \log_2 n}{\log_2 r}$ in case (ii). When P is arbitrary, one would expect that $\log_2 r$ should be replaced by the entropy of P. The remarkable results was, however, that while this expectation is correct in case (i), it is Rényi entropy of order 2 rather than Shannon entropy that enters in case (ii). This was the first operational characterization of a Rényi entropy of order α, though only for $\alpha = 2$.

Later, Rényi found modified versions of the above random search problem whose solution involved entropies of other (positive integer) orders α, see his 1965 survey paper [151, article 242]. In the same year, he introduced a general model of random search in his Invited Address [151, article 249]. Here, the role of the previous random labellings of the set H is played by functions $f : H \to \{1, \ldots, r\}$ randomly selected from a given class F of such functions, thus the labels (function values $f(x)$) need not be independently chosen for each $x \in H$. Under homogeneity conditions on the function class F, general results were obtained on the probability that k questions identify a fixed element of H, or all elements of H. These, in particular cases, lead to asymptotic results similar to those mentioned above.

Information theoretic methods in statistics

A basic problem in statistics is to infer an unknown distribution P from an observed sample $X^n = (X_1, \ldots, X_n)$ where $X_1, X_2 \ldots$ are independent random variables of distribution P. If the unknown P is assumed to belong to a known family of distributions, $\{P_\theta\}$, where θ is a (scalar or vector valued) parameter ranging over a given set, one has to estimate the true value of θ from the observed sample X^n.

Information measures relevant for this problem include Fisher's information and I-divergence. Still, the natural measure of the amount of information the sample gives for the unknown parameter is Shannon's mutual information $I(X^n, \theta)$, provided one is willing to adopt the Bayesian approach of assigning a prior distribution to the parameter, necessary for the definition of $I(X^n, \theta)$. The first to study the asymptotic behavior of $I(X^n, \theta)$ as $n \to \infty$ was Rényi, in the special case when θ had a finite number of possible values. Starting in 1964, Rényi treated this issue in 9 papers; the strongest results appear in [151, article 288] and [151, article 328]. He related this problem to that of the asymptotic behavior of the error probability of the Bayesian estimator of θ (called by him the "standard decision" Δ_n) and studied both problems in parallel.

Rényi proved that $I(X^n, \theta)$ converges to the entropy of θ or – equivalently – the missing information (Rényi's term for the conditional entropy of θ given X^n) converges to 0 exponentially fast, and so does the error probability $P(\Delta_n \neq \theta)$, with the same exponent. When θ had two possible values only, Rényi gave an exact asymptotic formula for $P(\Delta_n \neq \theta)$, using large deviations techniques. He also considered the case of (independent but) not identically distributed observations, and gave an upper bound to the missing information in terms of Hellinger integrals. This result is related to Kakutani's theorem that for two sequences $\{\mu_n\}$ and $\{\nu_n\}$ of probability measures, their infinite products are mutually absolutely continuous or singular according as the infinite product of the Hellinger integrals $\lambda_n = \int \sqrt{d\mu_n \, d\nu_n}$ is positive or 0.

Rényi's work gave substantial impetus to studying the interplay of statistics and IT. Bounds to error probability in terms of Shannon's and more general information measures are too numerous to cite here. Rényi's asymptotic formula for $P(\Delta_n \neq \theta)$ when θ has two possible values easily extends to θ with any finite number of values, see independent work of Vajda (Proc. Coll. Inform. Theory, Debrecen 1967, J. Bolyai Math. Society, Budapest, 1969, pp. 489–501). Rényi's result related to Kakutani's theorem actually gives an information theoretic proof of one half of that theorem, namely that $\Pi_{n=1}^{\infty} \lambda_n = 0$ implies singularity. An information theoretic proof of the other half was given by *Tibor Nemetz* {15}. Extensions of the study of the asymptotic behavior of $I(X^n, \theta)$ to the case of a continuous parameter θ, turned out of substantial interest for the theory of universal coding and Bayesian statistics, see Bertrand Clarke and Barron, J. Statist. Plan. Inference, Vol. 41, 1999, pp. 37–60. If θ is a k-dimensional parameter, $I(X^n, \theta)$ is asymptotically equal to $\frac{k}{2}\log_2 n$ plus a constant that depends on the prior distribution of θ (subject to regularity conditions); the constant is smallest for the so-called Jeffreys prior which is thereby distinguished as "least informative."

Vincze also devoted several papers to the interplay of IT and statistics, in particular to information-type measures of distance of probability distributions that belong to the class of f-divergences, see the subsection about Rényi informations. Members of this class of statistical significance include Hellinger distance and χ^2-distance, corresponding to $f(t) = 1 - \sqrt{t}$ and $f(t) = (t-1)^2$. Puri and Vincze in {17} introduced distances denoted by $I_N(P, Q)$ that correspond to $f(t) = \frac{1}{2}\frac{|t-1|^N}{(t+1)^{N-1}}$, $N \geq 1$. Their main result was that two sequences of probability distributions $\{P_n\}$ and $\{Q_n\}$ are mutually contiguous if and only if $I_{N_n}(P_n, Q_n) \to 0$ for all sequences of numbers $N_n \to \infty$.

Kafka, Ferdinand Österreicher and Vincze in {13} studied the problem for what f-divergences was some power of them a metric, and then what was the smallest such power. Previously, Csiszár and *János Fischer* in {6} showed that for $f(t) = 1 - t^\alpha$, $0 < \alpha < 1$, the corresponding f-divergence raised to the power $\min(\alpha, 1-\alpha)$ satisfied the triangle inequality (but was not a metric, for lack of symmetry, except if $\alpha = 1/2$, the case of Hellinger distance). The results of Kaffka, Österreicher and Vincze include that $\left(I_N(P,Q)\right)^{\frac{1}{N}}$ is a metric, the f-divergence corresponding to $f(t) = 1 + t^\alpha - t^{1-\alpha}$ raised to the power $\min(\alpha, 1-\alpha)$ is a metric, and so is the square-root of an f-divergence that gives the perimeter of the risk set for testing P against Q, studied earlier by Österreicher {16}; in neither case is any smaller power appropriate.

References

[5] Aczél, János – Daróczy, Zoltán, *On Measures of Information and their Characterizations* Academic Press (New York, 1975).

[98] Kullback, Solomon, *Information Theory and Statistics,* Wiley (New York, 1959), Dover (New York, 1978).

[110] Liggett, Thomas M., *Interacting Particle Systems,* Die Grundlehren der Mathematischen Wissenschaften in Einzeldarstellungen, Band 276, Springer-Verlag (New York, 1985).

[151] Rényi, Alfréd, *Selected Papers,* ed. Pál Turán, Akadémiai Kiadó (Budapest, 1976).

[152] Rényi, Alfréd, *Probability Theory,* Translated by László Vekerdi, North-Holland Series in Applied Mathematics and Mechanics, Vol. 10, North-Holland Publishing Company (Amsterdam–London); American Elsevier Publishing Co., Inc. (New York, 1970).

[154] Reza, Fazlollah M., *An Introduction to Information Theory,* McGraw-Hill (New York, 1961).

[188] Vajda, Igor, *Theory of Statistical Inference and Information,* Kluwer Academic (Boston, 1989).

[192] Wald, Ábrahám, *Sequential Analysis,* John Wiley and Sons (New York) – Chapman and Hall (London, 1947).

{1} A. Barron, Antropy and the central limit theorem, *Annals of Probability,* **14** (1986), 336–342.

{2} L. Campbell, A coding theorem and Rényi's entropy, *Information and Control,* **8** (1965), 423–429.

{3} I. Csiszár, Eine inrofmationtheoretische Ungleichung und ihre Anwendung auf den Beweis der Ergodizität von Markoffschen ketten, *Publ. Math. Inst. Hungar. Acad. Sci.,* **8** (1963), 85–108.

{4} I. Csiszár, Generalized entropy and quantization problems, *Trans. Sixth Prague Conference on Inform. Theory, etc.,* 1971, Academia (Praha, 1973), 299–318.

{5} I. Csiszár, Generalized cutoff rates and Rényi's information measures, *IEEE Trans. Inform. Theory,* **41** (1995), 26–34.

{6} I. Csiszár and J. Fischer, Informationsentfernungen im Raum der Wahrscheinlichkeitsverteilungen, *Publ. Math. Inst. Hungar. Acad. Sci.,* **7** (1962), 159–182.

{7} Z. Daróczy, Über Mittelwerte und Entropien vollstandiger Wahrscheinlichkeitsverteilungen, *Acta Math. Sci. Hungar.,* **15** (1964), 203–210.

{8} Z. Daróczy and I. Kátai, Additive zahlentheoretische Funktionen und das Mass der information, *Ann. Univ. Sci. Budapest, Sec. Math.,* **13** (1970), 83–88.

{9} P. Erdős, On the distribution function of additive functions, *Annals of Math.,* **17** (1946), 1–20.

{10} J. Fritz, An information-theoretical proof of limit theorems for reversible Markov processes, *Trans. Sixth Prague Conference on Inform. Theory, etc.,* 1971. Academia (Praha, 1973), 183–197.

{11} J. Fritz, An approach to the entropy of point processes, *Periodica Math. Hungar.,* **3** (1973), 73–83.

{12} P. Gács, Hausdorff-dimension and probability distribution, *Periodica Math. Hungar.,* **3** (1973), 59–71.

{13} P. Kafka, F. Österreicher and I. Vincze, On powers of f-divergences defining a distance, *Studia Sci. Math. Hungar.,* **26** (1991), 415–422.

{14} D. Kendall, Information theory and the limit-theorem for Markov chains and processes with a countable infinity of states, *Annals Inst. Statist. Math.*, **15** (1963), 137–143.

{15} T. Nemetz, Equivalence-orthogonality dichotomies of probability measures, *Limit Theorems of Probability Theory,* Colloquia Math. Soc. J. Bolyai, Vol. 11, North Holland (1975), 183–191.

{16} F. Österreicher, The construction of least favourable distributions is traceable to a minimal perimeter problem, *Studia Sci. Math. Hungar.*, **17** (1982), 341–351.

{17} M. Puri and I. Vincze, Measure of information and contiguity, *Statistics and Probability Letters,* **9** (1990), 223–228.

{18} M. Rudemo, Dimension and entropy for a class of stochastic processes, *Publ. Math. Inst. Hungar. Acad. Sci.*, **9** (1964), 73–87.

{19} C. Shannon, A mathematical theory of communication, *Bell System Technical Journal,* **27** (1948), 379–423 and 623–656.

{20} C. Shannon, Communication in the presence of noice, *Proc. IRE,* **37** (1949), 10–21.

Imre Csiszár

Alfréd Rényi Institute of Mathematics
Hungarian Academy of Sciences
P.O.B. 127
1364 Budapest
Hungary

csiszar@renyi.hu

BOLYAI SOCIETY
MATHEMATICAL STUDIES, 14

A Panorama of Hungarian Mathematics in the Twentieth Century, pp. 537–548.

Contribution of Hungarian Mathematicians to Game Theory

FERENC FORGÓ

Game theory, when defined in the broadest sense, is a collection of mathematical models formulated to study situations of conflict and cooperation. It is mainly concerned with finding the best actions for the individual decision makers in these situations and/or recognizing stable outcomes. Game theory attempts to provide a normative guide to rational behavior for individuals pursuing more or less different goals and make predictions about the outcomes thus realized.

The mathematical complexity of the models poses a real challenge to mathematicians who, if they want to be really successful, must possess not only the technical skills but have a deep understanding of the problems in a very wide variety of applications ranging from biology and human behavior to economics and warfare. János (John) von Neumann and János (John) Harsányi have left irremovable marks by their contributions to the very foundation of cooperative and noncooperative game theory. Jenő Szép, as an educator and textbook writer has helped generations of economists and decision scientists to keep abreast of the latest developments in this rapidly growing field. Their most important results, together with the context in which they are interpreted are briefly outlined below.

János (John) von Neumann, the last renaissance scientist of our time, was not only a brilliant mathematician but he also took interest in other sciences as well. His model of economic equilibrium has been a subject of study and a source of inspiration for generations of economists. While classical and neoclassical economic equilibrium models work with economic agents (producers, consumers) whose individual contribution in forming prices and producing goods is assumed to be negligible, there are several industries where concentration of capital creates situations in which a few major players decide on production volumes and/or prices. Strategic aspects of interaction on the market became an issue and raised questions that classical economic theory could not answer. So the need for a theory to study strategic behavior of participants in a conflict situation and the presence of the open mind of a genius came together in the thirties and forties of the 20th century to give birth to a brand new body of knowledge commonly known as game theory.

Von Neumann did not publish too much in game theory as far as the number of papers and books is concerned. His two major works, however, are landmarks in game theory. One is a paper in which the first, complete minimax theorem is stated and proved, the other is a book [117] in which the theoretical foundation of cooperative game theory is laid down. For the latter he found an economist, Oskar Morgenstern to help him reinforce the economic relevance of the model and make the monumental work the starting point of any research in cooperative game theory.

Though Emile Borel (1924) was the first who defined pure and mixed strategies for symmetric two-person games, he was unable to prove the existence of equilibrium in mixed strategies. In fact, he was in doubt about the validity of the minimax theorem, an equivalent to the existence theorem in case of two-person zero-sum games. It was von Neumann (1928) who first gave a complete proof of a minimax theorem which covers the special case of the mixed extension of finite, two-person zero-sum games.

Theorem 1 (von Neumann). *Let C and U be unit simplexes of finite dimensional Euclidean spaces, and f be a jointly continuous real valued function defined on $X \times Y$. Suppose that f is quasiconcave on X, that is to say, for all $y \in Y$ the upper level sets of f are convex, and f is quasiconvex on Y, that is to say, for all $x \in X$ the lower level sets of f are convex. Then*

$$\min_Y \max_X f = \max_X \min_Y f.$$

This result was later extended by von Neumann (1937) himself by replacing unit simplexes with nonempty compact, convex subsets of Euclidean spaces. In both cases, von Neumann used topological and fixed-point arguments. It turned out later on that fixed-point theorems are not necessary for the proof, convex analysis, in particular linear separation is enough to prove even more general results where the continuity of f is replaced by partial upper and lower semicontinuity in the respective variables, Sion (1958). Theorem 1 has opened a whole avenue of research about minimax theorems and their various generalizations and has been applied in many fields inside and outside of mathematics. For a good overview of minimax theorems we recommend Simons (1995).

Von Neumann was not only concerned with the existence of equilibria for two-person zero-sum games but also proposed a unique method for computing one in the case of symmetric matrix games.

The mixed extension of a finite, two-person zero-sum game can be completely defined by a matrix **A** of m rows and n columns with general element $a_{i,j}$ which represents the payoff player 1 gets from player 2 if player 1 plays his i^{th} and player 2 his j^{th} pure strategy. The players are allowed to mix their pure strategies by choosing probability vectors **x** and **y**, respectively, and are concerned with expected payoffs **xAy** player 1 gets on the average from player 2. It is easily seen that Theorem 1 ensures the existence of an equilibrium pair of strategies $\mathbf{x}^*$, $\mathbf{y}^*$ to satisfy

$$\mathbf{xA\,y}^* \le \mathbf{x}^*\mathbf{A\,y}^* \le \mathbf{x}^*\mathbf{A\,y}$$

for all probability vectors **x** and **y** of dimension m and n, respectively. We will briefly refer to the mixed extension of a finite two-person zero-sum game as a matrix game and define it by its payoff matrix **A**.

A matrix game is said to be symmetric if $\mathbf{A} = -\mathbf{A}^T$. The value (i.e., the payoff at equilibrium) of symmetric games is 0 and both players have the same equilibrium strategies. Von Neumann proposed the following method to find an equilibrium strategy of player 2 (which is also an equilibrium strategy of player 1).

Player 2's strategy $\mathbf{y}(t)$ is assumed to be a function of a continuous parameter t (time) and we suppose $t \ge 0$. Define the following functions:

$$u_i : \mathbb{R}^n \to \mathbb{R}, \quad u_i(\mathbf{y}) = \mathbf{e}_i\mathbf{Ay}, \quad (i = 1, \ldots, n)$$

$$\phi : \mathbb{R} \to \mathbb{R}, \quad \phi(a) = \max\{0, a\},$$

$$\Phi : \mathbb{R}^n \to \mathbb{R}, \quad \Phi(\mathbf{y}) = \sum \phi\big(u_i(\mathbf{y})\big)$$

where $\mathbf{e}_i$ denotes the ith unit vector.

Let $\mathbf{y}^0$ be any strategy of player 2. Consider the following system of differential equations:

$$\begin{aligned} y_j'(t) &= \phi\big(u_j(\mathbf{y}(t))\big) - \Phi\big(\mathbf{y}(t)\big)\, y_j(t) \\ y_j(0) &= y_j^0 \end{aligned} \qquad (j = 1, \ldots, n).$$

Since the right-hand side of the above system is continuous, it has at least one solution. Let $\mathbf{y}(t)$ be a solution and $t_1, t_2, \ldots$ be a positive monotone increasing sequence tending to ∞. Von Neumann's proved the following theorem:

Theorem 2. *Any limit point of the sequence* $\{\mathbf{y}(t_k)\}$, $(k = 1, 2, \ldots)$ *is an equilibrium strategy of player 2 in the symmetric matrix game* $\mathbf{A}$. *Furthermore, there exists a constant* C, *such that*

$$\mathbf{e}_i \mathbf{A} \mathbf{y}(t_k) \leq \sqrt{n}/(C + t_k), \qquad (i = 1, \ldots, n).$$

Numerically solving the differential equation system provides a good approximation of an equilibrium strategy of either player.

Using linear programming is a more efficient way of solving matrix games, but von Neumann's method remains an elegant, alternative approach which has also been an inspirational source for various kinds of learning processes.

Von Neumann's minimax theorem in the two-person zero-sum case was the precursor to the equilibrium concept developed by John Nash in 1951 for general-sum, n-person noncooperative games. A game G in normal (strategic) form is given as the $2n$-tuple

$$G = \{S_1, \ldots, S_n;\, f_1, \ldots, f_n\},$$

where for each $i = 1, \ldots, n$, S_i is the strategy set of player i and $f_i : S_1 \times \cdots \times S_n \to \mathbb{R}$ is his payoff function. Given the $(n-1)$-tuple of strategies $(s_1, \ldots, s_{i-1}, s_{i+1}, \ldots, s_n)$ of all players but i, a strategy $t \in S_i$ is said to be a best reply (to the strategy profile of the rest of the players), if

$$f_i(s_1, \ldots, s_{i-1}, t, s_{i+1}, \ldots, s_n) \geq f_i(s_1, \ldots, s_{i-1}, u, s_{i+1}, \ldots, s_n)$$

holds for all $u \in S_i$.

A strategy profile $s^* = (s_1^*, \ldots, s_n^*)$ is called a *Nash equilibrium point* of game G if

$$f_i(s_1^*, \ldots, s_n^*) \geq f_i(s_1^*, \ldots, s_{i-1}^*, s_i, s_{i+1}^*, \ldots, s_n^*)$$

holds for all $s_i \in S_i$ and $i = 1, \ldots, n$, or equivalently, s_i^* is a best reply to $(s_1^*, \ldots, s_{i-1}^*, s_{i+1}^*, \ldots, s_n^*)$.

Not only noncooperative game theory received the initiating impetus from von Neumann but the cooperative theory as well. In the seminal book written together with Oskar Morgenstern [117] he sets up the model of a cooperative game most commonly used for analysis ever since. Given a finite set of players, $N = \{1, \dots, n\}$, a pair $G = (N, v)$ is defined an n-person cooperative game in characteristic function form (with side payments) where v is a real valued function defined on the subsets (coalitions) of N. The function v assigns a real number $v(S)$ to every coalition, with the convention $v(\emptyset) = 0$. The value $v(S)$ represents the transferable utility coalition S can achieve on its own when its members fully cooperate. The theory is mostly concerned with how the utility $v(N)$ achievable by the grand coalition N can be distributed taking into account the power, as expressed by the characteristic function, the coalitions have. Von Neumann and Morgenstern consider only essential, constant-sum games in their book. A game $G = (N, v)$ is essential if

$$\sum_{i \in N} v(\{i\}) < v(N)$$

and it is constant-sum if

$$v(S) + v(N - S) = v(N)$$

holds for any coalition S. We call the game $G = (N, v)$ superadditive if

$$v(S) + v(T) \leq v(S \cup T)$$

for all disjoint coalitions S, T.

For a given game $G = (N, v)$, let S be a coalition and $\mathbf{x} = (x_1, \dots, x_n)$ a real n-vector.

Define

$$x(S) = \sum_{i \in S} x_i.$$

An n-vector $\mathbf{x} = (x_1, \dots, x_n)$ is called an imputation if it is individually rational, i.e.,

$$x_i \geq v\big(\{i\}\big) \quad \text{for all} \quad i \in N$$

and Pareto optimal or efficient, i.e.,

$$x(N) = v(N).$$

An imputation represents a distribution of $v(N)$ among players in such a way that no player will get less than his own value $v\big(\{i\}\big)$. We say that imputation $\mathbf{x}$ dominates imputation $\mathbf{y}$, written $\mathbf{x} \operatorname{dom} \mathbf{y}$, if there is a coalition S such that $x_i > y_i$ for all $i \in S$ and $x(S) \leq v(S)$. If $\mathbf{x} \operatorname{dom} \mathbf{y}$, then coalition S can block the imputation $\mathbf{y}$ since it can give more to its members and also has the power to achieve this.

Von Neumann and Morgenstern define a "solution" to a game $G = (N, v)$ as a subset V of the set all imputations which is both internally and externally stable, i.e.

(i) there is no $\mathbf{x}, \mathbf{y} \in V$ such that $\mathbf{x} \operatorname{dom} \mathbf{y}$,

(ii) if $\mathbf{y} \notin V$, then there is an $\mathbf{x} \in V$ such that $\mathbf{x} \operatorname{dom} \mathbf{y}$.

To distinguish from other solution concepts that have emerged since the ground breaking work of von Neumann and Morgenstern the above "solution" is usually referred to as *stable set* or *von Neumann–Morgenstern solution.* Von Neumann and Morgenstern interpreted stable sets as a standard of behavior within a society. Distribution of the commonly gained wealth is accepted and can be maintained if it belongs to a stable set. Within this set no distribution is both favorable and achievable by any segment of the society while any distribution outside of this set can be prevented from becoming socially acceptable by certain social groups which have both the motivation and power to do so. Von Neumann and Morgenstern left open the question of which imputation in a stable set V will actually realize. This is assumed to be determined by the bargaining ability of the players, outside forces, chance etc. They were not disturbed at all by the fact that in many games there is a multitude of different stable sets. They considered each as a standard of behavior and did not consider part of their model which one of these will realize. They were, however, deeply concerned with the existence of stable sets. They could prove in their book the following theorem.

Theorem 3 [117]. *Every superadditive, essential, three-person game has at least one stable set.*

Although some special classes of games were shown by von Neumann and Morgenstern to have stable sets, they were unable to prove a general existence theorem. To settle the existence of stable sets has proved to be a hard problem over the years. Lucas constructed a 10-person, nonconstant-sum game in 1969 which has no stable sets and Bondareva et al. (1974) proved that all 4-person games have stable sets. The question of existence for general games is unsettled for 5 to 9-persons. No one has proved or disproved the original conjecture of von Neumann and Morgenstern that every constant-sum game has at least one stable set. It is also conjectured that games with no solutions are "rare", known examples are unstable in the sense that minor changes in the characteristic function value causes them to have stable sets.

Stable sets have a surprisingly rich mathematical structure and give rise to extremely difficult problems. An excellent overview of the developments from von Neumann and Morgenstern to the early nineties is given by Lucas (1992).

One of the basic assumptions of classical game theory is complete information and common knowledge: rules of the game, the available strategies, the payoff functions are assumed to be known by each player together with the infinite hierarchy: "all players know it, each one knows that everyone knows it, and so on". If we want to come closer to reality, we have to find ways for analyzing conflict situations where players have only partial information about certain constituents of the game.

János (John) Harsányi, the Economics Nobel Laureate of 1994, was the first to provide a consistent model for games with incomplete information which became the most commonly applied approach to treat informational disparities of agents not only in game theory but in the new, rapidly developing field of Economics of Information as well. The very heart of the concept is usually referred to as the *Harsányi Doctrine* or *The Common Prior Assumption:* If C denotes the possible states of the world with generic element c which is a specification of all parameters that may be the object of uncertainty in a game G, then all the players share the same prior probability distribution on C which is common knowledge among them. This does not imply that all players have the same subjective probabilities since they may have different information about the true state of nature. The subjective probability of a player is his posterior (in the Bayesian sense) given his information. Posteriors may well be different but differences in probability estimates of distinct individuals should be explained by differences in information and experience.

We will demonstrate the power of the *Harsányi Doctrine* by a brief outline of his original model published in a series of articles in 1967. When asked, he himself thought that these articles had brought him the Nobel prize. It took thirty years for the scientific community to really appreciate the contribution of the original model and the underlying idea (The Harsányi Doctrine) to help better understand strategic behavior in conflicts with incomplete information.

Following Harsányi we will call a game of incomplete information in normal (strategic) form an *I-game* which is given as

$$G := \{S_1, \dots, S_n; C_1, \dots, C_n; R_1, \dots, R_n; V_1, \dots, V_n\}$$

where for player i, S_i is his strategy set, C_i is the (finite) set of the information vectors available to him, R_i is a function that assigns to every information vector $c_i \in C_i$ a (subjective) probability vector $R_i(c_{-i} \mid c_i)$ over the set $C_{-i} = C_1 \times \cdots \times C_{i-1} \times C_{i+1} \times \cdots \times C_n$, $c_{-i} = (c_1, \dots, c_{i-1}, c_{i+1}, \dots, c_n)$, V_i is his expected payoff function which assigns a real number (utility) to any n-tuple of strategies and information vectors using the probability distribution R_i. We can view the information vector c_i as representing certain physical, social, and psychological attributes of player i himself, in that it summarizes some crucial parameters of player i's own payoff function, as well as the main parameters of his beliefs about his social and physical environment including his beliefs about the beliefs of the other players. From this point of view, vector c_i may be called player i's *attribute vector* or *type.* R_i is player i's subjective probability distribution over the types of the other players conditioned on his own type.

Applying the Harsányi Doctrine, we assume that there is an objective probability distribution R^* over the product of the information sets (types) whose conditional probabilities coincide with the subjective conditional probabilities of the players, that is

$$R_i(c_{-i} \mid c_i) = R^*(c_{-i} \mid c_i)$$

holds for all attribute vectors and all $i = 1, \dots, n$. Then the I-game with the common prior R^* is called the *Bayesian game associated with* G and is denoted by

$$G^* = \{S_1, \dots, S_n; C_1, \dots, C_n; R^*; V_1, \dots, V_n\}.$$

With the help of the distribution R^*, we can now define the normal form $N(G^*)$ of an I-game G (and G^*) as

$$N(G^*) = \{S_1^*, \dots, S_n^*; W_1, \dots, W_n\}.$$

The strategy sets in $N(G^*)$ are the sets of the so-called *normalized strategies* which are functions from the range set of the information sets to the original strategy sets. In other words, a normalized strategy $s_i^* \in S_i^*$ tells player i what strategy to use from S_i for each possible value of his own information vector c_i. The payoff functions, $W_1, \dots, W_n$, are expected payoffs with respect to the information vector c using the objective probability distribution R^*. Notice that $N(G^*)$ does not include the c_i's and R^* any more.

If in the same game, instead of the distribution R^*, we use the conditional subjective distribution functions $R_i(c_{-i} \mid c_i)$ when defining the payoff functions, then we get the *semi-normal form* $SN(G^*)$ of G^* as

$$SN(G^*) = \{S_1^*, \dots, S_n^*; C_1, \dots, C_n; R^*; Z_1, \dots, Z_n\}$$

where the Z_i's are conditional expected payoffs obtained by using the conditional probabilities $R^*(c_{-i} \mid c_i)$ which are equal to $R_i(c_{-i} \mid c_i)$ by the Harsányi doctrine.

A strategy profile $s^* = (s_1^*, \dots, s_n^*)$ of game $SN(G^*)$ is said to be a *Bayesian equilibrium point* of the I-game G, if, for all $i = 1, \dots, n$, s_i^* is a best reply to the strategy profile $(s_1^*, \dots, s_{i-1}^*, s_{i+1}^*, \dots, s_n^*)$ of the rest of the players for all possible values of c_i (with the possible exception of a set of measure 0 in C_i).

Theorem 4 (Harsányi 1967). *Let G be an I-game and G^* the Bayesian game associated with G. A normalized strategy profile s^* is a Bayesian equilibrium point of G if and only if s^* is a Nash equilibrium point of the normal form $N(G^*)$.*

Harsányi's Bayesian approach is based on beliefs of players concerning certain parameters of the *I-game.* But it also involves beliefs about other players' beliefs and so on, leading to an infinite hierarchy of beliefs. The introduction of *type* has proved to be a useful tool to get around this difficulty operationally, but leaving a void for the formal mathematical treatment. Mertens and Zamir (1985) justified the validity of the Harsányi model with the strictest mathematical rigor.

Harsányi's model of games of incomplete information and the Harsányi doctrine has been so incorporated into modern economics that it is impossible to give a complete list of its applications. Harsányi himself used it in addressing other fundamental questions in game theory. We just mention two without going into any details.

One of the basic problems in noncooperative game theory is how to select a particular Nash equilibrium point of the mixed extension of a finite n-person noncooperative game if there are many equilibria. Additional assumptions and requirements have been proposed, as they say, to *refine* the equilibrium concept. The refinement due to Harsányi has become famous under the name: *tracing procedure,* Harsányi (1975). In this model, it is assumed that each player starts his analysis of the game situation by assigning a subjective prior probability distribution to the set of all pure strategies available to each other player. The Harsányi doctrine is also assumed: these distributions are conditionals of a common prior. Then the players will modify their subjective probability distributions in a continuous manner until all of these probability distributions converge to a specific equilibrium point of the game.

Another controversial issue is how to interpret mixed strategies for finite games. This is a problem when theory is confronted with practice: hardly anyone would randomize in the way it is assumed in classical models of game theory, that is, before each play of the game a lottery is performed and the players will do whatever is dictated by its outcome. In Harsányi's model, Harsányi (1973), a game with incomplete information is associated with the finite game under study. This game is obtained by a small perturbation of the payoffs. Then, as is proved by Harsányi, if the variations in payoffs are small, almost any mixed equilibrium of the finite game is close to a pure strategy equilibrium of the associated Bayesian game and vice versa. Thus even if no player makes any effort to use his pure strategies with the required probabilities, the random variations in the payoffs make every player choose his pure strategies with the right frequencies. The assumption of small variations is not very restrictive since payoffs are usually utilities of players whose estimations are never exact.

János von Neumann's and János Harsányi's contributions to modern game theory have had a great impact on the directions in which game theory has developed and their work is still a significant marker in contemporary research in the field. Hungarian mathematics, and science in general, must be very proud of their accomplishments and cherish the fame they have brought to Hungary. Though they got their first university degrees in Hungary, due to several reasons and special historical circumstances they lived their active life mostly abroad, thereby sharing the fame and publicity they earned between the homeland and the country they lived in.

Until the early sixties, any economic theory or methodology other than the Marxian was taboo in Hungary and completely missing from university curricula. A few professors at the University of Economics in Budapest realized in the early sixties that part of the methodology of modern economics, such as activity analysis with the mathematical programming underpinning and game theory when stripped of any ideology and used for the analysis of economic problems existing in modern societies, whether it be a free market economy or a (partially) planned and government controlled socialist economy, could be introduced in the curriculum of the university. They even selected a small, special group of students who were given a special curriculum heavily loaded with mathematics and "western-style" economics.

Professor **Jenő Szép,** head of the Mathematics Department at the University of Economics at that time and the late professor Béla Krekó were the main driving forces in this endeavor. Jenő Szép designed the first course in game theory which was based on Burger's book [20] and included both noncooperative and cooperative games. Out of these lectures, grew the first game theory book in Hungarian [176] co-authored by one of his former students, Ferenc Forgó. In addition to covering the standard topics, the introduction of the Nash equilibrium concept was embedded into a more general equilibrium model invented and studied by Jenő Szép (1970) himself.

The success of this book led to the German and the English version [176]. The English version served as text for game theory courses in several universities world wide. Later another author, Ferenc Szidarovszky of the University of Arizona joined Szép and Forgó to write an extended and improved text [45]. The specialty of this book is that it enhances the nonconvexities arising in various models of conflict situations.

Jenő Szép, as an educator, text-book author and inspirational source for generations of scientists and practitioners has done a lot to further the cause of meaningful and creative application of mathematics in economics in general and in game theory in particular.

References

[20] Burger, E., *Einführung in die Theorie der Spiele,* Walter de Gruyter (Berlin, 1966).

[45] Forgó, Ferenc – Szép, Jenő – Szidarovszky, Ferenc, *Introduction to the Theory of Games: Concepts, Methods, Applications,* Nonconvex optimization and its applications, Vol. 32, Kluwer Academic Publishers (Dordrecht–Boston, 1999).

[117] Morgenstern, Oskar – Neumann, John von, *Theory of Games and Economic Behavior,* Princeton University Press (Princeton, NJ, 1944); 3rd edition 1953.

[176] Szép, Jenő – Forgó, Ferenc, *Bevezetés a Játékelméletbe,* Közgazdasági és Jogi Könyvkiadó (Budapest, 1974). *Einführung in die Spieltheorie,* Akadémiai Kiadó (Budapest), Verlag Harri Deutsch (Frankfurt/Main, 1983). *Introduction to the Theory of Games,* Akadémiai Kiadó (Budapest), D. Reidel (Dordrecht, 1985).

Ferenc Forgó
Corvinus University
H-1093 Budapest
Fővám tér 8
Hungary

ferenc.forgo@bkae.hu

BOLYAI SOCIETY
MATHEMATICAL STUDIES, 14

A Panorama of Hungarian Mathematics in the Twentieth Century, pp. 549–554.

A Short Guide to the History of Hungary in the 20TH Century

JÁNOS HORVÁTH

These prefatory remarks are designed to help the reader with the biographical sketches which follow. They contain some information about the history of Hungary in the twentieth century.

For centuries, since the middle of the sixteenth century, the Kingdom of Hungary was part of the Habsburg Empire. In 1900 Hungary, or more precisely the Lands of the Holy Hungarian Crown, were part of the Austro–Hungarian Monarchy. Since the compromise of 1867 with Austria, Hungary was an independent kingdom united with Austria by the fact that the emperor of Austria was also the king of Hungary (thus Francis-Joseph, Emperor of Austria since 1848, became King of Hungary (1867–1916), followed by Charles IV). Each country had its own parliament and delegated a commission to discuss the common problems with a similar commission of the other country. However, the defense, finances and foreign affairs were common. Hungary covered the whole interior of the Carpathian basin and had a surface of 325,000 km^2 (125,500 square miles). This included Croatia: the kings of Hungary were also the kings of Croatia since 1102, with some interruptions, but Croatia had an autonomous status with an own assembly. Ethnic Hungarians (i.e., the people whose mother tongue was Hungarian) were a minority of about 48% in the country. The other nationalities, besides the Croats in the southwest, were Serbs in the south (and also further north along the Danube), Romanians in Transylvania in the east, Slovaks in the northwest (but also in many other parts of the country), Ruthenians (also called Carpatho–Ukranians) in the northeast, and Germans scattered in many parts of the country.

As World War I was ending, in October 1918 a revolution with a democratic liberal program broke out in Hungary. The country became an independent republic, separate from Austria, with count Mihály Károlyi as

President. At the same time Croatia and the Slovaks declared their independence from Hungary, and Romanian troops invaded the eastern part of the country. The Károlyi regime was replaced on March 21, 1919 by the Communist Hungarian Republic of Councils (Magyar Tanácsköztársaság), to which we will refer in the text by its better known names: "Hungarian Soviet Republic" or the "Commune". In the highly contagious climate of revolutionary euphoria, a surprising number of academics, intellectuals and artists lent their talents to the new administration's Utopian projects in education and the arts at the beginning of the communist rule; only to lose their enthusiasm after two or three months. E.g., László Dienes, brother of a mathematician mentioned in the text, was one of the commissars of Budapest. There were some atrocities committed; later the period was often referred to as "red terror", the number of victims is estimated to be about three hundred.

The Soviet regime was overthrown at the end of July when the Romanians invaded most of the country: the era of counterrevolution started. In November Admiral Miklós Horthy entered Budapest as the head of the "National Army". On March 1, 1920 Hungary was declared to be a kingdom, independent of Austria, but without a king: Horthy was appointed Regent. The peace treaty for Hungary was signed in Versailles on June 4, 1920. In Hungary it is referred to as the Trianon treaty (the act of signing took place in the Great Trianon Palace). Hungary lost over two thirds of its territory to the newly created countries of Czechoslovakia and Yugoslavia (then called Kingdom of the Serbs, Croats and Slovenes), to Romania, and a sliver to Austria: it became a small country with an area of 93,000 km^2 (36,000 square miles). Together with these territories, one third of ethnic Hungarians remained outside the borders of Hungary.

The counterrevolutionary period started with a "white terror" with considerably more victims than the "red terror". Some city councils were executed, and many innocent Jews were killed. A "numerus clausus" law introduced in 1920 restricted the admission of nationalities to the universities according to their proportion in the population but its purpose was in fact to limit the admission of Jews, who for the first time in the 20th century were considered an ethnicity and a religion. At that time, only people of Jewish religion were considered Jews, those who converted were not. There was a tacit antisemitic arrangement, according to which Jews were excluded from state employment; those already employed, with the very few exceptions of world-class professors, were sent into retirement or simply fired.

The next decade saw a gradual normalization of the country under the leadership of the conservative Prime Minister István Bethlen. First the murders stopped, then life slowly went back to almost normal. The "numerus clausus" law was modified: equal opportunity for Jews to enter university was restored. However, the great depression, the rise of Italian and German dictatorships, the fear of bolshevism, and the irresistible desire of territorial revision led the country to a fatal course leading to the losing side of a second world war, the repressive Jewish Laws of the late thirties, and the holocaust. First, seemingly, there were some successes.

After Czechoslovakia was dismembered, on November 2, 1938 the "first decision of Vienna" returned to Hungary the southern part of Slovakia inhabited mainly by Hungarians (12,000 km^2 = 4600 square miles). Then on March 15, 1939 Ruthenia (Carpatho–Ukraine) was incorporated into Hungary (another 12,000 km^2). On August 30, 1940, according to the "second decision of Vienna", the northern half of Transylvania (45,000 km^2 = 17,400 square miles) returned to Hungary. A large portion of this land was purely Hungarian, inhabited by the so-called "Secklers" (Székelyek). The decision was sponsored by Italy and curiously it was seconded by Germany, which had Romania as an even closer ally than Hungary. Finally on April 11–14, 1941 the Hungarians occupied the part of Yugoslavia called Vojvodina, which also had a large Hungarian population. However, by the end of the war all this was lost again.

Hungary declared war on the Soviet Union on June 27, 1941. (Later, in December 1941, Hungary declared war on the United States as well, under the initiative of then Prime Minister Bárdossy. The dubious glory of this act unique for a country of this size is only diminished by the fact that President Roosevelt did not take it seriously, and later in 1942 the US declared war on Hungary, Romania and Bulgaria at the same time.) A Hungarian army was mobilized and sent to the eastern front, and got almost annihilated in January 1943. Citizens who were not trusted to serve in the military, mainly Jews and Romanians, were drafted for labor service. Some were sent to the front, some worked in Hungary or in the occupied Yugoslav territory. Their treatment depended on the commanding officers. After the debacle of 1943, Hungary stopped sending soldiers to the front, and the government of Prime Minister Miklós Kállay started secret negotiations with the western allies concerning an armistice. The Germans got to know this, and in March 1944 summoned Horthy to Germany, required full participation of Hungary in the war, and occupied the country. Paradoxically, before the German occupation Hungary was an island of peace and safety in Europe, giving refuge, e.g., to many Poles or fugitive French prisoners of war. This abruptly ended. Again troops were sent to the east and the Germans managed to deport an estimated number of 440 thousand Jews of the provinces to concentration camps in Poland (Auschwitz) and Germany. (Those in the labor service were saved from deportation.) In July Admiral Horthy summoned a faithful motorized regiment to surround Budapest and to impede the deportation of the 200,000 Jews of Budapest. In spite of this belated effort, about two thirds of the Hungarian Jews were killed.

Horthy also sent secret negotiators to Moscow and an armistice was declared on October 15. Unfortunately most of the country was still in the hands of the Germans, who arrested Horthy and installed a Nazi ("arrow-cross") government of Hungary with Ferenc Szálasi as "nation leader". The Jews of Budapest started to be deported (several neutral embassies in Budapest issued passports for large numbers of Jews in Budapest, thus saving some of them from deportation; the Swedish diplomat Raoul Wallenberg, in 1945 deported by the Soviet army, is the best-known person leading such actions), a number were shot on the shore of the Danube, and the remainder transferred into a ghetto around the largest Budapest Synagogue. The arrow-cross had the plan to set fire to the ghetto and burn the Jews alive, but they were stopped from doing this by the German SS.

After a terrible siege of Budapest (when all the bridges over the Danube were blown up) and battles in the western part of the country (Transdanubia), the Soviet occupation ("liberation") of the country was completed on April 4, 1945. In the new peace treaty the territories regained in 1938–1941 were returned to the respective countries. Actually Czechoslovakia got a little more, but Ruthenia became part of Ukraine, then a republic of the Soviet Union. The Hungarian government tried to convince Stalin to leave Northern Transylvania in Hungarian possession, but Stalin answered that Romania had concluded an armistice before Hungary (namely on August 23, 1944, one day after the Soviet army had broken through the Romanian front).

For some time there was a restricted democracy in Hungary. The first elections were clean and fair for the parties admitted, with the Smallholders' Party winning absolute majority in the Parliament. Still the Communist Party was in a very strong position due to the presence of the Soviet army, and in 1947 the first signs were showing what was to come: several people (among them the son-in-law of President Zoltán Tildy) were executed on trumped-up charges or arrested and deported by the Soviet army (e.g., Béla Kovács, general secretary of the Smallholders' Party, who was among them, was imprisoned in the Soviet Union till 1955); the elections were cheated in order to get the Communists as the strongest party. The Social Democratic Party joined the Hungarian Communist Party to form the Hungarian Workers' Party (Magyar Dolgozók Pártja = MDP), and from 1948 this was the only party in the country. Hungary was declared to be a People's Republic, industry was socialized, and a large part of the agriculture reorganized by force into farmers' cooperatives. The reign of terror started in 1949. Many who belonged to the former bourgeois class were deported to small villages with just one suitcase, their apartments and belongings allotted to party members. When Marshall Tito of Yugoslavia quarrelled with Stalin, by some strange quirk of Marxist logic the authorities punished the "chained dog of the imperialists" by hanging several devoted Hungarian communists on preposterous charges. Thus László Rajk, a fighter in the Spanish civil war and minister of the interior who had organized the dreaded political police ("section for the protection of the state"), was accused, among others, of spying for the Gestapo. Many were sent into internment camps and concentration camps, of which the most infamous was in Recsk, in the wine-growing region around Eger (see: *"My Happy Days in Hell"* by the Hungarian poet György (George) Faludi, Andre Deutsch, London, 1962). Again, a kind of "numerus clausus" was introduced, this time not only at universities but also at secondary schools: children from the former upper-class families or families with a declared religious conviction were generally not admitted, and the number of students not from worker or peasant families became restricted.

In July 1956 the effective ruler of the country, Mátyás Rákosi was relieved of his duty as First Secretary of the MDP, and in October the Hungarian Revolution broke out (see: *"Ten days that shook the Kremlin"* by Tibor Méray, and *"The revolt of the mind"* by Tamás Aczél and Tibor Méray, Frederick A. Prager, New York, 1959). It seemed to be successful, but after 10 days, upon orders of the Presidium of the Soviet Communist Party, Soviet troops crushed the revolution, and its leaders (such as Imre Nagy, Pál Maléter) were executed a year and a half later. The MDP reformed itself under the name Hungarian Socialist Workers' Party (Magyar Szocialista Munkáspárt = MSZMP). The new regime of János Kádár became one of the longest (32 years) in the Hungarian history. The writer of this introduction did not live in the country during these years. After the retaliations it soon became clear that the regime was considerably milder than the preceding one. Thus in 1950–1956 travel to the west was restricted to the highest ranking party members, and while at the 1954 Amsterdam International Congress only two mathematicians from Hungary were present (Alexits and Rényi), at the 1958 Edinburgh Congress their number was 27. Beginning with the sixties, exit visas to leave the country for scientific or even touristic purposes were more and more easy to get, although they were by no means automatic. The relative freedom made it possible to sense the growing consensus in the country that "Existing Socialism does not work, and Working Socialism does not exist."

This led to the fact that, probably for the first time in history, in 1989–90 a ruling Communist party, the MSZMP handed over the power voluntarily to an emerging opposition.

Thereupon Hungary became a republic, there were clean elections, and the country has switched to a more or less successful market economy.

For further reading on the topic, see

- Ignác Romsics: Hungary in the 20th Century, Corvina, Budapest, 1999.
- Margaret Macmillan: Paris 1919, Random House, New York, 2002.

BOLYAI SOCIETY
MATHEMATICAL STUDIES, 14

A Panorama of Hungarian Mathematics in the Twentieth Century, pp. 555–562.

Education and Research in Mathematics

ÁKOS CSÁSZÁR

The purpose of this short introduction is to give a survey of the institutions in Hungary connected with research and the nurturing of young talents in mathematics, for readers who are not familiar with the subject.

1900–1920

After 1883, there were two kinds of secondary schools in Hungary which led to university studies: "gimnázium" and "reáliskola". Both started after four years of elementary school and had eight grades. In the humanistic gimnázium Latin and Greek were taught, while the reáliskola (like its model, the Austrian "Realschule") put emphasis on science, mathematics, descriptive geometry, and modern languages. Both schools ended with the "érettségi" examination (Reifeprüfung, Matura or Abitur in German, Baccalauréat in French), with written as well as oral examination. Passing the érettségi at a gimnázium gave right to enter any university, while the érettségi at a reáliskola gave right to studies at the Technical University, Faculties of Science, and to the Academies of Mining, Forestry, and Economics. The forming of secondary school teachers was directed by Teacher's Colleges attached to the universities. The College in Budapest was created by Mór von Kármán (father of Theodore von Kármán) and had a practice school called "Mintagimnázium" (minta = model), which was one of the strongest gimnáziums of Hungary (e.g., T. Kármán, E. Teller, D. König, P. Lax studied there). Otherwise, secondary schools were either state schools or church schools. Some of the church gimnáziums were famous top-standard schools; in bringing up talents in mathematics, the Lutheran Evangelical Gimnázium in Budapest (A. Haar, J. Harsányi, J. von Neumann, E. Wigner), the Benedictine Gimnázium in Győr (J. Farkas, J. König,

J. Pál, F. Riesz, M. Riesz), and the Piarist Gimnázium in Budapest (L. Grosschmid, G. Hajós, E. Makai) were the most successful among them.

At the beginning of the century, there were two universities in Hungary: one in Budapest founded in Nagyszombat [today Trnava, Slovakia] in 1635 and another in Kolozsvár [today Cluj-Napoca, Romania] established in 1872. There was also a Technical University in Budapest active in forming engineers and specialists in economics. As research centres in mathematics, the Budapest Technical University and the University of Kolozsvár both played an important role, while the University of Budapest gained considerable influence in mathematical activities only in the first decade of the century.

In the 1910's two more universities were established, one in Pozsony [today Bratislava, Slovakia] and one in Debrecen, with very modest activity in mathematics until the end of the period.

As in other countries, future researchers in mathematics usually took the degree "Doctor of Philosophy" at a university. After having written a "Habilitation" thesis and presented an inaugural lecture, one was awarded the title of "private professor" (magántanár, Privatdozent in German, the term we shall use in the Biographies). This title gave the right to teach at the university (venia legendi) but involved only a minimal financial remuneration. However, secondary school teachers of state schools who became Privatdozents had their teaching load of 18 weekly hours reduced to 12.

The Hungarian Academy of Sciences had about a dozen members engaged in research in mathematics, grouped in the Section for Mathematics and Physics. The Mathematical and Physical Society was founded in 1894 and had about 400 members. Both the Academy and the Society organized sessions with lectures on various subjects in mathematics and physics, mostly connected with the research results of the lecturer.

There were two periodicals in which research papers in mathematics appeared in Hungarian: Mathematikai és Természettudományi Értesítő [Mathematical and Scientific Bulletin] published by the Academy, and Mathematikai és Physikai Lapok [Mathematical and Physical Journal] published by the Society. The Academy also had a periodical containing papers in foreign languages: Mathematische und Naturwissenschaftliche Berichte aus Ungarn.

Beginning with 1894, the Mathematical and Physical Society organized every year a contest in mathematics for those who have just graduated from secondary school. After the death in 1919 of the physicist Loránd Eötvös, first president of the Society, this contest was named "Eötvös Competition". Every time, three problems were to be solved during a period of four hours, and any kind of resource was allowed to be used.

Many eminent Hungarian mathematicians began their mathematical careers by winning a prize at this competition. A few names of these famous winners are: L. Fejér (1897), Th. von Kármán (1898), D. König (1902), A. Haar (1903), M. Riesz (1904), F. Lukács (1909), G. Szegő (1912), T. Radó (1913), L. Rédei (1918), L. Kalmár (1922), E. Teller (1925). The solutions of the problems of the Eötvös Competitions were collected in the two volumes of the Hungarian Problem Book, first edited 1929 in Hungarian by J. Kürschák, and later published in English translation.

Every year, the Ministry of Education organized a competition, similar to the "Concours Général" in France, in every subject.

There was also a periodical (Középiskolai Mathematikai és Physikai Lapok [Mathematical and Physical Journal for Secondary Schools]), founded in 1894 by D. Arany, that published educational papers in mathematics and physics and, primarily, mathematical and physical problems for readers aged between 14 and 18. Solutions submitted by the readers constituted the major part of the content. Many future mathematicians started as authors of one or more solutions in this journal. The photos of the most successful problem solvers were published in the last issue of each year; therefore the best solvers had often known one another from these photos before they personally met at the university.

In order to train high level gimnázium teachers, the Eötvös Collegium (Eötvös College) was founded as early as 1895, following the example of the Ecole Normale Supérieure; it was a boarding-school for university students preparing themselves for the vocation of teacher. There were first class researchers in mathematics who had been members of this College (e.g. M. Riesz).

1920–1945

Beginning with 1924, a third kind of secondary schools was established, the so-called "reálgimnázium", a hybrid of gimnázium and reáliskola: beside mathematics and sciences, it had eight years of German, six years of Latin, four of a second modern language (French, English or Italian), and no Greek. And then, in 1934, the secondary school system was again reformed and made more uniform, by establishing theory-oriented gimnáziums, practice-oriented "líceum"-s and specific practice-oriented schools.

The peace treaty of Versailles that ended the first world war contained seriously damaging decisions for Hungary; about 2/3 of the territory of the country was detached in favour of other countries. Also three and a half million ethnic Hungarians became citizens of other countries.

In particular, Kolozsvár was annexed by Romania under the name Cluj, and the former University of Kolozsvár continued its activity in Szeged from 1921. Similarly, the city of Pozsony went to Czecho-Slovakia with the name Bratislava; the former University of Pozsony worked in Pécs from 1921, but, for lack of a faculty of science, it did not influence research in mathematics.

On the other hand, the universities in Budapest and Szeged and, to a certain extent, the University of Debrecen and the Budapest Technical University continued to be research centres in mathematics.

The list of periodicals publishing research papers in mathematics was enriched with an important new journal, the Acta Scientiarum Mathematicarum, published by the University of Szeged from 1922, containing papers written in foreign languages. Due to the activity of the editors, in the first place F. Riesz, A. Haar and B. Kerékjártó, it reached very soon the rank of an important periodical in mathematics. Its name was often changed: Acta Literarum ac Scientiarum Regiae Universitatis Hungariae Francisco-Josephinae, Sectio Scientiarum Mathematicarum; later: Acta Universitatis Szegediensis, Sectio Scientiarum Mathematicarum; after the second world war: Acta Scientiarum Mathematicarum (Szeged); now: Acta Universitatis Szegediensis, Acta Scientiarum Mathematicarum.

During the second world war from 1939 to 1945, Hungary regained a part of the territories lost in the Versailles treaty; therefore the University of Kolozsvár continued its activity for a few years as a Hungarian university, while the University of Szeged remained intact with about the same faculty as between 1921 and 1940.

The Academy, the Mathematical and Physical Society, and the earlier periodicals connected with mathematics continued their activity roughly in the same way as before 1920.

1945–2000

In 1948–49 the school system underwent a major change. The compulsory primary school was extended to eight years. A unified gimnázium had four years, with specialization within the school. Most of the church schools were taken over by state. Beside the gimnáziums, secondary technical schools in specific directions were also established; they also ended with a specialized "érettségi" examination. However, the érettségi no longer gave right to register at university. For that, entrance examinations had to be passed, in their oral part (which included also questions on political issues and the student's opinion on them) favouring students from worker or peasant families.

In 1966, the Fazekas Mihály Gimnázium in Budapest opened a class specialized in mathematics. In subsequent years, several other gimnáziums followed the example, still the Fazekas Gimnázium could maintain its leading position, many of the present leaders of Hungarian mathematics are its alumni.

The fall of the communist system in 1989 resulted also in major changes in schooling. The school landscape became rather chaotic, with four-year, six-year, and eight-year gimnázium curricula, sometimes even in the same school. Several church schools have been returned gradually to their previous owners.

After the second world war, Hungary was restricted again to the territory possessed in the 1920's.

The universities influencing mathematical research were, a few years after 1945, the same as before (University of Budapest, University of Szeged, University of Debrecen, Budapest Technical University). Beside them, in Cluj (Kolozsvár, Romania), the newly founded Bolyai University offered curricula in Hungarian until 1959 when it was unified with the Romanian Babeş University; after this, the number of courses in Hungarian was strongly reduced at the Babeş–Bolyai University.

Rather soon new technical universities were established in Miskolc and Veszprém, and the teaching of specialists in economics was split off from the Budapest Technical University and the Budapest University of Economics was founded. A number of colleges were also founded in order to train specialists in various fields (e.g. teachers or engineers); a part of these colleges had some contact with research in mathematics.

In 1950, scientific titles were reorganized following the Soviet system. The doctoral degree granted by universities was abolished (and reintroduced in the mid-sixties). The title Privatdozent was also abolished (but many more paid posts created at universities.) Two new degrees were created, granted by the Committee for Scientific Qualification, with members from universities and academical institutes. Requirements for the Candidate degree ("Candidate of Science") were higher than for a university doctorate. The highest degree was "Doctor of Science", for which usually a very substantial thesis was required, as well as taking active part in the life of the scientific community. In the mid-nineties issuing new Candidate degrees was stopped and, instead of the earlier "doctorates", Ph. D. became the official name of the degree granted by universities. The "Doctor of Science" degree was changed into the title "Doctor of the Hungarian Academy of Sciences", awarded by the Academy under conditions similar to the requirements for the earlier Doctor of Science degree.

The Hungarian Academy of Sciences founded a new research institute of applied mathematics in 1948 attached to the Budapest Technical University. In 1950 this institute became an independent Research Institute of Applied Mathematics, later renamed Mathematical Research Institute and, since 1999, Alfréd Rényi Institute of Mathematics.

In 1973, the Hungarian Academy of Sciences founded the Computer and Automation Research Institute, containing sections working in mathematical research.

The Section for Mathematics and Physics of the Academy was split in 1993 into a Section for Mathematics and another for Physics. The number of ordinary and corresponding members belonging to the Section for Mathematics is about 25; there are also external members, i.e. Hungarian scientists living abroad, and honorary members, i.e. foreign scientists having a sort of connection to Hungarian science. External members living in neighbouring countries have the option to spend longer periods of time in Hungary in the house of the foundation Domus Hungarica to carry out research.

The Mathematical and Physical Society split, and continued its activity from 1947 as János Bolyai Mathematical Society and Loránd Eötvös Physical Society, respectively. As to the former, it began in the 1960's to organize international conferences on various domains of mathematics and to publish proceedings of many of these conferences. The series Colloquia Mathematica Societatis János Bolyai, started in 1969 (from 1993 Bolyai Society Mathematical Studies), is a very important contribution to the publication of research papers in Hungary. Until 2000 more than 70 volumes appeared in these series.

The Academy and the Bolyai Society organized jointly two Hungarian Mathematical Congresses in 1950 and in 1960, with a number of foreign participants. The Proceedings of the first one was published in 1952.

The Hungarian Academy of Sciences started in 1974 a series of monographs with the title Disquisitiones Mathematicae Hungaricae (partly in Hungarian, but mostly in foreign languages).

The list of periodicals intended to publish new results in mathematics has been enriched by a large number of new journals. So from 1946 to 1949, four issues of Hungarica Acta Mathematica were edited, and the Hungarian Academy of Sciences started Acta Mathematica Academiae Scientiarum Hungaricae in 1950, since 1983 called Acta Mathematica Hungarica. The Mathematical Research Institute launched Studia Scientiarum Mathematicarum Hungarica in 1966. The János Bolyai Mathematical Society has been publishing Periodica Mathematica Hungarica since 1971. The University of Debrecen started Publicationes Mathematicae in 1949 and the University of Budapest the Annales Universitatis Scientiarum Budapestinensis de Lorando Eötvös Nominatae, Sectio Mathematica, in 1958, and Sectio Computatorica, in 1978. There are also two periodicals published in international cooperation: Analysis Mathematica (Hungarian Academy of Sciences and Academy of the USSR, from 1975 to 1991, Hungarian Academy and Russian Academy from 1992) and Mathematica Pannonica (Technical University of Miskolc till 1996, then University of Pécs [Hungary], Montanuniversität Leoben [Austria], University of Trieste [Italy]), started in 1990, and two journals with international Editorial Boards: Acta Cybernetica, launched by the University of Szeged in 1969, and Combinatorica, a journal of the Bolyai Society started in 1981.

All these journals publish articles in foreign languages. There are also a number of periodicals containing research and expository papers in Hungarian. The Academy published A Magyar Tudományos Akadémia Matematikai és Fizikai Tudományok Osztályának Közleményei [Communications of the Section for Mathematical and Physical Sciences of the Hungarian Academy of Sciences] between 1955 and 1977. The Academy launched also Alkalmazott Matematikai Lapok [Journal for Applied Mathematics] in 1975, which was taken over by the Bolyai Society in 1997. The Mathematical Research Institute published A Magyar Tudományos Akadémia Matematikai Kutatóintézetének Közleményei [Communications of the Research Institute for Mathematics of the Hungarian Academy of Sciences] from 1952 to 1964;

its continuation is Studia Scientiarum Mathematicarum Hungarica in foreign languages. The János Bolyai Mathematical Society publishes, as a continuation of Mathematikai és Physikai Lapok, from 1950 Matematikai Lapok [Mathematical Journal]. Mathematical research papers also appear in general (not specialized mathematical) journals of several universities.

The publication of Középiskolai Matematikai és Fizikai Lapok [Mathematical and Physical Journal for Secondary Schools] has been continued by the two societies (János Bolyai Mathematical Society and Loránd Eötvös Physical Society). After some issues published in English at special occasions, the journal has a regular English version since 2002. On the website www.komal.hu of the journal, much is accessible also in English.

The number of mathematical competitions was essentially increased. The former Eötvös Competition is called, since 1949, Kürschák Competition. (The Eötvös Competition is since that time the name of a similar competition in physics, organized by the Eötvös Society.)

There is, since 1949, a Miklós Schweitzer Competition, organized for university students; this is a homework competition in which ten (or more) problems are to be solved during ten days. The solutions of the problems from the competitions between 1949 and 1961 were published in 1968 in the book Contests in Higher Mathematics.

Since 1947, there is a Dániel Arany Competition, organized in several series for secondary school students of lower grades.

All of these competitions are organized by the Bolyai Society. Beside them, the Ministry still organizes the yearly competitions for the students of the last grades. There are also local competitions in mathematics, organized in several cities by the local section of the Bolyai Society.

Since 1992, secondary school students in Hungary and in neighbouring countries have an International Hungarian Mathematical Competition, organized more or less similarly to the International Mathematical Olympiads. As to the latter, Hungarian students always achieve impressive results.

Finally, it is worth mentioning that the Középiskolai Matematikai és Fizikai Lapok still runs its yearly problem-solving competitions, and continues to publish the photos of the best solvers. Hopefully, this competition will bring a rich supply of researchers in mathematics in the future generations as well.

BOLYAI SOCIETY
MATHEMATICAL STUDIES, 14

A Panorama of Hungarian Mathematics in the Twentieth Century, pp. 563–607.

Biographies

TÜNDE KÁNTOR-VARGA

A short note on the use of names of persons and institutions, and on pronunciation:

Quite a number of Hungarian mathematicians used several ways of writing their names on their publications during their career. In particular, till about 1950 it was a wide-spread habit to translate first names when writing in a foreign language, and use a generally accepted 'corresponding name' if the first name had no version in the given language. After the official (Hungarian) name of an author who followed this usage, we indicate other forms of the name used in publications which might be necessary for identifying the author, but we shall avoid, say, listing all the three of Georg, George, and Georges – we put only one of such easily recognisable variants.

Most universities in Hungary changed their names several times during the twentieth century. To cause less confusion, we always write University of Budapest, University of Debrecen, University of Szeged, and Budapest Technical University instead of the name valid in the given year. Likewise, we always write Mathematical Research Institute for the Mathematical Institute of the Hungarian Academy of Sciences, which also had three names during its history.

Finally, some words about how to pronounce Hungarian names. Writing is by and large phonetic, so knowing how to pronounce letters and some combinations of letters means that one is more or less on the safe side.

vowels:

a – as in n*au*ght but short; **á** – as in f*a*ther; **e** – as in l*e*t; **é** – as in c*a*pe or French th*é*; **i** – as in *i*t; **í** –as in sh*ee*t; **o** – as in n*o*t in Scottish pronunciation or French p*o*mme or German h*o*ffe; **ó** – as in *a*ll or French b*eau* or German B*oo*t; **ö** – as in French b*oeu*f or German L*ö*ffel; **ő** – as in French d*eu*x or German sch*ö*n; **u** – as in p*u*t; **ú** – as in m*oo*d; **ü** – as in French t*u* or German d*ü*nn; **ű** – as in French m*u*r or German fr*üh*; **y** (at the end of family names) – as in *i*t.

There are no diphthongs in Hungarian except **au** (as in n*o*w) in some words of foreign origin.

Notice that the accents ´ or ˝ on a vowel mean that the vowel is long. Stress is always on the first syllable of a word, irrespective of lengths – e.g., in the name Fejér the *e* is short and stressed whereas the *é* is long but not stressed.

consonants:

represented by single letters: **b**, **d**, **f**, **k**, **l**, **m**, **n**, **p**, **r**, **t**, **v**, **z** are pronounced as you would expect;

c – as in *ts*ar; **g** – as in *g*et; **h** – as in *h*ouse, except at the end of a word where it is not pronounced; **j** - as in *y*es; **s** – as in *sh*eet; **w** – as in *v*eal; **x** – as in a*x*e;

double letters, such as **ll**, **nn**, **ss** etc. represent consonants which are pronounced long like *n* in u*nn*atural;

diagraphs, i.e., combinations of (usually) two letters which represent a single speech sound as *gh* in lau*gh*: **cs** – as in *ch*ur*ch*; **gy** – as in *d*uty; **ly** – as in *y*es; **ny** – as in o*ni*on; **sz** – as in *s*in (notice the difference to the Polish pronounciation of sz! this can be the source of some embarrassment); **ty** – as in vir*t*ue; **zs** – as in French *j*our; notice that in their long version the first letter is doubled, hence **ssz** – as in di*ss*erve.

Finally, there are also some traditional diagraphs preserved only in family names: *cz* stands for *c* (so the name "Vincze" is pronounced vin*ts*e, not vin*ch*e as it would be in Polish); *ts* stands for *cs*, *eő* stands for *ő*.

BIOGRAPHIES

ALEXITS György (also George) (Budapest, January 5, 1899 – Budapest, October 14, 1978) began his studies at the University of Budapest in 1917. During the Communist rule in 1919 he joined the socialist student federation, and therefore emigrated to Austria later in the same year. He continued his studies in Graz, where he obtained the Ph.D. in 1924. Returned to Hungary in 1924 and worked at an insurance company till 1926. Then he moved to Romania to help found the Romanian Communist Party; in 1926–27 he taught at a secondary school in Giurgiu and lectured also at the University of Bucharest. In 1927 he returned to Hungary and was a schoolteacher till 1940. In 1940–41 he taught at the University of Kolozsvár, from 1941 till 1944 at the Budapest Technical University. He took part in the resistance against the German occupation and was therefore deported to Dachau in 1944. After returning in 1945, he was school director till 1947. He became Deputy Minister of Education (1947–48). From 1948 to 1967 he was a professor at the Budapest Technical University, from 1967 to 1970 he worked at the Mathematical Research Institute. He was elected corresponding member of the Hungarian Academy of Sciences in 1948, ordinary member in 1949, and in 1949–50 the first Secretary General of the reorganized Hungarian Academy of Sciences. He was awarded the Kossuth Prize in 1951 and 1970, the Tibor Szele Prize in 1976. The main area of his scientific interest was geometry and analysis, mainly orthogonal series.

ALPÁR László (Nagyvárad [now Oradea, Romania], January 29, 1914 – Budapest, September 18, 1991) won the Eötvös Competition in 1932 and entered the University of Budapest but was soon expelled as a Communist. After holding several jobs and being arrested for political reasons several times, in 1937 he emigrated to France and continued his studies at the Sorbonne in Paris. Suspect of being Communist, he was interned from 1939 to 1944, then he escaped and joined the Résistance. He returned to Hungary in 1945 and worked in the trade union movement till 1949 when he was imprisoned in connection with a political process and interned in labour camps till 1953. Returned to a job in industry in 1953, was rehabilitated in 1956, then completed his studies at the University of Budapest and obtained his teacher's degree in 1957. From 1956 to 1991 worked at the Mathematical Research Institute. Obtained the degree of Candidate in 1961, Doctor of Mathematical Sciences in 1978. Field of research: complex function theory.

ARANY Dániel (Pest, July 11, 1863 – Budapest, January, 1945) entered the University of Budapest. After receiving his teacher's diploma was an assistant at the Forestry School in Selmecz (now Banska Štiavnica, Slovakia). In 1893 returned to teach at the secondary school he had attended in Győr. During a visit in Paris with a scholarship became acquainted with the Journal de Mathématiques Élémentaires, which gave him the idea of starting the Középiskolai Matematikai Lapok (Mathematical Journal for Secondary Schools, abbr. KöMaL) in 1894, the second such journal in the whole world. He moved to Budapest in 1894 and passed the editorship of KöMaL to László Rátz, who edited it until 1914. Arany was a secondary school teacher, active in nurturing young mathematical talent. He wrote a series of mathematics textbooks for students aged 12 to 18. Was sent into retirement in 1919, after which he worked as an actuary.

BAKOS Tibor (Szeged, June 8, 1909 – Budapest, December 12, 1998) won first prize in the Eötvös Competition in 1926, studied at the University of Budapest from 1926 to 1931 and was a member of the Eötvös Collegium. He became a schoolteacher and taught at various secondary schools till 1944. After captivity as a prisoner of war, he returned to Szeged in 1947. Three more years of teaching at a Gimnázium followed, then he got a position at the University of Szeged. From 1958 to 1974 he was the Editor in Chief of the secondary school journal KöMaL (see above), and even after retirement he was an active member of the Editorial Board till his death. He was a constant member on the Committees of several mathematical contests. He twice got the Beke Manó Prize (1957, 1965).

BÁLINT Elemér (Budapest, December 4, 1888 – Budapest, August 20, 1967) studied at the Budapest Technical University and obtained his teacher's diploma in 1911. He worked as a teacher and also taught descriptive geometry at the Technical University. Obtained his Ph.D. with a dissertation on the roots of polynomials. Did military service during World War I, lost his teaching position in 1919, and then worked in insurance. In 1951 he was appointed professor first at the Technical College, then at the Budapest Technical University. Areas of scientific interest: polynomials, numerical and graphical approximation methods.

BALOGH Zoltán (Debrecen, December 7, 1953 – Oxford, Ohio, USA, June 19, 2002) studied at the University of Debrecen. He took positions at the University of Debrecen and, from 1988, at Miami University in Oxford, Ohio, USA. He obtained the title of Candidate in 1980, and the title of

Doctor of Mathematical Sciences in 1989. Field of research: set-theoretical topology.

BARANYAI Zsolt (Budapest, June 23, 1948 – Budapest, April 18, 1978) studied at the University of Budapest from 1967 to 1972, took his Ph.D. at the same university in 1975 and obtained the title of Candidate posthumously. Had a position at the University of Budapest from 1972 till his untimely death. Beside working in mathematics, he was a widely known concert player on the recorder. Area of research: combinatorics, his most famous result is on hypergraphs.

BARNA Béla (Máramarossziget [now Sighet, Romania], March 30, 1909 – Debrecen, June 9, 1990) entered the University of Budapest in 1926 and switched to Debrecen in 1928. Obtained his teacher's diploma in 1931 and a Ph.D. in 1932. Lectured without salary at the University of Debrecen on number theory until 1935, then taught at various secondary schools. In 1951 he joined the staff of the University of Debrecen. In 1957 he became Candidate and in 1967 Doctor of Mathematical Sciences. Scientific interest: mean values and zeros of functions.

BAUER Mihály (also Michael) (Budapest, September 20, 1874 – Budapest, March 2, 1945) started publishing at the age of l8, and obtained his teacher's diploma in Budapest. After graduating he held various positions at the Budapest Technical University. In 1922 he was the first winner of the Gyula Kőnig Prize of the Eötvös Loránd Mathematical and Physical Society. He was sent into early retirement in 1936. Main areas of interest: algebra, number theory.

BEKE Manó (Pápa, April 24, 1862 – Budapest, June 27, 1946) entered the Budapest Technical University but soon switched to the University of Budapest. He obtained his teacher's diploma in 1883 and his Doctorate in 1884. He taught at a secondary school in Budapest until 1895, and spent the year 1892/93 in Göttingen with a scholarship. He became aware of the activities of Felix Klein concerning the reform of the teaching of mathematics, and after his return he became the leader of the reformers of teaching mathematics in Hungary. In 1895 he started teaching at the Mintagimnázium in Budapest. In 1900 he was appointed professor at the University of Budapest. Between 1906 and 1909 he was the Chairman of the Commission to reform the teaching of mathematics in secondary schools. Became a corresponding member of the Hungarian Academy of Sciences in 1914. After World War I the Council of the University started a disciplinary investigation against him because of his political activity, and in 1922 he was stripped of his position at the University and of his membership in the Academy, following which he worked for a publisher. His book on Differential and Integral Calculus was the textbook for several generations of mathematics students. Areas of scientific interest: linear differential equations, determinants, problems from physics. In 1951 the Bolyai János Mathematical Society inaugurated the Beke Manó Prize for outstanding results in the teaching and the popularization of mathematics.

BIHARI Imre (Budapest, June 20, 1915 – Budapest, September 9, 1998) studied in Budapest between 1933 and 1938. Taught in secondary schools in Budapest in 1942–50, then had a position at the Budapest Technical University until 1962. Obtained the title of Candidate in 1962 and of Doctor of Mathematical Sciences in 1979. In 1962 he switched to the Mathematical Research Institute. Scientific area: non-linear differential equations: oscillation, inequalities.

BIRÓ Balázs (Budapest, September 21, 1955 – Budapest, December 27, 1997) graduated from the University of Budapest in 1979. First he worked as a programmer at an institute for computation, then from 1983 he was at the Mathematical Research Institute, first with a research scholarship and from 1986 in a regular position. He obtained the title of Candidate in 1989. Field of research: algebra, algebraic logic.

BORBÉLY Samu (also von Borbély) (Torda [now Turda, Romania], April 23, 1907 – Budapest, August 14, 1984). Studied at the Technische Hochschule Charlottenburg in Berlin from 1926. In 1929 he was a student-assistant at the Applied Mathematics Department of R. Rothe and later at the Mathematics Department of G. Hamel. Obtained the diploma of Engineer-Mathematician in 1933 and at the same time was appointed assistant. In 1934 he switched to the Institute for Aviation Technology of the Technical University. In 1938 he obtained the title of Doctor of Engineering (Dr. Ing.) with his work on airfoil theory. In 1940 a position was offered to him at the University of Kolozsvár, where he held various positions till 1944. As an expert in aviation, he was arrested May 4, 1944 by the Gestapo and transported to Berlin. He was set free in the middle of September, 1944. He returned to Kolozsvár in 1945, where he became a professor in the Department of Mathematics and Geometry at the short-lived Bolyai University. He returned to Hungary and had positions from 1949 to 1955 at the Technical University for Heavy Industry in Miskolc, from 1955 to 1977 at the Budapest Technical University, between 1960 and 1964 at the University of Magdeburg. He became a corresponding member of the Hungarian Academy of Sciences in 1946, and ordinary member in 1979. Scientific interest: mathematical physics, numerical and computerized integration.

BRINDZA Béla (Csongrád, September 18, 1958 – Debrecen, November 2, 2003) studied a year in Szeged and then in Debrecen. After his studies he worked mainly at the University of Debrecen but also spent 3 years with a research fellowship at Macquarie University in Sydney, Australia, and taught 4 years at the University of Kuwait. Obtained the title of Candidate in 1986 and of Doctor of Mathematical Sciences in 1999. Field of research: diophantine equations.

BUZÁSI Károly (Püspökladány, July 16, 1930 – Debrecen, August 7, 1988). His family moved to Ungvár [later Uzhgorod, Soviet Union, now Uzhhorod, Ukraine] in 1940. After graduating from secondary school he registered at the University of Uzhgorod as a student by correspondence because at the same time he taught as a schoolteacher. In 1953 he became a regular student. Obtained his degree and moved to Hungary in 1956. Taught at secondary schools in Debrecen until 1961. From 1961 to 1988 he held various positions at the University of Debrecen. Was a guest professor at the University of Plovdiv (Bulgaria) in 1963/64, obtained his degree of Candidate in Kharkov, Soviet Union, in 1968. The degree of Doctor of Mathematical Sciences was awarded to him posthumously. Areas of scientific interest: representations of finite groups, coding theory.

CSILLAG Pál (also Paul) (Budapest, April 17, 1896 – Budapest, December 24, 1944) studied in Budapest and obtained the Ph.D. from the University in 1921. Worked as a mathematician in industry. Areas of interest: harmonic analysis, complex function theory.

CZIPSZER János (Budapest, November 16, 1930 – Budapest, June 15, 1963) won first prize in the Kürschák competition in 1948, studied at the University of Budapest. In 1953 he joined the Mathematical Research Institute. He collaborated with several distinguished authors, but did not seek scientific degrees because of exaggerated modesty. Scientific areas: approximation, geometry, topology, applied mathematics.

DÁVID Lajos (also Ludwig von Dávid) (Kolozsvár [now: Cluj-Napoca, Romania], May 28, 1881 – Leányfalu, January 9, 1962) studied at the University of Kolozsvár. Obtained his Doctor's Diploma in 1904, after which he spent time in Göttingen and in Paris. Taught at various secondary schools. Habilitated in Kolozsvár in 1910. From 1914 taught in Budapest and in 1916 habilitated in Budapest. From 1919 to 1929 he was a professor at a Teacher's College in Debrecen, and from 1925 he also lectured at the University of Debrecen, where he became a professor in 1929. He organized the Mathematical Seminar and Library in Debrecen and was the founding editor of the journal Proceedings of the Mathematical Seminar of the University of Debrecen (in Hungarian, 1928–1940). Between 1940 and 1944 he was a professor at the University of Kolozsvár. In 1944 he returned to Budapest. His main scientific interest was the work of Gauss and of the Bolyais.

DEÁK Jenő (Budapest, March 1, 1948 – Budapest, February 8, 1995) studied at the University of Budapest. He obtained the degree of Candidate in 1980. Worked at the Mathematical Research Institute. Areas of interest: summability, general topology.

DÉNES József (Budapest, April 16, 1932 – Budapest, August 19, 2002) attended the University of Budapest. He obtained the degree of Candidate in 1961. Held various positions in applied mathematics and computer science in Budapest: 1954–1964 Ministry of the Interior, 1964–1969 Central Research Institute of Physics, 1969–1988 Institute for Coordination of Computer Techniques. Main fields of interest: combinatorics, combinatorial methods in algebra, coding theory.

DIENES Pál (also Paul) (Tokaj, November 24, 1882 – Turnbridge (England), March 23, 1952) studied at the University of Budapest but spent a year in Paris, where he came into contact with Jacques Hadamard, Émile Picard and Paul Appell. Starting with 1904 taught at a secondary school in Budapest. Returned to Paris in 1908–10, where he obtained the degree of "docteur ès sciences" at the Sorbonne. Dienes came from a politically left-wing family. As a Communist, he became the Political Comissar of the University of Budapest during the Hungarian Communist regime in 1919, and was charged with organizing the School of Science of the newly founded University of Debrecen. After the fall of the Soviet Republic, he fled Hungary and obtained a position at the University of Wales with the help of Hadamard in 1921. Moved later to Swansea and in 1929 to Birkbeck College in London. Became a professor in 1946 at the age of 63. Areas of interest: complex analytic functions, differential geometry, infinite matrices.

EGERVÁRY Jenő (also Eugen) (Debrecen, April 16, 1891 – Budapest, November 30, 1958) studied at the University of Budapest, from where he obtained his Doctor's Diploma in 1914. He worked first at the Seismological Observatory in Budapest and then at a secondary school. In 1932 he obtained the Gyula Kőnig Prize. In 1938 he habilitated at the

University of Budapest, and was appointed professor at the Budapest Technical University in 1941, where he organized a research group in applied mathematics. In 1950 he became the Chairman of the Scientific Council of the newly founded Research Institute for Applied Mathematics of the Hungarian Academy of Sciences, predecessor of the Mathematical Research Institute. He was elected a corresponding member of the Academy in 1943, ordinary member in 1946, and received the Kossuth Prize in 1949 and in 1953. He was the editor of the problems section of several journals. His scientific interest was very broad: classical function theory (polynomials and trigonometric polynomials), geometry, differential and integral equations, matrix theory and numerical methods, optimization, mathematical physics, applications of mathematics to engineering.

ELBERT Árpád (Kaposvár, December 24, 1939 – Budapest, April 25, 2001) studied at the University of Budapest, after which he became a member of the Mathematical Research Institute. Candidate in 1971, Doctor of Mathematical Sciences in 1989. Scientific interest: ordinary differential equations, semilinear differential equations, functions of a real variable, special functions.

ERDŐS Jenő (Hajdúszovát, June 7, 1931 – Debrecen, January 16, 2004) studied at the University of Debrecen, then taught at the same university. He took the Candidate degree in 1960. Field of research: abelian groups.

ERDŐS Pál (also Paul) (Budapest, March 26, 1913 – Warsaw, Poland, September 20, 1996). Both his parents were mathematics-physics teachers who tutored him at home. In secondary school, he was one of the most successful problem solvers of KöMaL, among whom his later group of friends was recruited: Tibor Gallai, Géza Grünwald, Pál Turán, Eszter Klein, Márta Wachsberger, György Szekeres, Endre Vázsonyi, László Alpár, Dezső Lázár. He studied in Budapest from 1930 to 1934, then with the help of L. J. Mordell obtained a scholarship to Manchester, UK, where he met the first ones of his more than four-hundred co-authors: György Pólya, Richard Rado, Stan Ulam, Chao Ko, Harold Davenport. In October 1938 he left Europe to spend a year and a half at the Institute for Advanced Study in Princeton, New Jersey, USA, then had temporary positions at various universities in the USA, and after World War II he spent also some time in Great Britain. In the United States he got into trouble because he corresponded with the Chinese mathematician Loo-Keng Hua, and when he wanted to go to the 1954 International Congress in Amsterdam,

Netherlands, he was told that once he left the United States, he would not be permitted to return. Erdős, who valued his liberty above everything else and was ready to make enormous financial sacrifices to travel where and when he wanted, could not accept this. He left and was allowed to return for a short visit only in 1959. After 1955 Budapest was his home base. He became corresponding member of the Hungarian Academy of Sciences in 1956, ordinary member in 1962. He travelled constantly, was in India one day, in Australia the next, and went, say, to California from there, since after a while he was free to enter the US. He died while attending a conference in Warsaw, Poland. He obtained the Kossuth Prize in 1958 and the State Prize in 1983, the Tibor Szele Prize in 1971, was member of eight academies on four continents, and was awarded fifteen honorary doctorates. In 1984 he got the Wolf Prize, and in 1991 the Gold Medal of the Hungarian Academy of Sciences. His interest covered mainly number theory, approximation, probability, combinatorics, graph theory, set theory but he has many publications in other fields as well: if anyone came to him with a problem, he was likely to be able to solve it. His ability to state open problems was legendary. He cared deeply about people, was extremely generous, and worried about the fate of mankind.

FARAGÓ Andor (Budapest, September 26, 1877 – 1944). He attended the University of Budapest, then taught at various secondary schools. His enormous merit was to restart in 1924 the Mathematical Journal for Secondary Schools (KöMaL), founded by Dániel Arany and László Rátz, and to edit it at a high level until 1939. He perished during World War II, details are not known.

FARKAS Gyula (also Julius) (Sárosd, March 28, 1847 – Pestszentlőrinc [now part of Budapest], December 27, 1930) attended the Benedictine Gimnázium in Győr. Under the influence of his teacher the famous physicist Ányos Jedlik he got interested in physics, and studied physics and chemistry at the University of Budapest. Taught at a secondary school between 1870 and 1874, and was the tutor of the children of count Géza Batthyány from 1874 to 1880, where he had the opportunity to perform experiments in the home laboratory. On a trip abroad he became acquainted with Charles Hermite. In 1880 he obtained his doctorate, and in 1881 became a "Privatdozent" in the field of complex function theory at the University of Budapest. In 1887 he became a professor of Theoretical Physics at the University of Kolozsvár. Corresponding member of the Hungarian Academy of Sciences in 1898, ordinary member in 1914. He was president of the science section of the Transylvanian Museum Society. His efforts made the

University of Kolozsvár achieve a distinguished level. In 1892 he obtained the honorary doctorate of the University of Padova. Areas of interest: mechanics, thermodynamics, linear inequalities, complex function theory, iteration of functions, work of Bolyai.

FÁRY István (Gyula, June 30, 1922 – Berkeley, California, USA, November 2, 1984) entered the University of Budapest and the Eötvös Collegium in 1940. He obtained his Ph.D. from the University of Szeged in 1948. The same year he moved to Paris where he got a position at the Centre National de la Recherche Scientifique. In 1955 he obtained the degree of "docteur ès sciences" under Jean Leray, after which he went to the Université de Montréal. From 1957 to 1971 he held a position at the University of California at Berkeley, USA. Areas of interest: graph theory, packing and covering, differential geometry, topology, algebraic geometry.

FEJÉR Lipót (also Leopold) (Pécs, February 9, 1880 – Budapest, October 15, 1959) won second prize in the Eötvös Competition. Entered the Budapest Technical University first to study mechanical engineering, then switched to the program preparing secondary school teachers, and finally transferred to the University of Budapest. He spent the year 1899-1900 at the University of Berlin, where Hermann Amandus Schwarz had an influence on him, and where he came into contact with Constantin Carathéodory and Erhardt Schmidt. He obtained his Ph.D. in 1902, spent some time in Göttingen and Paris, and in 1905 he habilitated and obtained a position at the University of Kolozsvár. In 1911 he moved to the University of Budapest, where he spent the rest of his life. He was the first mathematician in Hungary around whom a real school has emerged. Just a few prominent names from his pupils: Egerváry Jenő, Erdős Pál, Fekete Mihály, Lukács Ferenc, Pál Gyula, Pólya György, Riesz Marcel, Sidon Simon, Szász Ottó, Szegő Gábor, Turán Pál. He was elected a corresponding member of the Hungarian Academy of Sciences in 1908 and an ordinary member in 1930. In 1935 he was awarded an honorary doctorate from Brown University in Providence, Rhode Island, USA, in 1948 he was the recipient of one of the first Kossuth Prizes awarded. He was a member of the Bavarian and of the Polish Academies and of the Königliche Gesellschaft der Wissenschaften zu Göttingen. Fields of interest: Fourier series, interpolation, complex variables, mechanics.

FEJES TÓTH László (also Ladislaus Fejes) (Szeged, March 12, 1915 – Budapest, March 17, 2005) studied at the University of Budapest. From 1941 to 1944 he was an assistant at the University of Kolozsvár. He taught at a secondary school in Budapest from 1945 to 1948. He habilitated at the University of Budapest in 1946 and from 1949 he was a professor at the University for Chemical Industry in Veszprém. In 1964 he was invited to head the Chair of Geometry at the University of Zurich, Switzerland, but the Hungarian authorities did not allow him to accept the offer. From 1965 he worked at the Mathematical Research Institute in Budapest, of which he was the director between 1970 and 1983. Taught also at numerous foreign universities. He was elected corresponding member of the Hungarian Academy of Sciences in 1962, ordinary member in 1970. He was awarded the Kossuth Prize in 1957, the State Prize in 1973, the Tibor Szele Prize in 1977, and a Gauss Bicentennial Medal in 1977. He was elected corresponding member of the Sächsische Akademie der Wissenschaften and of the Wissenschaftliche Gesellschaft Braunschweig, and received an honorary doctorate from the University of Salzburg. In 2002 he was awarded the Gold Medal of the Hungarian Academy of Sciences. His main interest was geometry, in particular covering and filling problems.

FEKETE Mihály (also Michael) (Zenta [now Senta, Serbia], July 19, 1886 – Jerusalem, Israel, May 13, 1957). He attended the University of Budapest, obtained his Ph.D. in 1909 with a dissertation on number theory, after which he spent a year in Göttingen, where he worked with Landau. After returning to Budapest he was an assistant of Manó Beke, when he became acquainted with Fejér, who influenced his further research interests. Habilitated in 1914 but taught at various secondary schools in Budapest. In1920 he lost his job as a teacher and started to work in insurance. He was the tutor of John von Neumann. In 1925 he obtained a teaching position at the Reáliskola of the Budapest Jewish Congregation. He received an invitation from the Hebrew University of Jerusalem in 1928, where he held various positions, among others Director of the Albert Einstein Institute. In 1955 he was awarded the Israeli Prize for Exact Sciences. His interest was complex function theory, polynomials, approximation, summability of Fourier series.

FELDHEIM Ervin (Kassa [now Košice, Slovakia], September 21, 1912 – Bor, Yugoslavia, 1944). He studied in Paris and obtained the degree of "docteur ès sciences". After returning to Hungary worked in an insurance company. Scientific interest: interpolation, polynomials.

FÉNYES Tamás (Budapest, July 7, 1929 – Budapest, April 30, 2000). He studied engineering at the Budapest Technical University between 1947 and 1952. From 1951 to 1953 he worked in industry as an electrical engineer.

From 1953 he had a position at the Mathematical Research Institute. He became a Candidate in 1967. Field of research: differential equations, distributions, applications.

FENYŐ István (also Stefan) (Budapest, March 5, 1917 – Budapest, July 28, 1987). He obtained a mathematics-physics teacher's diploma at the University of Budapest in 1939, and in 1942 a diploma in chemistry. Worked as a chemist in industry between 1942 and 1945. In 1945 he obtained a position at the Budapest Technical University. Habilitated in 1950. He spent several years teaching in Rostock (then German Democratic Republic). Areas of interest: complex function theory, mean values, integral equations, applied mathematics.

FODOR Géza (Szeged, May 6, 1927 – Szeged, September 28, 1977) attended the University of Szeged. After graduating worked at the Szeged branch of the Mathematical Research Institute. From 1959 he was at the University of Szeged. In 1954 he obtained the degree of Candidate, in 1967 Doctor of Mathematical Sciences. In 1973 he was elected corresponding member of the Hungarian Academy of Sciences. Research interest: logic, set theory.

FREUD Géza (Budapest, January 4, 1922 – Columbus, Ohio, USA, September 27, 1979) obtained a degree in mechanical engineering from the Budapest Technical University in 1950 and then worked at the Institute of Physics first at the University of Budapest, then at the Budapest Technical University. Obtained the degree of Candidate in 1954, and in 1957 that of Doctor of Mathematical Sciences. From 1954 to 1974 he had a position at the Mathematical Research Institute. Was awarded the Kossuth Prize in 1959. In 1974 he left Hungary and, after some visiting professorships, in 1976 he took a position at the Ohio State University in Columbus, USA. Areas of interest: approximation, interpolation, Tauberian theorems, orthogonal polynomials, partial differential equations.

FREY Tamás (Budapest, June 23, 1927 – Budapest, April 6, 1978). Studied at the Budapest Technical University and obtained a diploma in electrical engineering in 1950, after which he took a position at the same University. He became a Candidate in 1956 and Doctor of Mathematical Sciences in 1970. From 1962 he was at the Institute of Computation of the Hungarian Academy of Sciences, of which he was the director between 1963 and 1969. In 1969 he returned to the Technical University. Fields of interest: approximation, qualitative theory of differential equations, simulation problems, programming, computer science, applications to technology and biology.

GALLAI Tibor (Budapest, July 15, 1912 – Budapest, January 2, 1992) won first prize at the Eötvös Competition in 1930 and then studied at the University of Budapest, where he became a member of the group of friends of Pál Erdős. After taking his diploma, he worked in insurance and in the industry until 1939. Under the influence of Dénes Kőnig he got interested in graph theory. In 1940 he took a Ph.D. from the University of Budapest. From 1945 to 1949 he taught at a secondary school. In 1949 he became a professor at the Technical University and in 1952 he obtained the degree of Candidate. For his work in mathematical education he received the Kossuth Prize in 1956. Being an extremely modest person, he resigned his professorship in 1958, joined the Mathematical Research Institute and at the same time taught at a secondary school. He became Doctor of Mathematical Sciences in 1988 and a corresponding member of the Hungarian Academy of Sciences in 1990. He was awarded the Tibor Szele Prize in 1972. His main research interest was the theory of graphs.

GEŐCZE Zoárd (Budapest, August 23, 1873 – Budapest, November 26, 1916) attended the University of Budapest. He taught at various secondary schools. While spending a year in Paris, he wrote a difficult paper on surface area. In 1910 he was again in Paris, obtained a doctorate at the Sorbonne. After that he taught at a secondary school in Budapest, and became in 1913 "Privatdozent" at the University of Budapest. His fundamental work on surface area got appreciation only later, mainly through the work of Tibor Radó.

GERGELY Jenő (also Eugen) (Kolozsvár [now Cluj-Napoca, Romania], March 4, 1896 – Cluj-Napoca, May 15, 1975) studied at the University of Kolozsvár, where Frigyes Riesz was one of his teachers. He became Riesz's assistant and obtained his Ph.D. under him. When the University moved to Szeged, he resigned his position and remained in Kolozsvár, where he taught at a secondary school. In 1948 he was appointed lecturer at the Bolyai University (later Babeş–Bolyai University) of Kolozsvár, where he remained until his death. Areas of interest: descriptive geometry, geometry of ovals, differential geometry, the geometry of Hilbert spaces, work of János Bolyai.

GOLDZIHER Károly (also Charles, Karl) (Budapest, February 26, 1881 – Budapest, November 6, 1955) was the son of Ignác Goldziher, an internationally known orientalist. Studied in Budapest and then in Göttingen, where under the influence of Felix Klein he became interested in applied mathematics. He taught in Budapest, first in a secondary school and from 1908 at a Teacher's College. In 1911 he habilitated at the Technical University. In 1920 he returned to teach in secondary school, in 1935 became an extraordinary professor at the Technical University, and after World War II an ordinary university professor. In 1952 he obtained the title of Doctor of Mathematical Sciences. Areas of interest: statistics, actuarial mathematics, mathematical education.

GROSSCHMID Lajos (also Louis de Grosschmid, Ludwig von Grosschmid) (Nagyvárad [now Oradea, Romania], April 21, 1886 – Budapest, June 13, 1940). Attended the University of Budapest, where he obtained his doctorate in 1911 and became the assistant of Manó Beke. In 1912/13 studied in Göttingen, Munich and Paris. Habilitated in 1918 at the University of Budapest in the theory of algebraic number fields. In 1919 he was appointed extraordinary professor of business mathematics at the College of Economics of the Technical University, and in 1924 ordinary professor. In 1916 he was elected member of the St. Stephen Academy, and in 1936 corresponding member of the Hungarian Academy of Sciences. Fields of interest: number theory, algebra (mainly the algebra of quadratic forms), ballistics, probability, business mathematics.

GRÜNWALD Géza (Budapest, October 18, 1910 – September 7, 1942) was a schoolmate and friend of Pál Erdős. Lajos Erdős, the father of Pál, supported him, because his father, a house painter, had problems in sending his two sons to school. Since he could not register at any University in Hungary, he started his studies in Italy. With the help of Alfréd Haar, he was finally admitted to the University of Szeged, where he won all the mathematical prizes. He obtained his teacher's diploma in 1936 and from 1937 on he worked as an applied mathematician at the "Tungsram" electric factory with the physicist Zoltán Bay. In 1952 the János Bolyai Mathematical Society established in his memory the Géza Grünwald Prize, given each year to young researchers. Scientific interest: Lagrange interpolation, strong summability of Fourier series, complex variables.

GRYNAEUS István (also Etienne) (Eperjes [now Prešov, Slovakia], March 21, 1893 – Budapest, September 28, 1936). He entered the University of Budapest and was a member of the Eötvös Collegium but was drafted at the beginning of World War I. Was taken prisoner in 1915 on the Russian front and returned home only in 1920. He obtained his mathematics-physics teacher's diploma in 1921, became an assistant and was awarded his Ph.D.

in 1922. Spent the year 1925-26 in Paris with a Rockefeller Fellowship, then continued in Delft (Holland). Habilitated to "Privatdozent" in 1932 and after that was an instructor at the Eötvös Collegium and at the Teacher's College. Scientific research areas: differential geometry, differential equations, Pfaff systems.

GYARMATHI László (Petrozsény [now Petroşani, Romania], June 24, 1908 – Debrecen, September 7, 1988) entered the University of Budapest and the Eötvös Collegium in 1926, obtained his teacher's diploma in 1931. Taught at various secondary schools in Debrecen. In 1948 he obtained a diploma as a teacher of descriptive geometry at the Technical University of Budapest. He obtained his Ph.D. in 1950 and the degree of Candidate in 1966. From 1951 to 1974 he taught at the University of Debrecen. Main areas of interest: descriptive geometry of several dimensions, projective and non-euclidean geometry.

GYÍRES Béla (Zagreb, Croatia, March 29, 1909 – Budapest, August 26, 2001) attended the University in Budapest (1928-1933), then taught at secondary schools. Obtained a Ph.D. in 1941 from the Budapest Technical University. In 1943 he became an assistant at the Commercial College in Kassa [now Košice, Slovakia], and starting with 1945 he was at the University of Debrecen. In 1946 he habilitated, became a Candidate in 1952, Doctor of Mathematical Sciences in 1962. He was elected corresponding member of the Hungarian Academy of Sciences in 1987, ordinary member in 1990. He obtained the State Prize in 1980. Fields of research: linear algebra, probability theory and statistics.

HAAR Alfréd (Budapest, October 11, 1885 – Szeged, March 16, 1933) won first prize in the Eötvös Competition in 1903. Started to study chemistry at the Budapest Technical University but soon transferred to the University of Budapest. Between 1905 and 1909 he studied in Göttingen, Germany. In 1909 he obtained his Ph.D. there under the direction of David Hilbert, and very soon after he became a "Privatdozent". Taught very briefly as a substitute at the Eidgenössische Technische Hochschule in Zurich, Switzerland, and in 1912 was appointed extraordinary professor at the University of Kolozsvár, and ordinary professor there in 1917. After World War I he spent some time in Budapest, and in 1920 together with Frigyes Riesz he founded the Mathematical Institute of the then new University of Szeged. In 1922 the two started the journal Acta Scientiarum Mathematicarum of the University of Szeged. He was elected corresponding member of the Hungarian Academy of Sciences in 1931. Scientific interests:

orthogonal functions, partial differential equations, calculus of variations. The Haar system of functions and the Haar measure on a locally compact group carry his name.

HAJÓS György (also Georg) (Budapest, February 21, 1912 – Budapest, March 17, 1972). He was the grandson of Adam Clark, builder of the Chain Bridge in Budapest. Obtained his teacher's diploma in 1929 from the University of Budapest and taught in secondary schools until 1935, when he got a position at the Budapest Technical University. Obtained his Ph.D. in 1938. In 1949 he became professor at the University of Budapest. Was elected corresponding member of the Hungarian Academy of Sciences in 1948, ordinary member in 1958. He was a member of the Romanian Academy of Sciences and of the German Leopoldina Academy, was awarded the Kőnig Gyula Prize of the Mathematical Society in 1942, the Kossuth Prize twice (1951, 1962). His most famous result is the solution of a conjecture of Minkowski in convex geometry, which he transformed into a problem in abelian group theory. Field of research: geometry, graph theory.

HARSÁNYI János (also John C.) (Budapest, May 29, 1920 – Berkeley, California, USA, August 9, 2000). Attended the Lutheran Evangelical Gimnázium in Budapest and was a problem solver for KöMaL. Studied philosophy, sociology and psychology at the University of Budapest and obtained his Ph.D. in philosophy in 1947. Emigrated to Australia in 1950. Studied economics at the University of Sydney in 1950-53, taught at the University of Queensland in 1953-56 and at the Australian National University (Canberra) in 1958-61. He moved to the USA in 1961, was first a professor at Wayne State University (Detroit, Michigan) in 1961/63, then at the University of California at Berkeley from 1965 to 1990, and a professor emeritus after that. He was awarded the Nobel Prize in Economics in 1994 jointly with John Nash and Reinhard Selten for their work on noncooperative games. Areas of interest: economics, game theory.

HOSSZÚ Miklós (Somogyszob, March 7, 1929 – Budapest, June 4, 1980) graduated from the University of Budapest in 1951. He obtained the degree of Candidate in 1957, Doctor of Mathematical Sciences in 1964. He taught at the Technical University for Heavy Industry in Miskolc, 1951–1972, and then at the University for Agriculture in Gödöllő, 1972–1980. Scientific interest: functional equations in algebraic structures, in particular in quasigroups, and mathematical programming.

HUHN András (Szeged, January 26, 1947 – Budapest, June 6, 1985) studied at the University of Szeged and after graduating took a position there. Obtained his Ph.D. in 1972 and the degree of Candidate in 1975. Spent the year 1978/79 at the University of Manitoba (Canada). In 1985 he went to Darmstadt with a Humboldt fellowship. Area of scientific interest: algebra, mainly the theory of lattices.

HUNYADY Jenő (also Eugene de Hunyady) (Pest, April 28, 1838 – Budapest, December 29, 1889) lived in the 19th century, still we decided to include his biography here because those who were active in algebra at the beginning of the 20th century (Beke Manó, Kőnig Gyula, Kürschák József, Rados Gusztáv) were under his influence. He attended secondary school in Pest. His father was a well-situated physician who could send his son to study mathematics in Berlin with Kummer, Kronecker, Clebsch. Hunyady obtained a Ph.D. in Göttingen with a dissertation on algebraic curves. Returning to Hungary in 1865, he habilitated at the Budapest Technical University in 1866, where he was appointed professor in 1869. In 1867 he was elected corresponding member of the Hungarian Academy of Sciences. Scientific interests: algebra, mainly linear algebra and algebraic geometry.

JOÓ István (Sárvár, September 19, 1948 – Budaörs, December 8, 1998). He graduated in 1973 from the University of Budapest and obtained a Ph.D in the same year. From 1975 to 1995 he taught at the University of Budapest. Became a Candidate in 1980. In 1980/81 he was a visiting professor at the Moscow State University, and in 1986/87 at the Ohio State University. Areas of scientific interest: approximation theory, number theory, convex analysis, differential equations with applications to physics, engineering and biology, control theory, game theory.

JORDÁN Károly (also Charles) (Pest, December 16, 1871 – Budapest, December 24, 1959). He studied abroad, in Paris, France, at the École Polytechnique, and obtained a diploma of Chemical Engineering at the Eidgenössische Technische Hochschule of Zurich, Switzerland, in 1893. Then worked one year in the Perkins Laboratory at Owen's College of the Victoria University of Manchester, England, and in 1894 became assistant at the University of Geneva, Switzerland, where he obtained a doctorate in physics and became a "Privatdozent" in physics. Between 1896 and 1899 he was employed as a chemical engineer at the Société d'Études Electrochimiques of Geneva and at a factory in Laibach, Austria [now Ljubljana, Slovenia]. Returned to Budapest in 1899 where he validated his engineer's diploma at the Technical University and his Ph.D. at the University of Budapest. Continued his studies at the University of Budapest attending courses in seismology, astronomy and mathematics. From 1909 to 1913 he was the head of the Seismological Institute. During World War I he worked as a meteorologist and taught mathematics and physics at the military school in Várpalota. In 1919/20 he lectured at the University on statistics, and between 1920 and and 1950 at the University of Economics of Budapest. In 1923 he became "Privatdozent", in 1933 extraordinary professor, in 1940 ordinary professor. Obtained the Gyula Kőnig Prize in 1928, the Kossuth Prize in 1956 and was elected a corresponding member of the Hungarian Academy of Sciences in 1947. Areas of scientific interest: probability, statistics, calculus of finite differences, interpolation.

KALMÁR László (Edde-Alsóbogátpuszta, March 27, 1905 – Mátraháza, August 2, 1976) won the Eötvös Competition in 1922. He attended the University in Budapest, was a member of the Eötvös Collegium, and obtained his Ph.D. in 1927, simultaneously with his graduation. Spent one year in Germany, where he became interested in logic under the influence of Hilbert. He was offered an assistantship at the Physics Institute of the University of Szeged but soon switched to the Mathematical Institute, where he remained the rest of his life. He was elected corresponding member of the Hungarian Academy of Sciences in 1941 and ordinary member in 1961. He was awarded the Gyula Kőnig Prize in 1936, the Kossuth Prize in 1950, the State Prize in 1975, the Tibor Szele Prize in 1970. The János Bolyai Mathematical Society named a competition for general school students after him. Areas of interest: interpolation, logic, computer science. He built the first computer in Hungary.

KALUZSAY Károly (Facskó [now Fačkov, Slovakia], December 28, 1889 – Nosowce [now Ukraine], August 10, 1916) studied at the University of Kolozsvár in 1908–1914. In 1913/14 he was an assistant at the Institute for Applied Physics and Electrotechnics. His Ph.D. dissertation on the converse of the Jordan theorem, which is his only publication, was written under Frigyes Riesz. From 1911 on he taught at a "higher elementary school" for girls, where Gyula Szőkefalvi Nagy was the principal. He was drafted on January 15, 1915, and disappeared in a battle. Frigyes Riesz wrote to his brother Marcel in 1917: "the most gifted of my former students, Dr. Kaluzsay Károly, disappeared on the Russian front more than a year ago".

KÁRMÁN Tódor (also Theodore von Kármán) (Budapest, May 11, 1881 – Aachen, May 7, 1963). His father Mór was responsible for the 1879 curriculum for secondary schools, and founded the Teacher's College and its practice school, the Mintagimnázium in Budapest. He was tutored at home in elementary school and then attended the Gimnázium founded by his father, from where he graduated in 1898 and won the Eötvös Competition. Studied mechanical engineering from 1898 to 1902 at the Budapest Technical University, where he obtained a prize in mechanics. After military service he was an assistant at the Budapest Technical University and worked as an engineer at the Ganz mechanical factory. In 1906 he traveled with a scholarship to Göttingen, Germany, where he got into contact with Felix Klein and L. Prandtl. It is there that he wrote his doctoral thesis in 1908 and became a "Privatdozent" in 1909. He spent some time in Berlin and Paris where he became interested in the theory of aviation. In 1912 he was appointed professor at the School for Mining and Forestry in Selmecbánya [now Banská Štiavnica, Slovakia], but since he was not able to pursue his research there, with the help of Felix Klein he was appointed University Professor at the Technical School of Aachen, Germany, where he founded the Aeronautical Research Institute. During World War I he served in the Austro-Hungarian Army and collaborated in the development of a helicopter. After the war he returned to Hungary, worked at the Budapest Technical University. Since he also worked at the Education Commisariat during the Hungarian Soviet Republic, he subsequently had to leave Hungary. From 1919 to 1930 he was again in Aachen. In 1934 he settled at the California Institute of Technology, USA, where he became the director of the Guggenheim Aeronautical Laboratory. In the 1940's he founded the Jet Propulsion Laboratory and the Aerojet Engineering Corporation. He was a member of twenty academies, in particular, of the Royal Society beginning in 1946. He had numerous honorary doctorates, obtained the National Medal of Science in 1963, and many other distinctions. He was the first president of the International Austronautical Academy. Scientific interests: hydro- and aerodynamics (von Kármán vortex street), structure of crystals, thermodynamics, rocketry.

KÁRTESZI Ferenc (Cegléd, February 13, 1907 – Budapest, May 9, 1989). His first article "The Tetrahedron", written while he was still in school, was published in KöMaL. Attended the University of Budapest. He obtained his diploma as a mathematics-descriptive geometry teacher in 1930, after which he was an assistant at the Budapest Technical University. He left because of his increasing interest in teaching, and taught at a secondary school from 1931 to 1940. He spent the year 1936/37 in Bologna associated with several Italian geometers. In 1939 he studied the Bolyai manuscripts in Marosvásárhely [now Tîrgu Mureş, Romania]. During World War II he was in military service and was a prisoner of war. After his return to Budapest he had the title of a secondary school teacher but was assigned to the University of Budapest. In 1947 he became a professor at the Teacher's College. Habilitated in 1948 in the field of projective and descriptive geometry and organized the teaching of descriptive geometry at the University of Budapest. In 1952 he became Candidate and in 1968 Doctor of Mathematical Sciences. Areas of interest: projective and descriptive geometry, finite geometry, didactics.

KERÉKJÁRTÓ Béla (also de Kerékjártó) (Budapest, October 1, 1898 – Gyöngyös, May 26, 1946) attended the University of Budapest between 1916 and 1920. He obtained his Ph.D. in 1920 and became "Privatdozent" in analysis and geometry in 1922. In 1922/23 he was a visiting professor in Göttingen; his lectures there became his famous Yellow Springer book. Then he spent two years in Princeton. He was appointed extraordinary professor at the University of Szeged in 1925 and ordinary professor in 1929. Lectured in Barcelona and in Paris. In 1938 he was appointed professor at the University of Budapest. He was elected corresponding member of the Hungarian Academy of Sciences in 1934, ordinary member in 1945, and in 1936 corresponding member of the Société Royale des Sciences de Liège. Areas of interest: topology, geometric group theory, foundations of geometry.

KERTÉSZ Andor (Gyula, February 19, 1929 – Budapest, April 3, 1974) studied at the University of Debrecen in 1947–1952. He obtained his diploma as a mathematics-descriptive geometry teacher in 1952 and became a research student first of Tibor Szele, after whose death László Rédei directed his research. Was awarded the degree of Candidate in 1954 and Doctor of Mathematical Sciences in 1957. He taught at the University of Debrecen from 1950. He was a professor also at the University of Halle in 1962–64 and in 1968–71. The German Leopoldina Academy elected him a member. Areas of interest: rings, modules, abelian groups, set theory.

KIS Ottó (Budapest, May 22, 1931 – Budapest, April 26, 2001). He obtained his diploma of mathematics-physics teacher from the University of Budapest in 1952, and was a graduate student in Leningrad [now St. Petersburg] from 1952 to 1955 under the direction of V. I. Krylov. He obtained the degree of Candidate in 1955. Worked from 1955 to 1968 at the University of Budapest, then at the Budapest Technical University. Areas of interest: approximation, interpolation, numerical analysis.

KISS Péter (Nagyréd, March 5, 1937 – Eger, March 5, 2002). He studied at the University of Budapest. During 12 years he taught at a secondary school, and in 1971 he was appointed to the Teacher's College in Eger. He became a Candidate in 1977 and habilitated at the Kossuth University of Debrecen in 1996. He obtained the title of Doctor of Mathematical Sciences in 1999, and in 1997 he received the Albert Szent-Györgyi Prize. Scientific interest: number theory.

KLUG Lipót (also Leopold) (Gyöngyös, January 23, 1854 – Budapest, 1944) studied at the Budapest Technical University from 1870 to 1874 and obtained a diploma for teaching mathematics and descriptive geometry. Between 1874 and 1893 he taught in Pozsony [now Bratislava, Slovakia], where he wrote his first books on geometry. In 1893/97 he was a teacher at a secondary school in Budapest and habilitated to "Privatdozent" at the University of Budapest. He started teaching descriptive geometry at the University of Kolozsvár in 1897, where he became a professor in 1900. After retirement he moved to Budapest. At the 50th anniversary of the Loránd Eötvös Society he offered a year's retirement pay for a foundation to promote geometry in Hungary. The Klug Lipót Prize was awarded only once, in 1943. The recipients were László Fejes (Tóth) and Ferenc Zigány. The 90-year old professor living alone disappeared during the siege of Budapest. Areas of research: descriptive geometry, synthetic geometry.

KOSIK Pál (Dolné Zahorany [Hungarian name: Magyarhegymeg], Czechoslovakia [now Slovakia], April 16, 1931 – Budapest, October 4, 1985) lost his sight in the first year of his life and grew up in asylums, attending only two classes till 1941. Then, after the South of Slovakia was returned to Hungary, he got to Budapest, finished elementary school still in asylums, worked a year in handicraft, and then completed secondary school from 1949 to 1953. Studied at the University of Budapest from 1953 till 1958 and then got a position at the Mathematical Research Institute. Scientific interest: differential equations, numerical analysis.

KOVÁCS Béla (Kissikátor, September 25, 1946 – Debrecen, September 14, 2001). He obtained a mathematics diploma from the University of Debrecen in 1970. Between 1970 and 1972 he had a research grant; following that he worked at the University of Debrecen. He obtained a doctorate from this university in 1972. Scientific interests: number theory, group theory.

KŐNIG Dénes (Budapest, September 21, 1884 – Budapest, October 19, 1944). Son of Gyula Kőnig. Attended the Mintagimnázium in Budapest, where Manó Beke was his teacher. Wrote the first volume of his book on popular mathematics ("Mathematical Recreations") while he was still in school, and the second volume as a student. Won the Eötvös Competition in 1902. He studied at the University of Budapest in 1902–04 and from 1904 spent five semesters in Göttingen. He obtained his Ph.D. in 1907 with a dissertation on geometry, and worked at the Technical University of Budapest until the end of his life. He was very active in the Mathematical-Physical Society, edited its Journal and judged the mathematical competition. In 1918 he and his brother established the Gyula Kőnig Prize for mathematics to commemorate their father. Specialities: graph theory, topology.

KŐNIG Gyula (also Julius) (Győr, December 16, 1849 – Budapest, April 8, 1913). He attended the Benedictine Gimnázium in Győr, and even before obtaining his "matura" he studied at the medical faculty in Vienna. He studied mathematics in Budapest and then in Heidelberg, Germany, where in 1870 he obtained his Ph.D. with a dissertation on elliptic functions. He spent a year in Berlin, where Kronecker had a great influence on him. He returned to Hungary and became a "Privatdozent" in 1871, professor at the Technical University from 1874 to 1905. His seminar with Kürschák was the center of mathematical life in Budapest. Was also active in the development of secondary schools and the preparation of teachers. He was one of the founders of the Mathematical-Physical Society and its honorary vice-president. Mathematical interests: algebra, number theory, logic, set theory.

KÜRSCHÁK József (also Josef) (Buda, March 14, 1864 – Budapest, March 26, 1933). He entered in 1881 the Teacher's College of the Budapest Technical University. He started research in mathematics under the influence of Gyula Kőnig, wrote his first paper on the polygons inscribed in and circumscribed about a circle while still a student. He taught at various schools before getting his diploma in 1888. He obtained his Ph.D. in 1890 with a dissertation on the calculus of variations, and then taught at the Technical University. He became corresponding member of the Hungarian Academy of Sciences in 1896, ordinary member in 1914. He was very active in the Mathematical and Physical Society, and published a collection of problems given at the Eötvös Competition. The mathematical competition organized by the Bolyai Society for high school students, the successor of the Eötvös Competition, is named after him. Areas of interest: geometry, number theory, algebra, calculus of variations, partial differential equations. He introduced the abstract notion of valuation of fields.

LAKATOS Imre (Debrecen, November 9, 1922 – London, February 2, 1974) studied mathematics, physics and philosophy at the University of Debrecen from 1941 to 1944. In 1945/46 he was a student at the Eötvös

Collegium in Budapest and at the same time an assistant of Ottó Varga at the University of Debrecen. He obtained his Ph.D. in Debrecen in 1947 with a thesis in in epistemology. Having joined the illegal Communist Party during the War, he became politically very active in 1945, participated in the destruction of the Eötvös Collegium, and as a recompense was sent to Moscow as a research student in theoretical physics. In 1950 he was expelled from the Party, arrested, and sent to a concentration camp. Freed in 1953, he joined (with the help of Alfréd Rényi) the Mathematical Research Institute, where he translated Pólya's book "How to Solve it". After the Revolution in 1956 he escaped and went with a Rockefeller fellowship to Cambridge, England. There he earned another Ph.D. with a dissertation which became his most famous book "Proofs and Refutations". He became the successor of Karl Popper at the London School of Economics. Area of research: philosophy of science, in particular of mathematics.

LÁNCZOS Kornél (also Cornelius) (Székesfehérvár, February 2, 1893 – Budapest, June 25, 1974) studied mathematics, physics and philosophy at the University of Budapest, and obtained his diploma in 1916. Between 1916 and 1920 he was an assistant at the physics department of the Technical University. He took a Ph.D. in 1921. In 1920 he left for Germany. Until 1924 he was an assistant in Freiburg, from 1924 to 1928 in Frankfurt, in 1928/29 in Berlin and then again in Frankfurt until 1931. Between 1931 and 1952 he lived in the USA. He was an engineer at the Boeing company. From 1949 to 1952 he worked at the Numerical Analysis Institute in Los Angeles. In 1952 he moved to Ireland. He was a visiting professor at Dublin University until 1954, when he became a member of the Dublin Institute for Advanced Studies and remained there until his retirement in 1968. He was awarded the Chauvenet Prize, was a member of the Royal Society of Ireland, and had several honorary doctorates. Areas of interest: relativity, quantum mechanics, integral equations, functional analysis, numerical methods.

LAX Péter (Budapest, May 1, 1926 –) was a pupil at the Mintagimnázium in Budapest. As a child prodigy he was tutored by Rózsa Péter. The family left Budapest for New York in November of 1941. He studied at New York University from 1944 and obtained his Ph.D. there in 1949. Worked in the Manhattan Project at the Los Alamos Laboratory in 1945/46, where he first came into contact with John von Neumann and with computers, and also from 1950 to 1958. From 1949 he was also a faculty member of New York University, Director of the AEC Computing and Applied Mathematics Center from 1964 to 1972, Director of the Courant Institute of Mathematical Sciences from 1972 to 1980, then Director of the

Courant Mathematics and Computing Laboratory. He was president of the American Mathematical Society in 1979/80. He is a member of the National Academy of Sciences of the USA, Foreign Associate of the French Academy of Sciences, of the Russian Academy of Science, honorary member of the Hungarian Academy of Sciences (1993), etc. Honorary doctor of the University of Paris VI. Recipient of the Chauvenet Prize (1974), the Norbert Wiener Prize (1975), the National Medal of Science (1986), the Wolf Prize (1987), the Steele Prize (1992), the Abel Prize (2005). Mathematical interests: nonlinear partial differential equations, shock waves, scattering theory, numerical mathematics.

LÁZÁR Dezső (Erzsébetfalva [later: Pestszenterzsébet, now part of Budapest], March 14, 1913 – Ukraine, 1943) graduated from the University of Szeged, in 1941 became a secondary school teacher. Died in the war. Area of interest: set theory.

LENGYEL Béla (Budapest, October 5, 1910 – Irvine, California, USA, November 1, 2002) studied at the University of Budapest from 1927 to 1933 and obtained his mathematics-physics teacher's diploma in 1933. Became an assistant of Kürschák at the Technical University but also performed experiments in the Department of Physics. He obtained his Ph.D. in 1934 with a dissertation on operator theory. Worked in insurance. Spent the year 1935/36 with a fellowship at Harvard University where he associated with M. H. Stone. He returned to Hungary but emigrated in 1939 to the USA where he taught at the Rensselaer Polytechnic Institute (Troy, New York) and at Brown University, physics at the City College of New York. He was appointed professor of physics in 1943 at the University of Rochester. From 1946 he had a position at the Naval Research Laboratory, and from 1950 he was an advisor of the Office of Naval Research. He moved to California in 1952, where he worked during eleven years at the Hughes Research Laboratories. Scientific interests: operators in Hilbert space, interpolation, statistics, lasers.

LIPKA István (also Stephan) (Budapest, May 9, 1899 – Budapest, September 24, 1990) entered the University of Budapest as a regular student in mathematics and physics, but attended also the Technical University. He obtained his teacher's diploma and his Ph.D. in mathematics in 1923, then taught at a secondary school until 1926. Became an assistant of Béla Kerékjártó in Szeged, spent a semester in Hamburg in 1929, became "Privatdozent" in 1933 and obtained the Gyula Kőnig Prize in 1938 for his work in algebra. From 1942 to 1945 he was a docent at the University of

Szeged. He was sent into retirement in 1946, then obtained positions in industry. In 1954 he obtained the title of Doctor of Technical Sciences. Areas of scientific interest: before 1945: algebra, the geometry of polynomials, complex function theory; after 1945: technical mathematics.

LUKÁCS Ferenc (Budapest, June 27, 1891 – Budapest, November 30, 1918). In secondary school, Frigyes Riesz was his teacher for a short time, who noticed his talent. Won second prize at the Eötvös Competition in 1909. He studied at the University of Budapest. After obtaining his Ph.D., he became an assistant of József Kürschák at the Technical University, and married Tekla Téri in 1914. His wife was not a mathematician but discovered a nice geometric theorem (see [129] I, Sec 3, No. 111). Areas of scientific interest: power series, Fourier series, polynomials.

MAKAI Endre (Budapest, November 5, 1915 – Budapest, November 8, 1987) won the Eötvös competition in 1933. He studied at the University of Budapest while a member of the Eötvös Collegium. Wrote his first paper while still a student, obtained his diploma in 1938, was unemployed until 1942, when he got a job at the Chinoin Chemical Factory. Received a Ph.D. in 1942. After World War II he worked at the research laboratory of the Tungsram company. In 1951 he got a position at the mathematics department of the College of Mechanical Engineering of the Technical University, moved in 1961 to the Mathematical Research Institute. He became Doctor of Mathematical Sciences in 1955, was awarded the Prize of the Academy in 1970 and the State Prize in 1973. Research interests: differential equations, special functions, vibration of membranes.

MEDGYESSY Pál (Egercsehi, October 10, 1919 – Budapest, October 8, 1977) studied at the University of Budapest while a member of the Eötvös Collegium. Then he got a position at the Institute for Physics of the Medical College of Debrecen University. Starting with 1955 he worked at the Mathematical Research Institute in Budapest. He obtained the degree of Candidate in 1955 and Doctor of Mathematical Sciences in 1973. Areas of interest: probability theory, statistics, and their applications.

MIKOLÁS Miklós (Celldömölk, April 5, 1923 – Budapest, February 2, 2001) entered the University of Budapest and the Eötvös Collegium in 1942, obtained his teacher's diploma in 1947, and his Ph.D. in 1948. He was an assistant of Lipót Fejér and then a docent at the University of Budapest until 1964. He obtained the degree of Candidate in 1955 and Doctor of Mathematical Sciences in 1992. From 1964 he was a professor at the Technical University. Areas of interest: analytic number theory, summability methods, orthogonal polynomials, fractional calculus, applied statistics, and applications of mathematics to technical problems.

MOGYORÓDI József (Nagyoroszi, October 2, 1933 – Budapest, March 27, 1990) studied from 1952 to 1957 at the University of Budapest, where he obtained a mathematics diploma in 1957. Starting with 1958 he taught at the University of Budapest and, from 1971, also at the University of Debrecen. Became Candidate in 1967, and Doctor of Mathematical Sciences in 1980. Areas of interest: probability, statistics, applications, computer science.

MOLNÁR Ferenc (1933 – Budapest, January 18, 1962) attended the University of Budapest from 1950 to 1954, and after receiving his diploma became an assistant of György Hajós in the Department of Geometry. Field of interest: geometry.

MOÓR Arthur (Budapest, January 8, 1923 – Sopron, August 26, 1985) started his university studies in Szeged in 1941 but because of the war graduated only in 1947. He was a member of the Eötvös Collegium, where László Kalmár was one of his teachers. From 1947 to 1950 he taught at the Teacher's College in Szarvas. His first works on Finsler geometry caught the attention of Ottó Varga, with whose help he transferred to a secondary school in Debrecen, where he taught from 1950 to 1952. In 1953 he became a research student under the direction of Varga and obtained the degree of Candidate in 1956. From 1956 to 1968 he worked at the University of Szeged and obtained the degree of Doctor of Mathematical Sciences in 1964. Moved to the University of Forestry in Sopron in 1968. Area of interest: differential geometry, mainly Finsler geometry.

NEUMANN János (also John von Neumann) (Budapest, December 28, 1903 – Washington, DC, USA, February 8, 1957). One of the most brilliant minds of the twentieth century. Attended the Lutheran Evangelical Gimnázium in Budapest, where László Rátz was his teacher, Jenő Wigner his friend. As a child prodigy, he was mentored by József Kürschák, Mihály Fekete, and Gábor Szegő. In 1921 he registered simultaneously at two universities: in Budapest for mathematics, and in Berlin, later Zurich for chemical engineering. He returned to Budapest at the end of each semester to pass the examinations and received his Ph.D. in mathematics in 1926. Spent one year in Göttingen, became "Privatdozent" in 1927 in Berlin and in 1929 in Hamburg. In 1930 he went to Princeton University as a guest professor and later as a professor, and from 1933 was a member of the Institute for Advanced Study. Between 1943 and 1955 he was a consultant of the Los

Alamos laboratory and in 1955 was appointed member of the Atomic Energy Commission. He was president of the American Mathematical Society from 1951 to 1953. He was member of seven academies, had honorary Doctor's degrees from seven universities and was awarded the following distinctions: Bôcher Prize (1937), Medal of Merit, Civilian Service Award, U.S. Navy (1947), Medal of Freedom (1956), Albert Einstein Commemorative Medal (1956), Enrico Fermi Award (1956). Areas of interest: logic, set theory, operator theory, measure theory, ergodic theory, game theory, computer science, quantum mechanics, applied mathematics.

OBLATH Richárd (Versec [now Vršac, Serbia], June 11, 1882 – Budapest, June 18, 1959) studied at the University of Budapest and obtained his mathematics-physics teacher's diploma in 1905. Taught at secondary schools in several towns between 1905 and 1919. After the fall of the Soviet Republic he lost his job. Worked first in insurance and from 1922 to 1945 as a mathematician of the General Mining Company. From 1946 he was a lecturer at the University of Budapest. In 1955 he obtained the title of Candidate. He was active in the organization of the Bolyai Society and gave lectures popularizing mathematics. Areas of interest: elementary number theory, actuarial mathematics, history of mathematics.

OLÁH Gyula (Kiskunfélegyháza, August 16, 1931 – Budapest, April 28, 1983) began his studies in mathematics-physics at the University of Budapest in 1949. After obtaining his diploma in 1953 he got a position at the same university. Between 1960 and 1965 he worked at the Mathematical Research Institute. He obtained the title of Candidate in 1971. From 1965 to 1973 he worked at the Ministry of Education, from 1973 at the Budapest Technical University. Areas of interest: graph theory. Later, his daughter Vera was the editor of KöMaL for several years.

PÁL Gyula (also Julius) (Győr, June 27, 1881 – Copenhagen, Denmark, September 6, 1946). He attended the University of Budapest, obtained his diploma in 1908 and taught until 1918 at a secondary school. In 1916 he received a Ph.D. at the University of Kolozsvár. In 1919 he moved to Copenhagen and started teaching at the Sankt Jørgens Gymnasium, where Børge Jessen was his student. From 1925 he taught at the Polytechnic School, between 1932 and 1938 also at the University of Copenhagen. Areas of interest: approximation, plane topology, Kakeya's needle problem.

PÁL László György (Nagyszénás, April 5, 1929 – Budapest, May 15, 2001) studied at the University of Budapest from 1947 to 1951 and was a member of the Eötvös Collegium. After graduating he worked at the same university till his retirement in 1996, with the exception of the period 1972–76 when he taught at the University of Lagos, Nigeria. He became a Candidate in 1967, Doctor of Mathematical Sciences in 1995. Scientific interest: orthogonal series, interpolation theory.

PAPP Zoltán (Debrecen, May 31, 1951 – Budapest, November 19, 1991) studied mathematics at the University of Debrecen. He worked at the Research Institute of the Postal Service, where he developed algorithms for the planning of optimal communication networks. Field of research: diophantine number theory.

PÉTER Rózsa (Budapest, February 17, 1905 – Budapest, February 16, 1977) started studying chemistry at the University of Budapest but soon switched to mathematics, where László Kalmár had a great influence on her. Her first paper, written while she was still a student, was on number theory, but she became interested very early in recursive functions. She earned her diploma in 1927, but obtained a teaching position only in 1933, and she had it until 1939. She received her Ph.D. in 1937. In 1945 she was appointed a secondary school teacher, then became a professor at the Teacher's College. Between 1955 and 1975 she was a professor at the University of Budapest. She was awarded the degree of Doctor of Mathematical Sciences in 1952 and was elected corresponding member of the Hungarian Academy of Sciences in 1973. She was awarded the Kossuth Prize (1951), the Beke Manó Prize (1953), the State Prize (1970). Areas of interest: recursive functions, computer science, mathematical linguistics, popularization of mathematics and mathematical education.

POLLÁK György (Budapest, April 26, 1929 – Pécs, June 29, 2001), winner of the Kürschák competition in 1947, started his university studies in Budapest in 1947 but between 1948 and 1953 he studied at the University of Kazan (Russia). He got a position first at the Bolyai Institute of the University of Szeged in 1953/54, from 1958 at the Szeged Department of the Mathematical Research Institute. He obtained the title of Candidate in 1961. Field of research: algebra, mainly semigroups.

PÓLYA György (also George) (Budapest, December 13, 1887 – Palo Alto, California, USA, July 7, 1985) started studies in Budapest, first law, literature, philosophy, and then switched to physics and mathematics. He spent the years 1910-1914 mostly in Vienna, Göttingen and Paris, obtained his Ph.D. in Budapest in 1912. He worked at the Federal Polytechnic School in Zurich from 1914 to 1940. His famous collection of problems in analysis written jointly with Gábor Szegő appeared in 1925. In 1940 he left for the United States. After two years at Brown University he became a professor at Stanford, becoming emeritus in 1953. Was a member of the National Academy of Science of the USA, of the Hungarian Academy of Sciences and other Academies, was awarded four honorary doctorates. Both the Mathematical Association of America and the Society for Industrial and Applied Mathematics have prizes named after him. Areas of interest: complex function theory, probability, combinatorics, number theory, psychology of mathematical discovery.

PUKÁNSZKY Lajos (Budapest, November 4, 1928 – Philadelphia, Pennsylvania, USA, February 19, 1996) studied at the University of Debrecen in 1947/51, then he was attached to the Mathematical Research Institute in Budapest although he worked in Szeged from 1952. He obtained the degree of Candidate in 1955. He left Hungary in January 1957 and went to the USA. He first obtained a fellowship in Chicago, then spent three years at the Research Institute for Applied Science in Baltimore. In 1960/64 he was at the University of Maryland, College Park. In 1963 he was appointed to the University of California in Los Angeles but spent 1964/65 at the University of Paris, France. In 1965 he was offered a professorship at the University of Pennsylvania, from where he retired in 1994. Areas of interest: operator algebras, in particular von Neumann algebras and quasi-unitary algebras, representations of Lie groups, in particular exponential and solvable Lie groups.

RADÓ Ferenc (also Francisc, François) (Timişoară, Romania, May 21, 1921 – Cluj-Napoca, Romania, November 27, 1990) started studies at the Technical University in Bucharest. In 1940/44 he did labor service in the Hungarian Army. From 1944 to 1946 he studied at the University of Kolozsvár. Between 1946 and 1949 he taught at a secondary school, but was also appointed to the Pedagogical Institute in Timişoară. From 1950 until his retirement in 1985 he taught at the Bolyai University and the Babeş–Bolyai University of Kolozsvár. He obtained his Ph.D. in 1959. He was an advisor of the Computing Institute of the Romanian Academy of Sciences. He spent the year 1969/70 as a visiting professor at the University of Waterloo, Canada. Fields of interest: functional equations, nomograms, algebraic foundations of geometry. He wrote a number of didactical works in Romanian and in Hungarian.

RADÓ Tibor (Budapest, June 2, 1895 – New Smyrna Beach, Florida, USA, December 25, 1965) won the Eötvös Competition in 1913. Started his studies at the Budapest Technical University but soon switched to the University of Budapest. During World War I he fought on the Russian front, fell into captivity and spent several years in Siberia. He escaped and returned to Hungary through China and India. Continued his studies in 1921 at the University of Szeged, where he obtained his Ph.D. and became an assistant of Frigyes Riesz and Alfréd Haar. In 1928 he went to Munich with a Rockefeller Fellowship and in 1929 emigrated to the USA. First he taught in Houston, but from 1930 until his retirement in 1948 he was a professor at the Ohio State University in Columbus. Afterwards, he worked a few more years at the University of Chicago. Areas of interest: Riemann surfaces, surface area, Plateau's problem, subharmonic functions.

RADOS Gusztáv (also Gustav) (Pest, February 22, 1862 – Budapest, November 1, 1942) studied at both the University and the Technical University in Budapest (1879–1883). In 1882 he published his first article on higher congruences. Spent the year 1884/85 with Felix Klein in Leipzig. From 1885 he had a position at the Budapest Technical University. Became corresponding member of the Hungarian Academy of Sciences in 1894, and ordinary member in 1907. He was awarded the Great Prize of the Academy in 1936 and the University of Kolozsvár awarded him an honorary doctorate. He was a founding member of the Mathematical and Physical Society, its Vice-President in 1913, its President in 1933. Areas of interest: algebra, in particular linear algebra, number theory, differential geometry.

RADOS Ignác (Pest, May 15, 1859 – 1944). Brother of Rados Gusztáv. He studied at the University of Budapest and the Budapest Technical University, and obtained his mathematics-physics teacher's diploma in 1883. Taught first at the Commercial Academy of Budapest, and then at secondary schools in Székelyudvarhely [now Odorhei, Romania] and Budapest. He published a number of articles on the work of great mathematicians. Translated the "Appendix" of János Bolyai and the book of Paul Stäckel on the Bolyais into Hungarian. Main interest: history of mathematics.

RAPCSÁK András (Hódmezővásárhely, December 12, 1914 – Debrecen, October 16, 1993) was admitted to the University of Szeged and the Eötvös Collegium in 1933. Because of health problems he had to interrupt his studies and obtained his diploma in 1942. Taught at secondary schools in several towns. Became interested in differential geometry under the influence of Ottó Varga, and starting with 1945 also lectured at the University of Debrecen. He obtained his Ph.D. in 1947. He also taught at the Teacher's

College in Debrecen, and starting in 1949/51 at the similar College in Eger. In 1951 he returned to the University of Debrecen. He became a Candidate in 1955, Doctor of Mathematical Sciences in 1960, corresponding member of the Hungarian Academy of Sciences in 1967, ordinary member in 1982. Research interest: differential geometry, mainly Finsler spaces.

RÁTZ László (Sopron, April 9, 1863 – Budapest, September 30, 1939) attended the University of Budapest, where he obtained his teacher's diploma, after which he studied in Strassburg and in Berlin. In 1890 he became a teacher at the Lutheran Evangelical Gimnázium in Budapest and taught there for 35 years. Among his students were Alfréd Haar, John von Neumann, Jenő Wigner, and János Harsányi. Between 1896 and 1914 he edited the Mathematical Journal for Secondary Schools (KöMaL). The Bolyai János Mathematical Society named after him a yearly meeting for mathematics teachers.

RÉDEI László (also Ladislaus) (Rákoskeresztúr [now part of Budapest], November 15, 1900 – Budapest, November 21, 1980) won second prize at the Eötvös Competition in 1918. Studied at the University of Budapest and obtained his diploma as a mathematics-physics teacher and also his Ph.D. in 1922. He taught at secondary schools in several towns. He habilitated at the University of Debrecen in 1932, spent the year 1934/35 in Göttingen and was awarded the Gyula Kőnig Prize in 1940. In 1940 he was offered a position at the University of Szeged. From 1967 to 1971 he worked at the Mathematical Research Institute in Budapest. He was elected corresponding member of the Hungarian Academy of Sciences in 1949, ordinary member in 1955, member of the German Leopoldina Academy in 1962, and was awarded the Kossuth Prize in 1950 and in 1955, an honorary doctorate from the Szeged University in 1971. He was awarded the Tibor Szele Medal in 1973. Areas of interest: algebraic number theory, algebra, geometry.

RÉNYI Alfréd (Budapest, March 20, 1921 – Budapest, February 1, 1970) attended the University of Budapest, and obtained his Ph.D. in 1945 in Szeged. In 1946/47 he was a research student in Leningrad [now St. Petersburg] under the direction of Yu. V. Linnik. Habilitated at the University of Budapest in 1947. He was appointed professor at the University of Debrecen in 1949, where with Ottó Varga he founded the journal Publicationes Mathematicae. In 1950 he became founding director of the Applied Mathematics Institute of the Hungarian Academy of Sciences, later Mathematical Research Institute, which now bears his name. From 1952 he also had a position at the University of Budapest. Elected corresponding member of the Hungarian Academy of Sciences in 1949, ordinary member in 1956. He received the Kossuth Prize in 1949 and in 1954. Areas of interest: number theory, probability and its applications, statistics, information theory, complex variables, combinatorics, popularization of mathematics.

RÉNYI Kató (also Catherine) (Budapest, October 24, 1924 – Budapest, August 31, 1969), wife of Alfréd Rényi. Started university studies in Budapest in 1942, continued in Szeged in 1945, then in Leningrad in 1946/47, and graduated in Budapest in 1949. Starting with 1950 she taught at the University of Budapest. She was an excellent teacher. Field of research: complex function theory. The Bolyai Society named after her a prize given for scientific achievements obtained by students before taking their Master's degree.

RÉTHY Mór (Nagykőrös, November 3, 1848 – Budapest, October 16, 1929) studied at the Technical Universities of Buda and Vienna. After obtaining his engineer's diploma he was an assistant at the Budapest Technical University, then a secondary school teacher. He studied in Heidelberg and also in Göttingen, where his first article, on the refraction of light, was presented. From 1874 to 1876 he was professor of theoretical physics and mathematics at the University of Kolozsvár [now Cluj-Napoca, Romania], and from 1886 at the Budapest Technical University. He was elected corresponding member of the Hungarian Academy of Sciences in 1878, and ordinary member in 1900. Research areas: work of Bolyai, complex function theory, mathematical physics.

RÉVÉSZ Gábor (Budapest, May 31, 1954 – Budapest, June 26, 1997) studied economics in Budapest and London, and graduated from the London School of Economics in 1978. Then he became a Ph.D. student with P. M. Cohn and took his degree in 1981. In 1981–84 he had research fellowships in London and Berlin, from 1984–87 a position at the University of Kansas at Lawrence, USA. Then he returned to Hungary, and was at the Technical University for Heavy Industry in Miskolc. In 1995 he gave up his position at the university for financial reasons, and took up work in economics. Field of research: algebra, mainly ring theory.

RIESZ Frigyes (also Frederic) (Győr, January 22, 1880 – Budapest, February 28, 1956) started studying engineering at the Eidgenössische Technische Hochshule in Zurich but after a couple of years returned to the University of Budapest to study mathematics and physics. He obtained his teacher's certificate and his Ph.D. in 1902. He was appointed teacher in Lőcse [now Levoča, Slovakia] in 1904 and in Budapest in 1908, but spent most of his time in Göttingen and in Paris with scholarships. In 1912 he was appointed professor at the University of Kolozsvár. In 1920 he was, with Alfréd Haar, one of the founders of the Mathematical Institute of the new University of Szeged. They were also the founders of the Acta Scientiarum Mathematicarum, journal of this Institute. In 1946 he moved to the University of Budapest. He was elected corresponding member of the Hungarian Academy of Sciences in 1916, and ordinary member in 1936. He was a corresponding member of the French Academy of Sciences (1948), of the Bavarian Academy and of the Royal Physiographical Society of Lund. He was awarded the Great Prize of the Hungarian Academy in 1945, and the Kossuth Prize in 1949 and 1953. He received honorary doctorates from the University of Szeged (1946), the University of Budapest (1950) and the University of Paris (1954). Areas of interest: integral equations, functional analysis (of which he was one of the creators), complex analysis, subharmonic functions (which were also his creation), ergodic theory.

RIESZ Marcel (Győr, November 16, 1886 – Lund, Sweden, September 4, 1969), brother of Frigyes Riesz. He won the Eötvös Competition in 1904, then studied at the University of Budapest and was a member of the Eötvös Collegium. He received his Ph.D. in 1909. Spent the academic years 1906/07 and 1909/10 in Göttingen, 1910/11 in Paris, where he got an invitation from Gösta Mittag-Leffler to give three lectures in Sweden. He accepted, and stayed in Sweden for the rest of his career. He was first at the University of Stockholm, and from 1926 until his retirement in 1952 at the University of Lund. He spent the year 1947/48 at the University of Chicago and after retirement until 1960 he was a visitor at Princeton University, the Courant Institute, Stanford University, the University of Maryland and Indiana University. He was a member of the Swedish Academy and an honorary doctor of the University of Copenhagen. Areas of interest: trigonometric series, complex function theory, in particular Dirichlet series, the moment problem, partial differential equations, relativistic quantum theory, algebra, number theory.

SALLAY Melánia (Kispest [now part of Budapest], June 7, 1934 – Budapest, September 10, 1981). She studied at the University of Budapest and obtained her mathematics-physics teacher's diploma in 1956. From the same year she had a position at the Mathematical Research Institute. Field of research: approximation and interpolation theory.

SARKADI Károly (Budapest, September 12, 1914 – Budapest, August 19, 1985) obtained his mathematics-physics teacher's diploma from the University of Budapest in 1937. Between 1937 and 1947 he did military service, taught at a secondary school, was sent to the front line, and fell into captivity. From 1947 to 1952 he was again a secondary school teacher. From 1952 on he worked at the Mathematical Research Institute. He obtained the degree of Candidate in 1958, Doctor of Mathematical Sciences in 1976. In 1976 he was at the University of California in Berkeley with a Ford Fellowship. In 1966 he obtained the State Prize. Field of interest: statistics.

SCHLESINGER Lajos (also Ludwig) (Nagyszombat [now Trnava, Slovakia], November 1, 1864 – Gießen, December 16, 1933) started his university studies in Heidelberg and continued in Berlin, where he obtained his Ph.D. in 1887 and habilitated in 1889. In 1897 he was extraordinary professor of the University of Bonn, and the same year became ordinary professor at the University of Kolozsvár [now Cluj-Napoca, Romania]. In 1911 he got a call from Budapest, but he went to Gießen, from where he retired in 1930. He was the son-in-law of Lazarus Fuchs. The Hungarian Academy of Sciences elected him corresponding member in 1902. Areas of interest: differential equations, automorphic functions.

SCHOPP János (1910 – Budapest, October 25, 1980). He obtained a teacher's diploma from the University of Budapest in 1934; he was also a member of the Eötvös Collegium. He worked as an actuary until 1948, taught at a secondary school until 1951, and then at the Budapest Technical University. Area of scientific interest: geometry.

SCHWEITZER Miklós (Budapest, February 1, 1923 – Budapest, January 28, 1945) obtained second prize at the Eötvös Competition in 1941. He was not admitted to the University, nevertheless learned mathematics and obtained his first result in 1942. His manuscripts were published posthumously by Pál Turán. Since 1949 the Bolyai János Mathematical Society has an annual competition for university students named after Schweitzer. Area of interest: infinite series and products.

SERES Iván (Budapest, December 15, 1907 – Budapest, February 25, 1966) studied in Budapest and obtained his mathematics-physics teacher's diploma in 1930. Until 1944 tutored, was assistant editor of KöMaL and worked in insurance. After 1945 he taught until 1949 at secondary schools. In 1949/51 he worked at the National Library and in 1951/52 in industry. From 1952 he had a position at the Mathematical Research Institute. He obtained the degree of Candidate in 1955. Area of research: irreducibility of polynomials.

SIDON Simon, see Szidon Simon.

SÓLYI Antal (1912 or 1913 – 1946). He studied at the University of Szeged, where he obtained his Ph.D. in 1941 with a dissertation on Haar's variational lemma and its applications.

SOMORJAI Gábor (Budapest, October 23, 1951 – Budapest, January 15, 1978) attended the University of Budapest, obtained his diploma in 1975 and got a position at the Mathematical Research Institute. Areas of interest: approximation, complex function theory.

SONNEVEND György (Szombathely, March 31, 1944 – Budapest, April 9, 1996) studied mathematics at the University of Budapest from 1962 to 1967 and then got a position at the Institute for Computation of the Hungarian Academy of Sciences. From 1969 he spent three years with L. S. Pontryagin at the Steklov Institute in Moscow, and there he obtained the degree of Candidate in 1973. From 1976 on he was at the University of Budapest but spent the period 1987–1992 at the University of Würzburg, Germany. In 1995 he obtained the title Doctor of Mathematical Sciences. Research interest: optimal control, numerical analysis.

STEINFELD Ottó (Szarvas, March 5, 1924 – Budapest, July 8, 1990). After secondary school he was not allowed to university studies, so he worked as a bricklayer. Then he was drafted for labor service, where his health was ruined. In 1945/50 he studied mathematics and physics at the University of Szeged, and after receiving his diploma, he got a position there. Became Candidate in 1955, then moved to Budapest, where he worked at the Mathematical Research Institute. He obtained the title of Doctor of Mathematical Sciences in 1969. Area of research: algebra, mainly semigroups and ordered algebraic structures.

SURÁNYI János (Budapest, May 19, 1918 –) studied at the University of Szeged from 1937 to 1941, was drafted for labor service from 1942 to 1945. He spent the years 1945/48 at the University of Szeged, where with Paula Soós he restarted the KöMaL. Between 1948 and 1951 he worked at the Ministry of Education and the National Institute for Pedagogy. From 1951 to 1988 he was a professor of the University of Budapest. In 1970/71 he was visiting professor at Sherbrook University, Canada. He became Candidate in 1953, Doctor of Mathematical Sciences in 1957. Received the Beke Manó Prize in 1952. For several decades, he was the head of the committee organizing the Kürschák competition. Areas of interest: logic, number theory, combinatorics, didactics of mathematics, competitions.

SUTÁK József (Szabadka [now Subotica, Serbia], November 5, 1865 – Budapest, July 19, 1954). He was a member of the Piarist order. Studied theology in Nyitra [now Nitra, Slovakia]. After obtaining his teacher's diploma he taught at the Piarist Gimnáziums first in Szeged, then in Budapest. He obtained his Ph.D. from the University of Budapest in 1892, habilitated in 1896, and translated the "Appendix" of János Bolyai in 1897. He was professor of higher geometry at the University of Budapest from 1912 to 1936. Areas of research: geometry, analysis and physics.

SZ.-NAGY Béla, see Szőkefalvi-Nagy Béla.

SZ.-NAGY Gyula, see Szőkefalvi-Nagy Gyula.

SZÁSZ Ferenc Andor (Kisújszállás, December 16, 1931 – Budapest, May 11, 1989). From 1950 to 1954 he studied at the University of Debrecen, and in 1954/55 was assistant of Tibor Szele there. After the death of Szele he taught at a secondary school and then got a research scholarship. From 1960 he worked at the Mathematical Research Institute in Budapest. He became Candidate in 1960, Doctor of Mathematical Sciences in 1973. Field of research: ring theory.

SZÁSZ Ottó (Alsószúcs [now Dolna Suča, Slovakia], December 11, 1884 – Montreux, Switzerland, September 19, 1952). Between 1903 and 1907 he studied at the University of Budapest and at the Budapest Technical University, then in Göttingen. He obtained his Ph.D. in Budapest in 1911, and continued his studies in Paris, Munich and Göttingen. Habilitated in 1914 in Frankfurt, in 1917 in Budapest. Taught at the University of Frankfurt from 1914 to 1933. He was awarded the Gyula Kőnig Prize in 1930. In 1933 he emigrated to the USA. He first taught at the Massachusetts Institute of Technology, then in 1933/35 at Brown University, and beginning with 1936 at the University of Cincinnati. He spent a year at the Institute of Numerical Analysis in Los Angeles. Areas of research: infinite determinants, continued fractions, trigonometric polynomials, series and summation, special functions.

SZÁSZ Pál (also Paul) (Budapest, July 11, 1901 – Budapest, February 12, 1978) studied in Budapest, obtained his mathematics-physics teacher's diploma in 1924, then became the assistant of Lipót Fejér. He obtained his Ph.D. in 1927, spent the year 1928/29 in Berlin at the Humboldt University, and in 1933 habilitated in Budapest. In 1957 he was awarded the title of Doctor of Mathematical Sciences. He had positions at the University of Budapest and at the Teacher's College in Budapest. He was an excellent teacher, lectured instead of Fejér Lipót on Differential and Integral Calculus and other topics in analysis at the University of Budapest. His more than 1300-page book "Elements of Differential and Integral Calculus" is not only a very successful textbook but also a true handbook in Analysis. Areas of research: Fourier series, interpolation, non-euclidean geometry.

SZEGŐ Gábor (also Gabriel) (Kunhegyes, January 20, 1895 – Palo Alto, California, USA, August 7, 1985) won the Eötvös Competition in 1912. He studied at the University of Budapest and was a member of the Eötvös Collegium. He spent the summers of 1913 and 1914 in Berlin and Göttingen. From 1915 to 1918 he was in military service. In Vienna at the air force he met von Kármán and von Mises, and obtained his Ph.D. in 1918. In 1919/20 he was an assistant of József Kürschák at the Budapest Technical University. In 1920 he moved to Berlin, where he first worked in a bank and as an editor of the Fortschritte der Mathematik. He habilitated at the University of Berlin in 1921, obtained the Gyula Kőnig Prize in 1924. In 1926 he was appointed professor at the University of Königsberg [now Kaliningrad, Russia] from where he emigrated to the USA in 1934. He first taught at Washington University in St. Louis and in 1938 went to Stanford, from where he retired in 1960. He was elected corresponding member of the Austrian Academy of Sciences in 1960, honorary member of the Hungarian Academy of Sciences in 1965. He was a member of the National Academy of Science of the USA. Areas of interest: classical analysis, Toeplitz determinants, orthogonal polynomials, mathematical physics.

SZEKERES György (also George) (Budapest, May 29, 1911 – Adelaide, Australia, 28 August, 2005) studied chemical engineering at the Budapest Technical University (1928–1932) but belonged to the group of friends of Pál Erdős, with whom he collaborated and published mathematics. One member of this group, Eszter Klein, became his wife. Until 1939 he worked in a leather factory and in 1939 he emigrated to Shanghai, where he started research in group theory. In 1948 he received an offer from the University of Adelaide (Australia), where he stayed until 1963. Then he became a professor of mathematics at the University of New South Wales (Sydney), where he got an honorary Ph.D. in 1976. He is a founding member of the Australian Mathematical Society, its president in 1972/74, a member the Australian Academy of Science (1963), recipient of the Lyle medal (1968), honorary member of the Hungarian Academy of Sciences (1986). He initiated mathematical competitions in Australia. Areas of research: combinatorics, number theory, group theory, diophantine approximations, general relativity.

SZELE Tibor (Debrecen, June 21, 1918 – Szeged, April 5, 1955) won the Eötvös Competition in 1936. He started to study mechanical engineering in Budapest but after one semester he switched to study mathematics and physics at the University of Debrecen. He obtained his diploma in 1941 and got a position at the Institute for Theoretical Physics in Szeged. He wrote his Ph.D. dissertation under the influence of László Rédei but because of military service defended it only in 1946. From 1946 to 1948 he was an assistant at the University of Szeged. In 1948 he returned to Debrecen and habilitated in algebra and combinatorics. In 1952 he was awarded the Kossuth Prize and got the title of Doctor of Mathematical Sciences. He inspired a large school of students to do research in algebra. The Bolyai Society named after him a prize given to those who created a mathematical school. Areas of interest: abelian groups, rings, modules.

SZÉLPÁL István (Szeged, August 9, 1917 – Szeged, June 22, 1984) attended the University of Szeged. He taught at a secondary school but was dismissed for political reasons in 1949. After the death of Tibor Szele in 1955 he withdrew from mathematics and devoted himself to farming. Area of interest: algebra (groups, rings).

SZÉNÁSSY Barna (Ungvár [now Uzhhorod, Ukraine], December 11, 1913 – Debrecen, November 12, 1995) studied mathematics-physics in Debrecen and obtained his diploma in 1936. He taught at secondary schools and obtained in 1937 his Ph.D. in Debrecen with a dissertation on the infinitesimal ideas of Farkas Bolyai. He spent the year 1942/43 in Berlin. After military service and captivity, he was a teacher again until 1951. From 1951 until retirement in 1977 he worked at the University of Debrecen. He was awarded the degree of Candidate in 1962, Doctor of Mathematical Sciences in 1991. Area of interest: history of Hungarian mathematics.

SZENTMÁRTONY Tibor (Budapest, September 22, 1895 – Budapest, July 18, 1965). He obtained his diploma in 1921 from the University of Budapest. He held various positions at the Budapest Technical University from to 1950. After that he worked at the Mathematical Research Institute. Areas of interest: operator calculus, tensor calculus, probability.

SZÉP Jenő (Budapest, January 13, 1920 – Budapest, October 18, 2004) attended the University of Budapest from 1938 to 1943, and was assistant there from 1941 to 1946. From 1946 to 1961 he taught at the Teacher's Colleges in Budapest and Szeged. Became Candidate in 1952 and Doctor of Mathematical Sciences in 1957. From 1961 to 1993 he was a professor at the University for Economics in Budapest. In 1957 he received the Prize of the Academy, in 1993 the Albert Szent-Györgyi Prize. Fields of research: groups and semigroups, game theory, applications of mathematics.

SZIDON Simon (in non-Hungarian publications he spelled his name **Sidon**) (1892 – Budapest, April 27, 1941) acquired a mathematics-physics teacher's diploma but because of his extreme shyness worked as an actuary. Area of research: trigonometric and power series, orthogonal systems. Sidon sets are named after him.

SZILÁRD Károly (Győr, September 26, 1901 – Budapest, April 5, 1980), brother of the famous physicist Leó Szilárd, studied in Germany between 1919 and 1925 in Jena and also in Göttingen, where he obtained his Ph.D in 1927. From 1925 to 1932 he worked in Berlin as an engineer and mathematician mainly in the industry, but in 1926–27 at the Kaiser Wilhelm Institut für Physikalische Chemie. Since 1925 he was a member of the German Communist Party and in 1933 he emigrated to the Soviet Union. From 1933 till 1948 he worked as a physicist at the Central Aero-Hydroinstitute in Moscow. In 1948 he was interned and worked on the development of missiles. In 1953 (still before Stalin's death!) he got a Stalin Prize. He was set free in 1956 and became an assistant editor of the Referetivnyi Zhurnal (Mathematics). He obtained the title of Candidate in Moscow in 1960. In the same year he returned to Hungary, where he worked at the Mathematical Reserarch Institute. He obtained the title of Doctor of Mathematical Sciences in 1976. Areas of interest: classical analysis, differential equations.

SZŐKEFALVI-NAGY Béla (in most of his publications he used the shorter form **Sz.-Nagy**) (Kolozsvár [now Cluj-Napoca, Romania], July 29, 1913 – Szeged, December 21, 1998). Son of Gyula Szőkefalvi-Nagy. He studied at the University of Szeged, where he got his Ph.D. on orthogonal systems in 1937. The next two years he spent some time in Leipzig, Grenoble and Paris, and from 1939 to 1948 he taught at the Teacher's College in Szeged. He habilitated in 1940 at the University of Szeged, where he became professor in 1948. He was awarded the Gyula Kőnig Prize in 1942, was elected corresponding member of the Hungarian Academy of Sciences in 1945, ordinary member in 1956. He was an honorary member of several academies: Soviet (1971), Irish (1973), Finnish (1976), and had several honorary doctorates (Dresden, Turku, Bordeaux, Szeged). He was awarded the Kossuth Prize in 1950 and 1953, the State Prize in 1978, the Tibor Szele Prize in 1978, the Lomonosov Gold Medal of the Russian Academy in

1980, and the Gold Medal of the Hungarian Academy of Sciences in 1987. Research areas: linear operators, approximation theory, Fourier analysis.

SZŐKEFALVI-NAGY Gyula (also **Sz.-Nagy, Sz. Nagy, von Sz. Nagy, Julius Sz. Nagy**) (Erzsébetváros [now Dumbrăveni, Romania], April 11, 1887 – Szeged, October 14, 1953). He started his studies at the University of Kolozsvár [now Cluj-Napoca, Romania] in 1905 and obtained his Ph.D. in 1909. He taught in secondary schools in Transylvania from 1909 till 1929. He spent the year 1911/12 in Göttingen, Germany, with a scholarship. He habilitated in Kolozsvár in 1915 and in Szeged in 1922, obtained the Gyula Kőnig Prize in 1926. In 1929 he moved to Szeged, where he first taught at the Teacher's College, and in 1939 became professor of geometry at the University. Between 1940 and 1945, when northern Transylvania returned to Hungary, he was a professor in Kolozsvár. In 1945 he returned to the University of Szeged. Became corresponding member of the Hungarian Academy of Sciences in 1934, ordinary member in 1946. Research areas: polynomials, algebraic curves, geometric constructions.

SZŰCS Adolf (Budapest, November 29, 1884 – Budapest, February, 1945) studied in Budapest and Paris (France). He obtained his teacher's diploma and his Ph.D. in 1907. Taught ten years at a secondary school, but was from 1912 also an assistant at the Budapest Technical University. Habilitated in 1913 and in 1920 he transferred completely to the Budapest Technical University. Areas of interest: partial differential equations, calculus of variations, algebra, diophantine approximation.

TAKÁCS Lajos (Maglód, August 21, 1924 –). He was placed second in the Eötvös Competition of 1943. Obtained his Ph.D. at the Budapest Technical University in 1948. He worked at the research laboratory of Tungsram (1945/55) and at the Mathematical Research Institute (1950/58). In 1953/58 he had a position also at the University of Budapest. In 1957 he obtained the title of Doctor of Mathematical Sciences. He left Hungary in 1958. In 1958/59 he was a visiting professor of the University of London, UK, in 1959/66 professor at Columbia University, New York, USA, and since 1966 at the Case Western Reserve University, USA. He was elected exterior member of the Hungarian Academy of Sciences in 1993. Area of research: probability, in particular stochastic processes and their applications.

TANDORI Károly (Novi Sad, Yugoslavia [now Serbia], August 23, 1925 – Szeged, January 24, 2005) studied at the University of Szeged (1944–1948), where he became an assistant at the Bolyai Institute in 1949. He obtained the degrees of Candidate in 1953 and Doctor of Mathematical

Sciences in 1957. He became a professor at the University of Szeged in 1962. He was elected a corresponding member of the Hungarian Academy of Sciences in 1965, and an ordinary member in 1976. He was awarded the Kossuth Prize in 1961, the Szele Tibor Prize in 1983, the Széchenyi Prize and Szent-Györgyi Albert Prize in 1992, the Pro Urbe Prize of Szeged in 1994, the title Honorary Doctorate of the University of Szeged in 1997, and the Szőkefalvi-Nagy Béla Memorial Medal of the Acta Scientiarum Mathematicarum in 2001. Research areas: general orthogonal series, Fourier series, strong laws of large numbers.

TARGONSKI György (Budapest, March 27, 1928 – München, Germany, January 10, 1998) studied at the University of Budapest from 1947 to 1952, and then taught at the Budapest Technical University. Left Hungary in 1956, and for several years had only short-term positions. He took a Ph.D. in theoretical physics in Cambridge, UK, in 1963. From 1963–1974 he was at the Fordham University in New York, USA, then 1974–1993 at the University of Marburg, Germany. Areas of research: iteration theory, functional equations, operator theory, theoretical physics.

TURÁN Pál (also Paul) (Budapest, August 18, 1910 – Budapest, September 26, 1976) studied at the University of Budapest, where in 1933 he obtained his teacher's diploma and in 1935 his Ph.D. He lived from tutoring until 1938 when he obtained a position at a secondary school. He habilitated in 1945, spent some time in Copenhagen (Denmark) and Princeton (USA) in 1946/47 and in 1949 was appointed professor at the University of Budapest. He was a visiting professor at many universities. He was elected corresponding member of the Hungarian Academy of Sciences in 1948, ordinary member in 1953. Was awarded the Kossuth Prize in 1948 and 1952, and the Tibor Szele Prize in 1975. His wife, Vera T. Sós, is the "grande dame" of Hungarian mathematics. Areas of interest: number theory, approximation and interpolation, complex function theory, graph theory. He developed the "power-sum method".

VÁLYI Gyula (also Julius) (Marosvásárhely [now Tîrgu Mureş, Romania], January 25, 1855 – Kolozsvár [now Cluj-Napoca, Romania], October 13, 1913) entered the University of Kolozsvár in 1873 to study mathematics and physics. He obtained his diploma in 1877 after which he spent two years in Berlin, Germany, with a scholarship. In 1880 he defended his doctoral dissertation on a partial differential equation coming from engineering. Kapteyn published in 1910 a revised version of Vályi's thesis. In 1881 Vályi became "Privatdozent" and started teaching at the University of Kolozsvár, where in 1884 he became professor of theoretical physics and in 1885 professor of mathematics. In 1891 he was elected corresponding member of the Hungarian Academy of Sciences. Areas of research: partial differential equations, projective and analytic geometry, number theory, Bolyai studies.

VARGA Ottó (Szepetnek, November 22, 1909 – Budapest, June 14, 1969). As he was a child, his family moved to Késmárk [now Kežmarok, Slovakia], where he attended secondary school. He started to study construction engineering at the Technical University of Vienna, Austria. In 1928 he moved to Prague, Czechoslovakia, where he was a regular student at the Charles University but attended also the Technical University. In 1933 he obtained a mathematics and physics teacher's diploma and also a Ph.D. on Finsler spaces under the direction of L. Berwald. In 1934/35 he worked in Hamburg, Germany, with W. Blaschke and did research in integral geometry. In 1936 he returned to Prague and habilitated in 1937. In 1941 he returned to Hungary, first to Kolozsvár, and in 1942 to Debrecen. From 1958 to 1967 he was a professor at the Budapest Technical University and from 1967 he had a position at the Mathematical Research Institute. In 1944 he received the Gyula Kőnig Prize, in 1950 he was elected corresponding member of the Hungarian Academy of Sciences, in 1952 he was awarded the Kossuth Prize and in 1965 he was elected ordinary member of the Academy. Areas of research: differential geometry, integral geometry.

VÁZSONYI Endre (also Andrew) (Budapest, November 4, 1916 – Santa Rosa, California, USA, November 13, 2003) won the Eötvös Competition in 1934. He studied at the University of Budapest, where under the influence of Dénes Kőnig he got interested in the theory of graphs. He obtained his Ph.D. in 1938, then emigrated to the USA. Studied at Harvard University between 1942 and 1948, then worked as an aircraft engineer. In 1969/72 he was a professor of computer science at the University of California, and in 1972/79 at the University of Rochester. Areas of research: graph theory, operations research, differential equations, supersonic flight, mathematical economics.

VERESS Pál (also Paul) (Kolozsvár [now Cluj-Napoca, Romania], July 19, 1893 – Budapest, January 27, 1945) studied at the Universities of Budapest, Göttingen (Germany) and Kolozsvár. In Budapest he was a member of the Eötvös Collegium. Obtained in 1917 his Ph.D. in Kolozsvár, in 1919 his teacher's diploma, after which he taught in secondary school. He spent the year 1925/26 in Berlin, Germany. From 1928 he was a professor at the Teacher's College in Budapest and was also an instructor at the

Eötvös Collegium. He habilitated in Budapest in 1929, and between 1936 and 1938 he substituted for the professor of geometry. Between 1940 and 1944 with György Alexits, György Hajós and Ferenc Kárteszi he edited the "Mathematical and Didactical Journal" which replaced the KöMaL, suspended for racial reasons. Areas of interest: real analysis, actuarial mathematics, teaching of mathematics.

VERMES Pál (Újpest [now part of Budapest], July 31, 1897 – London, UK, February 26, 1968) was a pupil of György Pólya in secondary school. He studied at the Budapest Technical University. In 1919 he interrupted his studies and worked in business in Hungary and Austria until 1938, when he emigrated to England. While teaching in a secondary school, he finished his mathematical education at Birkbeck College, where Pál Dienes was an instructor. He was awarded the Armitage-Smith Prize in 1945, received a doctorate in 1947, and was appointed lecturer in 1948. Areas of interest: summability, infinite matrices, complex function theory, graph theory.

VINCZE István (also Stephan) (Szeged, February 26, 1912 – Budapest, April 12, 1999) studied at the University of Szeged (1930/35) and then worked as an actuary. He obtained the degree of Candidate in 1952 and Doctor of Mathematical Sciences in 1972. In 1949 he became one of the founders of the Mathematical Research Institute in Budapest. He also taught at the Budapest Technical University in 1949/52, and at the University of Budapest in 1952/82. Areas of interest: statistics, inequalities, geometry, complex function theory.

WALD Ábrahám (Kolozsvár [now Cluj-Napoca, Romania], October 31, 1902 – India, December 13, 1950). He attended secondary school in Kolozsvár. When he graduated there in 1921, instruction at the local University was in Romanian, a language he was not familiar with, so he first got some private tutoring in mathematics, and in 1926 went to study in Vienna, Austria. There he attended the Technical University for a year and only then was accepted at the University, where he got into contact with Karl Menger. He soon had to return to Romania for his military service, but in 1930 he was again in Vienna, obtained his Ph.D. in 1931 and in 1933 he obtained a position at the Economics Institute led by Oskar Morgenstern. In 1938 he emigrated to the USA, where he worked at Columbia University. Areas of research: geometry, mathematical statistics.

WIGNER Jenő (also Eugene) (Budapest, November 17, 1902 – Princeton, New Jersey, USA, January 3, 1995) wanted to study physics but at the request of his father registered at the Budapest Technical University from where he soon transferred to the Technische Hochschule in Berlin, Germany. There he attended the lectures of Einstein, Max Planck, Max von Laue, Werner Heisenberg, Wolfgang Pauli, etc. As a third year student he directed exercises at the Kaiser Wilhelm Institut. He obtained his Ph.D. in chemistry in 1925 under Mihály Polányi. In 1925/26 he worked in his father's factory in Budapest, and in 1926 he was called back to the Kaiser Wilhelm Institut. Then he was an assistant at the University of Berlin and in Göttingen. In 1929 he habilitated in Berlin. In 1930 he left Germany for Princeton University, USA, where he did not get tenure, so in 1936/38 he taught at the University of Wisconsin, USA. Then he returned to Princeton. He was awarded the Franklin Prize (1950), the Fermi Prize (1958), the Atoms for Peace Medal (1960), the Max Planck Medal (1961), the Nobel Prize in physics (1963), the Albert Einstein Prize (1972), the Leó Szilárd Medal of the Hungarian Nuclear Society (1994). He was elected to the National Academy of the USA, fellow of the Royal Society (1970), honorary member of the Loránd Eötvös Physical Society (1983), Honorary Doctor of the University of Budapest (1987), Honorary Member of the Hungarian Academy of Sciences (1988). His sister was the wife of P.A.M. Dirac. Scientific areas: group theory, representations of Lie groups, quantum mechanics.

WINTNER Aurél (Budapest, April 8, 1903 – Baltimore, Maryland, USA, January 15, 1958) studied at the University of Budapest from 1920 to 1924 but did not graduate there. In 1927 he went to Leipzig, Germany, where he got a Ph.D. in 1929. He spent 1929/30 in Rome, Italy, where he worked with T. Levi-Cività. In 1930 he married the daughter of Otto Hölder, and moved to Baltimore, USA, where he was on the faculty of Johns Hopkins University. In 1937/38 he was a visitor at the Institute of Advanced Study in Princeton, USA. Areas of research: astronomy, celestial mechanics, linear operators in Hilbert space, almost periodic functions, probability, number theory, ordinary differential equations, differential geometry.

ZÁNYI László (Budapest, May 5, 1905 – ?) attended the University of Budapest between 1925 and 1929, where he obtained a mathematics-physics teacher's diploma. Followed simultaneously theological studies, was ordained as a priest and joined the Piarist Order in 1929. He obtained a Ph.D. in 1933. He taught in secondary schools of the Piarist Order. After the secularization of church schools in 1948, he worked in 1949/50 as a curate, then left the country and lost contact with Hungary and mathematics. Area of interest: algebra, number theory.

A Panorama of Hungarian Mathematics in the Twentieth Century, pp. 609–621.

References

[1] *Abelian Groups, Rings and Modules,* ed. A. V. Kelarev, R. Göbel, K. M. Rangaswamy, P. Schultz, C. Visonthaler, Agram 2000 Conference, July 9-15, 2000, Perth, Western Australia. Contemporary Mathematics, Vol. 273, American Mathematical Society (Providence, RI, 2001).

[2] Achieser, N. I., *Theory of Approximation,* Ungar (New York, 1956).

[3] Aczél, János – Gołąb, Stanislaw, *Funktionalgleichungen der Theorie der geometrischen Objekte,* PWN (Warszawa, 1960).

[4] Aczél, János, *Lectures on Functional Equations and their Applications,* Mathematics in Science and Engineering, Vol. 19. Academic Press (New York–London, 1966).

[5] Aczél, János – Daróczy, Zoltán, *On Measures of Information and their Characterizations* Academic Press (New York, 1975).

[6] Aczél, János, *A Short Course on Functional Equations based upon recent applications to the Social and Behavioral Sciences,* Reidel–Kluwer (Dordrecht–Boston, 1987).

[7] Aczél, János – Dhombres, Jean G, *Functional Equations in Several Variables: with applications to mathematics, information theory, and the natural and social sciences,* Encyclopedia of Mathematics and its Applications, Vol. 31. Cambridge University Press (Cambridge–New York, 1989).

[8] Alexits, György, *Convergence Problems of Orthogonal Series,* International Series of Monographs on Pure and Applied Mathematics, Vol. 20. Pergamon Press (New York–Oxford–Paris, 1961).

[9] *Analytic Number Theory,* Proceedings of a Conference in Honor of Paul T. Bateman, ed. Bruce C. Berndt, Harold G. Diamond, Heini Halberstam, Adolf Hildebrand, Progress in Mathematics, Volume 85. Birkhäuser (Boston, MA, 1990).

[10] *Analytic Number Theory,* Proceedings of a Conference in Honor of Heini Halberstam, ed. Bruce C. Berndt, Harold G. Diamond, Adolf J. Hildebrand, Progress in Mathematics, Volumes 138–139, Birkhäuser (Boston, MA, 1996).

[11] Asanov, G. S., *Finsler geometry, relativity and gauge theories,* Reidel (Dordrecht, 1985).

[12] Beke, Manó, *Determinánsok* (Determinants), Természet és Technika, Volume 1, Atheneum (Budapest, 1926).

[13] Bernstein, Vladimir, *Leçons sur les progrès récents de la théorie des séries de Dirichlet,* Collection des Monographies sur la Théorie des Fonctions, Gauthiers-Villars (Paris, 1933).

[14] Bieberbach, Ludwig, *Analytische Fortsetzung,* Ergebnisse der Mathematik und Ihrer Grenzgebiete, Neue Folge, Heft 3, Springer-Verlag (Berlin–Göttingen–Heidelberg, 1955).

[15] Boas, Ralph Philip, Jr, *Entire Functions,* Pure and Applied Mathematics, Vol. V, Academic Press (New York, 1954).

[16] Bohr, Harald, *Fastperiodische Funktionen,* Ergebnisse der Mathematik und ihrer Grenzgebiete, Band 1, No. 5, Springer-Verlag (Berlin, 1931)./*Almost Periodic Functions,* Chelsea Publishing Company (American Mathematical Society) (New York, 1947).

[17] Bollobás, Béla, *Random Graphs,* Academic Press (London–New York, 1985).

[18] Borel, Armand, *Seminar on Transformation Groups,* Annals of Mathematics Studies, No. 46, Princeton University Press (Princeton, NJ, 1960).

[19] Borwein, Peter B. – Erdélyi, Tamás, *Polynomials and Polynomial Inequalities,* Graduate Texts in Mathematics, Volume 161, Springer-Verlag (New York, 1995).

[20] Burger, E., *Einführung in die Theorie der Spiele,* Walter de Gruyter (Berlin, 1966).

[21] Carathéodory, Constantin, *Gesammelte Mathematische Schriften,* C.H. Beck'sche Verlagsbuchhandlung (München, 1954–57).

[22] Cartan, Henri, *Collected Works,* ed. R. Remmert and J-P. Serre, Springer-Verlag (Berlin–Heidelberg–New York, 1979).

[23] Császár, Ákos, *Fondements de la topologie générale,* Akadémiai Kiadó (Budapest–Gauthiers-Villars (Paris, 1960).

[24] Császár, Ákos, *Foundations of General Topology,* Pergamon Press (Oxford–London–New York–Paris, 1963).

[25] Csiszár, Imre – Körner, János, *Information Theory; Coding Theorems for Discrete Memoryless Systems,* Disquisitiones Mathematicae Hungaricae, Vol. 12, Akadémiai Kiadó (Budapest, 1986).

[26] De Vore, R. A., *The Approximation of Continuous Functions by Positive Linear Operators,* Lecture Notes in Mathematics, No. 293, Springer-Verlag (Berlin–Heidelberg–New York, 1972).

[27] Dienes, Pál, *Leçons sur les singularités des fonctions analytiques,* Gauthier-Villars (Paris, 1913).

[28] Dienes, Pál, *Taylor Series,* An introduction to the theory of functions of a complex variable, Oxford University Press (1931).

[29] Dieudonné, Jean, *La théorie analytique des polynomes d'une variable (à coefficients quelconques),* Mémorial des Sciences Mathématiques, Fasc. XCIII, Gauthiers-Villars (Paris, 1938).

[30] Ditzian, Zeev – Totik, Vilmos, *Moduli of Smoothness,* Springer series in computational mathematics, Vol. 9, Springer-Verlag (Berlin, 1987).

[31] *Entire Functions and Related Parts of Analysis,* eds. Jacob Korevaar, S. S. Chern, Leon Ehrenpreis, W. H. J. Fuchs, L. A. Rubel, Proceedings of Symposia in Pure Mathematics, Vol. XI, American Mathematical Society (Providence, RI, 1968).

[32] Erdős, Pál, *The Art of Counting,* Selected Writings, ed. Joel Spencer, Mathematics of Our Time, Vol. 5, The MIT Press (Cambridge, Mass. 1973).

[33] Erdős, Pál – Joel, Spencer, *Probabilistic Methods in Combinatorics,* Academic Press (New York, 1974).

[34] Erdős, Pál – Hajnal, András – Máté, Attila – Rado, Richard, *Combinatorial Set Theory: Partition Relations for Cardinals,* Studies in Logic and the Foundations of Mathematics, Vol. 106, North Holland Publishing Co. (Amsterdam–New York–Oxford); Akadémiai Kiadó (Budapest, 1984).

[35] Erdős, Pál – Gruber, P. M. – Hammer, J., *Lattice Points,* Longman (Harlow, 1989).

[36] Erdős, Pál – Surányi, János, *Válogatott Fejezetek a Számelméletből* (Selected Chapters from the Theory of Numbers), Polygon (Szeged, 1996). *Topics in the Theory of Numbers,* Undergraduate Texts in Mathematics, Springer-Verlag (Berlin–New York, 2002).

[37] Faraut, Jacques – Korányi, Ádám, *Analysis on symmetric cones,* Oxford University Press (Oxford, 1994).

[38] Farkas, Miklós, *Periodic Motions,* Applied Mathematical Sciences, Vol. 104, Springer-Verlag (New York, 1994).

[39] Farkas, Miklós, *Dynamical Models in Biology,* Academic Press (New York, 2001).

[40] Fejér, Lipót, *Összegyűjtött Munkái = Gesammelte Arbeiten,* ed. Pál Turán, Akadémiai Kiadó (Budapest, 1970).

[41] Fejes Tóth, László, *Lagerungen in der Ebene, auf der Kugel und im Raum,* Grundlehren der Mathematischen Wissenschaften in Einzeldarstellungen, Band LXV, Springer-Verlag (Berlin–Göttingen–Heidelberg, 1953).

[42] Fejes Tóth, László, *Regular Figures,* Pergamon (London), The MacMillan Co. (New York, 1964). *Reguläre Figuren,* Akadémiai Kiadó (Budapest, 1965).

[43] Feldheim, Ervin, *Théorie de la convergence des procédés d'interpolation et de quadrature mécanique,* Mémorial des Sciences Mathématiques, Fasc. XCV, Gauthier-Villars (Paris, 1939).

[44] Foiaş, Ciprian – Sz.-Nagy, Béla, *Harmonic Analysis of Operators on Hilbert Space,* North-Holland Publishing Co. (Amsterdam, 1970).

[45] Forgó, Ferenc – Szép, Jenő – Szidarovszky, Ferenc, *Introduction to the Theory of Games: Concepts, Methods, Applications,* Nonconvex optimization and its applications, Vol. 32, Kluwer Academic Publishers (Dordrecht–Boston, 1999).

[46] Frank, Philip – Mises, Richard von, *Die Differential- und Integralgleichungen der Mechanik und Physik,* Friedr. Vieweg and Sohn (Braunschweig, 1925), 2nd edition 1930.

[47] Freud, Géza, *Orthogonale Polynome,* Akadémiai Kiadó (Budapest), Birkhäuser (Basel, 1969). *Orthogonal Polynomials,* Pergamon Press (London–Toronto–New York, 1971).

[48] Frobenius, Georg, *Gesammelte Abhandlungen,* ed. J.-P. Serre. Springer-Verlag (Berlin–New York, 1968).

[49] *From Frege to Gödel: a Source Book in Mathematical Logic, 1879–1931,* ed. Jean Van Heijenoort, Harvard University Press (Cambridge, Mass., 1967).

[50] Fuchs, László, *Partially ordered algebraic Systems,* Pergamon Press, (Oxford–London–New York–Paris, 1963). *Teilweise geordnete algebraische Strukturen,* Akadémiai Kiadó (Budapest) – Vandenhoeck and Ruprecht (Göttingen, 1966).

[51] Fuchs, László, *Infinite Abelian Groups,* Pure and Applied Mathematics, Vol. 36, Academic Press (New York–London, 1970, 1973).

[52] Fuchs, László – Salce, Luigi, *Modules over Non-Noetherian Domains,* Mathematical Surveys and Monographs, Volume 84, American Mathematical Society (Providence, RI, 2000).

[53] *Functional Analysis and its Applications* (Nice, 1986), ICPAM Lecture Notes. World Sci. Publishing (Singapore, 1988).

[54] Furstenberg, Hillel, *Recurrence in Ergodic Theory and Combinatorial Number Theory,* R. Porter Lectures, Rice University, Princeton University Press (Princeton, NJ, 1981).

[55] Gál, István Sándor (Gaal, Steven Alexander), *Linear Analysis and Representation Theory,* Grundlehren der Mathematischen Wissenschaften in Einzeldarstellungen, Band 198. Springer-Verlag (Berlin–Heidelberg–New York, 1973).

[56] Gilbarg, David – Trudinger, Neil S., *Elliptic Partial Differential Equations of Second Order,* Grundlehren der Mathematischen Wissenschaften in Einzeldarstellungen, Band 224, Springer-Verlag (Berlin–New York, 1983).

[57] Grätzer, György, *Universal Algebra,* Springer-Verlag (New York, 1979).

[58] Grätzer, György, *General Lattice Theory,* Birkhäuser-Verlag (Basel–Boston, 1998).

[59] Grenander, Ulf – Szegő, Gábor, *Toeplitz Forms and their Applications,* University of California Press (Berkeley and Los Angeles, 1958)/Chelsea (New York, 1984).

[60] Gyires, Béla, *Linear approximations in convex metric spaces and the application in the mixture theory of probability theory,* World Scientific Publ. Co., Inc. (River Edge NJ, 1993).

[61] Győri, István – Ladas, Gerasimos, *Oscillation Theory of Delay Differential Equations,* Oxford Mathematical Monographs, Clarendon Press (Oxford, 1991).

[62] Haar, Alfréd, *Összegyűjtött Munkái = Gesammelte Arbeiten,* ed. Béla Sz.-Nagy, Akadémiai Kiadó (Budapest, 1959).

[63] *Haar Memorial Conference,* ed. J. Szabados and K. Tandori, Bolyai János Mathematical Society – Elsevier (Amsterdam–New York, 1987).

[64] Hadamard, Jacques – Mandelbrojt, Szolem, *La série de Taylor et son prolongement analytique,* Scientia, No. 41, Gauthiers-Villars (Paris, 1926).

[65] *Handbook of Game Theory with Economic Applications,* Vol. I, ed. R. J. Aumann, S. Hart, North-Holland (Amsterdam, 1992).

[66] Hardy, G. H. – Riesz, Marcel, *The General Theory of Dirichlet Series,* Cambridge Tracts in Mathematics and Mathematical Physics, No. 18, Cambridge, University Press (London, 1915), reprint: 1952.

[67] Hardy, G. H. – Littlewood, J. E. – Pólya, György, *Inequalities,* Cambridge University Press (London, 1934); Second Edition, 1951.

[68] Hardy, G. H, *Divergent Series,* Oxford University Press (London, 1949).

[69] Hayman, W. K., *Multivalent Functions,* Cambridge Tracts in Mathematics and Mathematical Physics, No. 48, Cambridge University Press (London, 1958).

[70] Helson, Henry, *Harmonic Analysis,* Addison-Wesley (Reading, MA, 1983).

[71] Hewitt, Edwin – Stromberg, Karl, *Real and Abstract Analysis,* Springer-Verlag (New York, 1965).

[72] Hilbert, David – Ackermann, W., *Grundzüge der theoretischen Logik,* Grundlehren der Mathematischen Wissenschaften in Einzeldarstellungen, Band 27. Springer-Verlag (Berlin–Göttingen–Heidelberg, 1928), 3rd edition 1949.

[73] *Hungarian Problem Book III,* Based on the Eötvös Competitions 1929–1943, compiled by Gy. Hajós, G. Neukomm, J. Surányi, translated and edited by Andy Liu. Anneli Lax New Mathematical Library, vol. 42, Mathematical Association of America (Washington, D.C., 2001).

[74] Hurwitz, Adolf – Courant, Richard, *Funktionentheorie mit einem Anhang von H. Röhrl,* Grundlehren der Mathematischen Wissenschaften in Einzeldarstellungen, Band 3, 4th edition, Springer-Verlag (Berlin–Heidelberg–New York, 1964).

[75] Iserles, A., *Numerical Analysis of Differential Equations,* Cambridge Texts in Applied Mathematics, Cambridge University Press (Cambridge–New York, 1996).

[76] Isidro, José M. – Stachó, László, *Holomorphic Automorphism Groups in Banach Spaces: An Elementary Introduction,* North-Holland Mathematics Studies, No. 105, Elsevier (New York, 1985).

[77] Jetter, Kurt – Lorentz, George G. – Riemenschneider, Sherman D., *Birkhoff Interpolation,* Encyclopedia of Mathematics and Its Applications, Vol. 19, Addison-Wesley Publishing Co. (Reading, Mass., 1983).

[78] Jordán, Károly, *Matematikai Statisztika (Mathematical Statistics),* Természet és Technika, Volume 4, Athenaeum (Budapest, 1927, in Hungarian), and *Statistique mathématique,* Gauthier-Villars (Paris, 1927, in French).

[79] Jordan, Charles, *Statistique mathématique,* Gauthier-Villars (Paris, 1927).

[80] Jordan, Charles, *Calculus of Finite Differences,* Chelsea (New York, 1950).

[81] Jordan, Charles, *Chapters on the Classical Calculus of Probability,* Akadémiai Kiadó (Budapest, 1972).

[82] Kahane, Jean-Pierre, *Séries de Fourier absolument convergentes,* Ergebnisse der Mathematik und Ihrer Grenzgebiete, Band 50, Springer-Verlag (Berlin–Heidelberg–New York, 1970).

[83] Kahane, Jean-Pierre and Lemarié-Rieusset, Pierre Gilles, *Séries de Fourier et Ondelettes,* Cassini (Paris, 1998). *Fourier Series and Wavelets,* Gordon and Breach (New York, 1995).

[84] Kármán, Tódor, *Collected Works of Theodore von Kármán,* Volumes 1–4, Butterworths Scientific Publications (London, 1956); Volume 5, Von Kármán Institute for Fluid Dynamics, Rhode-St. Genese (Belgium, 1975).

[85] Kárteszi, Ferenc, *Introduction to Finite Geometries,* North-Holland (Amsterdam–Oxford–New York, 1976).

[86] Katznelson, Yitzhak, *An Introduction to Harmonic Analysis,* Dover (New York, 1976).

[87] Kerékjártó, Béla, *Vorlesungen über Topologie. I. Flächentopologie,* Grundlehren der Mathematischen Wissenschaften in Einzeldarstellungen, Band VIII, Springer-Verlag (Berlin, 1923).

[88] Kerékjártó, Béla, *A geometria alapjairól. I. Az euklidesi geometria elemi felépítése,* A szerző kiadása a Magyar Tudományos Akadémia támogatásával, Szeged, 1937; II. Projektív geometria, Magyar Tudományos Akadémia (Budapest, 1944).

[89] Kerékjártó, Béla, *Les fondements de la géométrie. I. La construction élémentaire de la géométrie euclidienne,* Akadémiai Kiadó (Budapest, 1955): *II. Géométrie projective,* Akadémiai Kiadó (Budapest, 1966).

[90] Kertész, Andor, *Vorlesungen über Artinsche Ringe,* Akadémiai Kiadó (Budapest) – B. G. Teubner (Leipzig, 1968). Lectures on Artinian Rings, Akadémiai Kiadó (Budapest, 1987).

[91] Kertész, Andor, *Einführung in die transfinite Algebra,* Akadémiai Kiadó (Budapest, 1975).

[92] Kiss, Elemér, *Mathematical gems from the Bolyai chests;* János Bolyai's discoveries in number theory and algebra as recently deciphered from his manuscripts, Akadémiai Kiadó–Typotex Ltd. (Budapest, 1999).

[93] Kőnig, Dénes, *Az analysis situs elemei, Első rész, Felületek* (The elements of analysis situs. First part, Surfaces), Magyar Tudományos Akadémia (Budapest, 1918).

[94] Kőnig, Dénes, *Theorie der endlichen und unendlichen Graphen,* Kombinatorische Topologie der Streckenkomplexe, Akademische Verlagsgesellschaft M.B.H. (Leipzig, 1936)/ Chelsea Publishing Company (American Mathematical Society) (New York, 1950).

[95] Kőnig, Gyula, *Einleitung in die allgemeine Theorie der algebraischen Größen,* Teubner (Leipzig, 1903). *Az algebrai mennyiségek általános elméletének alapvonalai,* Magyar Tudományos Akadémia (Budapest, 1903).

[96] Kőnig, Gyula (Julius), *Neue Grundlagen der Logik, Arithmetik und Mengenlehre,* Veit (Leipzig, 1914).

[97] Koosis, Paul, *Introduction to Hp Spaces,* London Mathematical Society Lecture Note Series, 40, Cambridge University Press (Cambridge–London–New York, 1980).

[98] Kullback, Solomon, *Information Theory and Statistics,* Wiley (New York, 1959), Dover (New York, 1978).

[99] Kürschák, József, *Hungarian Problem Book based on the Eötvös Competitions 1894–1928,* revised and edited by G. Hajós, G. Neukomm, J. Surányi, translated by Elvira Rapaport. New Mathematical Library, vols. 11–12, Random House (New York), L.W. Singer Co. (Syracuse); Anneli Lax New Mathematical Library, Mathematical Association of America (1963).

[100] Lánczos, Kornél (Cornelius), *Collected Published Papers with Commentaries,* 6 volumes, ed. William R. Davis et al., Foreword by George Marx. North Carolina State University (Raleigh, 1998).

[101] Lánczos, Kornél (Cornelius), *The Variational Principles of Mechanics,* Mathematical Expositions, No. 4, 2nd edition. University of Toronto Press (Toronto, 1962).

[102] Lánczos, Kornél (Cornelius), *Albert Einstein and the Cosmic World Order,* Interscience Publishers (New York, 1965).

[103] Lánczos, Kornél (Cornelius), *Space through the ages,* The evolution of geometrical ideas from Pythagoras to Hilbert and Einstein, Academic Press (London–New York, 1970).

[104] Landau, Edmund, *Darstellung und Begründung einiger neuerer Ergebnisse der Funktionentheorie,* Julius Springer (Berlin, 1929).

[105] Lax, Péter – Phillips, R. S., *Scattering Theory.* Pure and Applied Mathematics, vol. 26, Academic Press (New York–London, 1967).

[106] Lax, Péter – Phillips, R. S., *Scattering Theory for Automorphic Functions,* Annals of Mathematics Studies, No. 87, Princeton University Press (Princeton, N.J., 1976).

[107] Lax, Péter, *Hyperbolic Systems of Conservation Laws and the Mathematical Theory of Shock Waves,* CBMS-NSF Regional Conference Series in Applied Mathematics, No. 11, Society for Industrial and Applied Mathematics (Philadelphia, PA, 1973), Second printing, 1984.

[108] Leindler, László, *Strong Approximation by Fourier Series,* Akadémiai Kiadó (Budapest, 1985).

[109] Lévy, Paul, *Théorie de l'addition des variables aléatoires,* Monographies des Probabiltés; Calcul des Probabilités et ses Applications, fascicule 1, Gauthier-Villars (Paris, 1937).

[110] Liggett, Thomas M., *Interacting Particle Systems,* Die Grundlehren der Mathematischen Wissenschaften in Einzeldarstellungen, Band 276, Springer-Verlag (New York, 1985).

[111] Lorentz, George G., *Approximation of Functions,* Holt, Rinehart and Winston (New York, 1966); 2nd edition, Chelsea (New York, 1986).

[112] Magyar, Zoltán, *Continuous Linear Representations,* North-Holland Mathematics Studies, No. 168, Elsevier (Amsterdam–London, 1992).

[113] Marden, Morris, *Geometry of Polynomials,* Mathematical Surveys and Monographs, Volume 3, American Mathematical Society (Providence, RI, 1989).

[114] Medgyessy, Pál, *Decomposition of Superpositions of Distribution Functions,* Akadémiai Kiadó (Budapest, 1961).

[115] Medgyessy, Pál, *Decomposition of Superpositions of Density Functions and Discrete Distributions,* Halsted Press [John Wiley and Sons] (New York, 1977).

[116] *Minimax and Applications,* ed. D. Z. Du, P. Pardalos, Kluwer Academic Publishers (Dordrecht, 1995).

[117] Morgenstern, Oskar – Neumann, John von, *Theory of Games and Economic Behavior,* Princeton University Press (Princeton, NJ, 1944); 3rd edition 1953.

[118] Neumann, János (John von Neumann), *Collected Works,* 6 volumes, ed. A. H. Taub, Pergamon Press (New York, 1961).

[119] Neumann, Johann von, *Mathematische Grundlagen der Quantenmechanik,* Die Grundlehren der Mathematischen Wissenschaften in Einzeldarstellungen, Band XXXVIII, J. Springer (Berlin, 1932).

[120] Neumann, John von, *1903–1957,* ed. J. C. Oxtoby, B. J. Pettis, G. B. Price, Bull. Amer. Math. Soc., Vol. 64, No. 3, Pt. 2, May 1958. Reprinted as monograph, American Mathematical Society (Providence, RI, 1988).

[121] Neumann, *The Legacy of John von Neumann,* ed. James G. Glimm, John Impagliazzo, Isadore Singer. Proceedings of Symposia in Pure Mathematics, Vol. 50, American Mathematical Society (Providence, RI, 1990).

[122] Nevanlinna, Rolf, *Eindeutige Analytische Funktionen,* Die Grundlehren der Mathematischen Wissenschaften in Einzeldarstellungen, Band XLVI, Springer-Verlag (Berlin–Göttingen–Heidelberg, 1953).

[123] Newman, D. H., *Approximation with Rational Functions.* CBMS Regional Conference Series in Mathematics, Vol. 41, American Mathematical Society (Providence, RI, 1979).

[124] Newman, M. H. A., *Elements of the topology of plane sets of points,* Cambridge University Press (1939).

[125] Obrechkoff, Nikola, *Quelques classes de fonctions entières limites de polynomes et de fonctions méromorphes limites de fractions rationnelles,* Actualités Scientifiques et Industrielles, No. 891, Hermann (Paris, 1941).

[126] Péter, Rózsa, *Rekursive Funktionen,* Akadémiai Kiadó (Budapest, 1951). English Translation: *Recursive Functions,* 3rd edition, Academic Press (New York, 1967).

[127] Péter, Rózsa, *Playing with Infinity,* Mathematical Explorations and Excursions, Dover (New York, 1976).

[128] Pólya, György, *Collected Papers, Vol. I, Singularities of Analytic Functions,* ed. R. P. Boas. Mathematics of Our Time, Vol. 7 (1974); *Vol. II, Location of Zeros,* ed. R. P. Boas, Mathematics of Our Time, Vol. 8 (1974); *Vol. III, Analysis,* ed. J. Hersch and G.-C. Rota, Mathematics of Our Time, Vol. 21 (1984); *Vol. IV, Probability, Combinatorics, Teaching and Learning in Mathematics,* ed. G.-C. Rota, Mathematics in Our Time, Vol. 22 (1984), The MIT Press (Cambridge, Mass).

[129] Pólya, György – Szegő, Gábor, *Aufgaben und Lehrsätze aus der Analysis,* Grundlehren der Mathematischen Wissenschaften in Einzeldarstellungen, Band 19–20, Springer-Verlag (Berlin, 1925), Dover (New York, 1945), 2nd edition, Springer, 1954, 3rd edition, Springer 1964, 4th edition, Heidelberger Taschenbücher, Band

73–74, Springer, 1970. English translation: *Problems and Theorems in Analysis,* Die Grundlehren der Mathematischen Wissenschaften, Band 193, 216, Springer (Berlin–New York–Heidelberg, 1972, 1976).

[130] Pólya, György – Szegő, Gábor, *Isoperimetric Inequalities in Mathematical Physics,* Annals of Mathematics Studies, No. 27, Princeton University Press (Princeton, NJ, 1951).

[131] Pólya, György, *How to solve it, a new aspect of mathematical method,* 2nd edition, Princeton Paperbacks, No. 246, Princeton University Press (Princeton, N.J., 1971).

[132] Pólya, György, *Mathematics and Plausible Reasoning. I. Induction and Analogy in Mathematics, II. Patterns of Plausible Inference,* Princeton University Press (Princeton, N.J., 1954).

[133] Pólya, György, *Mathematical Discovery,* On Understanding, Learning and Teaching Problem Solving, Two volumes, John Wiley and Sons (New-York–London, 1962/1965).

[134] Pólya, György, *Mathematical Methods of Science.* Revised Edition, ed. Leon Bowden, New Mathematical Library, vol. 26, The Mathematical Association of America (Washington, D.C., 1977).

[135] Pólya, *Studies in Mathematical Analysis and Related Topics - Essays in Honor of George Pólya,* ed. Gábor Szegő, Charles Loewner, Stefan Bergman, Menahem Max Schiffer, Jerzy Neyman, David Gilberg, Herbert Solomon, Stanford University Press (Stanford, California, 1962).

[136] Post, *Solvability, provability, definability: the Collected Works of Emil L. Post,* ed. Martin Davis, Birkhäuser (Boston, 1994).

[137] *Proceedings of the International Conference organized to Commemorate the 100-th Anniversary of Leopold Fejér and Frederic Riesz,* Budapest, August 22–28, 1980, ed. B. Sz.-Nagy, J. Szabados, Colloquia Mathematica Societatis János Bolyai, 35. North-Holland (Amsterdam–New-York, 1983).

[138] Pukánszky, Lajos, *Leçons sur les représentations des groupes,* Monographies de la Société Mathématique de France 2, Dunod (Paris, 1967).

[139] Pukánszky, Lajos, *Characters of Connected Lie Groups,* With a preface by J. Dixmier and M. Duflo, Mathematical Surveys and Monographs, Volume 71, American Mathematical Society (Providence, RI, 1999).

[140] Rademacher, Hans – Toeplitz, Otto, *Von Zahlen und Figuren,* Springer-Verlag (Berlin, 1930).

[141] Rademacher, Hans – Grosswald, Emil, *Dedekind Sums,* Carus Mathematical Monographs, No. 16, The Mathematical Association of America (1972).

[142] Radó, Tibor, *On the Problem of Plateau,* Ergebnisse der Mathematik und ihrer Grenzgebiete, Band 2, No. 2, Springer-Verlag (Berlin, 1933).

[143] Radó, Tibor, *Subharmonic Functions,* Ergebnisse der Mathematik und ihrer Grenzgebiete. Band 5, No. 1, Springer-Verlag (Berlin, 1937)/Chelsea (New York, 1949).

[144] Radó, Tibor, *Length and Area,* Amercian Mathematical Society Colloquium Publications, Vol. XXX. (1948).

[145] Radó, Tibor – Reichelderfer, P. V., *Continuous Transformations in Analysis with an introduction to algebraic topology,* Die Grundlehren der Mathematischen Wissenschaften in Einzedarstellungen, Band LXXV, Springer-Verlag (Berlin–Göttingen–Heidelberg, 1955).

[146] Rédei, László, *Algebra, I,* Akadémiai Kiadó (Budapest, 1954). German translation: *Algebra, I,* Akademische Verlagsgesellschaft, Geest Portig (Leipzig, 1959). English translation: *Algebra, I,* Akadémiai Kiadó (Budapest) – Pergamon Press (Oxford, 1967).

[147] Rédei, László, *Theorie der endlich erzeugbaren kommutativen Halbgruppen,* Hamburger Mathematische Einzelschriften, Heft 41, Physica-Verlag (Würzburg, 1963), *The theory of finitely generated commutative semigroups,* Akadémiai Kiadó (Budapest) – Pergamon Press (Oxford, 1965).

[148] Rédei, László, *Begründung der euklidischen und nichteuklidischen Geometrien nach F. Klein,* Akadémiai Kiadó (Budapest) – Teubner Verlag (Leipzig, 1965). *Foundation of Euclidean and non-Euclidean geometries according to F. Klein,* Akadémiai Kiadó – Pergamon Press (Oxford, 1968).

[149] Rédei, László, *Lückenhafte Polynome über endlichen Körpern,* Akadémiai Kiadó (Budapest) – Deutscher Verlag der Wissenschaften (Berlin) – Birkhäuser (Basel, 1970). *Lacunary polynomials over finite fields,* Akadémiai Kiadó (Budapest) – North Holland (Amsterdam, 1973).

[150] Rédei, László, *Endliche p-Gruppen,* Akadémiai Kiadó (Budapest, 1989).

[151] Rényi, Alfréd, *Selected Papers,* ed. Pál Turán, Akadémiai Kiadó (Budapest, 1976).

[152] Rényi, Alfréd, *Probability Theory,* Translated by László Vekerdi, North-Holland Series in Applied Mathematics and Mechanics, Vol. 10, North-Holland Publishing Company (Amsterdam–London); American Elsevier Publishing Co., Inc. (New York, 1970).

[153] Rényi, Alfréd, *Foundations of Probability,* Holden-Day, Inc. (San Francisco–Cambridge–London–Amsterdam, 1970).

[154] Reza, Fazlollah M., *An Introduction to Information Theory,* McGraw-Hill (New York, 1961).

[155] Richtmyer, R. D., *Difference Methods for Initial-Value Problems,* Interscience Tracts in Pure and Applied Mathematics, No. 4, Interscience Publishers (New York–London, 1958).

[156] Riesz, Frigyes, *Összegyűjtött Munkái = Œuvres complètes = Gesammelte Arbeiten,* ed. Á. Császár, Akadémiai Kiadó (Budapest, 1960).

[157] Riesz, Frigyes – Sz.-Nagy, Béla, *Leçons d'Analyse Fonctionnelle,* Akadémiai Kiadó (Budapest, 1952). English translation: *Functional Analysis,* Ungar (New York, 1955).

[158] Riesz, Marcel, *Collected Papers,* ed. L. Gårding and L. Hörmander, Springer-Verlag (Berlin–Heidelberg–New York, 1988).

[159] Riesz, Marcel, *Clifford Numbers and Spinors,* ed. E. Folke Bolinder and Pertti Lounesto, Kluwer Academic Publishers (Dordrecht–Boston–London, 1993).

[160] Saff, Edward B. – Totik, Vilmos, *Logarithmic Potentials with External Fields,* Grundlehren der Mathematischen Wissenschaften, Vol. 316, Springer-Verlag (Berlin–New York, 1997).

[161] Sarkadi, Károly – Vincze, István, *Mathematical Methods of Statistical Quality Control,* Probability and Mathematical Statistics, Vol. 18., Academic Press (New York–London, 1974).

[162] Savage, L. J., *The Foundations of Statistics,* John Wiley (New York, 1954).

[163] Schur, Issai, *Gesammelte Abhandlungen,* ed. Alfred Brauer and Hans Rohrbach, Springer-Verlag (Berlin–New York, 1973).

[164] Schwartz, Laurent, *Étude des Sommes d'Exponentielles,* Hermann (Paris, 1959).

[165] Stahl, Herbert – Totik, Vilmos, *General Orthogonal Polynomials,* Encyclopedia of Mathematics and its Applications, Vol. 43, Cambridge University Press (New York, 1992).

[166] Stein, Elias M., *Harmonic Analysis: Real Variable Methods, Orthogonality and Oscillatory Integrals,* Princeton Mathematical Series, 43, Princeton Unversity Press (Princeton, NJ, 1993).

[167] Steinfeld, Ottó, *Quasi-ideals in Rings and Semi-groups,* Akadémiai Kiadó (Budapest, 1978).

[168] Surányi, János, *Reduktionstheorie des Entscheidungsproblems im Prädikatenkalkül der ersten Stufe,* Akadémiai Kiadó (Budapest); VEB Deutscher Verlag der Wissenschaften (Berlin, 1959).

[169] Szabados, József – Vértesi, Péter, *Interpolation of Functions,* World Scientific (Singapore–Teaneck, NJ, 1990).

[170] Szász, Ferenc Andor, *Radikale der Ringe,* Akadémiai Kiadó (Budapest), Deutscher Verlag der Wissenschaften (Berlin, 1975). *Radicals of Rings,* Akadémiai Kiadó (Budapest) - John Wiley and Sons (Chichester–New York, 1981).

[171] Szász, Gábor, *Bevezetés a hálóelméletbe,* Akadémiai Kiadó (Budapest, 1959). *Einführung in die Verbandstheorie,* Akadémiai Kiadó (Budapest, 1962). *Introduction to Lattice Theory,* 3rd enlarged edition revised by R. Wiegandt, Academic Press (New York–London), Akadémiai Kiadó (Budapest, 1963). *Théorie des treillis,* Monographies universitaires de mathématiques, 36, Dunod (Paris, 1971).

[172] Szász, Ottó, *Collected Mathematical Papers,* ed. H. D. Lipsich, University of Cincinnati (1955).

[173] Szegő, Gábor, *Collected Papers,* ed. R. Askey, Birkhäuser (Boston–Basel–Stuttgart, 1982).

[174] Szegő, Gábor, *Orthogonal Polynomials,* American Mathematical Society Colloquium Publications, Vol. XXIII., 1939, revised 1959, 3rd edition 1967, 4th edition 1975.

[175] Szénássy, Barna, *History of Mathematics in Hungary until the 20th Century,* Springer-Verlag (Berlin, 1992).

[176] Szép, Jenő – Forgó, Ferenc, *Bevezetés a Játékelméletbe,* Közgazdasági és Jogi Könyvkiadó (Budapest, 1974). *Einführung in die Spieltheorie,* Akadémiai Kiadó (Budapest), Verlag Harri Deutsch (Frankfurt/Main, 1983). *Introduction to the Theory of Games,* Akadémiai Kiadó (Budapest), D. Reidel (Dordrecht, 1985).

[177] Sz.-Nagy, Béla, *Spektraldarstellung linearen Transformationen des Hilbertschen Raumes,* Ergebnisse der Mathematik und ihrer Grenzgebiete, Band 5, No. 5. Springer-Verlag (Berlin, 1942).

[178] Sz.-Nagy, Béla, *Unitary dilations of Hilbert space operators and related topics,* Conference Board of the Mathematical Sciences, Regional Conference Series in Mathematics, No. 19, American Mathematical Society (Providence, RI, 1974).

[179] Takács, Lajos, *Introduction to the Theory of Queues,* Oxford University Press (New York, 1962) – Greenwood Press (Westport, CT, 1982).

[180] Takács, Lajos, *Combinatorial Methods in the Theory of Stochastic Processes,* Wiley (New York, 1967).

[181] Titchmarsh, E. C., *The Theory of the Riemann Zeta-Function,* Clarendon Press (Oxford, 1951).

[182] Totik, Vilmos, *Weighted Approximation with Varying Weight,* Springer Lecture Notes in Mathematics, No. 1569, Springer-Verlag (Berlin–New York, 1994).

[183] *Transactions of the Sixth Prague Conference on Information Theory, Statistical Decision Functions, Random Processes,* Academia, Prague, Academic Press (New York, 1973).

[184] Turán, Pál, *Collected Papers,* ed. Pál Erdős, Akadémiai Kiadó (Budapest, 1990).

[185] Turán, Pál, *List of Publications of Paul Turán,* J. Approx Theory 29, No. 1 (1980), 14–22.

[186] Turán, *To the memory of Paul Turán,* Studies in Pure Mathematics, ed. Paul Erdős, László Alpár, Gábor Halász, András Sárközy, Akadémiai Kiadó (Budapest) – Birkhäuser-Verlag (Basel–Boston–Stuttgart, 1983).

[187] Turán, Pál, *On a New Method of Analysis and its Applications,* Published posthumously with the assistance of Gábor Halász and János Pintz, Pure and Applied Mathematics. A Wiley-Interscience Series of Texts, Monographs and Tracts, John Wiley and Sons (New York–Chichester–Brisbane, 1984).

[188] Vajda, Igor, *Theory of Statistical Inference and Information,* Kluwer Academic (Boston, 1989).

[189] Veress, Pál, *Valós Függvények* (Real Functions), Studium (Budapest, 1934).

[190] Vincze, István, *Mathematische Statistik mit industriellen Anwendungen,* Akadémiai Kiadó (Budapest, 1971); 2nd expanded edition in two volumes, Bibliographisches Institut (Mannheim–Wien–Zurich), Akadémiai Kiadó (Budapest, 1984). *Matematikai statisztika ipari alkalmazásokkal,* Műszaki Könyvkiadó (Budapest, 1968, 1974).

[191] Vincze, István – V"rbanova, Mariya, Непараметрични статистически методи, Series in Modern Mathematics, No. 17, Nauka i Izkustvo (Sofia, 1981). *Nemparaméteres matematikai statisztika, elmélet és alkalmazások* (Nonparametric mathematical statistics, theory and applications), Akadémiai Kiadó (Budapest, 1993).

[192] Wald, Ábrahám, *Sequential Analysis,* John Wiley and Sons (New York) – Chapman and Hall (London, 1947).

[193] Wald, Ábrahám, *Statistical Decision Functions,* John Wiley and Sons (New York) – Chapman and Hall (London, 1950).

[194] Walsh, Joseph L., *Bibliography of Joseph Leonard Walsh,* J. Approx. Theory 5, No. 1 (1972), xii–xxviii.

[195] Webster, A. G. – Szegő, Gábor, *Partielle Differentialgleichungen der mathematischen Physik,* B. G. Teubner (Leipzig, 1930).

[196] Weil, André, *L'intégration dans les groupes topologiques et ses applications,* Hermann (Paris, 1940).

[197] Weisz, Ferenc, *Martingale Hardy spaces and their applications to Fourier Analysis,* Lecture Notes in Mathematics, Vol. 1568, Springer-Verlag (Berlin–New York, 1994).

[198] Weyl, Hermann, *Selecta,* Birkhäuser-Verlag (Basel–Stuttgart, 1956).

[199] Wigner, Jenő (E.P.), *Collected Works,* ed. A. Wightman, Springer-Verlag (Berlin–New York, 1993).

[200] Wigner, Jenő, *Gruppentheorie und ihre Anwendung auf die Quantenmechanik der Atomspektren,* Die Wissenschaft, No. 85, F. Vieweg, Braunschweig, 1931; J. W. Edwards, 1944.

[201] Wintner, Aurél, *Spektraltheorie der unendlichen Matrizen,* Einführung in den analytischen Apparat der Quantenmechanik, S. Hirzel (Leipzig, 1929).

[202] Wintner, Aurél, *The Analytical Foundations of Celestial Mechanics,* Princeton Mathematical Series, No. 5, Princeton University Press (Princeton, NJ, 1941).

[203] Zygmund, Antoni, *Trigonometric Series,* Cambridge University Press (London–New York, 1959).

Name Index